# HANDBOOK FOR BUILDING CONSTRUCTION

# HANDBOOK FOR BUILDING CONSTRUCTION

## Administration, Materials, Systems, and Safety

**Clifford J. Schexnayder, P.E., Ph.D.**
Eminent Scholar Emeritus
Del E. Webb School of Construction
Arizona State University
Tempe, Arizona

**Christine Fiori, Ph.D.**
Program Director of the Construction Management Program
Drexel University
Philadelphia, Pennsylvania

Mc Graw Hill

**New York Chicago San Francisco Athens London**
**Madrid Mexico City Milan New Delhi**
**Singapore Sydney Toronto**

**Library of Congress Control Number: 2021931762**



**Handbook for Building Construction: Administration, Materials, Systems, and Safety**



1 2 3 4 5 6 7 8 9 CCD 26 25 24 23 22 21

ISBN 978-1-260-45688-2
MHID 1-260-45688-9

**Sponsoring Editor**
Ania Levinson

**Editing Supervisor**
Donna M. Martone

**Acquisitions Coordinator**
Elizabeth Houde

**Project Manager**
Jyoti Shaw

**Copy Editor**
Bindu Singh

**Proofreader**
Kirti Dogra

**Indexer**
Edwin Durbin

**Production Supervisor**
Pamela A. Pelton

**Composition**
MPS Limited

**Art Director, Cover**
Jeff Weeks

This book is dedicated to my father, Peter A. Fiori who taught me to love and appreciate all aspects of building and to respect every trade. The truly great men and women are those who actually execute the work.

Christine M. Fiori
Clifford J. Schexnayder

# CONTENTS

## Chapter 8. Risk Management 181



## Chapter 9. Building Materials 197

## ABOUT THE AUTHORS

**Clifford J. Schexnayder** is a Del E. Webb School of Construction, Arizona State University Eminent Scholar Emeritus. He has authored five editions of Construction Planning, Equipment & Methods, two editions of *Construction Management Fundamentals*, and the latest edition of *Moving the Earth*, all with McGraw-Hill, Inc.

He received his Ph.D. in civil engineering (construction engineering and management) from Purdue University, and a Masters and Bachelors in civil engineering from Georgia Institute of Technology. A construction engineer with over 50 years of practical experience, Cliff has worked with major construction contractors as field engineer, estimator, and corporate chief engineer. In addition, he served with the U.S. Army Corps of Engineers on active duty and in the reserves, retiring as a colonel. His last assignment was as Executive Director, Directorate of Military Programs, Office of the Chief of Engineers, Washington, D.C.

He has taught construction management, estimating, and planning at Arizona State University, Louisiana Tech University, Northeastern, and the University of New Haven. As a visiting professor he taught these courses at Virginia Tech, the Universidad de Piura, in Peru, the United States Air Force Academy, and at the Israel Institute of Technology, Technion City, Haifa, Israel.

Cliff served as a consultant to the Autoridad del Canal de Panama reviewing the cost estimate for the 3rd Lane project and for the Secretary of the Business, Transportation & Housing Agency of California appraising issues associated with constructing the main east span of the San Francisco-Oakland Bay Bridge. He is a member of the National Academy of Construction and the Academia Panamericana de Ingeniería.

**Christine Fiori** is the Program Director of the Construction Management Program at Drexel University where she teaches courses in Project Controls, Equipment Applications and Economics, and Strategic Management. Prior to joining the faculty at Drexel University, she served as the Preston and Catharine White Fellow and Associate Director of the Myers-Lawson School of Construction at Virginia Tech. She received her Bachelors, Masters, and Ph.D. in Civil Engineering with a concentration in Geotechnical Engineering from Drexel University. She was a Civil Engineering officer in the United States Air Force and taught at both the United States Air Force Academy and Arizona State University.

Christine led a National Science Foundation sponsored team to explore the construction techniques of the Inca, specifically the Inca road throughout Peru. This research supported development of the Smithsonian exhibit at the Museum of the Native American Indian "Engineering an Empire." She was recognized as an Engineering News Record Top 25 Newsmakers of 2010 for her research on the Inca Road. In addition, Dr. Fiori was featured on the Science Channel in an Episode of Strip the City pertaining to Machu Picchu. She has led diverse groups of student teams to Vietnam, Kenya, Belize, Guatemala, and Haiti to complete construction projects and community engagement programs. Currently her work is focused in Belize and Africa. She also serves as a Bridges to Prosperity Construction Mentor, serves on the Board of Directors and as a mentor for the ACE Mentoring program of Southeastern Pennsylvania, was elected as the first Affiliate member of the Carpenters' Company of the City and County of Philadelphia, and is an instructor at the United Brotherhood of Carpenters and Joiners Training Center in Las Vegas, Nevada.

## LIST OF CONTRIBUTORS

**Lester M. Hunkele, III,** P.E., F. CMAA, Hunkele Consulting, Myrtle Beach, South Carolina. Lester managed the technical, contractual, and administrative aspects of the development and design-build of both the four million-square-foot Ronald Reagan Building in Washington, D.C. and the six million-square-foot Pentagon Renovation surrounding 9/11. (Chapter 2: The Project Team.)

**Meredith G. Kinkle** is the founder of Kinkle Consulting, LLC. She has more than 14 years of experience in planning, and scheduling of infrastructure, roadway, commercial, and bridge projects. She was named ENR's Southeast Top Young Professional of 2017. Kinkle Consulting provides technical and Project Management consulting using 3D Modeling integration with 4D capabilities. They also provide schedule auditing and analysis as well as delay and claim management. (Chapter 7: Scheduling Projects.)

**Daniel D'Angelo,** P.E., Clifton Park, New York. Principal Civil Engineer with Applied Research Associates, Inc., Risk Manager to the Executive Steering Committee, technical adviser, and federal liaison for the Tappan Zee Hudson River Crossing Project. He provides consultant services on risk management and project management. (Chapter 8: Risk Management.)

**Robert Moreno,** Agoura Hills, California. Sr. Vice President, Southern California for Bergelectric. Projects of note include the Children's Hospital of Orange County and St. John's Santa Monica Hospital. (Chapter 12: Electrical Construction.)

**Lantz Holtzhower,** Ph.D. Kennesaw State University, Marietta, Georgia, Assistant Professor Construction Management, College of Architecture and Construction Management. (Chapter 13: Mechanical Piping Systems.)

**Aaron Cohen,** CPC, Phoenix, Arizona. Director at InEight Inc. for building estimating and project management software solutions and AGC Lecturer at Arizona State University. (Chapter 16: Software Support.)

**Aviad Shapira,** D.Sc., Professor (Emeritus) of Construction Engineering and Management, Technion–Israel Institute of Technology, Haifa, Israel. Over the years he was a Visiting Professor at the University of New Mexico in Albuquerque and at the University of Wisconsin–Madison. Dr. Shapira accrued his practical experience as a project engineer and project manager in a general contracting firm prior to pursuing an academic career. He specializes in building construction equipment selection, operation, management, economics, productivity, and safety. (Chapter 19: Equipment for Building Construction.)

# PREFACE

While many construction task operations have not changed, the industry has moved away from design-bid-build project delivery to Design/Build and Construction Manager/General Contractor (CMGC) methods. Only 19% of the construction put in place in the next 3 years will be delivered by the old design-bid-build approach. Sustainability is the political/social buzzword; however, a better statement of what can and is now being delivered is low-energy or zero-energy use buildings. New materials and designs allow creation of energy-efficient homes and buildings. The concepts are not new; in fact the Greek philosopher Socrates studied how to build energy-efficient structures 2,500 years ago.

Other critically important construction project delivery issues are: (1) schedule to meet the owner's need, time to market, (2) comprehensive work safety, and (3) delivered quality. To accomplish these goals, builders are moving to coordination of tasks using Building Information Modeling (**BIM**) and other linked software systems. These tools provide all members of the project team with access to current project drawings and related documents, including specifications. "The future of construction is connecting the field to the office."[1]

**BIM**
*BIM models are different from 2D or even 3D CAD drawings. They are made up of intelligent objects, so when design changes are made, the model is updated throughout design no matter who is working with it. BIM is a process, not an application*

The challenge, even on small projects, is to ensure that everyone—not just the engineers in the office but the superintendents and foremen in the field directing the work—is working with the latest set of plans. When a project is constructed using paper documents and plans, someone is going to be left out of the data-loop. The result will be delays and rework.

> Software tools such as BIM and PlanGrid are the digitized future of construction processes. They serve to accelerate project delivery, improve productivity, and enhance delivered quality.

Always be aware of available tools but never forget; construction is a people business and success goes to those who spend time personally observing and interacting with people. Human relationships are critically important and must be developed. The computer and the apps on the iPads are tools, but people lead and perform the work of construction. This imperative is discussed in Chapter 2: The Project Team.

Many individuals and firms have supplied information and illustrations for this work, and we owe them a great debt of gratitude. Others have freely given their time to explain specialty areas of construction. Their support and help is appreciated. However, full responsibility for all content rests with our author team.

We solicit comments on this edition.

—***CLIFF SCHEXNAYDER***
*Construction Engineer*
*Frisco, Texas*

—***CHRISTINE FIORI***
*Program Director of the Construction*
*Management Program*
*Drexel University*

[1] Tracy Young, PlanGrid's co-founder and CEO. PlanGrid is now (20 Dec. 2018) part of Autodesk.

# HANDBOOK FOR BUILDING CONSTRUCTION

# CHAPTER 1
# THE CONSTRUCTION INDUSTRY

*The construction industry is the second largest goods-producing industry in the United States. It employs more than 6.4 million people in craft and management positions. Construction work can be broken down by type: residential, commercial/institutional, high-rise building, industrial, and heavy/highway. Most contracts are awarded to a general contractor who awards subcontracts to specialty contractors. Today, Alternative Project Delivery methods are used for the majority of projects, with the most common project delivery method being design–build where the designer and contractor are a team from the beginning of a project. This is a change from the design–bid–build project delivery method of the past.*

## CONSTRUCTION

The use of Alternative Project Delivery (APD) methods, sometimes referred to as Alternative Contracting Methods (ACMs), including variations of Design Build (DB), Construction Manager/General Contractor (CM/GC), Construction Management-at-Risk (CM-at-Risk), and Alternative Technical Concepts (ATCs) has changed the roles and responsibilities of owners, designers, and contractors. To further complicate the situation, many of these delivery methods include a private financing component, Public-Private Partnership (PPP or P3s), adding a fourth participant interested in timely delivery of a project—the concessionaire. Almost all sectors of the construction industry, building, industrial, and heavy-highway, have embraced the use of APD methods. Such alternative delivery methods are changing how a project owner or a concessionaire, the designer, and the contractor:

1. Achieve project success
2. Manage project delivery
3. Allocate Quality Assurance (QA) activities
4. Allocate project risks

> APDs are about identifying opportunities and allocation of risk.

## GLOSSARY OF PROJECT DELIVERY METHOD TERMS

**A + B bidding (cost-plus-time bidding).** A bidding procedure used to select the low responsive bidder based on a monetary combination of the traditional bid price (A) and the time (B) needed to complete the project or a critical portion of the project.

**Alternative Technical Concepts (ATCs).** Proposing teams are allowed to submit innovative, cost-effective solutions for achieving project success. The offered solutions must be equal to or better than the owner's design and/or construction criteria. The ATC process is often used when a project is being delivered with the design–build method.

**Construction Manager (CM).** Sometimes referred to as the "construction manager—general contractor (CMGC)" method.

**Construction Manager at Risk (CMAR).** When the CM makes a commitment to deliver the project within a guaranteed maximum price (GMP).

**Design–Bid–Build (DBB).** A project delivery method where a completed project design is provided to the bidding contractors. The project owner has separate contracts with the designer and the construction contractor. Construction cost (bid price) is a significant factor in contractor selection. The method is often termed the "traditional method."

**Design–Build (DB).** More project owners are entering into a single contract with one entity, the design-build team, to deliver their projects.

**Design–Build–Finance–Operate–Maintain (DBFOM).** This method requires the contractor to design, construct, finance, operate, and maintain an owner's facility. It is a type of Public–Private Partnership (P3) concession.

**Multiparameter bidding.** A bidding procedure used to select a project contractor based on a monetary combination of the traditional bid price (A), the completion time (B), and other elements (C) such as construction quality, safety, and life-cycle costs. Quantification of the elements and bidder evaluation methodology are included in the procedure. The method is sometimes termed A + B + C bidding.

## A GOODS-PRODUCING INDUSTRY

Construction is a goods-producing industry. Still, certain segments of the industry are service businesses: for instance, designers, who prepare studies and designs for clients, provide a service. On the other hand, contractors, who build facilities, produce a product. The most significant differences between the construction industry and most other product industries are the size and cost of the product, and in most cases custom-design, one-of-a-kind features. The product to be delivered by any construction contract is also a manufactured product: a building, road, factory, church, residence, or office building. These products are constructed in response to the specific needs of a customer. Contractors service the needs of their customers, but the product of construction is not a service: it is a product, properly built in accordance with the expectations of an owner.

Construction in the United States can be described as a single industry. But it includes more than just the work of building structures on a project site. It includes engineers and architects who create project designs, the manufacturers and distributors of the materials and equipment for the project, the construction managers who lead the on-site work, unions and tradespeople who perform the actual work, the municipal officials who review plans and enforce the building codes, the workers at concrete batch plants where concrete is proportioned for delivery to the project site, the workforce in manufacturing facilities assembling prefabricated components used in many buildings, and many others in supporting roles. The industry is in reality a sector of the economy.

Construction can be an easy-in, easy-out business. It is easy to get started because the licensing requirements are not uniform from state to state and not difficult to fulfill in most states. Nearly anyone with a pickup truck and a cell phone can start a small construction business. The number of people keeping their accounts receivable tallied on one visor of the pickup and the accounts payable tallied on the other is probably quite large.

For very small companies, getting out of the construction business is not difficult either. All it takes is one bad project. Most construction companies are small, having only one or two permanent employees. Such companies are generally not able to withstand a single unsatisfactory or deficient project or a down year, so the number of construction companies experiencing failure each year is quite large when compared to other industries. About 10,000 construction companies, somewhere around 1–5% of the industry, fail every year with about $2 billion in liabilities. According to many studies, profit margins in the industry can be very minimal, only about 1–3%.

## PROJECT SCOPE

The first and most important project planning document is the owner's definition of facility need, the Project Scope. Producing a clear and comprehensive description of project purpose and desired features is not an easy task. To simply tell an architect, "We need a school building for six hundred students" will not produce the desired facility. The result of such direction will be multiple change orders, and change orders cost an owner extra money.

> Inadequate scope definition is a leading cause of project failure.

When APD methods are used, the contractor becomes involved in supporting the project owner's effort to craft a project scope document. The time and effort devoted to working closely with an owner to clearly describe goals—deliverables, significant milestones, facility performance, expectations/commitments to external stakeholders—will lead to accurate predictions of project cost and schedule, and to delivery of a facility having the desired operational characteristics. The need for full and clear project scope definition is imperative even for small projects. Nevertheless, in many cases projects suffer from inadequate and imperfect scope definition. The development of a clearly articulated project scope using standardized terminology requires participation by a wide cross-section of individuals and, in many cases, active participation by external stakeholders. Without the effort of carefully developing a project scoping document, significant items can be overlooked. These could include:

- Flood plain encroachment (project constraint)
- Fire water pressure needs for multistory building (a mandatory deliverable)
- Facility effect on surrounding area traffic flow (effect on external stakeholders)
- Electrical wiring to take advantage of solar energy (owner-directed inclusion)
- Permitting and building code requirements (regulatory requirements)

In addition, be forewarned, many project owners are sensitive to things like:

- Sustainability objectives
- Long-term carbon footprint
- Energy independence systems
- Use of natural lighting

Get everyone to sign off on the final statement of project scope. With an agreed-to scope there will be minimum cost and schedule changes required during the construction phase of project delivery.

## PROJECT QUALITY

With the DBB contracting method, contractors perform Quality Control (QC) sampling and testing, and the results from the QC tests may or may not be used by the owner for work item acceptance decisions. The contractor's focus is on meeting the contract specifications; this is a minimum requirement for quality. Quality management becomes much more important when APD methods are used by a project owner. With an APD method, the responsibility roles of the owner and contractor for achieving quality are changed. In many cases the project designer will not have a direct contractual relationship with the owner. With the shift to an APD method, the contractor has the ability to leverage and introduce new approaches to providing quality.

The use of an APD contracting method usually leads to greater owner satisfaction with delivery of the project. A comparative study of public school projects in the Southeastern United States completed between January 2006 and December 2012 found the CM at Risk Alternative contracting method superior to DBB in all areas of owner satisfaction: overall service, planning, cost control, schedule control, quality control, communications, and cooperation.

## ORGANIZATION OF THE INDUSTRY

There are many ways to describe the organization of the construction industry, but it may be best to visualize it as organized around delivery of five types of projects: (1) residential, (2) commercial/institutional, (3) high-rise building, (4) industrial, and (5) heavy/highway. Owners who expect to earn a profit from the constructed facility build most projects. To fulfill a social need, government agencies build the last category.

### Residential Construction

This category includes contractors and developers who build individual homes, apartment complexes, and assisted-living facilities. Builders of individual homes generally fall into two groups: those who build custom homes individually suited to the buyer's needs and builders of tract homes, offering standard designs and floorplans with a limited number of options or upgrades. Residential construction is generally not bid work. Builders of custom homes and apartments negotiate a construction price with the owner. Developers of tract homes set the price of the finished house based on market conditions.

### Commercial/Institutional Construction

Commercial construction includes offices, manufacturing plants, medical centers, and retail shopping centers, whereas institutional construction refers to healthcare, education, recreation, or public works projects. Projects in this category range in size from small convenience stores to multimillion-dollar office buildings.

Both private owners and government owners contract for buildings in this category. Commercial projects can be very different compared to residential work, having different rules, regulations, and building standards. Private owners generally negotiate the price of the buildings, while the government agencies typically use a low-bid or best-value process.

### High-Rise Building Construction

High-rise structures pose particular design and construction challenges (Fig. 1.1). There is probably no universal definition of high-rise buildings, but one government planning agency uses:

- Low-rise: 1–6 stories (in the order of 10–76 ft)
- Mid-rise: 7–12 stories (in the order of 88–150 ft)
- High-rise: 13 stories and higher (greater than 165 ft; and a mechanical floor is needed for every 25 stories)

Emporis Standards (a global provider of building information) defines a high-rise building as "a multi-story structure of unknown height from 12–39 floors" and a skyscraper as "a multi-story building whose architectural height is at least 330 ft." In New York City, Central Park Tower will have a height of 1,550 feet and the skylines of many other cities are changing with new high-rises. Steel frame construction is used for these tall buildings.

**FIGURE 1.1** Large commercial project site with multiple high-rise buildings.

## Industrial Construction

Manufacturing plants, petrochemical refineries, pipelines, electricity-generating facilities, and high-tech facilities, such as hospitals and clean rooms, are examples of industrial facilities. Contractors who perform this category of work are specialists in the kind of projects they undertake, and they normally negotiate a price for the delivered facility.

**FIGURE 1.2** Major bridge project.

### Infrastructure Construction

In addition to highways, this category includes structures such as dams and levees, canals, water and wastewater treatment projects, bridges (Fig. 1.2), railroads, and tunnels. Such work, which represented about 25% of all construction spending in 2017, is usually awarded by a low-bid or a best-value process, with design-build more prevalent for large complex projects.

In the case of transportation projects, the owner is usually a government agency. For some complex projects, owners prefer to prequalify bidders. This means a contractor will be allowed to bid only after completion of a prequalification process to demonstrate experience, qualifications, technical and managerial expertise, and financial capacity for the project.

## *PUBLIC AND PRIVATE WORKS*

There are many ways to categorize the construction industry besides project type. One of the most convenient is to consider public work (government) projects and private work. Public sector work accounted for less than 25% of the construction dollars spent in 2017. The largest share of the construction market is private work. The type of owner, public or private, is significant because of the differences in procurement procedures and regulations.

The public bidding process is encumbered by rules, which have no application to private projects. Public work is defined here as all work performed using the public's money. This includes government agencies at all levels: federal, state, and local. It also embraces small agencies such as local water districts and school boards. Public agencies are not spending their own money; the money expended to complete a project belongs to the taxpayers. Such organizations have bureaucratic rules and procedures, thousands of rules and policies. Projects must be awarded to the lowest responsible and responsive bidder. Exceptions are not allowed. There are similarly strict rules governing how projects are advertised. Anyone who can provide a bid bond is allowed to bid unless the project requires contractors to prequalify.

Private persons and corporations, on the other hand, are spending their own money. They are not as encumbered by bureaucracy and rules. They may open bids (or simply take proposals) in any way they choose. They may decide who will be allowed to submit proposals. They may choose not to advertise a project at all, but to award it to their favorite contractor who has built a reputation of delivering quality. The majority of construction is privately funded. Public projects receive most of the publicity, but there is more private work than public work.

## *PROJECT PARTICIPANTS*

A description of the industry would not be complete without including the other professions, government organizations, and stakeholders who are part of the construction industry.

### Owners

No project is ever undertaken without an owner who defines the need, provides the financing, makes the decision to build the facility, and manages the construction process. Owners can be either public or private. Most public owners, such as the Corps of Engineers or departments of transportation, are experienced **construction managers**. The majority of private owners, such as small manufacturing companies, have little or no construction management expertise and depend on consultants to help them manage their projects. Besides providing the project funding, the primary responsibility of the owner is to define the scope of the work.

**construction manager**
*A person or entity who coordinates the construction process on behalf of an owner.*

### General Contractors

Contracts are typically awarded to a **general contractor** (GC) or prime contractor, who is the party signing the contract with the owner to deliver the project. The GC could be an individual, a partnership, a corporation, or other business entity capable of performing construction work as a contractor. While there are differences, many states have specific licensing requirements for general contractors. Typically, requirements include:

- Completion of a trade exam together with a business and law exam
- Documentation of work experience
- Proof of general liability insurance

**general contractor**
*A general contractor has the primary contact with a project owner and supervises the project while self-performing some tasks and hiring specialty subcontractor for others.*

States without specific General Contractor licensing requirements often have other stipulations such has holding a business license, or registration with a Contractors Board or State Division of

Labor. States without licensing regulations usually advise contractors to check the licensing requirements of local governments before beginning a project.

The general contractor manages project construction and delivered quality, keeps the work on schedule, controls costs, and ensures coordination of work tasks and their performance in a safe manner. Because the GC holds the contract to deliver the project, this entity/person handles all coordination with the owner. The GC may hire subcontractors skilled at a specific type of work and knowledgeable about the code requirement, and other regulations pertaining to their work specialty.

The general contractor is hired to oversee and coordinate all project work, and to deliver a completed project, whereas a specialty contractor specializes in the execution of a particular type of work.

## Subcontractors or Specialty Contractors

The construction of even a small building may require 20 or more **subcontractors**, also known as **specialty contractors**. Normally the subcontractor has a contractual agreement with the GC to perform a specific item of work and be paid by the GC.

Subcontractors or specialty contractors make up the largest portion of the construction industry. They offer a specialized set of skills such as electrical or mechanical, steel erection, acoustical, drywall and painting, and carpeting. A subcontractor holds a contract with the GC, not with the project owner. In most cases subcontractors are selected by the GC but in some cases a subcontractor is specified for a project by the owner. General contractors evaluate and select subcontractors based on offered price, but other important considerations are:

**subcontractors**
*Subcontractors work under a prime or general contractor performing specialty work tasks as independent contractors.*

- Safety record, typically the company's accident Experience Modification Rate (EMR)
- Financial strength
- Specialized equipment
- Past performance, including responsiveness to schedule and quality

> Subcontractors tend to work for contractors with whom they have an established relationship.

## Designers, Architects, and Engineers

In the case of DBB projects, owners select an **architect** or **engineer** (A/E) to design the project. The A/E is selected based on demonstrated ability to design the project in a specified time frame and to provide a design buildable at a cost acceptable to the owner. The designer will prepare the construction documents (plans and specifications) for the owner. Architects are usually the lead designers for building projects—vertical construction. They subcontract some of the work, such as the structural frame and mechanical/electrical/plumbing (MEP) systems, to engineers who specialize in such work. Engineers are normally the lead designers for industrial facilities and heavy civil projects. They may subcontract part of the work, such as train stations or office buildings, to an architect.

The designer's involvement during construction will depend on the project delivery method used by the owner. In the case of a **DBB** project, the owner may employ the designer to oversee the performance of the construction contractor. In such a capacity, the designer will evaluate the quality of the contractor's work and approve progress payments. If it is

**architect**
*An architect, working for a project owner, designs and reviews the construction of buildings.*

**engineer**
*Engineers collaborate with architects to design the structural, electrical, and mechanical systems of a building.*

necessary to issue contract change orders, the designer will assist the owner in negotiating with the contractor the magnitude of the resulting cost and time adjustments to the contract.

## Construction Manager

The term **construction manager** has many meanings. Some contractors call themselves construction managers. A professional construction manager provides a service—usually by serving as an advisor to an owner and overseeing the owner's construction contracts.

Frequently, construction managers subcontract 100% of the project work. Normally they prefer to be involved in both the design and construction processes because they bring construction expertise to the design process. Some engineering and architectural firms offer construction management services, to represent the owner during project construction. To make the definition even more complex, not all people who call themselves construction managers work for a construction contractor: there are some government agencies that have employees classified as construction managers.

**construction manager**
*Construction managers plan, coordinate, budget, and supervise projects.*

## Trades

The trades consist of both union and **open shop** workers. The rules of employment vary by state and locality. Unions are organized by craft, with specific unions representing workers within a specific craft trade. Take for example, electricians: those who belong are members of an electricians' union. The carpenters' union is somewhat more diverse. It can, based on local practice, represent workers who are carpenters, piledrivers, millwrights, shipwrights, exterior/interior specialists, scaffold erectors, insulators, and related craft workers.

**open shop**
*A contracting business that does not require craft employees to have union membership.*

## Labor Unions

The purpose of **labor unions** is to improve the lives of their members. They use the power of collective bargaining to address wages, benefits, and working condition issues. A union acts as an intermediary between its members and the employing business. Unions develop skilled workers by their apprenticeship training programs.

While unions are generally formed around crafts and trades, today many unions represent several trades, not only the trade used in its name. Unions are operated by the members, have constitutions, and are made up of locals. The carpenters' union, "The United Brotherhood of Carpenters and Joiners of America," has over 2,500 locals. The locals serve as the contract-negotiating body for their members in their local area, negotiating labor agreements presented to and accepted by the signatory companies. The signatory companies employ union workers and agree to the terms of the union's contract. Employers of union members benefit from the union training programs and hiring halls.

**labor unions**
*Unions organize workers to influence work conditions. They also provide their members construction tasks skill training.*

## Insurance Companies

Contractors are often required to provide a bid bond as a condition of being allowed to offer a price to construct a project. A bid bond provides the project owner assurance about a bidder's financial resources to undertake the job under the terms of the offered contract and at the price quoted in the bid. If a contractor fails to provide the required bond, the bid will automatically be rejected.

Usually, contractors must provide insurance, performance bonds, and payment bonds prior to contract award. Insurance companies also provide these other required bonds. Correspondingly, they arrange liability and property insurance for contractors. Most public construction contracts require bonding because the bonds provide the project owner legal and financial protection.

## Banks

Banks provide the necessary working capital to contractors so expenses for materials, labor, equipment, and overhead can be paid in the interval between performance of the work and progress payments by the owner. Typically, owners make progress payments monthly. Banks also provide both short-term and long-term financing to project owners.

## Suppliers

All material used for construction projects, from concrete to paint, comes from suppliers. Many suppliers assist designers with material and installed equipment selection. Suppliers provide prices to contractors for preparing bids, prepare shop drawings, and fabricate items specifically for individual projects. Designers rely heavily on standard specifications and standards such as those published by the American Society for Testing and Materials (ASTM). It is important for designers to understand the standard specifications and design standards they are using because there are design standards for nearly every level of product quality. When the owner wants a high-quality product, it is important to use a specific quality standard. It is equally important to ensure the quality of a construction project by using quality suppliers.

## Permitting Agencies and Building Authorities

Permitting agencies and building authorities represent the interests of public safety. This group includes federal agencies such as the Federal Highway Administration and the General Services Agency, state departments of transportation, and local agencies such as municipal zoning boards and even homeowners' associations. They administer publicly funded construction projects, and they ensure private construction projects comply with zoning laws and building codes.

The intent of building codes is ensuring reasonable standards of construction quality and safety for the public. Building codes usually require a contractor to obtain a building permit from a public agency prior to any construction, repair, or alteration of a building or structure. An application for a building permit together with any necessary supporting data are reviewed by the building department to confirm compliance of the proposed work with the applicable building laws/codes of the jurisdiction. During progress of the work, a building department official will examine the areas of work subject to inspection. In the case of commercial buildings, final approval is needed before issuance of a Certificate of Occupancy. A Certificate of Occupancy is not a guarantee of quality or even of conformance to the local Building code. Because of the legal doctrine of sovereign immunity, there can be no redress against the building department if defects are discovered later.

## The Public

The public is impacted by every construction project. Those effects can be good or can have a negative impact on an area. Voters approve bond issues for new schools, roads, and other public facilities. But the public is often inconvenienced while a project is being constructed. Individual members of the public residing close to a project can be irritated by construction noise, vibrations, dust, or lighting for nighttime work. These can be especially bothersome when work occurs early in the morning, after nightfall, or on weekends. Adjacent structures can be damaged by construction project activities. Significant settlement or vibration damage may occur due to dewatering, pile driving, and heavy equipment movement.

## Industry Associations

Most participants in the construction industry belong to one or more industry or professional association. Membership in some of these associations is by the firm or corporation, whereas in other cases membership is by the individual.

***American Concrete Institute (ACI)*** is a technical and educational organization. The mission of ACI is to develop, share, and disseminate the knowledge and information needed to utilize concrete to its fullest potential. It is a leading authority and resource of consensus-based standards and other technical resources addressing concrete design, construction, and materials. The Institute's publication *Building Code Requirements for Structural Concrete*, ACE 318-14 is an important document for contractors responsible for the placement of building concrete.

***American Institute of Constructors (AIC)*** is a professional society for practicing constructors. It is the sponsoring organization for the Constructor Certification Program. The AIC Constructor Certification Commission was organized to expand the constructor-qualifying process to include a written examination and to offer an internationally recognized certification process. The certification process was peer developed and is intended to set high standards for skills, knowledge, education, and conduct for those recognized as a Certified Professional Constructor (CPC).

***The American Institute of Architects (AIA)*** is a professional organization for architects in the United States. AIA provides model contract documents for the design and construction industry. It attempts to raise public awareness of the value of architecture and the importance of good design.

***American Institute of Steel Construction, Inc. (AISC)*** is a nonprofit trade association and technical institute established to serve the structural steel industry in the United States. AISC publishes AISC 360 *Specification for Structural Steel Buildings*. This specification is the guidance document for steel building structural design and is referenced in all U.S. building codes.

***American Society of Civil Engineers (ASCE)*** represents members of the civil engineering profession worldwide. It is the world's largest publisher of civil engineering content and an authoritative source for codes and standards to protect the public. The Society's discipline-specific institutes and technical groups bring together volunteers from around the world to advance the profession's technical expertise. The Construction Institute within ASCE specifically addresses design skills, and supervision and management of projects.

***Associated Builders and Contractors (ABC)*** is a national trade association representing contractors, subcontractors, material suppliers, and related firms. ABC is the only national association devoted exclusively to the merit shop philosophy of providing the best management techniques, the finest craftsmanship, and the most competitive bidding and pricing strategies in the industry, regardless of labor affiliation.

***Associated General Contractors of America (AGC)*** was the first major association of general contractors. Today the AGC has over 26,000 members, including over 6,500 general contractors and over 9,000 specialty contractors, plus suppliers, and service providers. Historically, the association has handled wage and contract negotiations with local building trade unions throughout the United States.

***Construction Industry Institute (CII)*** is a research organization with a singular mission: improving the competitiveness of the construction industry. The CII is a unique consortium of leading facility owners and contractors who have joined together to find better ways of planning and executing capital construction programs.

***National Society of Professional Engineers (NSPE)*** is a national society of licensed professional engineers (PEs) across all disciplines. It promotes the ethical and competent practice of engineering and licensure. NSPE serves its members and the public through state and territorial societies and chapters.

## CONSTRUCTION LABOR FORCE

Twenty-five states have right-to-work laws, making it illegal to require union membership as a condition of employment. On the average, 8.9% of the construction workers in the right-to-work states belong to a union. The other states are union states where the state laws permit union organization of employees and union contractors agree to hire only through the union hiring halls.

### Project Labor Agreements

Government agencies have on occasion utilized collective bargaining **project labor agreements (PLAs)** for their projects. A PLA is for a specific construction project and is applicable only for the project's duration. The agreements terminate at completion of the project. PLAs can have many different sets of agreements, but the basic agreement commits contractors to hire workers only through union hiring halls.

In return, the unions agree not to **strike** for the duration of the project. By law, a PLA has no impact on a contractor's existing employees, only on hiring new employees. According to recent court decisions, this kind of requirement does not exclude open shop contractors from bidding work. However, the number of union versus nonunion contractors who successfully bid PLA projects does not support the claim.

**project labor agreement (PLA)**
*Agreement between the contractor/s and unions for a single large project to hire workers through the union hiring hall in exchange for a no-strike agreement.*

**strike**
*A work stoppage by a body of workers to enforce compliance with demands made on an employer.*

### Construction Crafts

Trade unions all have very valuable training programs, and these are an important contribution to the industry. But these programs are able to reach only a limited number of people. Craft training requires a great deal of "hands-on" work. For some tools, the instructor needs to put his or her hands onto the hands of the student to demonstrate and teach the proper use of the tool. Because of the nature of the training, the size of classes must be limited. The industry needs more trained craftspeople than the unions can provide. Construction company owners believe the top challenge facing the construction industry is a shortage of trained labor. Labor and management need to work together to solve this problem. Here are short descriptions of the tasks each construction craft performs.

***Brickmasons, Blockmasons, and Stonemasons.*** The work of a mason varies in complexity, from laying a simple masonry walkway to installing an ornate exterior on a high-rise building. Brickmasons and blockmasons are the trowel trades. They build and repair walls, floors, partitions, fireplaces, chimneys, and other structures with brick, precast masonry panels, concrete block, and other masonry materials. Stonemasons work with natural-cut stone and artificial stone, building walls or creating exterior stone surfaces and floors.

***Carpenters.*** The tasks of cutting, fitting, and assembling wood and other materials for the construction of buildings, highways, bridges, docks, industrial plants, and many other structures are the responsibility of carpenters. Builders are increasingly using specialty trade contractors who, in turn, hire carpenters who specialize in just one or two types of work. Some of these activities are setting forms for concrete construction; erecting scaffolding; or doing finishing work, such as installing interior and exterior trim. However, a carpenter employed by a general building contractor often must perform a variety of the tasks associated with new construction, such as framing walls and partitions, putting in doors and windows, building stairs, laying hardwood floors, and hanging kitchen cabinets.

***Cement Masons and Concrete Finishers.*** One of the most common and durable materials used in construction is concrete. Cement masons and concrete finishers place and finish the concrete. They may also color concrete surfaces; expose aggregate (small stones) in walls and sidewalks; or fabricate concrete beams, columns, and panels. In preparing a site for placing concrete, cement masons check the forms (formwork) for holding the liquid concrete and properly aligning it according to the plan dimensions. They then direct the placing of the concrete and supervise laborers who use shovels or special tools to spread the liquid concrete. Concrete finishers will guide a straightedge back and forth across the top of the forms to "screed," or level, the freshly placed concrete. Immediately after leveling the concrete, finishers carefully smooth and/or texture the concrete surface.

***Construction Equipment Operators.*** Skilled operators, also called operating engineers, maneuver or control the heavy machinery (cranes, bulldozers, front-end loaders, rollers, backhoes, graders, dredges, hoists, drills, pumps, and compressors) used to construct roads, bridges, buildings, and other structures. The operation of these machines is becoming more complex as a result of computerized controls. Equipment operators may set up and inspect equipment, make adjustments, and perform minor repairs and maintenance to their machines. Some states require construction equipment operators to have an occupational license. When equipment operation involves moving a machine from one site to another, a commercial driver's license (CDL) is needed. An operator of piledriver equipment working in some states must have a crane license.

***Construction Laborers.*** Although the term "*laborer*" implies work requiring relatively low levels of skill or training, many tasks performed by these workers require a high level of training and experience. Construction laborers perform a wide range of physically demanding tasks. They may be employed to operate hand and power tools, air hammers, earth tampers, cement mixers, small mechanical hoists; set braces to support the sides of excavations; and erect scaffolding. Construction laborers help other craft workers including carpenters, plasterers, and masons.

***Electricians.*** The electricians perform electrical construction, fabrication, installation, and erection work. They install electrical systems and equipment for a variety of purposes, including lighting, climate control, security, and communications. Electricians work inside or outside buildings.

***Glaziers.*** Glass has many uses in modern buildings. Insulated and specially treated glass keeps in warmed or cooled air and provides good condensation and sound-control qualities; tempered and laminated glass makes doors and windows more secure. In large commercial buildings, glass panels give office buildings a distinctive look while reducing the need for artificial lighting. Glaziers are responsible for selecting, cutting, installing, replacing, and removing all types of glass.

***Painters.*** A painter prepares surfaces for paint and then paints the surface with the appropriate number of coats. The work requires bending, kneeling, reaching, and climbing. Some projects require a painter to be high above the ground or floor; therefore, safety equipment can be necessary. Others apply sealers to protect outside walls from wear caused by exposure to the weather and air pollution.

***Pipelayers, Plumbers, Pipefitters, and Steamfitters.*** Although pipelaying, plumbing, pipefitting, and steamfitting sometimes are considered a single trade, workers generally specialize in one of the four areas. Pipelayers place clay, concrete, plastic, or cast-iron pipe for drains, sewers, water mains, and oil or gas lines. Plumbers install and repair the water, waste disposal, drainage, and gas systems in homes and commercial and industrial buildings. Plumbers also install plumbing fixtures, bathtubs, showers, sinks, toilets, and appliances such as dishwashers and water heaters. Pipefitters install and repair both high- and low-pressure pipe systems used in manufacturing, in the generation of electricity, and in heating and cooling of buildings. They likewise install automatic controls to regulate these systems. Some pipefitters specialize in only one type of system. Steamfitters, for example, install pipe systems to move liquids or gases under high pressure.

***Sheet Metal Workers.*** The installation and maintenance of air-conditioning, heating, ventilation, and pollution control duct systems; siding; rain gutters; downspouts; skylights; restaurant equipment; outdoor signs; and many other products made from metal sheets is the responsibility of the sheet metal workers. They may work with fiberglass and plastic materials. While some workers specialize in fabrication, installation, or maintenance, most do all three jobs.

***Structural and Reinforcing Iron- and Metalworkers.*** Many structures have steel frames of columns, beams, and girders. In addition, reinforced concrete—concrete containing steel bars or wire fabric—is an important material. In addition, metal stairways, catwalks, floor gratings, ladders, window frames, and decorative ironwork increase the functionality and attractiveness of structures. Metalworkers fabricate, assemble, and install these products.

Even though the primary metal involved in this work is steel, workers often are known as *ironworkers.* They install structural steel and reinforcing bars used in concrete. The structural metal arrives at the construction site in sections and is then lifted. Ironworkers connect the sections and set the cables to do the hoisting.

## SAFETY

The work environment of a construction project is dynamic. It is affected by weather and the changing conditions created by construction of the structural frame of a facility. It is a place where work teams executing very different tasks are confined in a common area. The constant change to the physical environment and interacting activities increase worker safety risks. Each worker within a specific trade is exposed to a unique pattern of risks. Roofers have higher mortality rates (55.5 per 100,000 workers) compared to other trades because they work at height. Reinforcing-iron workers suffer days off from work injuries because they typically perform their work on unstable and uneven surfaces. Ladders or scaffolding are used extensively by painters and create exposure to falls.

As a result, project managers need to think about project safety from both a comprehensive work site perspective and with recognition of specific trade-related hazards. Every trade has its own complement of safety risks driven by sources of possible injuries and accident scenarios. To achieve a safe project, accident risks must be considered by trade occupations and work tasks.

## ORGANIZATION OF THE CONSTRUCTION BUSINESS

Architects and design engineers accomplish design work. Normally for buildings, the architect will be the lead designer, which means it is the architect who has a contract with the owner. Structural engineers, mechanical engineers, or others may provide some design services to the architect; these engineering companies have a contract with the architect, not the owner. Usually the construction project itself is awarded to a general contractor. This is a contractor who manages the entire job, probably constructs part of the project, and subcontracts a large portion of the work to specialty contractors such as mechanical, electrical, masonry, and roofing subcontractors. The subcontractors have a contract with the general contractor and no contractual relationship with the owner or each other.

Some states and municipalities have laws permitting contracting using multiple prime contractors. Because there is no general contractor to perform the day-to-day management and coordination of the project when multiple prime contractors work on a project, the owner must perform those functions. The owner must likewise perform the general conditions responsibilities such as providing project access, clean-up services, office space, and utilities. Owners prefer multiple primes either to save the general contractor's markup on subcontractor work or to finish the work sooner by starting specialty contractor work early. However, the owner must intensively manage the specialty contractors, a function normally done by the general contractor. Many of these owner-managed contracts finish late because of the owner's limited experience in controlling construction work. This is not a widespread contracting system.

## OPPORTUNITIES

Many young engineers begin their careers by working as a project engineer or assistant project manager. Project engineers are assigned a great variety of tasks such as tracking **requests for information (RFIs)**, managing shop drawings, maintaining the daily records of the project, calculating pay estimates, updating project schedules, and resolving errors in plans and specifications. As a project engineer gains experience, they can become estimators, schedulers, and project managers. Larger companies have a greater ability to meet bonding requirements, which leads to a greater likelihood of working on a large project. This provides the opportunity to work with a larger project team. In small companies, project engineers work on smaller projects as members of smaller project teams.

**Requests for information (RFIs)**
*Formal documentation used by contractors to clarify contract requirements.*

Construction is a team endeavor. The job of each individual on the project team is to make sure everyone else on the team succeeds, because everyone is essential, and the project may fail if one person is allowed to fail. There are no jobs in construction where a person can work alone. In construction, people skills are an imperative key to success.

> Teams succeed or fail as a group.

General contractors tend to move their employees around more than specialty contractors because specialty contractors self-perform work and are regionally focused. General contractors tend to follow the work. There is always mechanical construction or electrical construction, for instance, in every city in the country, so specialty contractors have less need to move people.

A growing number of large companies are publicly held, with their stock traded in the stock market; some companies are privately held by a few individuals or a family; and some construction companies make ownership of company stock available as people are promoted up the management ladder. The impact of company ownership on an individual employee relates to areas such as potential for bonuses or profit sharing, potential ownership, or promotion limitations. All reputable companies are incorporated and licensed.

## DELIVERING A BETTER PRODUCT

New materials and management support tools are constantly becoming available. Therefore, for both company and personal success it is necessary to participate in one of the organizations listed in this chapter and to read trade literature. What is your awareness of the following?

**Building Information Modeling (BIM)**—Owners use BIM, a digital representation of physical and functional characteristics of a facility, as a management tool throughout a facility's life cycle. Therefore, project owners are including contract language directing contractors to use BIM. These requirements ensure that the owner's BIM needs are met. They also serve to give the entire project team a common understanding of how project components fit together. The use of BIM is an investment, an investment paying benefits by the elimination of project component conflicts and a reduced need for costly rework.

**High-Strength Reinforcement (HSR)**—The use of high-strength reinforcement can reduce the volume of reinforcement needed for the construction of a typical concrete building. This will reduce construction time, material costs, and, ultimately, the total cost of construction.

**Onsite nonpotable water**—With careful planning, wastewater, stormwater, and rainwater can be treated and reused in a building or for nonpotable needs such as irrigation, toilet flushing, and cooling. On-site nonpotable water systems contribute to sustainable water management and reduced the need for valuable potable water.

Always remember: people build projects. The most important ingredient for success is people, so build your team. At one location, this was painted on the wall in 2-foot-high letters:

## Our Vision

Safety—We do the Safe Thing

Excellence—We do the Best Thing

Caring—We do the Kind Thing

Integrity—We do the Right Thing

The following chapters will provide insight about these and other important project delivery techniques and tools.

# CHAPTER 2
# THE PROJECT TEAM

*Success in delivering a construction project is determined by how well the "project team" performs. The team must come together and be managed in a manner that facilitates its efficiency and effectiveness. The skills needed to manage such a team are not the typical hard skills familiar to builders, but the necessary skills can be learned and applied by those directing construction teams.*

## TEAM

What is a team? Most importantly teams are made up of people, who may or may not represent other organizations or companies. Typically, effective teams have their own organizational structure, with their own norms, goals, culture, and expectations, as well as assigned roles and responsibilities for each of the team members. Teams typically expect some degree of loyalty to the team and to the other team members.

In construction, teams (Fig. 2.1) may be formed with individuals from various trades, or from various project management organizations such as the CM, AE, owner, and even attorneys. The individuals on the team come with their own personalities, skills, shortcomings, and personal lives. They are not from a cookie cutter and so one size does not fit all when attempting to manage or lead a construction team.

While a construction project requires the orchestration of many different "assets," materials, equipment, time, money, land, and design, **the most important "asset" is people.**

Without people there would be no one to operate the equipment, install the materials, or develop the designs. Someone, and often "someones," needs to orchestrate all of these people and the other assets to accomplish the right things, at the right time, in the right way. These are the managers—foremen, superintendents, project managers, and task leaders. Some of these folks will be simply managers and some will be managers AND leaders. What's the difference?

## MANAGER VERSUS LEADER

Both managers and leaders use management skills, but leaders go a step further. An easy way to differentiate between managers and leaders is to view managers as those who "do things right," while leaders "do the right things!" Managers *often* go by the book and perform repetitive tasks. They know the right way to do certain tasks. Leaders have a vision of what all the tasks will accomplish, how to employ team members to accomplish the tasks, how to incorporate the tasks in a new way, and an

**FIGURE 2.1** A project team on the project site.

ability to think "out of the box" when needed. Leaders will redirect efforts when the situation changes; they take initiative. They understand when the totality of the project is veering off course or a new situation requires a change to the original plan. They lead execution of the required change to the plan. They know when things are off course because they have a vision of how the project should proceed and the important goals.

Leaders appeal to the motivation driver of their team members. Motivation is usually internal to the individual team members and not something externally created by the leader. Good leaders know what motivates the individual team members and appeals to their internal motivation drivers. According to psychologists, people do what they perceive to be in their self-interest. Motivation could include personal recognition for a job well done or done in unusual circumstances. Motivation could be earning time off, recommendation for a bonus, or an appeal to the project focus, such as patriotism on a Veteran's cemetery project. Leaders keep things going well or make changes if the task or project is not going well. Leaders lead—they do not, cannot, delegate their leadership responsibilities. Leaders understand the need for action; good leaders have a "bias for action." In an infantry example, when ambushed you should attack the ambushers if they are close, or you should run away from the ambushers if they are distant, but in no case should you do nothing and get destroyed! To paraphrase Lou Holtz, the famously successful football coach and philosopher: life is 10% what happens to you and 90% how you respond to it. Projects will not always go according to plan. Leaders ensure project accomplishment, in spite of what happens. To be effective at constructing a project both the leaders and managers have to be effective at dealing with people. Something important to remember:

| If you take care of your people, your people will take care of you and the project. |
| --- |

To be a leader, you must meet, know about, and support your people. A leader is there to support the team members; the team members are not there just to support the leader. You must look out for their safety. You need to provide the best working conditions possible based on the situation. You must not waste people's time. You need to facilitate.

Leader tasks include developing plans, assigning personnel, acquiring the needed assets (materials, equipment, budget, and personnel), ensuring personnel are properly trained or providing additional training, and empowering people to grow. Personnel development is very important for the future of an organization. Mentoring, counseling, and task assignments are key to personnel development. Personnel can be developed to learn new things for the benefit of the job at hand and to improve their overall capabilities for the future. Learning new things, development, is a great motivator for the majority of people.

Leaders communicate up, down, and sideways. They keep their boss(es) informed of the good and the bad; they are not "Lone Rangers." They manage expectations. They raise issues and present solutions for consideration. They communicate to their team and to other teams that affect the project, such as the owners, stakeholders, media, or city traffic control. Finally, real leaders exhibit *courage* to change direction and try new things; and they have a high level of *honesty* and *integrity*. Honesty is telling the truth; integrity is doing what you said you would do, properly, when no one is watching. *Trust* is an outcome of honesty and integrity, which will be discussed later. In addition, leaders practice the Golden Rule when dealing with people: Treat others as you would want to be treated. This is not the golden rule often encountered on construction projects: where it is often, he who has the gold, rules. For example: if you are doing something incorrectly and need to be counseled, would you want the counseling done in private or in front of your peers? There is an old adage for good managers and leaders: praise in public, condemn in private.

## MANAGEMENT

Managers are engaged with five management functions: (1) Plan, (2) Organize, (3) Staff, (4) Direct, and (5) Control.

### Plan

Planning is determining how to accomplish the project, usually it involves committing the plan to writing, and communicating it to the team members. This is how all the team members literally "get on the same page," know what and when they are to perform, and how they relate to other team members. It is not reasonable or effective, to simply bring in carpenters and have them "wing it" without the construction plans. They need plans with specific requirements. They also need the right materials, on time, and space to perform their work. A good plan anticipates what conditions will be encountered and how to operate within the constraints of the project. Construction planning is like sailing a boat. Sailboats have no reverse under sail, so you have to think way ahead of the boat's bow, both where it is now and where it is headed. You need to know the conditions such as water depth, wind direction and speed, the current's direction and speed, and the location of other boats and obstructions. The same is true of a project.

| Project work is not performed in vacuum. |
|---|

It is necessary to anticipate and allow for the environment of the work task. In the case of a construction project, it is good practice to first "build the project in your head." A leader visualizes the conditions, challenges, and activities. Then the pen is put to paper and the written plan prepared. Plans must be realistic. While one may know what ***should*** be done (in a perfect world), it is necessary to temper the plan based on the environment and resources available and develop a realistic plan—an accomplishable plan considering all impacts to project tasks.

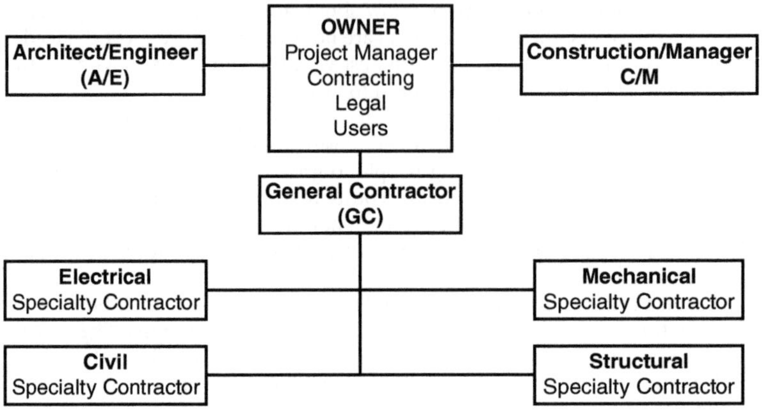


**FIGURE 2.2** A project organization chart.

## Organize

The organization is often depicted on a chart (Fig. 2.2) as hierarchical and horizontal relationships. This helps team members understand how they relate to one another. Developing the chart serves to delineate individual roles and responsibilities of the team members.

## Staff

Project *staff* are people. They are assigned to a project to fill specific positions on the organization chart and to perform specific functions to achieve the plan. Often the skill set of a particular individual may not match exactly all of the requirements of a job as indicated on the organization chart. Sometimes the individual may also have skills necessary for a different position. Such real-life situations often result in changes to the initial organization chart, making it more effective, given the skill sets of the actual staff member filling a position. In addition, training to fill-in knowledge or skill gaps can be very useful. Some staff may be needed and are affordable, only on a part-time basis. A part-time contributor with special skills is far preferable to trying to go it alone and potentially miss something important.

## Direct

Although the other four less obvious functions are also critical, directing is the most obvious day-to-day job of telling others what to do and how to do it. Directing can occur on the job site or during meetings. Directing requires good communication skills to ensure the course forward is clear and understood. It is the responsibility of the one communicating direction to ENSURE that the information provided is received and understood. Correspondingly, direction must be timely. The Plan should ensure that necessary meetings are held before certain activities are undertaken, not after the fact. "Chewing people out" when things go wrong, as a result of a failure to communicate the plan clearly in a meeting, is a poor form of directing. Ideally, directing takes place at planning meetings and again on the day of execution of the tasks. This is done so everyone knows what is required, knows their role and responsibility, and as a result things go smoothly.

### Control

Control consists of observing results and using those observations to modify the Plan, Organization, Staff, and Direction. If things are NOT going smoothly, someone needs to be on the spot directing and correcting. But the manager or leader must also assess "why" things did not go smoothly and what changes to one or more of the first four management functions should be made to avoid a repetitive deficiency. The deficiency could be in training, in communicating the plan, or in providing needed support.

> Success comes when there is a great emphasis on people.

Typically, people come from companies, and companies provide many things useful for projects. Companies provide: marketing forces used to find and secure the projects, insurance and bonding capacity without which most owners would not let the people work on the project, equipment, training, and staff. Even with all of these necessary "assets," ***companies do not build projects—people perform the work***. People in companies harness assets and make them available to the project as necessary. People from these companies become part of the project team and work in the team. These people report back to their companies and are paid by their companies, but have the project as their focus.

Management and leadership of people require soft skills which project engineers, construction managers, other supervisors are seldom exposed to in school. School and training routinely focus on the technical issues, to the exclusion of soft skills. This is a mistake. The technical issues are often not solved directly by the manager or leader, but by a manager or leader engaging the right technical experts, by communicating the issues to be solved, and by getting the best out of their team members. Again, these are soft skills. Hard skills (such as scheduling, safety, reinforced concrete, and air flow) are required to effectively construct the project, but to be effective and efficient and profitable the soft skills are critically important.

## *COMMUNICATION*

Effective communication is needed throughout the management process. Communication is the clear transmission and reception of ideas. While it is sometimes frustrating to the communicator, it is the responsibility of the communicator to ensure that the idea has been received and understood. Since all members of the team have diverse experiences and skills, different means of communication may be required so all members of the team understand. Some people learn through words, some through pictures, and some through body motions or hand movements. How will the manager or leader know if the idea has been received and understood? It is necessary to always seek feedback!

After presenting an idea to a group, ask if someone can articulate the critical points of the idea in their own words. A manager or the leader should seek to identify those team members who often do not grasp ideas readily. The manager or leader should then take them aside individually (not in front of the group) and ask them what they plan to do as a result of the direction.

IMPORTANT: if you just ask:

> "Do you understand?"
> Most people will say "yes" whether they really understand or not!

In such a situation the manager/leader does not know if all team members really understand the message. Another way to ensure the directions are understood is to create a mock-up. Using such a process both the "hands-on learners" and the "visual learners" will grasp the information.

Communication is a person-to-person engagement. Directions can be transmitted in many ways, some more effective than others. Face-to-face meetings, in-person meetings, and electronic audio/visual methods are preferable to communicating only in writing. Why? Because two-way communication is the most efficient method of transmitting ideas.

Experienced communicators will look for the reactions of those to whom the message is presented, looking at team member's body language and providing an opportunity for questions can help dramatically in providing clarification.

With multiple time zones and people not always available for face-to-face meetings, written communications certainly have their place. In fact, many things need to be committed to writing. Meeting agendas and minutes should be communicated in writing. The minutes should detail responsibilities resulting from the meeting: who, what, and when. These responsibilities are then reviewed at the next meeting. Such a process ensures accountability in the team, including the team leader.

To ensure understanding, follow-up meetings are important to provide opportunities to ask questions. A simple written directive with no meeting as a follow-up provides little opportunity for clarifications and can result in multiple interpretations and consternation among the team members. To be clear about today's world of communication methods: seldom is a text message sufficient direction!

One caution about meetings. All meetings should have a purpose and specific information to be transmitted or discussed. If not, do not waste people's time with a meeting! Publishing the agenda well in advance of the meeting lets people know if they should attend and provides an opportunity to add to the agenda if an issue needs input from multiple individuals. Fewer meetings, with real objectives, and a specified duration, ensure people bring energy and enthusiasm to the discussion—making the meetings more effective. Again, do not waste people's time.

Team leader accountability is important. It demonstrates integrity and reinforces the concept of the leader being part of the team effort in "doing," not just saying. Here is a real-life example: on a very challenging project the team leader ran the weekly coordination meeting—it was a large project with many meeting participants. Upon reaching an agenda item, the team leader noted how he had been responsible for accomplishing the item. Then he openly apologized for not having accomplished the task and then and there the leader said it would be done by the next week. You could have heard a pin drop! After an awkward pause, the leader went on to the next item on the agenda. Though the team leader was fairly new to the project, the work had been ongoing for 3 years. Following the meeting, the team leader asked a senior person from the meeting: "Why the silence?" and was told that his admission was the first time the individual in the team leader position had ever apologized for such a failure. The previous team leader had always blamed someone else. To make a simple apology and accept accountability was an important step in gaining the trust of the team. This single act of accepting ownership and responsibility was the beginning of bringing the team together.

Ideally all managers and leaders would be extroverts, and be schooled in oral, written, and body language communications. More likely, most engineers are introverts, so it is necessary to *act* like extroverts! Project managers and leaders need to take the initiative to meet people, learn about them, and be with them rather than hiding in an office and remaining in a closed "comfort zone." Management is a people skill. More practice makes for better leaders. A good test for determining if you are an extrovert or an introvert is to think of being invited to a marketing meeting. You will represent your company at a "meet and greet" reception with many people, all unknown to you when you enter. Do you dread the reception (INTROVERT reaction) or are you enthused to meet new people and "work the room" (EXTROVERT reaction)? All are not born extroverts, but if we want to be successful managers and leaders, we can learn extrovert behavior and become more comfortable in situations. Toastmasters is an organization in which you can learn how to meet people, practice "small talk," ask questions, and make presentations before groups. These are all useful manager/leader skills.

One technique for starting small talk is to ask easy questions. Instead of worrying about appearing ignorant, ask simple questions. Simple questions have two benefits. First, the other person should be able to answer the question easily, so the conversation feels like a social conversation, not an inquisition. Second, you know what you know, you want to know what the other person knows.

When problems arise, good leaders are HARD on the PROBLEMS, and EASY on the PEOPLE. People will solve the problems. They do not need additional stress imposed by the leader. The leader engages the right people, provides as much insight as possible, creates a collaborative environment, and supports the knowledge of the staff and workers. Unless the situation was caused intentionally or by negligence, there is no benefit in getting upset and taking it out on individuals. Rather, it's time to get going on a solution, and the solution will probably come from the staff or workers close to the situation. Note: stakeholders above the level of the leader may be upset at the situation, so the leader may have to provide a buffer between the project stakeholders' angst and those team members solving the problem.

> The leader represents the team to the stakeholders, while providing time and distance for the staff and workers to accomplish the work.

Leaders make the hard calls and **take responsibility**. If someone needs to be disciplined or removed from the team for intentional or negligent action, then the leader must step up and take the appropriate action. Such action cannot be delegated or the leader will forfeit respect. Nor can the leader afford to sacrifice the team or the project by inaction. Leaders engender trust.

## TEAM BUILDING

The best teams are those comprised of individuals who have worked together before. They know each other and how their tasks relate. To use a sports analogy, they are well practiced. On a sports team, when a play is called, detailed explanations are not needed. Players know and execute their assignments, based on their past practicing together. In construction we are sometimes fortunate enough to have familiar teammates from a previous project. To create efficiency and effectiveness, some companies even form strategic relationships and assign the same personnel to work with their strategic partners. More often, however, some or all team members find themselves working on a project with new teammates. Then the team relationships have to be developed—often times the team must come together as the project progresses. The building of a project team can be as challenging as the technical complexity of the project work.

> Team building (people working together) is critically important to the success of the project.

Again, using a sports analogy, there are many examples of highly skilled team mates ("all stars") who work poorly together and fail to achieve expected results. Such was the case when one of the NBA "Dream Teams" failed to win Olympic Gold. Sure, the talent (Staff) was there—but the team work (Plan, Coordination, Control) was not. The coaches (leaders) failed to establish a winning team culture. Letting individual stars "do their own thing" instead of knitting them together as a team was a failure in team building. Then enter "Coach K," who is known for leadership; he has turned teams around and brought individual stars together (Control), resulting in many Olympic and World gold medals.

On construction projects we can have lots of individually talented performers who operate in their own "silos," who *meet* together, but who do not *work* together. The results are almost always suboptimized and are often characterized by a lack of trust and a lack of team work. Worse, antagonistic relationships develop. Antagonistic situations often slow the project and increase its cost, including claims. At the heart of these failures is frequently a lack of trust among team members who come to believe individual profit motives are the reason for the lack of teamwork. The individual skills of team members will get the project only so far. Team member motivations and attitudes are key to forging a successful, working team. Trust must be developed.

In team building and in accomplishing projects effectively, a leader must ensure creation of an openness where everyone can be heard, even those with differing opinions. Sometimes a group of similar minded people will overlook an important item, leading to problems. This is not to say the leader wants to create discord. But the leader must work to get everyone on the same page. Before an important decision, it is best for the leader to gather all relevant input. At the Pentagon Renovation project, prior to 9/11, an engineer who was often considered to be out of the main stream of thought asked an important question: "Will the existing walls be strong enough to restrain the blast forces on the blast proof windows being installed?" The project had already been designed. Someone must have previously considered the question, right? Wrong! The walls did not have sufficient strength. As a result, the walls were reinforced with steel concurrently with the blast proof window installation. On 9/11 when the plane hit this exact section of the building with reinforced walls and blast proof windows, the reinforced walls held up for about 20 minutes and then collapsed, allowing people to escape, people who would have been crushed had the wall collapsed immediately. The engineer's question, and the freedom to ask it, saved lives.

## TRUST

Trust is a requisite for leadership, for team work, and for saving time. Again, trust is a requisite. Trust is born of honesty and integrity. Trust is why projects have "close-in inspections." Once the wall is built, the installation behind the wall is no longer visible. We all trust the work was done right and inspected properly before being closed in, and before we take an elevator in a skyscraper! The Corps of Engineers has a three-step quality program. It provides inspection at the appropriate level, thereby building trust in the construction quality. Some checking is always required. As Ronald Reagan said about nuclear weapons with the Russians, "Trust but verify." However, trust means less checking is required. Trust leads to speed, less over inspection, and many other benefits.

Lou Holtz became head coach of several major college football teams. One of those teams had been plagued by a lack of trust among players just before his arrival. The defensive line was not stopping the running game of the opposition. So the linebackers decided to move up toward the defensive line to help stop the run. By doing so they left an opening between their new position and the defensive backs, so the defensive backs moved up closer to the linebackers. Soon opposition coaches noticed the defensive team compression and sent wide receivers past the defensive backs for easy passing touchdowns. Once Holtz arrived, he took individual names off the back of jerseys, and gave each team member a T-shirt. On these T-shirts was printed "Trust me to play my position." Those T-shirts were worn in practice and on game days. Within 3 years the team was undefeated and a national champion. The talent level was about the same; much improved trust and focus on a team (vs. individual) goal were key to the transformation.

Another aspect of trust is using information properly. Many times the CM, AE, or GC will have access to owner-sensitive information. The information has been provided by the owner to make the project more effective, but not to be shared with others. The information could be physical characteristics, proprietary processes, or other technical data. Real harm can come to the owner (a team member) if such information makes its way to the owner's competitors or opponents. When the New York World Trade Center was bombed in 1996, there were articles and photographs in the public domain explaining why the bombing had failed to take down the structure! Armed with this information, 9/11 attack was effective, hurting not only the owner but also resulting in the loss of many lives. Owners trust the team to keep information confidential.

So how do you develop trust and build a team on an ad hoc basis? How about considering Partnering?

## PARTNERING

The Corps of Engineers had been plagued by project claims. Their Chief Counsel, Lester Edelman, and their Chief of Construction, Richard Hanson, strategized on how to reduce claims, and how to stop issues from festering into claims. They developed the concept of "Partnering" to be used on construction projects. The result of considering team members as partners led to remarkable project delivery success. The project environment under Partnering is far less adversarial, more conducive to working together, and results in far fewer claims and legal actions. Under Partnering, the project gets accomplished better, faster, more efficiently, less expensively, with greater job satisfaction for all team members, and more repeat business. It's hard to beat those outcomes.

Partnering starts at the top, with the owner, AE, CM, GC, and other major players. These partners in the project venture commit to common goals such as an on-time project and everyone making a profit within the project budget. While everyone should make a profit, no one should maximize their profit at the expense of another team member. Partnering is not just "feel good" talk, but is a realistic commitment to work together, to be fair and reasonable, to solve problems, and for all members of the team to be successful. Following a meeting of the top members of each organization, Partnering usually results in a subsequent meeting of team members one level lower in each organization's hierarchy—those who will focus on this specific project and who will work daily assuring project success. Partnering meetings are usually facilitated by an experienced third party and the team members develop a written charter listing the project goals. The charter is signed by all the partners. Because the top leaders of the various organizations associated with the project have committed to work together with the owner, it is much easier for the next level down to also commit. The process of developing the charter allows all entities to express their hopes and concerns so other members know the expectations of each partner. These expectations can be addressed during the life of the project. Quarterly follow-up Partnering meetings are held until the project is completed. The follow-up meetings assess how the team is doing in all respects and provide a forum to discuss how to implement improvements where needed. The meeting is self-directed, with the partners working out process details and the facilitator attending for the sole purpose of helping to keep things moving.

A key aspect of Partnering is the commitment to solve issues at the lowest level and within a fixed amount of time. If an issue is not resolved in the specified time, it is raised one level on a management "ladder" developed at the partnering meeting and specified in the charter. The issue is hopefully resolved at the next level or again raised up the ladder in each organization. Issues not resolved in this manner then go to mediation—not arbitration. Mediation keeps the team members engaged, working together, and avoids the necessity of going to court. The mediator provides advice concerning how he or she believes a court would decide the various parts of the issue. The purpose is to educate the partners and help ***them*** settle the issue amicably. Partnering is based on relationships, not on transactions—people working with and trying to understand and respect the values and interests of the other partners. Each of the affected partners is seeking a workable solution, satisfactory to all, rather than resorting to expensive litigation which distracts from the project's goals.

Partnering works because the focus is project success, success for all members of the team. Team members support one another and do not make it difficult for each to succeed. If one team member notices a problem developing for another team member, the team member who observes the situation raises the issue seeking a solution rather than letting it develop badly ... for everyone. While Partnering has been employed successfully many times, it cannot be mandated in a contract. It can only be offered. A team member cannot partner with an entity who refuses to participate in the partnering process. Partnering is not Pollyanna; it does not mask real-life considerations and problems. A partner does not just give in to everything and lose money. All are supposed to be successful.

In 1989, the Ronald Reagan Building and International Trade Center in Washington D.C. was in turmoil, with the owner revising 40% of the building design after award of the construction contract. The relationship between the owner and the developer and the project's GC was quite poor. The fixed budget project was 4 years behind schedule. As a change to the status quo, Partnering was attempted and resulted in the building being completed within budget, in spite of being still 4 years behind schedule. In addition, there were no claims. Partnering started when the owner met with the developer and they agreed to speak with one voice. The six major entities on the project met in the first partnering session. The first meeting was followed by a next level down partnering meeting of the six management entities. A charter was developed and the issue resolution process ("ladder") was developed. Subsequently, the resolution process was used many times. A few issues went to mediation and all were resolved at mediation; none went to court.

**FIGURE 2.3** A dedicated team of iron workers publicly demonstrated their project focus.

## PROJECT FOCUS AND CONCLUSION

When all team members focus first on the project, other "ankle biter" issues tend to disappear or dissipate and are more easily handled. The larger goal raises the level of conversation and attitudes. Potential adversaries become friends. Engineering professor German Gallardo Zevallos in Peru observed:

> "With friends, you can do the impossible."

Once there was an inspector who spoke to two masons, asking each the same question: "What are you doing?" The first said "I'm setting stones." The second said "I'm building a cathedral." The second mason understood project focus and probably sought to deliver better construction quality and had better job satisfaction. In a similar vein, a dedicated team of iron workers on a Seattle Veterans Affairs hospital project publicly brought team focus to their work, as shown in Fig. 2.3.

Project focus (our common goal) is critical, but it is achieved by remembering: people are our most important asset.

### People Working Together

The engineers of a joint venture construction team, building the Pearl Harbor Memorial Bridge ("Q-Bridge") in New Haven, CT made it a tradition to fry turkeys for the trade workers before the Thanksgiving holidays (Fig. 2.4). The engineers then got behind the tables and served the workers.

**FIGURE 2.4** Construction engineers preparing and serving Thanksgiving dinner for their teammates.

Finally, if you are to have a successful construction career, you must also care for ***yourself*** as an important asset. To avoid burnout, you need to balance many things in addition to the project: sleep, nutrition, exercise, family, and your form of spirituality. Yes, you sometimes need to work extra hours and, yes, you need to work hard, but your career is a marathon not a sprint, and your own "care and feeding" is as critical as taking care of your people. It will pay long-term benefits in both longevity and job satisfaction.

# CHAPTER 3
# CONTRACT ADMINISTRATION

*Everyone involved in the delivery of construction projects must understand contract terms and structure, and know contract administration processes. Successful execution of a project requires knowledge of the agreement and the project specifications plus other contractual requirements such as bonds and insurance. No two projects are the same size or have the same complexity but all contracts require similar administrative processes. A contract for a large complex project might require a complicated monthly pay estimate system, while a simple contract might have a more basic monthly pay estimate system. But they both require a monthly pay estimate matched to the project owner's specified process and format.*

## DESCRIPTION OF A CONTRACT

Contracts are the instruments used for the procurement of everything in the construction business, both goods and services. The makeup of a contract, whether the owner is a public agency or a private corporation, is essentially the same. The form of the documents may change, but the elements of the contract remain. There are many federal, state, and local laws affecting construction contracts. Consequently, the successful contractor must be well initiated in the legal requirements of construction contracting.

### Definition

In its simplest, most essential form, a construction contract is defined as an agreement between two parties. It requires a "meeting of the minds" and there must be a service and consideration. One party must agree to perform work (service) for the other party and receive payment (consideration) for the executed work. The contract must be "enforceable by law," which for practical purposes means the contract must be for performance of a legal service. Although there is no requirement for a contract to be written, realistically it must be. Oral contracts are essentially impossible to enforce in construction because of the lack of evidence regarding the scope of the agreement. Generally, a written contract eliminates problems by removing doubt concerning the agreed-upon terms.

For a contract to be valid and enforceable by law, it must have four essential elements.

**Offer**—One of the parties makes a promise to perform a service. An offer can normally be withdrawn up until the time it is accepted, though, in the case of many government projects the offer (the bid) may not be withdrawn once submitted. Always carefully read the bid documents for stipulations concerning offer withdrawal.

**Consideration**—In return for the service the other party provides something of value, usually a monetary payment but it could be something else.

**Acceptance**—The acceptance completes the "meeting of the minds." In construction, the acceptance is award of the contract to the offering contractor.

**Mutuality**—The parties understand and agree to the substance and terms of the contract.

These four elements define the contract. In addition, the subject matter of the contract must be lawful and the contracting parties must have the legal capacity to enter a contract. A contract to commit a crime is not legal or enforceable and a contract with a minor is not lawful. Contracts are signed by representatives of both the owner and contractor who have the legal authority to sign for their organizations.

## ESSENTIAL CONTRACT DOCUMENTS

While many documents make up a construction contract, the four *essential* documents are:

Agreement

General Conditions

Special Conditions

Drawings and Specifications

### Agreement

The agreement represents and reflects the legal contract between two parties. This can be between the owner and the contractor or between the owner and the designer or between the general contractor (GC) and the subcontractors or between the contractors and material suppliers. The purpose of an agreement is to record in written form those items agreed to by the two parties. One common agreement used in the construction industry is AIA Document A101 published by the American Institute of Architects (AIA). The full title of the current version of the document is AIA Document A101-2017 *Standard Form of Agreement Between Owner and Contractor*. It is structured for use when the basis of payment is a stipulated sum (fixed price). The Engineers Joint Contract Documents Committee (EJCDC) C-520 2018 *Agreement between Owner & Contractor Stipulated Price* is another commonly used agreement. EJCDC is a joint endeavor of The American Council of Engineering Companies (ACEC), The National Society of Professional Engineers (NSPE), and The American Society of Civil Engineers–Construction Institute (ASCE).

A simple letter can constitute legal evidence of an agreement and be enforceable if it contains:

- Date of the agreement
- Names and addresses of the contracting parties
- Description of the scope of work
- Contract considerations (price)
- Payment conditions
- Time limitations (schedule)
- Reference to other documents (document list)
- Signatures

### General Conditions

The "General Conditions" part of the contract defines the rights, responsibilities, and relationships of each party to the contract—the owner and the general contractor; general contractor and a subcontractor; general contractor and suppliers. It sets out the guidelines for administration of the contract. The General Conditions are often referred to as "boilerplate," because the same wording is standard to many contracts. Contractors must know exactly what is contained in the "boilerplate."

Various standard forms of General Conditions have been developed by different organizations. These forms are familiar to all parties concerned and the wording has been tested in the courts. One commonly used general conditions for an agreement between owners and contractors is **AIA Document A201**. The full title of the document is *A201 General Conditions of the Contract for Construction*. Other standard general conditions forms are available for use in other types of contracts, such as the contract between a general contractor and subcontractor (AIA Document A401) or the contract between the owner and architect (AIA Document B141). A complete listing can be found on the AIA website at www.aia.org. Many other agencies have standard forms of General Provisions, including the General Services Agency (GSA), U.S. Army Corps of Engineers (CE), and state departments of transportation (DOTs).

**AIA Document A201** *Provides the terms and conditions under which the Owner, Contractor, and Architect work together during a construction project.*

Standard documents, such as AIA Document A201, are intended to be used in their entirety, and not in a "cut and paste" format. A201 is updated periodically. Necessary changes to the *General Conditions* are made by writing in the *Supplementary Conditions,* not by making changes to the general conditions themselves. The AIA Document A201 contains the following 14 articles:

1. General Conditions
2. The Owner
3. The Contractor
4. Administration of the Contract
5. Subcontractors
6. Construction by Owner or by Separate Contract
7. Changes in the Work
8. Time
9. Payments and Completion
10. Protection of Persons and Property
11. Insurance and Bonds
12. Uncovering and Correction of Work
13. Miscellaneous Provisions
14. Termination or Suspension of the Contract

Most experts feel it is not wise to use nonstandard general conditions. This is because all of the published standard general conditions have been tested in court and their legal interpretations are well known. Nevertheless, if the contract does not incorporate standard general conditions, such as those written by the AIA or the Associated General Contractors of America (AGC), and instead the designer or the owner writes the document, as a minimum, the nonstandard general conditions must include:

- A definition of the contract documents. This should list and provide a brief description of the documents forming the contract.
- Document precedence, a clear statement of which documents will have precedence over the other in case of a discrepancy (Fig. 3.1).
- Duties and responsibilities of the owner and contractor while construction is in progress.
- A definition of how a portion of the work can be awarded to a subcontractor and the working relationships between subcontractors.
- The rights of the owner to self-perform construction or to award work to separate contractors.
- The time for completion of the work.

**GENERAL CONDITIONS TO THE CONSTRUCTION CONTRACT**

**ARTICLE 1 GENERAL PROVISIONS**

**1.1 DEFINITIONS, CORRELATION AND INTENT**

1.1.7 Contract Document Order of Precedence. The Drawings, Specifications, and others Contract Documents will govern the Work. The Contract Documents are intended to be complementary and cooperative and to describe and provide for a complete Project. Anything in the Specifications and not on the Drawings, or on the Drawings and not in the Specifications shall be as though shown or mentioned in both.

1.1.7.1 If there is a conflict between Contract Documents, the document highest in precedence shall control. The precedence for the Contract Document shall be:

1.1.7.1 Permits from other agencies required by law
1.1.7.2 Change Orders, Supplemental Agreements, and approved revisions to Contract Specifications and Contract Drawings.
1.1.7.3 Contract Specifications
1.1.7.4 Contract Drawings—Material/Equipment Schedules
1.1.7.5 Contract Drawings—Detailed Plans
1.1.7.6 Contract Drawings—General Plans
1.1.7.7 Standard Plans, i.e., standard structural details, devices, or instructions referred to on the Plans or Specifications by title or number.
1.1.7.8 Reference Specifications, i.e., Building Codes, Test References, etc.

**FIGURE 3.1** General Conditions clause addressing precedence.

- The mode and the frequency of payment or the stages of completed work when the contractor will be paid. This section also contains the rights of both parties with respect to **retainage** and the definition of completion.
- The requirements and penal value of bonds and insurance. This section includes **bid bonds** as well as payment and performance bonds.
- The conditions constituting a change to the contract's scope of work must be clearly defined and the steps to be taken when a change order is required must be detailed.
- The system for dispute resolution, such as arbitration or mediation, must be defined. Also, the rights of each party to terminate the contract must be fully described.
- There are always some miscellaneous provisions such as the governing laws, delegation of work, requirements for inspections and tests, approvals during the work, and statutory limits.

**retainage**
*A portion (%) of the contract price kept by the project owner until project completion. It is protection against the contractor not completing the work.*

**bid bond**
*A bond to assure contractor acceptance of the offered contract and submittal of the required performance and payment bonds*

## Supplementary Conditions

The supplementary conditions are sometimes identified as special provisions or special conditions. The purpose of the supplementary conditions is to provide an extension of the general provisions of the contract to fit the circumstances of a specific project. They serve as amendments or augmentation to the general conditions and take precedence over the general conditions. Items included in the supplementary conditions are entirely subject to the discretion of the parties to the contract and may include topics such as:

- The number of copies of contract documents to be received by the contractor
- Survey information to be provided by the owner

- Materials provided by the owner
- Changes in insurance requirements
- Project phasing requirements
- Site visits
- Start date of the construction work
- Requirements for security and temporary facilities
- Procedures for submittal and processing of shop drawings
- Cost and schedule reporting requirements
- Traffic control and street cleaning requirements
- Responsibilities for testing of materials
- Actions to be taken in the event of discovery of artifacts or items of historical value

## Drawings

Drawings are the means by which the designer conveys a physical, quantitative, and visual description of the project to the contractor. The drawings are a two-dimensional representation of the physical structure designed to meet the objectives of the owner. The use of two-dimensional drawings requires detailed review, interpretation, and integration so construction tasks can be coordinated. Such drawings are also known as the plans. Two-dimensional computer-aided design (CAD) drawings have been the norm but the use of three-dimensional computer models is steadily growing. The linkage of three-dimensional building components (structural, piping, electrical) is a way to better-coordinated construction documents and placement of systems. This approach is the basis of building information modeling (BIM). However, printed and stamped construction drawings are still expected by many owners, code officials, financial stakeholders, and even construction contractors.

The contract drawings are organized into sections. The architectural drawings show the layout of the project. They are numbered sequentially beginning with page A-1 (mechanical drawings begin with M-1, electrical drawings begin with E-1, structural drawings begin with S-1, etc.). For a building project, the architectural drawings are the core drawings of the contract. They show the floor plans, exterior elevations, interior elevations, details, windows, doors, and finish schedules. The structural drawings show the load-carrying systems. They show how the structural members will support the building and transmit the loads to the foundation and the ground. The mechanical drawings show the plumbing; heating, venting, and air-conditioning (HVAC); and fire protection. The electrical drawings show the various electrical and communications systems in the building. Site drawings show the relation of the site to adjoining property, the drainage plan, and systems such as sanitary sewer and utilities.

But architecture/engineers can have their own systems for ordering the sheets. However, there is a United States National CAD Standard® (NCS). The NCS provides a consistent system for classifying

| Discipline Designators | |
|---|---|
| A | Architectural |
| B | Geotechnical |
| C | Civil |
| E | Electrical |
| F | Fire Protection |
| G | General |
| I | Interiors |
| L | Landscape |
| M | Mechanical |
| S | Structural |

electronic building design data. The Uniform Drawing System (UDS) of NCS is a standardized system for organizing building design information. The UDS format has three parts beginning with a discipline designator either primary or primary and secondary (C for civil, S for structural, and E for electrical are examples of primary designators). If only the primary is used, a hyphen is placed in the second position. If a primary and secondary are noted, the second character identifies the secondary; ET would identify electrical telecommunications. The third space is used as a sheet-type designator. The sheet type is indicated by single numerical character: 0 for general, 1 for plans (horizontal view or a combination of plan and profile), and 2 for elevations. There are 10 sheet-type designators, 0 to 9. The last two spaces in the system designate the sheet sequence number, see examples below.

| Discipline Designator | Sheet Type | Sequence Number |
|---|---|---|
| **P-** | **3** | **07** |
| Plumbing | Section | Sheet 7 |
| MP | 5 | 12 |
| HVAC Piping | Details | Sheet 12 |

## Specifications

Specifications may also be known as Technical Provisions. They are written conditions used in conjunction with the drawings. Together the drawings and the specifications describe in detail the requirements of the contract, to include expected quality. The specifications supplement the drawings and provide nongraphic descriptive data or information of such detail to be too lengthy for placement on the drawings.

> Specifications guide bidders in the preparation of cost proposals and the selected contractor in execution of the work.

The specifications provide the information a contractor needs for ordering materials, construction, and installation of the components. Specifications provide information regarding:

- The quality of materials
- The quality of workmanship
- Erection and installation methods
- Test and inspection requirements and methods

Specifications have restricted application, usually to a specific item or work operation. The designer or specification writer is able, therefore, to assign responsibility for each provision of the specifications to the desired specific party. Specifications must satisfy these basic criteria:

- Technical accuracy and adequacy
- Definite and clear stipulations
- Fair and equitable requirements
- Information formatted for ease of use during bidding and construction
- Legal enforceability

There are several general types of specifications. Design specifications, also known as materials, and workmanship specifications or prescriptive specifications state how work is to be performed. For example, a wall specification requiring the studs be placed 16 in. on center, and even specifying the nailing pattern. The contractor is not free to make changes to the spacing of the studs. By using a design specification, the designer assumes liability for the performance of the product. Looked at another way, the designer warrants the performance of the product.

Performance specifications state how the finished product must perform without dictating how the contractor is to do the work. A performance specification would specify that the concrete for a slab should have a 28-day strength of 3,500 psi. It is the contractor's responsibility to provide concrete meeting the requirement.

Open specifications allow any product meeting the requirement to be used. Closed specifications allow only products of a certain type. Proprietary specifications allow only one product, without any allowance for substitutions. Equal specifications normally specify one product, but allow the substitution of others considered as "equal." Be warned: read the "or equal" language carefully, what property of the item must be equal? Or equal clauses vary from one set of project specifications to another.

In case of a conflict between specifications and drawings, the specifications usually have precedence. The other option is to require the architect to resolve any conflicts between plans and specifications.

### Addenda

Any change to the bid documents after they are released for bidding but before bids are actually received requires the issuance of an addendum. This formal document, the addenda, changes the original bid documents and becomes a part of the bid package. At the time of **bid opening**, bidders must acknowledge all addenda. Addenda may be issued to change the bid opening date, to modify the original design, to delete or add items, or to correct errors. Addenda should not be issued close to the bid opening unless the bid date is also extended accordingly.

**bid opening**
*The opening and tabulation of bids at the prescribed time and place.*

### Documents Included by Reference

Contracts may make reference to other documents, which become part of the contract by reference. Most common among documents of this nature are laws, regulations, codes, covenants, and ordinances. Federal laws such as the Davis-Bacon Act (prevailing wage) and the Miller Act (**payment** and **performance bonds**) are usually included by reference within the General Conditions of project financed with Federal moneys. National or regional codes, such as the Uniform Building Code (UBC), and local building codes for the specific city or the local of a project are commonly included by reference. There may be occasions during construction when it is necessary to refer to local codes and ordinances. It is wise to have copies of referenced documents available at the project site and they must have the "exact name" and be the correct version of referenced document.

**payment bonds**
*Guarantee the contractor will pay subcontractors and material suppliers.*

**performance bonds**
*Guarantee the contractor will complete the work in accordance with the plans and specifications.*

## *CONTRACTUAL RELATIONSHIPS*

The principal project relationships are the owner, designer, general contractor, subcontractors, material suppliers, and construction managers, though there can be others. Not all of these will be contractual.

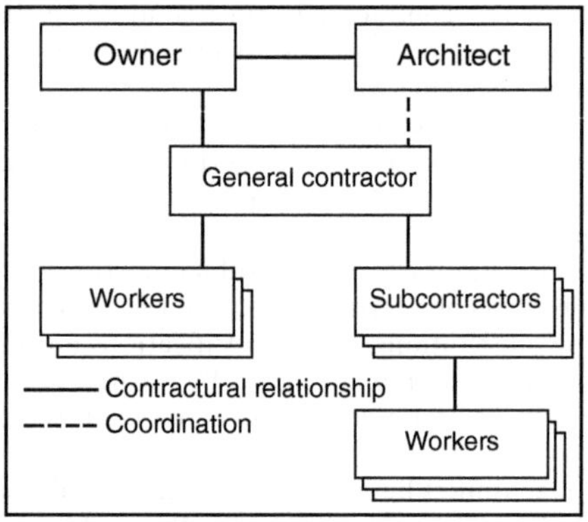


**FIGURE 3.2** Design–bid–build contract relationships between the owner, architect, general contractor, and subcontractors.

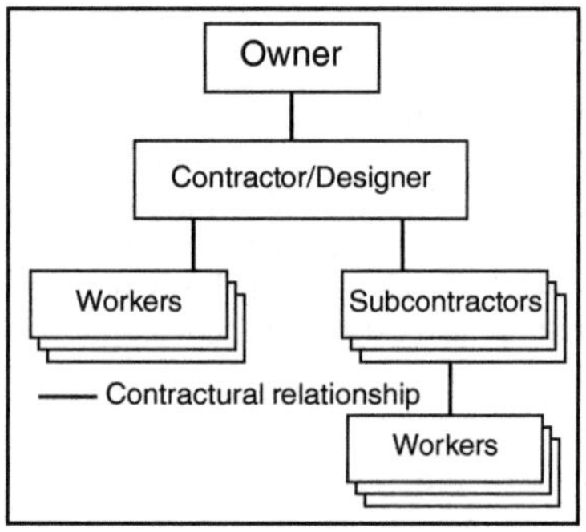


**FIGURE 3.3** Design–build project procurement contract relationships.

## Owners

In the case of a Design–Bid–Build (DBB) project, there is a contractual relationship between the owner and the general contractor (Fig. 3.2). The general contractor will usually have contracts with the specialty subcontractors. There is no contractual relationship between the owner and the subcontractors; this is important to understand. The owner hires the GC to perform the work and the GC hires necessary subcontractors. In the DBB relationship, the owner has a contract with the architect/engineer, but there is no a contractual relationship between the general contractor and the architect/engineer. The architect/engineer has many active functions on the job, such as processing and approval of shop drawings, approval of progress schedules, approval of periodic pay estimates, and acceptance of the work. The general contractor and the architect/engineer have a working relationship based on their separate contracts with the owner. Conversely, for a Design–Build (DB) contract, the contract structure is different (Fig. 3.3).

The owner may be represented on the job site by the architect/engineer, by a construction manager (CM), in-house staff, or a consultant team. The manner in which the owner is represented has no impact on the owner's responsibilities. Under the provisions of AIA A201, the owner has the responsibility to provide a safe workplace, respond to the contractor's questions, secure adequate funding, and pay the contractor promptly. The normal period of payment is monthly. The owner can reduce project cost by developing a reputation for prompt payment. With the availability of electronic information transfer, there is no reason owners cannot pay contractors within 3 days, and require the general contractor to pay the subcontractors and material suppliers promptly. But many project owners do not pay promptly.

### Designers

In the case of DBB projects, designers (the architect or engineer) have no contract with the contractor, but they have project functions impacting contractor performance. To deliver a successful project the designer and contractor must work together. The designers review the shop drawings. Nevertheless, to protect themselves from unnecessary liability, designers have begun to move away from approving shop drawings. AIA A201 *General Conditions of the Contract for Construction* limits the design professional's review of shop drawings to "checking for conformance with information given and the design concept expressed in the Contract Documents." Some attorneys believe the phrase "design concept" is undefined and so vague that it precludes the idea of an Architect's approval constituting incorporation of shop drawing into the Contract Documents.

> Read the contract carefully; make sure you understand the process for submitting shop drawing and the legal ramifications of shop drawing approval.

Designers review and approve monthly pay requests and have the potential to increase the owner's costs by delaying the process. Pay request process delays result in higher-priced change orders and a higher incidence of claims.

When there are disputes and claims, the designer will be called upon to provide comments and recommendations to the owner.

## GENERAL CONTRACTORS

The general contractor has a direct contractual relationship with the owner, subcontractors, and suppliers. In the case of a DBB project there is no contractual relationship between the general contractor and the designer. The GC has no contractual relationship with second-tier subcontractors or materials suppliers serving the subcontractors. However, the general contractor interacts with all these entities and as members of the team they depend on each other for mutual success.

The general contractor incurs an obligation to pay the subcontractors promptly when paid by the owner. Inflation and the cost of borrowing money will take all the profit out of a project for a subcontractor who is not paid promptly by the general contractor. Subcontractors soon learn if a general contractor has a reputation of holding their money. This can be a serious problem and some states have enacted legislation to require contractors to promptly pay their subcontractors.

### Subcontractors

The primary subcontractors, or specialty contractors, have contracts with the general contractor. They have no contractual relationship with any other party. All their communications must be routed through the GC. There is usually only one monthly project pay estimate and it is submitted by the GC to the owner. Each subcontractor's progress, and hence payment, is contained within the GC's pay estimate. Subcontractors are totally dependent on the GC for timely processing of their pay estimate and timely payment once the owner has paid the GC. Change orders often have a larger impact on a subcontractor than they have on the GC, simply because one change order can represent a higher

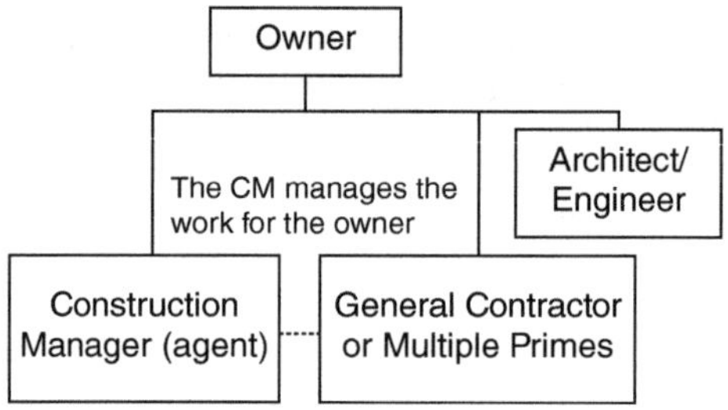


**FIGURE 3.4** Construction Manager arrangement.

percentage of a subcontractor's total earnings on the project. Subcontractors rely on the GC to conduct change order negotiations on their behalf.

### Material Suppliers

Material suppliers, such as electrical wholesalers, lumberyards, concrete batch plants, plumbing supply stores, and other businesses, sell their products directly to general contractors and subcontractors. Because contractors purchase large quantities, often for multiple projects, they receive bulk discounts and may have commercial trade deals in place to lower their material costs. The supplier's contract is with the contractor or sub-subcontractor with whom they are dealing. Suppliers may even be required to provide warranties to the owner or technical training to the owner's staff, but their contract is only with the contractor or a subcontractor. They can attach a lien to private property if they are not paid and the Miller Act protects them on public contracts.

### Construction Managers

Most construction managers (CMs) act as advisors directly to the owner. Typically they do not perform construction tasks like a GC would, but oversee construction contracts as the owner's agent (Fig. 3.4). It is a good idea to bring the CM onto the project team in time to help select the architect. The CM serves as a professional consultant during project design and brings predictability to the project's estimates and schedule. The project owner executes separate contracts with the CM and the designer. When the design is complete, the CM manages the construction for the owner.

There are large general contractors who refer to their companies as CMs. They may bid projects and then subcontract 100% of the work, performing none of the work with their own forces. This form of construction management is a variation of the general contractor role.

Some CMs operate "at risk" and are contractually obligated for delivering the project at a fixed price. They commit to deliver the project within a Guaranteed Maximum Price based on the construction documents and specifications at the time of the guarantee.

## *BID AND AWARD PROCESS*

Government projects must be advertised and are normally awarded to the lowest responsive bidder (Fig. 3.5). Private projects do not need to be advertised and are usually negotiated, with only a limited number of selected contractors involved in the negotiation. Advertisement is at the discretion of the private owner.

The public advertisement of government projects is required. This provides all contractors an opportunity to prepare a bid and submit an offer. Federal projects are advertised in the *Commerce*

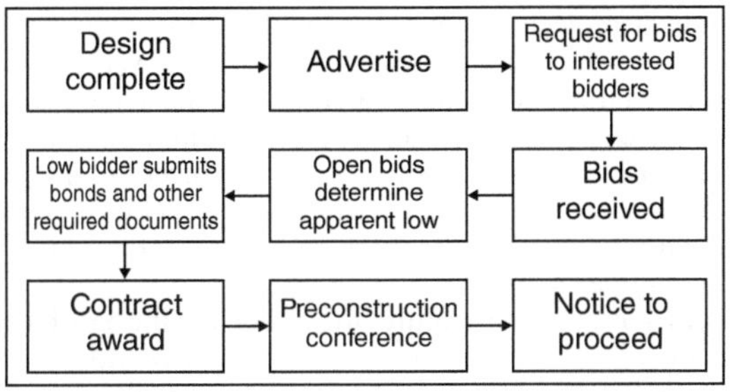


**FIGURE 3.5** The advertise, bid, and award process.

*Business Daily* (http://cbd-net.com/) which is The FedBizOpps (FBO) database. It is the source for federal procurement bidding opportunities, contracts awarded, and special notices. Dodge Construction Central (www.construction.com/) is another source of construction project information. Contractors seeking project bidding opportunities check such sources daily.

Construction project advertisements contain sufficient technical information to enable contractors to make a preliminary decision regarding their interest in bidding. As a minimum, the project name, owner, location, required start and completion dates, a short description of the project scope, estimated cost range, and the bid opening location and date are included in the advertisements. Advertisements contain the contact information for those desiring to obtain copies of the Instructions to Bidders.

Bid documents can now be obtained in more than one format. For many years, interested bidders had to go to the office of the owner or the designer to obtain the bid package, which consisted of the invitation to bid, specifications, drawings, and bid documents. Usually there is a fee involved in obtaining these documents. To receive drawings and specifications for Federal projects, contractors must be registered in the System for Award Management (SAM) system (www.sam.gov/SAM/) and in FedBizOpps, as drawings and specifications are usually issued electronically from those sites. Being able to obtain the documents in this manner is quicker and easier. There is cost saving for the agency because fewer copies of the documents must be printed and shipped. Potential bidders can download the bid documents as they are needed.

## Bidding

Historically, bids must be submitted on time, at the location specified, on the correct forms, with acknowledgment of all addenda, and (if required) a valid bid bond. Otherwise, the bid is considered nonresponsive and returned unopened. When the time for accepting bids is reached, an announcement is made and no other bids are accepted. Any bids turned in late will be returned unopened. Still the use of digital methods is rapidly gaining acceptance. Many government agencies have gone to electronic bidding. The use of electronic documents and bidding processes has been embraced by 47 State Transportation Agencies (STAs).

If it is a Federal Government project the Federal Acquisition Regulation (FAR) Title 48—Part 49 Sealed Bidding states: Bidders are responsible for submitting bids, and any modifications or withdrawals, so as to reach the government office designated in the invitation for bid (IFB) by the time specified in the IFB. They may use any transmission method authorized by the IFB (i.e., regular mail, electronic commerce, or facsimile).

The low "apparent" bidder may be announced at the bid opening, but the low bidder cannot be determined until the bid is checked and the bonds and certificates of insurance examined. Many agencies accept bids by mail and it is possible for a bid to be mailed on time, but not be received at the time of bid opening. It is also necessary to check the bids for mistakes and completeness of the bid documents. Only after an agency is certain about receipt of all bids and has examined all bids will the low bidder be announced.

Government projects are approved with a *programmed amount* plus *contingency funds.* It is very difficult to obtain approval to award projects above the programmed amount. Therefore, bids are sometimes advertised with a **base bid** and alternates, which can be additive or deductive. This is done only in cases when the adequacy of the project funding is questionable and the agency wants to maximize the probability of receiving an awardable bid by accepting alternates to bring the total price below the project programmed amount.

**base bid**
*Provides a mechanism for the owner to accept bids on a portion of the facility and not construct features whose cost would cause the budget to be exceeded.*

Alternates may consist of features to be added to the base bid, at an additional cost for each additive, or they may be deductive with a cost reduction for each alternate. Alternates are considered sequentially until the low bidder has provided all possible project features and the bid still remains below the programmed amount. Alternates must be accepted in numerical order, without skipping, and the total bid for each bidder must be considered with the addition of each alternate. In the case of additive alternates, the alternates are added in sequence until accepting the next alternate would result in the lowest bid being too high. In the case of deductive alternates, the alternates are accepted until a sufficient amount has been subtracted from the base bid to bring the total bid below the project budget.

**Alternative Bid Deductive Example:** A school district has determined there is a need to add a new gymnasium to an existing high school. The programmed amount for the project, which is the amount of funds available, based on the enabling legislation, is $12.5 million. The architect's estimate is $12.9 million, but the school district desires to retain as many of the features in the gymnasium as possible. The following deductive alternates and their estimated values have been identified:

1. Bleachers for the basketball court—$240,000
2. Racquetball courts—$125,000
3. Additional parking—$55,000
4. Replace brick exterior with metal siding—$98,000
5. Replace air-conditioning—$135,000

Five bids were received and evaluated to determine the low bidder.

| Bidder | A | B | C | D | E |
|---|---|---|---|---|---|
| Base bid | $12,879,384 | $13,045,900 | $13,243,765 | $12,999,001 | $13,684,700 |
| Alternative 1 | −224,318 | −237,982 | −264,280 | −274,460 | −312,238 |
| Cumulative | 12,655,066 | 12,807,918 | 12,979,485 | 12,724,541 | 13,372,462 |
| Alternative 2 | −97,210 | −114,000 | −107,340 | −145,600 | −99,000 |
| Cumulative | 12,557,856 | 12,693,918 | 12,872,145 | 12,578,941 | 13,273,462 |
| Alternative 3 | −43,500 | −62,300 | −65,000 | −68,500 | −73,600 |
| Cumulative | 12,514,356 | 12,631,618 | 12,807,145 | 12,510,441 | 13,199,862 |
| Alternative 4 | −97,500 | −89,450 | −99,650 | −92,450 | −89,650 |
| Cumulative | 12,416,856<br>**Low bidder** | 12,542,168 | 12,707,495 | 12,417,991 | 13,110,212 |
| Alternative 5 | Does not need to be considered. | | | | |

Using the total of the base bid and all alternates, the low bidder is A, with base bid and alternates 1 through 4 subtracted from the contract. The alternate 5 deduction, replace air-conditioning, was not accepted. It remains part of the contract. The contract will be awarded with air-conditioning, but without bleachers, racquetball courts, additional parking, and with a metal siding exterior.

The contracting officer could have decided to advertise the project with a base bid and then add the alternates to the base bid until the next one could not be included without exceeding the project budget. In such a case, the base bid would be reduced in scope and then the alternates would be added to the scope until adding the next alternate would cause the total bid to exceed the project budget. Both approaches (additive or deductive alternates) are acceptable and legal. Construction contractors are used to dealing with both methods.

**Alternative Bid Additive Example:** A new research laboratory building is to be constructed on a university campus. The programmed amount is $5.6 million. The architect's estimate is $5.8 million. The contract will not be awardable above $5.6 million. The decision is made to advertise the building with a base bid and four additive alternates. The four additive alternates and their estimated values are:

1. Additional installed testing equipment—$74,000
2. Upgraded administrative area—$88,000
3. Additional climate control areas—$48,000
4. Landscaping—$76,000

The following table shows the bids and the evaluation to determine the low bidder.

| Bidders | A | B | C | D | E |
|---|---|---|---|---|---|
| Base bid | $5,440,200 | $5,195,165 | $5,380,760 | $5,582,685 | $5,234,678 |
| Alternative 1 | 56,480 | 69,485 | 71,900 | 46,425 | 68,945 |
| Cumulative | 5,496,680 | 5,264,650 | 5,452,660 | 5,629,110 | 5,303,623 |
| Alternative 2 | 82,450 | 79,675 | 56,900 | 82,000 | 80,450 |
| Cumulative | 5,579,130 | 5,344,325 | 5,509,560 | 5,711,110 | 5,384,073 |
| Alternative 3 | 40,800 | 43,580 | 44,687 | 49,784 | 42,136 |
| Cumulative | 5,619,930 | 5,387,905 | 5,554,247 | 5,760,894 | 5,426,209 |
| Alternative 4 | 74,690 | 79,300 | 72,345 | 70,985 | 76,890 |
| Cumulative | 5,694,620 | 5,467,205<br>**Low bidder** | 5,626,592 | 5,831,879 | 5,503,099 |

The low bidder is B at the total bid price of $5,467,205 for the base bid and all four alternates. The total bid is below the programmed amount, and will be awarded.

Owners do not all use the same method for determining the low bidder when there are alternates, but the method must be stated in the bid documents and adhered to in awarding the contract. Some owners compare the total of the base bid plus all accepted alternates to determine the low bidder. Other owners determine the low bidder based only on the base bid and then accept as many of the alternates as funding allows. The method of determining the low bidder will affect a bidder's strategy. In the case of the low bidder determined by the base bid only, a contractor can be tempted to bid the base project low and increase the amount of the alternates to offset the low base bid. However, the tactic can be costly if only the base bid and none of the alternates are awarded. In every case of base bid and alternates, bidders must carefully read the contract documents so they understand the procedure for selecting the low bidder.

## Bid Depository

**bid depository**
*An organization used by owners, general contractors, subcontractors, and materials suppliers to ensure fair bidding.*

A **bid depository** provides a service to general contractors and subcontractors by receiving subcontractor bids and distributing them to general contractors. There are many types of bid depositories. Historically, bid depositories have been created by the trade associations in an effort to protect member companies from unscrupulous contractors. The main objective of any bid depository is the prevention of bid shopping. Bid shopping can occur after award of the contract when a general contractor requests a lower price from a specialty contractor. The request may or may not involve the disclosure of the bids already submitted to the general contractor by other specialty contractors. The practice is not ethical, but it does happen.

There are only a few bid depositories still operating because many were sued under antitrust laws and most of the depositories lost their cases with the courts ruling the practice to be a trade restriction.

## Award

Most government agencies still use the DBB contracting method and award the contract to the low-responsive, responsible bidder. However, DBB represents less than 30% of the construction market. DB represented about 40% of the market during the last decade (2010–2020) and some form of Construction Manager contracting captured slightly more than 30%.

When DBB or DB contracting methods are used, the contractor must provide the required payment and performance bonds, and certificates of insurance for all specified coverages to the project owner before the contract will be awarded. Once the owner has all required documents two copies of the contract will be sent to the contractor for signature. The contractor returns both copies to the owner who then signs both copies, keeps one, and returns the other signed copy to the contractor. With a signed contract, the general contractor will begin to finalize contracts with all subcontractors. As a general rule the subcontracts contain the same contract requirements as the contract between the owner and the general contractor, including but not limited to bonds, insurance, completion date, **liquidated damages**, and claims requirements or other articles punitive to the general contractor.

**liquidated damages**
*Daily amount, specified by the contract, paid by the contractor to the owner as compensate for late project completion. Liquidated damages are not penalties. They represent compensation for owner expenses and lost income.*

## Notice to Proceed

The notice to proceed (NTP) establishes the contractor's right of access to the owner's property. Sometimes the contract will specify the date when the contractor can proceed with the work. In such a case an NTP will not be required. If the contract requires the issuance of an NTP, it fixes the date when the work can commence and, if applicable, it will also fix the date for project completion.

It is usually not issued until after the contract between the owner and the general contractor is executed. The date of the NTP establishes the official start date of the contract. Contracts normally require the work be complete in some specified number of days; these could be calendar or work days, after NTP.

> Read the contract documents carefully, do they specify a 7-day calendar week or a 5-day work week for calculating time?

Change orders may alter the required completion date by adding days for the accomplishment of the additional work. Some contracts establish a maximum number of days after issuance of the NTP by which the contractor must begin physical work on the site. The owner and contractor should use the preconstruction conference to discuss and agree on the contractor's plan for beginning work. Open and efficient communications on such matters are essential.

## STANDARD CONTRACT DOCUMENTS

Standard contract documents for construction projects serve to reduce conflicts and avoid litigation. These have been developed by owner and government agencies who regularly engage in project contracting. Others are the result of combined efforts by professional groups from the design and construction contractor community.

### Government

Just as the AIA Document A201 is used for many private contracts, government agencies have their own standard general provisions. Their general provisions enjoy the same advantages as A201. The requirements are known and understood by all and have been court tested, so their meaning is well established. Owners should be discouraged from using locally written or "homegrown" contract forms because the parties cannot be as sure of how the courts would interpret those documents.

Agencies, from local school districts to state departments of transportation or state general services offices, also have standard specifications or technical provisions. The agency contracting personnel are intimately familiar with such standard provisions because they use them on a regular basis.

At the federal level, there are many agencies responsible for contract construction. The more commonly known are the U.S. Army Corps of Engineers, the Navy Facilities Engineering Command, and the General Services Administration. All have their own standard contract provisions and forms or use standard government documents. Contractors doing business with a government agency for the first time must quickly and thoroughly familiarize themselves with the agency's standard documents.

### American Institute of Architects

The American Institute of Architects is the professional voice of the architecture profession. AIA publishes many standard contract forms, all of which are offered for sale. Their general provisions, AIA Document A201, is designed for a contract between the owner and general contractor, but the AIA similarly provides standard forms for design contracts, construction management contracts, and others.

### Engineers Joint Contract Documents Committee (EJCDC)

EJCDC develops standard documents for engineering design and construction projects. Each family of documents is prepared by a subcommittee: Construction, DB, Engineering, and Procurement, but the General Committee of EJCDC is responsible for approving new documents and reviewing, updating, and modifying existing documents. The EJCDC documents conform to the CSI MasterFormat and have been tested and interpreted in courts of various jurisdictions. EJCDC offers some wise advice about changes to a part of standard documents: seek a legal review.

### Construction Owners Association of America

The Construction Owners Association of America (COAA) is a national organization of public and private owners. COAA believes the forms provided by other construction industry interest groups are biased in favor of the producing group. To provide a balanced alternative, COAA has produced a group of interrelated contracts for use by owners who want an alternative. The Documents Committee of COAA reviews industry contract documents and prepares comments for its membership to guide them in their contract negotiations.

## TYPES OF CONSTRUCTION CONTRACTS

There are many types of construction contracts and the type of contract selected depends on the category of work being performed and the conditions under which it will be performed. A summary of common contract types is presented in Table 3.1.

### Lump Sum

When a project is defined in detail and the contractor can accurately price the work, it may be bid "lump sum." These fixed price contracts are commonly used for uncomplicated or small projects. Lump-sum contracts apportion a greater amount of risk to the contractor. But even with stipulated price contracts there can be changes if the owner issues instructions to change the design, increase the quantities, or causes the sequence and working conditions to be modified. Lump-sum contracts are not appropriate where speed is important or when the required work is not well defined.

### Unit Price

Unit-price contracts are used for work where it is not possible to calculate the exact quantity of required materials. Unit-price contracts are commonly used for heavy/highway work. The designer may calculate 1,000,000 cy of earth to be moved, but the owner and contractor both know the amount could be more or less. The fair solution is for the owner to tell the contractor what the estimated quantity is and then to pay the contractor for the measured amount after the work is completed.

Contractors submit a price for each work item on a unit-price contract. Unit prices are multiplied by the engineer's estimated quantities and totaled. The low bidder is the bidder with the low total of all items. After award of the contract, items whose actual quantity varies from the estimated quantity by more than 15 or 20%, either above or below the estimated quantity, are sometimes subject to renegotiation of the unit price. When the actual quantity is low, the contractor may request a renegotiated higher price because the anticipated amount of earned overhead has been reduced. When the actual quantity is high, the owner may request a renegotiated unit price for the opposite reason. The goal of unit-price contracts is fairness to both parties.

### Cost Plus

Cost plus (cost reimbursable) contracts are used in situations where it is difficult or impossible for either the owner or the contractor to predict their costs during the negotiation, bid, and award

**TABLE 3.1** Common Types of Contracts and Their Normal Application

| Contract type | Application | Contract type | Application |
|---|---|---|---|
| Lump sum | Well-defined work where quantities and costs can be estimated accurately | Unit price | Projects where it is not possible to accurately estimate material quantities |
| Cost plus | Any type of facility with conditions making it impossible for either the owner or the contractor to compute costs accurately; an example would be work in remote areas | Incentive | Contracts with a financial incentive to contractors with the expertise to complete the project quickly, by rewarding early completion |

process. The calculation of costs may not be possible because of unpredictable and extreme weather conditions such as would be encountered in the Antarctic, unknown transportation requirements to remote locations, combat or war, or contracts where the amount of effort required depends on another contractor's work. A cost-plus agreement can also be good option for an owner when design plans are still being developed and time is a crucial factor in delivering a project.

Cost-plus contracts take many forms, the most common being cost plus a fixed fee paid on top of the materials and installation cost or cost plus a percentage. Most owners prefer cost plus fixed fee because then the amount of profit the contractor will earn cannot increase, thereby removing any incentive for the contractor to be less than thrifty or to produce poor-quality work. In very difficult situations or when the time to complete the work is not known with any certainty cost-plus percent contracts may be an appropriate contracting method.

When cost-plus contracting methods are used, the contractor is required to present evidence for the costs related to the project. But inclusion of incentives can be advisable to maintain productivity. A profit suitable to 6 months of effort may not be attractive if there is a possibility that the work may require more than a year to complete.

## Incentive Contracts

The most common types of incentive contracts are those offering a bonus for early completion but incentives could be based on cost and/or quality. If a contract has a penalty for late completion by law in many cases it must also offer a bonus for early completion. If contract definition is complete, and there is a full design and a set of specification, a fixed price incentive contract is possible.

In California, the eastbound I-580 connector bridge over I 880 west of the Oakland-San Francisco Bay Bridge collapsed due to the intense heat from a tanker truck fire. The California Department of Transportation (Caltrans) used an incentive contract to repair the 165-foot long bridge. The contract included an incentive of $200,000 per day for reopening the bridge early and a penalty of $200,000 per day for exceeding the 50-day deadline. The construction contractor completed the repairs in just 18 days, 32 days early, and collected the contract's maximum $5 million bonus. But this was not all added profit as the contractor spent extra money to accelerate the work.

Penalty for late finish and liquidated damages are not the same thing and should not be confused if used in an incentive contract. A penalty is a punishment prescribed by the contract. Penalties may be any amount the contractor is willing to accept. Liquidated damages are designed to pay the cost an owner experiences if the construction continues past a preset completion date.

Cost plus time or "A + B" contracts were developed to reduce construction time. Many projects involve renovation, repair, or improvements. The purpose of the A + B contract is to provide a way to compensate the contractor for the expertise in sequencing and staging to complete the work quickly and efficiently. Owners assign a monetary value to each day the work is being performed and the contractors will then bid a price to complete the work plus a number of days. The low bidder is the low total of performance cost plus the contractor's stated duration times the monetary value.

For example, the estimated cost to the owner by additional days and inconvenience might be estimated at $10,000 per day. Contractors are required to bid an A component for performing the work, plus a number of days, the B component. The low bidder is determined by converting the days to dollars, and adding the two components. It is possible to bid a construction cost higher than other bidders and still be the low bidder if a plan can be developed to complete the project in fewer total days.

Consider this example using $10,000 per day for the B component:

Bidder 1: $\$5{,}000{,}000 + 200 \text{ days} \times \$10{,}000/\text{day} = \$7{,}000{,}000$

Bidder 2: $\$5{,}350{,}000 + 160 \text{ days} \times \$10{,}000/\text{day} = \$6{,}950{,}000$

Bidder 2 is the low bidder in spite of a higher A construction cost component.

Seeking quality, the Kentucky Transportation Cabinet let an I-275 project using an A + B + C contract. The "A" part was the sum of the unit prices to perform the construction work. The "B" part was the number of contract days the bidder stated were required to complete the work. The price per contract day was established at $25,000. The contract additionally called for penalties of $25,000 per day for every day the contract used beyond the contract time. The "C" part was a project warranty. The contractor had to provide a 5-year warranty for the work but for every year beyond the 5 years up to a 10-year maximum, the bidder was given a $500,000 bidding credit.

Contracts of these types offer the owner an opportunity to buy and pay for the contractor's knowledge and as a result they are gaining acceptance. Contractors must focus sequencing of activities and minimizing construction time. Those who can "get in and get out" the soonest will be successful.

## Negotiated Contracts

Negotiated contracts are awarded through a negotiation process rather than through a bid process. An owner elects to negotiate because it affords contractor selection based on qualifications and secondly it creates a partnership instead of an adversarial relationship. In their purest form, an owner will ask a contractor for price. A negotiation process, which might address price, delivery speed, and constructability of the design, will ensue, culminating in an agreement to build the project at an agreed-upon, negotiated price. This process works especially well when private owners are negotiating the expenditure of their own money, and the owner and contractor have experience working together, that is, a relationship was previously established. It does not work as easily when the government is trying to do the same thing with public money.

The process called competitive negotiation involves the taking of bids, after which the final price is then negotiated. There are normally two or more contractors involved in the negotiation process. Negotiated contracts may be fast-tracked, with work beginning before all design work is complete. In such a case, the drawings needed first, such as site drawings and foundation design, will be completed and given to the contractor immediately.

## Job Order Contracts

**job order contracts** *Contracts used mainly for maintenance work. Contractors submit a bid in the form of a decimal multiplier for the owner-published book of standard rates for work. The contractor then performs work against a series of work orders. The process is designed to avoid a design contract for each work order.*

**Job order contracts** are used by large organizations such as universities and industries with major facilities. The owners of these facilities have a need for repair and maintenance work on a continuing basis. There are two basic methods for accomplishing all the ongoing painting, electrical repairs, clean-up, and other minor projects. One method is to have the owner's in-house maintenance crew perform the work. The other way is to hire a contractor. The purpose of the job order contract is to be able to hire the contractor quickly and to eliminate the need to hire a designer to prepare a design for each small item of work.

For small construction projects under $10,000, there is little time or need to have a designer, draw up plans and specifications, and then advertise the work for bid. The job order contract (JOC) avoids those processes and accomplishes the needed work quickly. The contractor assumes responsibility for completing the design work, if any design is required, and for completing the construction task, all on the same work order.

Three documents are provided to the contractors to bid the projects: (1) standard specifications, (2) general conditions, and (3) a book of owner cost estimates. Usually contractors are required to bid by submitting two coefficients, one for normal work hours and one for overtime work. To bid, contractors are given a number of theoretical projects. Contractors first review the specifications and study the owner's cost book. Then a contractor calculates its prices for the work items of the projects and divides its calculated cost by the owner cost. The ratio will produce a "coefficient" such as 0.967 for estimates below the owner's estimate or a number such as 1.0224 for estimates slightly above the owner's estimate. The bid process is low bid, with the coefficient replacing a dollar cost. The contract is awarded to the contractor submitting the lowest coefficient.

The contractor who is awarded the contract will receive work orders or "call orders" for each required task. JOC is not appropriate for large tasks. JOC provides a mechanism for the owner to obtain quality work quickly. Most owners award more than one JOC at a facility and then award more call orders to the contractors doing the best work.

JOC is a competitive bid process with specifications and cost data provided by the owner. Design is the responsibility of the contractor. The process provides a continuing incentive for quality work, while reducing project duration and the owner's administrative and work task costs.

## Design–Build

Seeking to bring better coordination between a project's design and construction phases, owners are moving to DB contracting. Contractually, the DB method offers the project owner a single point of responsibility for both the design and construction phases of a project. Having a seamless project team—designer and builder—allows each to provide insights into the costs and constructability of different designs. The DB team can also easily overlap project delivery activities to accelerate the schedule. Construction Industry Institute project delivery studies have repeatedly found DB to be a significantly faster delivery method than the traditional DBB. If the owner provides a well-defined scope and timely decisions, DB will be the fastest delivery method.

The fundamental contracting difference between DBB and DB is the single contract, but the performance difference is driven by the method's inherent culture of collaboration. Collaboration reduces the risk to all parties and is making DB the fastest growing delivery method in America.

## Construction Management-At-Risk (CMAR)

A construction manager (CM) is hired by an owner to act as a consultant during the development of project design. During the project development phase, the CM provides advice on cost, schedule, and constructability. Once the important design features are defined, the CM often assumes the role of a general contractor, the construction manager/general contractor (CMGC) model, and begins hiring subcontractors for project construction. The CM may hire key subcontractors sooner to assist in reviewing and commenting on designs ideas.

Guidance on the use of CMs published by The Associated General Contractors of America[1] encourages hiring of the CM before the selection of a design professional. When the CM is involved with the owner in selection of the design professional, a more collaborative relationship develops.

In some cases, the CM will provide the owner a Guaranteed Maximum Price (GMP) before final construction plans and specification are available. This contractual relationship is termed a CM-at-Risk (CMAR). It centralizes responsibility for construction under a single contract with the CM (usually a general contractor) committed to delivery of the project at the GMP. The CMAR has assumed the risk of subcontractor prices being higher than assumed based on the incomplete plans available at the time the guarantee was given.

## Alternative Technical Concepts (ATCs)

An owner can allow contractors to propose alternative technical concepts (ATCs) to a DBB request for pricing. A responding bidder offers a proposal during the bidding process to modify a contract requirement and gain a competitive advantage. The offered proposal must provide an equal or better technical solution to the design requirements of the bid invitation. In turn, the project owner must guarantee confidential consideration and pre-award approval of the ATC solution. The use of ATCs serves to enhance innovation and achieve efficiency.

[1] *CM/GC Guidelines for Public Owners*, second ed., 2007, Associated General Contractors of America and the National Association of State Facilities Administrators.

### Public Private Partnership (PPP or P3s)

The United States' Government Accountability Office has defined public–private partnerships (PPPs) as "a contractual arrangement that is formed between public and private-sector partners." PPPs involve a government agency (federal, state, or local) at any level establishing a long-term contractual relationship with a private partner to construct and operate a facility for a fixed duration. The government agency retains ownership of the public facility and to build the facility the private party must invest its own money. If the project is a road, tolls would be used to provide the income to repay the private partner for its investment. While often used for transportation projects PPPs have also been the delivery method for schools and hospitals. PPP projects, no matter the type of construction or even renovation, require a dedicated revenue stream to pay the concessioner.

Because the P3 approach is typically used for complex nonstandard projects, there are no standard contract documents. To provide the upfront money a concessionaire may be the holder of the agreement with the public entity and the construction contractor is only hired to build the project. In such case, it is a question of profit for delivering the project. But if the contractor is providing the financing, then the contract should be assessed considering allocation of risk and long-term return on investment (ROI).

## *BONDS (GUARANTEES)*

Bonds are a guaranteed loan waiting to happen. Unlike an insurance policy, no losses are expected, and if there are losses, they have to be repaid by the contractor. The mechanism for repayment is called an indemnity agreement. In the case of a performance bond, contractors sign indemnity agreements agreeing to reimburse the surety company if it must provide funds to complete the project. There are provisions made for the stockholders and even spouses to individually indemnify the surety.

There are three parties to a bond: (1) the principal (contractor), (2) the surety (bonding company), and (3) the obligee (owner) (Fig. 3.6). The surety and the principal make a joint and several guarantee to the owner, both parties are obligated to the owner.

### Bid Bonds

Bid bonds have two purposes: (1) to guarantee the contractor will enter into a contract if determined to be the lowest responsible bidder and (2) to guarantee the contractor will provide the required payment and performance bonds, and insurance policies. When the performance and payment bonds have been submitted and the contract signed, the contractor is released from the bid bond obligations. Most bid bonds are spread bid bonds: they require only forfeiture of the difference between the low bid and the next bidder's price if the guarantees are not met. There are also penal sum bonds. A penal sum bond is stated in the form of actual damages up to a limiting amount, usually 10–20% of the bid amount.

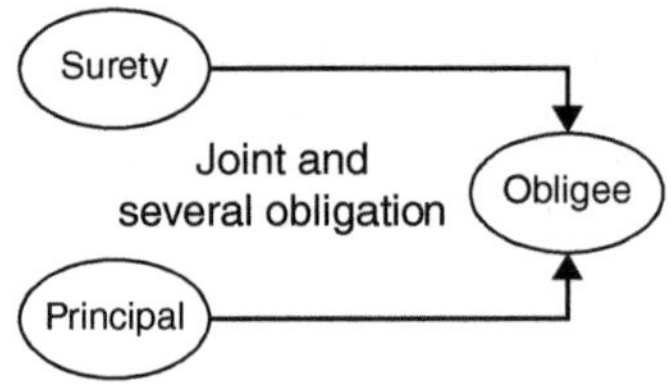


**FIGURE 3.6** Relationship between the surety, principal, and obligee.

## Performance Bonds

Performance bonds guarantee the contractor will perform to the contract requirements at the stated bid price. The surety is in the position of being asked to guarantee the contractor's performance. In effect, the surety guarantees the performance of the contractor, or if the contractor fails to perform, the surety will complete the project as described in the plans and specifications. Therefore, the contractor must demonstrate an ability to perform before the surety is willing to issue the payment and performance bonds. The surety will visit the contractor's home office and job sites, and will contact the owners of recently completed contracts. In addition, contractors can be expected to provide the surety with the following information:

- Résumés of key personnel in the firm
- List of work successfully completed
- Names and addresses of owners, designers, suppliers, subcontractors, general contractors, and others in the industry with whom the contractor has done business
- Organization chart and responsibilities of key individuals in the company
- Business continuity plan
- Explanation of how the project for which it is seeking a bond fits with the company's other work
- Strategic plan of the company
- Risk management and safety programs
- Safety performance
- List of business relationships with accountants, attorneys, consultants, and insurance companies
- Financial information such as balance sheets, statement of earnings, owner's equity, schedule of indirect costs, general and administrative expenses, and a cash flow statement
- Progress schedules of current work, list of completed contracts, and a list of current contracts

Contractors seek a surety who will underwrite projects with specific constraints of project type, location, and size. They present their company in its most favorable light when applying for payment and performance bonds but must be completely honest. Over time, the contractor develops a working relationship with its surety.

If a contractor defaults on performance of the contract, the surety has three basic choices:

- Buy back the bond. This amounts to giving the owner a check for the amount of the penal value of the bonds.
- Replace the contractor. Negotiate or advertise for bids for the purpose of retaining another contractor to finish the work.
- Finance the contractor. The bonding company runs the risk of spending more than the value of the bond, but this is still a common option because the contractor is familiar with the project.

Today most sureties try to replace the contractor.

> Be forewarned: carefully investigate both the supposedly completed work and the quality of the work when taking over a project in midstream, as in the case of a default.

## Payment Bonds

The Miller Act of 1935 applies to federal projects of over $200,000. It requires the contractor to furnish payment and performance bonds in a satisfactory amount. Most states have passed similar legislation, commonly termed Little Miller Acts. Government agencies require bonds because contractors

**TABLE 3.2** Summary of Construction Contract Bonds

| Bond type | Purpose |
|---|---|
| Bid | Ensures the low bidder will accept the contract if offered and will submit the required insurance and payment and performance bonds |
| Performance | Guarantees the contractor will complete the work in accordance with the plans and specifications |
| Payment | Guarantees the contractor will pay the subcontractors and suppliers |

cannot place a lien on government-owned property. The Miller Act and state-level Little Miller Acts provide the mechanism to guarantee the subcontractors, labor, and suppliers will be paid. Table 3.2 presents a summary of construction bonds.

### Bonding Limits

There are many factors impacting the bonding capacity a surety is willing to provide to a contractor. Bonds cost only about ¾–4% of the contract amount, so the failure rate must be kept very low. Consequently, bonding companies evaluate contractors carefully. Bond cost is a function of the risk assumed by the surety. There is not enough premium in bonds to cover the potential exposure. Sureties have to be correct in their evaluation of contractor performance and financial capability. As a rule of thumb, contractors can be bonded up to about 10 times their working capital (working capital = current assets minus current liabilities) or 4 times their net worth. No more than half of a contractor's bonding capacity should be used on any one project. These rules are further discussed in Chapter 4.

## *CONTRACTOR INSURANCE*

The project owner, as the party ultimately responsible for project completion, is seeking to enhance control over project safety and risk. Requiring contractors to have certain types of insurance is one way owners manage risk. Contractors, also seeking to control risk, use insurance to protect against possible events. A summary of construction insurance is presented in Table 3.3.

### Workers' Compensation Insurance

All 50 states have workers' compensation legislation requiring employers to insure workers for injuries on a no-fault basis. Workers' compensation insurance awards to a worker can be denied for willful misconduct, drug use, and intoxication.

Workers' compensation insurance protects workers who are injured on the job by covering medical and hospitalization expenses, plus a percentage (usually 75%) of the hourly wage during the time the employee is out of work. Serious injuries, such as loss of a limb or death, are compensated by a lump sum at a predetermined monetary amount. Correspondingly workers' compensation protects the contractor by limiting the workers' remedy from the employer to the amount covered by the workers' compensation law. Workers who are covered by Workers' Compensation Insurance cannot sue their employer for additional compensation, but some subcontractor employees, who cannot sue their employer because workers' compensation is their sole remedy from their employer, have sued the general contractor under the safe-place-to-work doctrine, which says it is the general contractor's responsibility to provide a safe place to work. These suits are known as "employee overclaims." Unlike other kinds of insurance, contractors cannot name other additional insureds on their workers' compensation policy.

**TABLE 3.3** Types of Construction Insurance

| Type of insurance | Coverage |
|---|---|
| Workers' compensation | Pays claimant in case of injury, disability, or death of employees resulting from work on the job. |
| General liability | Protects owners and contractors from the financial consequences of various risks, such as hazardous operations, or accidents during construction and after work is completed. The insurance pays for a variety of benefits, including legal defense expenses, injuries to people, and damage to property. |
| Builder's risk | Pays for damages and losses to a facility occurring while it is being built. |
| Excess liability | An umbrella policy pays for losses exceeding the primary policy limits, such as general liability, automobile liability, and employers' liability on workers' compensation. |
| Pollution liability | Pays for environmental losses associated with accidental chemical spills and the leakage or disbursement of dangerous vapors. |
| Design professional liability | Pays for architects' and engineers' professional liability for errors and omissions. This coverage is usually purchased by the architectural and engineering firms but could be needed for a design–build project. |
| Design–build errors and omissions | For companies working in the design–build arena, provides coverage for contractor errors. |
| Railroad protective | Liability insurance coverage for railroads, purchased by those who conduct operations (construction) on or adjacent to railroad property. |
| Longshoreman/maritime | Liability insurance is similar to workers' compensation; it provides coverage for workers, including construction workers, on the water (working on barges) or those working over water. |
| Automobile liability | Pays for damage caused by the policyholder's vehicles. Also pays medical costs of persons injured in or by the vehicles. This insurance is typically not included in wrap-up insurance because vehicles are operated outside the confines of the project. |
| Tools and equipment | Pays when a contractor's tools, equipment, field offices, or other property are destroyed, damaged, or stolen. This insurance is not included in wrap-up insurance because these items are considered mobile and, therefore, difficult to manage. In addition, the premium costs for these policies are not material and would be difficult to isolate from bids. |

Workers' compensation is an expensive required benefit contractors are obliged to maintain for their workers. Premiums are based on total payroll by trade. Rating bureaus retained by the state set the premium rates according to trade or craft. In addition, each company's experience modification ratio (EMR or E-mod) impacts the cost of workers' compensation insurance. Effectively, a company's EMR is a statistic used to calculate its workers' compensation insurance rate. Companies with a higher accident rate history pay more for their workers' compensation insurance than companies with better safety records. A work category EMR of 1 is average and it has no effect on the company's insurance premium. Companies with a high EMR have higher rates and, therefore, are less competitive at bid time. Many general contractors will not use subcontractors with an EMR above 1.0. Likewise, many owners will not hire GCs with an EMR greater than 1.0. This is especially true in private sector industrial construction.

Many believe the use of EMR ratings to qualify contractors penalizes smaller companies with good safety records. This is because smaller contractors have fewer hours worked so a single claim has a greater effect on their EMR.

## Commercial General Liability Insurance

There are four coverages provided in commercial general liability insurance policies:

1. **Premises/operations liability insurance** provides protection from risks associated with on-going construction operations. Most liability claims result from on-going construction operations such as when a member of the public is hit by a falling tool or someone's car is covered by overspray of paint. The contractor is protected in the case of such events by premises/operations liability. All contractors carry premises/operations liability insurance.
2. **Completed operations and product liability insurance** protects the contractor from liability resulting from construction defects after a building has been completed. If the ceiling in a building falls on a crowded room several years after the building is complete, the contractor is protected through completed operations and product liability insurance. This insurance begins where premises/operations insurance ends.
3. **Contractor's protective insurance** policies provide the general contractor coverage for occurrences caused by the subcontractors. When an accident is caused by a subcontractor, the injured party will sue the subcontractor and almost always the general contractor will also be sued. This is an insurance coverage protecting the general contractor.
4. **Contractual liability insurance** covers liabilities the contractor accepts by contract. Contractual liability results from the indemnification or hold harmless clauses in the construction contract; assessing the contractor's contractual risk requires an examination of the contract. There are many forms of indemnification clauses in construction contracts, with varying degrees of liability transferred to the contractor. Most of these clauses begin, "The contractor shall indemnify and save harmless. . . ."

A general contractor should require all subcontractors to name the general as an additional insured on the subcontractor's commercial general liability insurance policy. This must be done on a primary (the subcontractor's policy is the primary policy) and noncontributory (the general contractor's policy will not contribute to the claim) basis to protect the general contractor from subcontractor employee overclaims. This situation occurs when an employee of a subcontractor is injured on the job and collects workers' compensation from the subcontractor but sues the general contractor for additional compensation.

## Builder's Risk

A project owner should provide for a comprehensive all-risk "Builder's Risk" property insurance policy on a "no-fault" basis to insure project participants against a casualty loss to the work. The policy should be written for the value of the completed construction project. It should be in force at the start of construction and should continue until the project is certified for occupancy.

> A project owner may delegate, by the contract requirements, responsibility to procure the builder's risk insurance to the contractor.

Builder's risk covers risks such as fire, weather events, falsework, and vandalism to the work during the construction period. Some builder's risk insurance policies are "all-risk" to protect the owner, the general contractor, the subcontractors, and the material suppliers while the facility is under construction. All-risk means the policy covers an array of loss events except for those explicitly excluded. Floods and earthquakes are normally excluded from coverage. If the facility is in a flood- or earthquake-prone area special coverage must be purchased separately.

## Equipment Insurance

Equipment insurance covers theft of or damage to construction equipment.

## Umbrella Insurance

An umbrella policy is purchased by a contractor to add to the limits of coverage by the Commercial General Liability Policy, the automobile policy, and the employer's liability section of the workers' compensation policy. An umbrella policy comes into effect if the contractor suffers a loss greater than the limits of the other policies.

## Owner-Controlled Insurance Programs—OCIP

Owner-controlled insurance programs (OCIP), often referred to as "wraps," or wrap-up insurance can reduce project cost. They have traditionally been used for large, commercial projects with construction costs of over $50 million but are gaining usage with smaller residential projects.

In theory, contractors have no need for general liability and workers' compensation insurance because the owner covers all parties for those coverages under the terms of a wrap-up policy. Contractors are able to eliminate the need to buy insurance for the project so there is a cost saving to the owner. Projects covered by wrap-up insurance have extensive safety requirements. These help to control losses.

## Certificates

To demonstrate required insurance is in force, most contracts require the contractor to provide a certificate of insurance to the owner. When required, the certificate will have the owner as an additional insured on the contractor's policy for the project. It is not adequate to just receive a certificate of insurance stating the owner is an additional insured party. The actual endorsement naming the owner as additional insured, which changes the policy, is needed to have a third party named as an additional insured. The GC must ensure subcontractors provide certificates of insurance or the GC can be liable for the subcontractors' losses.

## Subrogation

Subrogation is the right of the insurance company to recover money from the responsible third party who is at fault in an accident. Subrogation is best described as the right to stand in the shoes of another. The insurance company stands in the shoes of its insured party to recover damages from the insurance company representing the third party who caused the accident. When your car gets hit in the rear, your insurance company will pay to repair the damages, but will recover its expense from the insurance holder of the at-fault driver. This is the right of subrogation. Another example would be a building damaged by a truck running into it. The builder's risk insurance would pay for the damage and would subrogate against the truck's owner.

Under the terms AIA Document A201, General Conditions, "the owner and the contractor waive all rights against (1) each other, any of their subcontractors, sub-subcontractors, agents and employees, each of the other, and (2) the Architect, Architect's consultants, separate contractors described in Article 6, if any." To create an effective working relationship on a project, the parties cannot allow their insurance companies to have the ability to recover from members of the contraction team for damage on the site occasioned by one of the team members. Each party is required to carry commercial general liability insurance. Subcontractors are required to name the GC as an additional insured to make sure the subcontractor provides the GC with subcontractor's liability insurance covering its operations on the project. As an additional insured of the subcontractor, the GC would send any claims resulting from the subcontractor's negligence to the subcontractor's policy.

## ADMINISTERING A CONSTRUCTION CONTRACT

Attention to contract administration results in a successful project completed on time and within budget. It is the best way to elimination disputes and claims. With the award and signing of the contract, the process begins and continues until all work is completed and has been accepted by the owner with all disputes completely resolved.

### Preconstruction Conference

Nearly all construction projects begin with a preconstruction conference. This is a formal meeting, with an agenda distributed by the owner before the meeting and minutes of the meeting published by the owner and distributed to all attendees following the meeting. Attendees should include all stake holders: the owner's representative, the designer, the general contractor with its superintendent and the estimator for the project, and as many subcontractors as possible, municipal authorities, utility companies, emergency organizations such as fire and police, and others who may be impacted by the work.

> Attendance by the estimator is important. The estimate was created based on a construction plan and the field leaders need to understand the plan.

The preconstruction conference is a business meeting and is not open to the public. Its main purposes are to open the lines of communication between owner and contractor, to establish processes for doing business, and to address any technical questions by the general contractor and the specialty subcontractors. Questions concerning acceptable "or equal" substitutes can be posed and the process for obtaining answers explained. Clearly establish QA/QC roles and responsibilities. The owner's representative should answer all questions in writing as soon as possible but not later than a specific agreed-upon date.

> Each trade, specialty subcontractor, involved with the project should determine their top issues and bring them to the preconstruction conference for discussion.

Important policies are reviewed, such as restrictions on work hours and days, critical progress dates, and other issues that are relevant to the project team. The safety is always discussed. Location of utilities can be a major concern both in terms of connection and those below excavation areas on the site. The accuracy of provided location information should be verified.

The one topic important to all contractors is the processing of pay estimates. At the preconstruction conference, the monthly progress cutoff date and the date of submission for the pay estimates are selected or the requirements already in the specifications are noted. The most important issue to the contractor is the number of days the owner will require before making payment to the general contractor, and the maximum number of days the general contractor has to pay the subcontractors.

There are many legislative efforts to shorten the time between completion of the work and receipt of payment. For example, the State of Arizona requires owners to pay the general contractor within 10 days of receipt of a monthly pay estimate and the general contractor to pay subcontractors within 10 days after receipt of their funds. If for any reason the general contractor cannot or will not pay the subcontractors, the general contractor is obliged to return the progress payment to the owner. The Arizona law applies to both public and private owners.

### Subcontracts

A subcontractor is a construction firm hired by the general contractor to perform work in a specialty area such as electrical or mechanical construction. They contract directly with the general contractor and usually do not have any contractual relationship with the owner. The general contractor remains responsible to the owner for the quality of the subcontractor's performance. However, the general contractor's contract with a subcontractor clearly states the subcontractor's responsibility for performance and correction of substandard work. If approval of the subcontractor by the owner is required, the approval must be obtained before the subcontract is signed, even though disapproval of subcontractors is a rare event. Some projects, such as buildings, may have many different subcontractors working on the site at the same time. Subcontractors are generally able to provide a better-trained and properly equipped workforce in their area of specialization.

Some subcontracts are small enough in scope to allow the use of an informal contract, such as a letter of proposal and an acceptance, but general contractors should use a formal written document for all subcontracts. There are standard subcontract forms available from organizations such as The Associated General Contractors of America, The American Institute of Architects, or the American Subcontractors Association (ASA) or contractors can develop their own forms. Preparation of a subcontract is an area where contractors should seek the advice and assistance of legal counsel.

Use of a standard subcontract form helps the subcontractors because they gain experience with the general contractor and know what contract terms to expect when they submit a subsequent proposal. In addition, change orders require subcontracts to also be changed to reflect the new requirements or work. Standard contract forms contain clear change order language understandable to both the general contractor and the subcontractor.

## Submittals

To verify the materials being ordered and subsequently installed in a project conform to the specifications, owners use a submittal process. By the process, the design professional has an opportunity to confirm or reject the proposed product. The submittal process serves to confirm compliance of the proposed product with the intent of the contract documents.

Submittals contain information concerning products to be incorporated in a project. The owner or designer must approve the submittals before products are ordered or incorporated into the work. The drawings, diagrams, catalog cuts, data sheets, and other submittal data prepared by the contractor, subcontractor, manufacturer, or supplier illustrates what items will be installed or how specific portions of the work will be performed. The designer should also seek to ensure the proposed product will function as intended.

The submittals also provide data allowing the designer to confirm the correctness of dimensional information. The submittal and the submittal approval process are commonly incorporated into the quality assurance efforts for the project and these serve to ensure the use of proper methods and materials. As a general rule, everything to be installed in the facility requires a submittal. Submittals are normally processed as shown in Fig. 3.7.

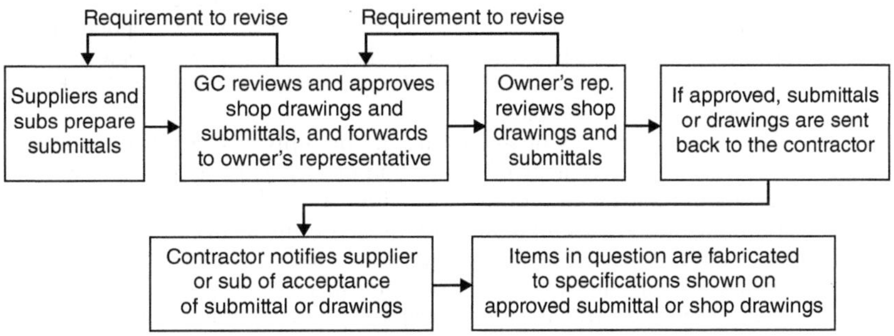


**FIGURE 3.7** Processing submittals and shop drawings.

The required routing of formal communications such as submittals is very important because incorrect routing can cause delays. For instance, submittals must be sent up the chain from the materials supplier to the specialty contractor to the general contractor to the designer for review and approval. If the designer rejects the submittal and requires a resubmittal, the submittal must make its way back down the chain to the originator, and then be resubmitted and sent back up the chain through the general contractor to the designer. This can be a time-consuming process and has the potential to delay project completion by delaying the material procurement process.

> The preconstruction conference is the opportunity for the contractors to work out an efficient use of technology to track submittals.

It is becoming common for owners, designers, general contractors, and subcontractors to design the submittal process so the originator of a submittal may route it directly to the designer for review, with an information copy to the general contractor and others needing notification. The designer then returns an information copy to the general contractor and the action copy directly to the originator of the submittal. This system has two advantages:

- Efficiency (saves time)
- Involvement by the general contractor only when required

There are some disadvantages to such a system. A possible disadvantage is the general contractor not being informed of all possible concerns in a timely manner.

## Request for Information

A request for information (RFI) is a specific type of written communication used to document questions stemming from the contract documents and to record the subsequent answers. The subject matter addressed by RFIs in many cases relates to design intent, affects the project schedule and budget, and may or may not be urgent. RFIs serve to document the question and require the other party to return a written/documented answer. The party requesting the information should always ensure that the designer issues the appropriate pricing documentation (change order) if the response to the RFI results in extra work or a cost change. Most general contractors have a standard form they use for all RFIs. The GC will require their subcontractors to use the same form. GCs also maintain a log of RFIs, sorted by urgency, to track the RFIs and the responses.

## Pay Estimates

Contractors are normally paid monthly; however, it is necessary to read the contract to understand the details. One of the most important aspects of any owner's reputation with contractors is the time required to process a payment after submission of the pay estimate by the contractor. Slow payment or **delayed payment** leads to higher bids. Fast payment leads to lower bids.

**delayed payment**
*Because of credit or cash flow problems, owners sometimes fail to pay contractors for services according to the schedule set by the contract. To justify the practice they may claim defective or substandard work.*

The actual process of submitting the monthly pay estimate involves estimating the percentage of completion of each job activity, calculating the amount due for each of those activities, and subtracting the retainage from the total. A portion of the monthly pay estimate belongs to the general contractor and a portion belongs to the subcontractors. Therefore, the obvious second question relates to the time taken by the general contractor to pay the subcontractors after receipt of payment from the owner. The same formula applies—general contractors who pay their subcontractors quickly receive lower bids than those who hold subcontractors' money.

The payment clauses in subcontracts can be different from the payment clauses in general contracts. In subcontracts, payment clauses fall into one of two categories, either "pay when paid" or "pay if paid." Subcontractors prefer a "pay when paid" clause because such a contract clause creates a contractual obligation for the general contractor to pay the subcontractor. "Pay if paid" clauses do not. Subcontractors should read the payment provisions carefully.

## Change Orders

Change orders are part of every construction contract because each project is a new prototype. No two projects are exactly alike and owners do not fund the design process at a quality level to produce a "perfect" design. Even houses in a large development, with each being very similar, have differences caused by the terrain, existing utilities, or other factors such as subsurface conditions. Change orders are a normal part of the contract administration process.

A change order does not change the scope of work; in fact, the requirements of the change order must be within the original scope of work. Changes outside the scope of work require a supplemental agreement and are legally quite different from change orders. Change orders change the details or conditions of the work and are used to add "extra" work or to delete work. The contract does not allow the contractor the option of refusing change order work (they are not change *options*). But most change orders entitle the contractor to additional time and compensation if the required work significantly modifies the original requirements and involves changes to the existing work, as well as additional work. Change orders must be written and must be issued by a party with the proper authority to issue the change; nevertheless, the parties involved can effectively waive this requirement by their actions. Owners normally provide a letter to the contractor to explain who has authority to issue and approve changes, and the exact monetary limits on each person's authority. The designer, for example, usually has no authority to obligate the owner and cannot approve change orders. Many construction companies use field orders for minor no-cost changes having no impact on the contract provisions. Field orders change the work but do not modify the contract; they simply provide a record of the change.

There is an important difference between *extra* work and *additional* work. Extra work is added by change order and was not included in the original requirements of the contract, but was within the original scope of work. A simple example of extra work would be the addition of a new door opening not shown on the original plans, but it is added by change order. The contractor is entitled to additional compensation. An example of *additional* work could occur in a project requiring the installation of a fire sprinkler system to an existing building. Additional work would be the need to remove a ceiling to gain access to the work area and then restore it to its original condition, but if this was known it should have been included in the original estimate for the work. There would be no additional compensation for removal and replacement of the ceiling.

Change orders are subject to negotiation and the contractor is rightfully interested in gaining as much compensation as possible for the extra work. Change order work is more expensive than the original contract work because it tends to be out-of-sequence work and there is additional overhead to track and schedule changes to implement the change. Most owners feel a contractor is entitled to fair compensation for the change order work. The contractor's problem is to convince the owner of what is fair and reasonable. Some contracts are written with prenegotiated markup and profit for change order work. This is part of the original contract. Large change orders on most government projects are subject to audit. The contractor agrees to open the company's books to the auditors to determine the actual costs associated with the change order work.

Change orders can be very simple or they can become very complicated. A change order to add a new door opening is very simple. A change order to protect a project from the effects of the weather can be very difficult.

## Claims

Claims are the unfortunate result of change orders not resolved or job conditions the contractor considers to be a material change condition but the owner does not agree. Often claims result from a disagreement on the meaning of the terms of the contract. Causes often causing claims are those addressing late progress payments, delays caused by the owner, or rejection by the designer of "or equal" substitutes. Claims often result from poor communications and a claim will be compounded by poor communication.

Contracts require the contractor to notify the owner of a claim within a specified number of days and then proceed with the work. The contractor is required to rely on the claims process for fair recovery. The contractor cannot refuse to proceed with work within the original scope of work, but it is sometimes difficult to determine whether work is within the original scope. In the case of federal contracts, the contractor is required to file a claim with the contracting officer, and if the contracting officer's decision is not favorable to pursue the claim in the judicial system. Contractors must read the contract to determine what process must be followed to file a claim. In all cases, public or private, the administrative remedies specified must be exhausted before any judicial remedy can be sought.

Successful settlement of a claim depends on adequate documentation. Contractors have learned to maintain high-quality daily records of the events in the field. Good records include meeting minutes, records of telephone conversations, project photographs taken regularly, and other records to reconstruct an accurate record of the events on which the claim is based.

## Dispute Resolution

Everyone in the industry recognizes the courts are not the best way to resolve disputes. Still the predominant method of dispute resolution is litigation. There is, however, a growing trend to find ways to resolve disputes without litigation. The results are mixed, with some methods more successful than others. Disputes are unresolved claims, unresolved disagreements, or unresolved change orders causing the parties to resort to legal action—the courts. They always involve money. The key word is unresolved. Disputes are not caused by illegal actions. They are caused by unresolved issues about compensation for the contractor and what the owner sees as an unnecessary extra expense because it should have been included in the original contract price.

> What should a contractor do when a situation appears to be headed to a claim situation?
>
> ➢ Keep detailed and accurate records.
>
> ➢ Seek to maintain and/or improve communications with the owner.

Construction is, by the nature of the contract, an "arm's length" business characterized by adversarial relationships and broken communications. Contractors need to communicate. When there are problems, the problems need to be communicated immediately and clearly. Bad news will not improve with time. Open and honest business practices together with good communications help to avert situations leading to an unresolved dispute. In a potential dispute situation, seek to resolve the dispute as rapidly as possible and at the lowest possible management level. Project management by an owner with a "do it my way and sue me when the project is over if you don't like it" philosophy is not good business, but it happens.

Contractors should try to resolve disputes when they occur. Seek to resolve them in the field and not in the courts. Many alternative dispute resolution methods are being tried, all with the objective of resolving disputes before they advance to the litigation stage.

## Alternative Dispute Resolution

Common alternative dispute resolution methods are:

- Partnering
- Dispute Review Boards

- Mediation
- Arbitration
- Mini trials
- Project neutrals

**Partnering** has gained favor as a philosophy of construction management based on trust. The basis of partnering is trust. Partnering meetings are held before the start of a project in an effort to develop a level of trust between the owner and contractors. Common goals and objectives are identified. A charter, in which all parties resolve to operate in an atmosphere of trust, is written and signed. A dispute resolution framework is put in place to require problems to be solved in the field within a very limited number of days, may be only 3 days, or elevated to the next level of management. Monthly partnering meetings are held to evaluate the partnering process. All meetings are chaired by an outside facilitator to create a "level playing field" for the contract partners. The move toward partnering appears to be very successful.

**Dispute Review Boards** (DRBs) are used to both prevent disputes and to settle disagreements rapidly. The members of the board bring construction experience and technical expertise specific to the project into their decision making. A DRB usually has neutral members who have been selected by the parties to the contract, the owner and the contractor. The board members serve from commencement of the project until the end. When called upon, the board issues "Advisory Opinions." Because the DRB is familiar with the project its recommendations are hard to reject.

**Mediation** is a system of selecting a neutral third party to assist the parties in reaching a mutually acceptable agreement. The mediator serves as a "go-between" to improve communications and does not impose a settlement on the parties. There is nothing binding about a mediation process. It is undertaken by mutual agreement of the parties involved, in an attempt to resolve the disagreement before it becomes a dispute. Often it is the last step before litigation and represents the efforts of both parties to avoid litigation.

**Arbitration** is slightly closer to litigation in philosophy than mediation because the arbitrator can in some cases impose a solution on the parties to the dispute or the parties can agree to accept the arbitrator's solution. Arbitration is not court and is not binding except by prior agreement between the parties to accept the findings of the arbitrator or when court directed. The arbitrator is selected by mutual agreement. The parties to the dispute can make the arbitration process binding by mutual agreement because both sides want to avoid the expense of a lawsuit or it can be ordered by a court. There are no arbitration records as there would be court transcripts, so the parties maintain a higher degree of privacy. The process is less costly than a lawsuit. The arbitrators are usually experts in the kind of construction issue in dispute, providing a more informed decision than might be obtained in court where the judge cannot possibly be an expert in every area of litigation presented. Most judges are not as knowledgeable about construction matters as arbitrators.

**Mini trials** allow each side to present its position in the dispute, and then a judge renders a decision. They are judicial proceedings. They have not proven to be very successful because the opposing sides have a tendency to escalate the proceedings into a normal hearing and leave the "mini" description behind. The level of effort for a "mini" tends to grow and as a result they fail to accomplish their intended purpose.

**Project neutral** is an approach to dispute resolution being used on very large, highly technical projects. The project neutral is a person (or team) hired before the construction contract is awarded. The team participates in the project throughout the construction period. The project neutral is technically knowledgeable and extremely well informed in matters of construction and the contract requirements. As the name implies, the project neutral does not represent the position of any of the project parties and does not advocate positions taken by any party. The project neutral is analogous to an official in an athletic contest providing expert on-the-spot decisions when they are needed. Typically, the project neutral, to remain well informed of the status of work and of the beginnings of any disputes or potential claims, will visit the project regularly. The function of the project neutral is to reach a solution to disputes. The finality of the decisions made by the project neutral depends on the provisions of the contract.

## Liquidated Damages and Substantial Completion

Liquidated damages are paid by the contractor to the owner if such amounts are required by the contract and if the contractor exceeds the allowed contract time period. A contract may require liquidated damages of a few hundred dollars per day or it may require liquidated damages of thousands of dollars per day. The amount is determined before the contract is advertised and it is based on an estimate of actual costs the owner or the public would incur by the late completion of the work. The amount is stated in the contract documents.

To avoid the imposition of liquidated damages, contractors must actively manage the project schedule. The project manager must have a realistic schedule to complete the work. In addition, change order work must always include additional time for the added work as part of the negotiation process.

Most contracts require liquidated damages for failure to reach "substantial completion" by a specified time. Contractors must know the requirement of each contract and the definition of substantial completion. The general rule (not a legal rule) is: the facility must be ready for the owner to move into and use. There can still be a "punch list" of items the contractor needs to correct to reach final completion and those items can be completed while the owner is occupying the facility. Punch list items can be anything from touching up paint to completing the landscaping. Punch list items cannot be substantial items of work such as installing the windows and doors or completing the air-conditioning system. If those items remain, the facility is not "substantially complete."

## Final Inspection

When the contractor feels the facility is substantially complete and the owner agrees the facility is ready for acceptance, as defined in the contract documents, a "final inspection" will be scheduled. Contractors should treat the final inspection as a formal contract requirement. It is the owner's last formal opportunity to walk the facility and identify work items requiring attention before formal acceptance by the owner. The final inspection generates the final punch list. Owners should prepare for the final inspection by working proactively with the contractor as the work progresses and staying knowledgeable of the status of the facility. This should be a "one bite out of the apple" event, meaning the final inspection is the last opportunity for the owner to identify remaining work items. The items agreed to and written into the punch list permit the contractor to proceed with the knowledge that when the punch list items are done, the contract requirements are fulfilled. The contractor should then be required to return to the site only for possible corrective work covered in the warranty.

# CHAPTER 4
# CONSTRUCTION ACCOUNTING

*A company's balance sheet and income statement are important accounting documents used to monitor financial performance. The balance sheet presents a snapshot of the company's fiscal health at a particular moment whereas an income statement shows all of the money a company has earned (revenues) and all of the money it has spent (costs and expenses) during a performance period, usually the fiscal year. Balance sheet analysis can identify potential liquidity and leverage challenges. The income statement provides a map of how a company is changing and the pace of change. It is essential for management to know how to analyze different elements of these documents. The summarized information of these documents serves company management as the basis for important decisions affecting company capability. These documents also guide decisions made by outside parties such as banks and sureties.*

## ACCOUNTING PRACTICES

Accounting is a structured method of recording, classifying, reporting, and interpreting the financial data of an organization. There are three types of accounting:

1. Generally Accepted Accounting Principles (GAAP accounting) for financial statements (lots and lots of rules)
2. Management Accounting (no rules) used in running the firm
3. Tax Accounting

This chapter discusses the first two and leaves the tax aspect to the accounting professionals who persistently work to understand the tax laws politicians dream up and constantly change.

Accounting is the process of recording, classifying, reporting, and interpreting the financial data of an organization. This is the role of financial accounting, with all the GAAP rules and "whatnot." Beyond GAAP and its rules there is management accounting used to make decisions about operational aspects of a business. The main purpose of management accounting is to aid decision makers in choosing among alternative courses of action. Effective management of every company requires a good **cost accounting** system, and the availability of current and accurate cost information. For information to be useful, there must be consistent application of accounting principles (e.g., a method of recording and reporting information) from one period to another. Accounting documentation using a consistent basis provides information needed to make business decisions for guiding a company on a profitable course.

**cost accounting**
*The accounting phase used to determine the cost of producing a product or provide a service.*

> Standing alone, no one financial statement provides sufficient information to summarize the activities of a company, and no one item or part of a statement can summarize the information.

Financial statements are the primary means of communicating useful financial information. Each statement is the result of

- Simplifying
- Condensing
- Aggregating transactions

The most important accounting document used to monitor a company's performance (earnings) is the income statement, while the **balance sheet,** another important document, demonstrates the company's overall financial health and potential for future growth. The income statement exhibits the sources of a company's production costs, expenses, and income.

**balance sheet**
*Shows the assets, liabilities, and company owner's equity on a specific date.*

## INCOME STATEMENT

A company's income statement is a summary of the revenues earned by a business, the expenses incurred in earning the revenues, and the resulting net income or loss. It is a record of earnings or losses for a given period. It is a picture of the company's performance for a specific period, usually the fiscal year. It shows all the **revenue** a company earned and all of the expenditures (expenses and costs) it incurred. *Expenses* are the direct outlays required to produce revenue; for example, rent, payroll, and materials. *Costs* refers specifically to depreciated assets; it is the depreciation portion charged to revenue during a period.

**revenue**
*An inflow of assets, not necessarily cash, in exchange for goods and services sold.*

The income statement provides a record of a company's performance across time. Most importantly, the income statement communicates if the business is profitable. It provides a map (Fig. 4.1) of how a company is changing and the pace of change. Company management must both understand the income statement and know how to analyze different elements of this important document.

Operating income (revenue) does not include interest earned. Nor does it include income generated outside the normal activities of the company, such as income on investments. Operating income is particularly important because it is a measure of profitability based on a company's operations. The income statement presents an assessment of whether or not the "core operations" of a company are profitable. It ignores income or losses outside of a company's normal domain. Earnings before taxes are the sum of operating and nonoperating income. Net earnings or net income (loss) after taxes is the proverbial "bottom line." It is the amount of profit a company makes after all of its income and all of its expenses are collected.

### Retained Earnings

**Retained earnings** is the amount of equity from profits a company has accumulated considering all of its transactions since its commencement of operations. If the company continually earns substantial profits, it is a stable company, or if it is like the Low Bid Construction Corporation (Fig. 4.1—"Income (loss) from operations"), there are major problems and the viability of the company is threatened. In addition to very thin gross profits, it appears Low Bid is carrying significant general and administrative expenses. These could include an unnecessarily fancy office, a company airplane, or unnecessary estimating cost caused by attempting to bid everything offered.

**retained earnings**
*The amount of net income retained by a company.*

**LOW BID CONSTRUCTION CORPORATION**
**STATEMENT OF INCOME AND RETAINED EARNINGS**

| | For the Years Ended December 31, | | |
|---|---|---|---|
| | **2019** | **2018** | **2017** |
| Revenue (Notes 10 and 16) | $174,063,148 | $163,573,258 | $210,002,272 |
| Cost of revenue (Note 10) | 169,404,787 | 158,935,103 | 200,070,826 |
| Gross profit | 4,658,361 | 4,638,155 | 9,931,446 |
| General and administrative expenses | 8,062,623 | 777,840 | 671,035 |
| Income (loss) from operations | (3,404,262) | (2,139,685) | 3,260,411 |
| Other income (expense): | | | |
| Interest income | 319,797 | 646,480 | 668,928 |
| Interest expense | (485,937) | (250,996) | (209,872) |
| Other income (expense) | 377,840 | (115,246) | 211,119 |
| | 211,700 | 280,238 | 670,175 |
| Income (loss) before income taxes | (3,192,562) | (1,859,447) | 3,930,586 |
| Income tax benefit (expense) | 668,631 | 284,861 | (1,590,480) |
| Net income (loss) (Note 17) | $(2,523,931) | $(1,574,586) | $2,340,106 |
| BEGINNING RETAINED EARNINGS | $3,090,214 | $4,664,800 | $4,664,800 |
| ENDING RETAINED EARNINGS | $566,283 | $3,090,214 | |

**FIGURE 4.1** Sample income statement. (The numerous notes accompanying the statement are not included here.)

## *BALANCE SHEET*

A company's balance sheet (Figs. 4.2 and 4.3) describes its financial position on a specific date, a single point in time. This is a snapshot of the company's financial condition. Because it is a single-point-in-time representation, it is sometimes referred to as a position statement. A balance sheet is prepared at least once per year, but it may also be presented quarterly, semiannually, or even monthly. This document provides information documenting what the company owns, what it owes (liabilities), and the value of the business to its owners (**equity**). The name, balance sheet, is used because the accounts must always be in balance. Assets must always equal the sum of liabilities and **owner's equity**.

**equity**
*A right, claim, or interest in property.*

**owner's equity**
*The equity of the owner or owners of a business in the assets of the business.*

### Assets

Resources expected to produce economic benefits for their owners are termed **assets**. They are the opening information presented on the balance sheet (Figs. 4.2 and 4.3). The assets of a construction company are usually money (cash), receivables, equipment and plant, property, and materials used to construct projects. Compared to other types of businesses, construction companies typically have fewer physical assets. A manufacturing business will have a large investment in buildings and machinery and may even at times carry significant inventory of raw materials with which to produce the product or an inventory of completed products in the warehouse ready for shipment. These are physical assets and a lender can rely on them as collateral against a loan to the company.

**assets**
*Things used to produce value.*

| **CONSOLIDATED BALANCE SHEET**<br>(Thousands of dollars) | | |
|---|---|---|
| **December 31,** | **2019** | **2018** |
| **ASSETS** | | |
| **Current assets** | | |
| Cash | $11,330 | $10,025 |
| Short-term investments, at cost, which approximates market | 8,620 | 14,738 |
| Accounts receivable including retentions $44,943 and $65,064 | 208,007 | 182,319 |
| Refundable federal income taxes | – | 11,225 |
| Costs and earnings in excess of billings on uncompleted contracts | 192,727 | 164,232 |
| Equity in joint ventures | 96,655 | 82,311 |
| Real property held for development and sale at the lower of cost or market | 45,163 | 33,000 |
| Other | 11,462 | 7,790 |
| Total current assets | 573,964 | 505,640 |
| **Investments and other assets** | | |
| Equity in affiliated enterprises | 9,815 | 16,349 |
| Marketable securities, at cost, market $9,667 and $10,741 | 259 | 259 |
| Other investments, at cost, which approximates market | 41,318 | 15,896 |
| Notes receivable and other assets | 18,024 | 8,959 |
| Total investments and other assets | 69,416 | 41,463 |
| **Property and equipment, at cost** | | |
| Land | 2,596 | 2,566 |
| Buildings, ways, and wharves | 164,322 | 148,900 |
| Construction and other equipment | 223,144 | 212,186 |
| Total property and equipment | 390,062 | 363,652 |
| Less accumulated depreciation | (140,169) | (123,702) |
| Property and equipment—net | 249,893 | 239,950 |
| Total assets | $893,273 | $787,053 |
| *The accompanying notes are an integral part of the financial statements.* | | |

**FIGURE 4.2** Sample consolidated balance sheet, assets part.

A construction company, however, conducts its business on property owned by someone else and only needs to actually possess a small office for corporate management. A construction company performing heavy/highway work could have a fleet of equipment, which represents substantial asset holdings, but many times such equipment is leased or rented so it does not belong to the company.

Exceptions would be companies engaged in the production of concrete and asphalt or owning quarries. Quarries with their aggregate reserves represent significant asset holdings. Therefore, because the value of actual physical assets possessed by a construction company is usually minimal, lenders must carefully look to the continuing operational performance of the company when evaluating the risk of a loan.

> Property and equipment are recorded on the balance sheet at cost and not the price at which an asset may be sold or the cost to replace the asset.

| LIABILITIES AND STOCKHOLDERS' EQUITY<br>(Thousands of dollars) | | |
|---|---|---|
| **December 31,** | **2019** | **2018** |
| **Current liabilities** | | |
| Short-term and current portion of long-term debt | $72,742 | $29,350 |
| Accounts payable and accrued expenses | 228,115 | 200,761 |
| Billings in excess and earnings on uncompleted contracts | 66,982 | 70,846 |
| Advances from clients | 19,687 | 32,076 |
| Income taxes | | |
| Currently payable | 11,462 | 3,476 |
| Deferred | 6,047 | 46,866 |
| Dividends payable | 3,236 | 2,672 |
| Total current liabilities | 408,271 | 386,047 |
| **Noncurrent liabilities** | | |
| Deferred income taxes | 23,666 | 20,611 |
| Deferred income and compensation | 15,513 | 15,602 |
| Accrued workmen's compensation | 27,973 | 23,189 |
| Long term debt | 103,611 | 76,063 |
| Total noncurrent liabilities | 190,763 | 135,465 |
| **Commitments** | | |
| **Stockholders' equity** | | |
| Common stock, par value $3.33½, authorized 20,000,000 shares | | |
| Issued 10,616,391 shares | 35,388 | 353,886 |
| Capital in excess of par value | 74,056 | 73,273 |
| Retained earnings | 196,764 | 170,421 |
| | 306,208 | 279,082 |
| Less cost of treasury stock, 802,878 and 899,744 | (11,969) | (13,541) |
| Total stockholders' equity | 294,239 | 265,541 |
| Total liabilities and stockholders' equity | $893,273 | $787,053 |

**FIGURE 4.3** Sample balance sheet, liabilities, and stockholders' equity part.

## Current Assets

If an asset can be converted into cash within 1 year it is classified as a current asset.

- *Cash* is the basic current asset.
- *Cash equivalents* are not cash, but they are considered equal to cash because they can be easily converted into cash. Cash equivalents are generally highly liquid, short-term investments such as U.S. government securities, commercial paper, and money market funds. To be considered cash, short term is generally defined as converting or maturing into cash within 90 days.

**FIGURE 4.4** Owned equipment, at cost on the consolidated balance sheet.

- *Accounts receivable* represent money clients (project owners) owe to the firm for services rendered or for goods sold (e.g., producers of asphalt and concrete provide material on credit).
- *Inventory*, in the case of a construction firm, is the material kept on hand for use in constructing projects.

## Long-Term Assets

Many economic resources, assets, cannot be readily converted into cash. These long-term or "other assets" are expected to provide benefit for future operations. Within the year after the balance sheet date they will not be sold or consumed.

- *Fixed assets* are tangible assets. Generally, fixed assets refer to items such as equipment (Fig. 4.4), vehicles, plant, buildings, and property. On the balance sheet, these are valued at their cost less depreciation, except land.

## Liabilities

Liabilities are obligations a company owes to outside parties. They represent rights of others to money or services of the company. Examples include bank loans, debts to suppliers, debts to employees, debts to subcontractors, and both deferred taxes and taxes currently payable to government entities. On the balance sheet, liabilities are generally broken down into current liabilities and long-term liabilities.

**liabilities**
*Legal financial debts or obligations.*

*Current liabilities* are obligations due to be paid within the year (12 months), such as accounts payable, interest on long-term debts, and taxes payable. The most pervasive item in the current liability section of the balance sheet is accounts payable. Billings in excess of costs and earnings on uncompleted contracts "overbillings" are also recognized as a current liability (Fig. 4.3).

- *Accounts payable* are debts owed to subcontractors and to suppliers for the purchase of goods and services on an open account.

Because current liabilities are usually paid with current assets, it is important to examine the degree to which current assets exceed current liabilities. This difference is called *working capital*.

**Current Obligations Example:** Review the balance sheet asset and liability information contained in Figs. 4.2 and 4.3. Does this company appear to be in a good position in regard to meeting current obligations?

| | Dec. 31, 2019 | Dec. 31, 2018 |
|---|---|---|
| Total current assets | $573,964 | $505,640 |
| Total current liabilities | 408,271 | 386,047 |
| Working capital | 165,693 | 119,593 |

The company's current assets are greater than its current liabilities. It should be able to meet current obligations.

- *Long-term debt* is a liability of a period greater than 1 year. It usually refers to loans the company has acquired from banks. These debts could include loans to purchase vehicles and equipment, and mortgage payments for buildings and land, as such they are often paid in installments. If this is the case, the portion of the principal to be paid in the current year is considered a current liability.

## Owner's Equity

**Owner's equity**
*The owner's investment in the business*

**Owner's equity** or the net worth of the company is the excess of assets over liabilities. It represents what the owners have invested in the business plus accumulated profits, which are retained in the company, from ongoing operations. Losses will reduce owner's equity. The formulation of the report's balance is shown by

$$\text{Equity (net worth)} = \text{Total assets} - \text{Total liabilities} \tag{4.1}$$

Considering the balance sheet asset and liability information contained in Figs. 4.2 and 4.3, it is possible to calculate the net worth of the company without knowing exactly the types of equity the company owners possess (Table 4.1).

**TABLE 4.1** Net Worth Calculations

| | Dec. 31, 2019 | Dec. 31, 2018 |
|---|---|---|
| Total assets | $893,273 | $787,053 |
| Total current liabilities | 408,271 | 386,047 |
| Total noncurrent liabilities | 190,763 | 135,465 |
| Total liabilities | 599,034 | 521,512 |
| Net worth | 284,239 | 265,541 |

At the bottom of the balance sheet in the "stockholders' equity" section (Fig. 4.3), the composition of the equity is presented. *Retained earnings* are the part of equity resulting from earnings in excess of losses.

## Liquidity and Working Capital

Management must study the balance sheet to measure a firm's liquidity, financial flexibility, profit generation ability, and debt payment ability. Financial flexibility is the capability of taking effective actions to alter the amounts and timing of cash flow in response to unexpected needs and opportunities.

Surety companies will carefully study the company's balance sheet before issuing bid, payment, and performance bonds for a project. The balance sheet reveals the company's financial stability and will, therefore, affect the surety's decision concerning the bonding limit (volume of work) of the construction company. Without the ability to secure a bond, the amount of work the company can undertake can be limited.

The analysis of a balance sheet will identify potential debt payment problems if they exist. These may signify the company's inability to meet financial obligations. *Liquidity* is the quality an asset has of being readily convertible into cash, and cash is necessary to keep the business operating—to pay the bills. *Working capital*, the funds available for operating the business [Eq. (4.2)] is the positive difference between current assets and current liabilities (Current Obligations Example). The relative amount of working capital is an indication of short-term financial strength:

$$\text{Working capital} = \text{Current assets} - \text{Current liabilities} \tag{4.2}$$

Working capital consists of funds used to pay operating obligations. These funds flow into, through, and out of a construction company. They flow in the manner of a circulatory system, as operations are conducted. This flow of working capital through the company is a lifeblood cycle, and without the "blood," the company dies. The flow of cash includes (Fig. 4.5)

1. Clients (project owners) input *contract revenues*.
2. A major portion of the revenue stream flows out of the company as direct costs to pay *direct project costs*: subcontractors, material suppliers, project labor, and other costs.
3. Part of the remainder is used for job *overhead costs* (this cost is allocated to and absorbed by all jobs).

FIGURE 4.5 Flow of funds through a construction company.

4. The portion remaining is classified as *gross profit* (operating profit).
5. *Nonjob overhead*, general and administrative (G&A), expenses of the company, are then deducted.
6. Finally, a small portion makes it through as *income* to the company. But the company does not get to keep all of this portion because taxes must be paid on income.
7. However, if there is even one bad project (total cost is greater than the contract amount), profits from profitable projects are used to complete the work (pay direct cost).

If the company is operating profitably, it should generate a cash surplus. If it does not generate surpluses, it will eventually run out of cash and expire. The Consolidated Balance Sheets for Low Bid Construction Corporation are shown in Fig. 4.6.

From the presented balance sheet data, it is clear Low Bid developed a liquidity and working capital problem in 2019 (Table 4.2) and was forced to use past profits (retained earnings).

Recognizing its problem, Low Bid made a provision in the assets part of the 2019 balance sheet to generate needed working capital. Examine the "Assets held for sale" row in the Assets section of Low Bid's balance sheet (Fig. 4.6). A note accompanying the balance sheet states "in order to improve working capital, the Company executed a definite agreement to sell certain assets. Accordingly, $3,213,484 of assets consisting of inventories and equipment has been classified as assets held for sale on the balance sheet."

## Current Ratio

While the magnitude of a firm's working capital, as calculated using Eq. (4.2), provides information about the ability of a firm to prosecute its workload, it is also an indication of the firm's ability to pay short-term obligations. There are other financial ratios designed to measure a company's ability to meet these obligations. Ratios provide a means to identify changes in position. Changes in position are not always apparent when reviewing simply the magnitude of values. The **current ratio,** Eq. (4.3), is a tool to identify changes in a company's ability to meet short-term obligations:

**current ratio**
*The relation of a company's current assets to its current liabilities.*

$$\text{Current ratio} = \frac{\text{Current assets}}{\text{Current liabilities}} \tag{4.3}$$

Financial analysis is the process of tracking changes in a company's financial health. Ratios are one method of analysis. When a ratio is not in conformance with the norm for the industry, the variance is a warning flag announcing the need to look deeper into a company's financial statements.

As a rule of thumb, the current ratio for a construction company should be above 1.3:1.

Checking the financial ratios for Fast Buck Construction Co., it appears the company has problems (refer to the following Working Capital Example). The current ratio gives an indication of trouble.

**Working Capital Example:** Fast Buck Construction Company's working capital situation.

**Fast Buck Construction Co. Balance Sheet Data (Thousands)**

| Year | 2019 | 2018 | 2017 | 2016 | 2015 |
|---|---|---|---|---|---|
| Current assets | $8,354 | $7,539 | $3,587 | $1,649 | $819 |
| Current liabilities | 10,559 | 6,514 | 3,127 | 1,089 | 467 |
| Working capital | –2,205 | 1,025 | 460 | 560 | 352 |
| Current ratio | 0.79 | 1.15 | 1.14 | 1.51 | 1.75 |

**Low Bid Construction Corporation**
**Consolidated Balance Sheets**

| | December 31, 2019 | December 31, 2018 |
|---|---|---|
| Assets (Note 9): | | |
| Current Assets: | | |
| Cash and cash equivalents (Notes 1 and 2) | $2,228,506 | $1,822,598 |
| Restricted cash (Notes 1, 2, and 16) | 2,401,548 | 1,783,005 |
| Accounts receivable, net (Notes 1, 3, and 16) | 21,377,904 | 14,297,564 |
| Prepaid expenses and other | 404,780 | 749,708 |
| Inventory (Note 1) | 3,365,750 | 4,288,235 |
| Income tax receivables (Notes 1 and 11) | – | 74,000 |
| Costs & estimated earnings in excess of billings | 5,294,054 | 9,828,009 |
| Total Current Assets | 35,072,542 | 33,543,119 |
| Property and equipment, net (Notes 1, 5, 8, and 12) | 15,267,791 | 18,111,506 |
| Assets held for sale (Note 1) | 3,213,484 | – |
| Deferred tax asset (Note 1 and 11) | 1,957,923 | 873,441 |
| Refundable deposits | 55,110 | 176,565 |
| Goodwill, net (Note 1) | – | 1,500,733 |
| Mineral rights and pit development, net | 533,608 | 1,180,666 |
| Claims receivable (Notes 1, 4, and 14) | 5,968,026 | – |
| Other assets | 80,558 | – |
| Total Assets | $ 62,149,042 | $ 55,386,030 |
| Liabilities and Stockholders' Equity | | |
| Current Liabilities: | | |
| Accounts payable (Notes 6 and 10) | $ 27,025,984 | $ 17,606,113 |
| Accrued liabilities (Notes 7 and 10) | 1,811,998 | 2,289,698 |
| Notes payable (Note 8) | 1,685,634 | 1,604,399 |
| Obligations under capital leases (Note 12) | 1,118,055 | 1,041,921 |
| Billings in excess of costs on uncompleted contracts | 4,625,657 | 6,054,814 |
| Total Current Liabilities | 36,267,328 | 28,596,945 |
| Deferred tax liability (Notes 1 and 11) | 2,718,734 | 2,272,700 |
| Notes payable, less current portion (Notes 8 and 9) | 9,484,479 | 7,674,608 |
| Obligations under capital leases, less current portion | 2,964,195 | 3,603,540 |
| Total Liabilities | 51,434,736 | 42,147,793 |
| Stockholders' Equity | | |
| Preferred stock—$0.001 par value; 1,000,000 shares authorized, none issues and outstanding | – | – |
| Common stock—$0.001 par value; 15,000,000 shares authorized 3,559,438 issued and outstanding | 3,601 | 3,601 |
| Additional paid-in capital | 10,943,569 | 10,943,569 |
| Capital adjustments | (799,147) | (799,147) |
| Retained earnings | 566,283 | 3,090,214 |
| Total Stockholders' Equity | 10,714,306 | 13,238,237 |
| Total Liability and Stockholders' Equity | $ 62,149,042 | $ 55,386,030 |

**FIGURE 4.6** Consolidated balance sheets, Low Bid Construction Corporation (the numerous notes normally accompanying a balance sheet are not shown here).

**TABLE 4.2** Liquidity Analysis for Low Bid Construction Corporation

| | Dec. 31, 2019 | Dec. 31, 2018 |
|---|---|---|
| Current assets | $35,072,542 | $33,543,119 |
| Current liabilities | 36,267,328 | 28,596,945 |
| Working capital | −1,194,786 | 4,946,174 |

**TABLE 4.3** Cash to Accounts Receivable Analysis for Low Bid Construction Corporation (Fig. 4.6)

| | Dec. 31, 2019 | Dec. 31, 2018 |
|---|---|---|
| Current assets—cash | $2,228,506 | $1,822,598 |
| Current assets—accounts receivable | 21,377,904 | 14,297,564 |
| Proportion cash to accounts receivable | 1:9.6 | 1:7.8 |

Looking at the change in working capital across the years 2015–2018, it would appear the company has the liquidity to meet its obligations. The amount of working capital actually grew from $352,000 to $1,025,000 between 2015 and 2018. The amount of working capital is growing, but the current ratio provides a warning. It alerts management to the fact that the growth is not keeping pace with the increase in current liabilities. The significant drop in the current ratio value in 2016 and 2017 provides a warning about the company's operations.

When the adequacy of working capital is studied, the composition of the company's current assets should be considered. A company with a high proportion of cash to accounts receivable is in a better position to meet its current obligations. The company with a high proportion of accounts receivable must turn those into cash before it can pay the bills. Again, using the Low Bid balance sheet (Fig. 4.6), the liquidity deterioration is also reflected by the cash to accounts receivable proportional change (Table 4.3).

Working capital *increases* when a company *makes a profit on a job,* sells equipment or assets, or borrows money on a long-term note. Contractors typically resist selling equipment or engaging in long-term borrowing.

Working capital *decreases* when the company *loses money on a job,* buys equipment, or pays off long-term debt. Contractors are typically eager to buy equipment and pay off debt.

## CONSTRUCTION CONTRACT REVENUE RECOGNITION

In many businesses, revenue is recognized at the point of sale, because most of the uncertainties related to the earning process are removed and the exchange price is known. The exception to the general rule of recognition at point of sale is caused by the long-term nature of construction projects.

> In construction accounting, everything follows accounting practices used in other industries except for revenue recognition.

The accounting measurements of revenue associated with long-term construction projects are a bit more difficult because events and amounts must be estimated for a period of years and even good estimates are modified because of owner-desired change orders or unexpected conditions.

**TABLE 4.4** OK Project Financial Data (Sample Data)

| Project financial data | Amount |
|---|---|
| Contract amount | $1,000,000 |
| Original estimated cost | 900,000 |
| Billed to date | 700,000 |
| Payments received to date | 630,000 |
| Costs incurred to date | 450,000 |
| Forecasted cost to complete | 400,000 |
| Costs paid to date | 400,000 |

There are several different methods a construction company can use to identify revenue including the cash, straight accrual, completed contract, and percentage-of-completion methods, but for financial statements, the only recognized method is the percentage-of-completion method.

## Cash Method of Revenue Recognition

With the *cash method*, a contractor records revenue when payments for services are received (actual receipts to date) and records costs as bills are paid (costs paid to date). As a result there is never an accurate picture of the company's financial condition or how individual projects are progressing (a profit or loss). Considering the project data in Table 4.4 and using the cash method of revenue recognition, the revenue to date for the project would be

| | |
|---|---|
| Payments received to date | $630,000 |
| Costs paid to date | $400,000 |
| Revenue to date | **$230,000** |

## Straight Accrual Method of Revenue Recognition

The *straight accrual method* considers cost as costs to date (even if not yet paid) and revenue as what has been billed to date. Considering the project data in Table 4.4 and using the straight accrual method of revenue recognition, the revenue to date for the project would be

| | |
|---|---|
| Billed to date | $700,000 |
| Costs incurred to date | $450,000 |
| Revenue to date | **$250,000** |

## Completed-Contract Method of Revenue Recognition

The *completed-contract method* is almost never used, but might be when

- The contractor's business is composed predominantly of very short-term contracts.
- The conditions for using the percentage-of-completion method cannot be met.
- There are inherent hazards in the contract beyond normal, recurring business risks.

Considering the project data in Table 4.4 and using the completed-contract method of revenue recognition, the revenue to date for the project would be zero at this time, because the project has not been completed.

## Percentage-of-Completion Method of Revenue Recognition

The *percentage-of-completion method* should be used when estimates of progress toward completion, revenues, and costs are reasonably dependable, and all of these conditions exist:

- The contract clearly specifies the enforceable rights regarding services to be provided and received by the parties, the consideration to be exchanged, and the manner and terms of settlement.
- The owner (buyer) can be expected to satisfy all obligations under the contract.
- The contractor can be expected to perform contractual obligations.

When using the percentage-of-completion method to calculate revenue, it is first necessary to calculate the percent of completed work:

$$\text{Percentage of completion} = \frac{\text{Cost to date}}{\text{Cost to date} + \text{Cost to complete}} \times 100\% \tag{4.4}$$

Considering the project data in Table 4.4 the percentage of completion for the project would be

$$\frac{\$450{,}000}{\$450{,}000 + \$400{,}000} \times 100\% = 52.9\%$$

The revenue to date for the project would be

$$\text{Revenue to date} = \text{Percentage of completion} \times \text{Contract price} \tag{4.5}$$

Therefore, revenue is 52.9% × $1,000,000 = $529,000

$$\text{Gross profit} = \text{Revenue} - \text{Cost} \tag{4.6}$$

| | |
|---|---|
| Revenue to date | $529,000 |
| Cost incurred to date | $450,000 |
| Gross profit to date | **$79,000** |

## Over/Underbilling

When the OK Project was originally bid, the anticipated profit was $100,000 (Table 4.4, contract amount minus original estimated cost). As the work progressed, the cost to complete the work has been continually revised; at the point in time when the financial data in Table 4.4 were prepared the new estimated final cost was $850,000 (Table 4.4, cost incurred to date plus forecasted cost to complete, $450,000 + $400,000). The construction company has billed the owner for $700,000 and has experienced an actual cost to date of $450,000; this leaves the contractor with $250,000 in cash. The contractor has not received all $700,000, because the owner is holding 10% retainage, but for the discussion here this fact will be ignored. Part of the $250,000 is true gross profit (what the contractor has really earned, $79,000) and part is overbilling, $171,000. To calculate overbilling, subtract the revenue-to-date (percentage-of-completion method) from the billed-to-date amount. This reveals the excess of the billing over the revenue. Such an undertaking demonstrates whose money is whose, for example what belongs to the contractor and what truly belongs to the owner.

| | |
|---|---|
| Billed to date | $700,000 |
| Revenue to date (percentage-of-completion method) | $529,000 |
| Overbilling or job borrow | **$171,000** |

Most contractors are not in the banking business; they do not have the funds to finance construction of the owner's project. Contractors, therefore, allocate more profit to those items of work they expect will be completed early in the project. In the case of a lump-sum project, this is done when the schedule of values is prepared. The total amount of revenue for the work is the identical; but more of the revenue is received with completion of the early items of work. The formula for calculating overbilling is given by

$$\text{Overbilling} = \text{Billed to date} - \text{Revenue} \tag{4.7}$$

Early in a project, a contractor is usually in an underbilling situation where actual earned revenue is greater than the amount billed to the owner:

$$\text{Underbilling} = \text{Revenue} - \text{Billed to date} \tag{4.8}$$

Assuming the overbillings have been paid, the contractor is effectively "borrowing" money from the owner by billing ahead of the actual amount of earned revenue. While overbilling is preferable to significant underbilling, it can pose difficulties. Consider a contractor who decides to use the borrowed cash for other purposes or who does not understand the true situation. At some point in the future the funds will be needed to complete the work and there may not be any left to pay the bills.

Managers should look for underbilling late in a project. When an underbilling situation occurs, the contractor is effectively allowing the project owner to "borrow" money, borrow the contractor's money because the contractor has performed the work (and incurred expenses) without timely payment. Underbillings reduce cash flow for the individual contract and the general operations of the company. An underbilling can be characterized as an unbilled receivable.

> Significant underbillings indicate problems and the possibility of unapproved change orders, and unapproved change orders may never be approved or paid.

## CONTRACT STATUS REPORT

A *contract status report* (work-in-progress accounting) summarizes the financial position of each unfinished project. It presents a comparison of work accomplished (revenue earned) against costs and a projection of the final financial outcome of each project. Table 4.5 presents the base data required to develop a contract status report. In preparing a contract status report, the project manager is usually responsible for evaluating the accuracy of the estimated cost to complete a project. The contract price is the sum of the original contract amount and all approved change orders.

The first step in developing a complete contract status report is to calculate a new or current (at the point in time when the report is prepared) estimate of the total project cost. Summing the cost to date of each project with its re-estimated cost to complete yields the new total cost estimate (col. 2, Table 4.6). Once the revised estimate of project cost is completed it is possible to use Eq. (4.4) (cost to date as a ratio of projected total cost) to calculate the percentage complete of each project (col. 5,

**TABLE 4.5** It Will Cost You a Million Construction Co. Contract Financial Data

| Contract | Contract price ($) | Cost to date ($) | Total amount billed to date ($) | Estimated cost to complete ($) |
|---|---|---|---|---|
| School 1 | 1,000,000 | 450,000 | 700,000 | 400,000 |
| School 2 | 1,000,000 | 450,000 | 400,000 | 400,000 |
| Store 1 | 1,000,000 | 450,000 | 600,000 | 500,000 |
| Store 2 | 1,000,000 | 450,000 | 450,000 | 460,000 |
| Warehouse | 1,000,000 | 475,000 | 525,000 | 575,000 |
| | 5,000,000 | 2,275,000 | 2,700,000 | 2,335,000 |

**TABLE 4.6** It Will Cost You a Million Construction Co. Contract Status

| Contract | Contract price ($) | Current estimated total cost ($) | Estimated profit at completion | Cost to date ($) | Percent complete | Total amount billed to date ($) | Estimated cost to complete ($) |
|---|---|---|---|---|---|---|---|
| | (1) | (2) | (3) | (4) | (5) | (3) | (4) |
| School 1 | 1,000,000 | 850,000 | 150,000 | 450,000 | 53 | 700,000 | 400,000 |
| School 2 | 1,000,000 | 850,000 | 150,000 | 450,000 | 53 | 400,000 | 400,000 |
| Store 1 | 1,000,000 | 950,000 | 50,000 | 450,000 | 47 | 600,000 | 500,000 |
| Store 2 | 1,000,000 | 910,000 | 90,000 | 450,000 | 49 | 450,000 | 460,000 |
| Warehouse | 1,000,000 | 1,050,000 | –50,000 | 475,000 | 45 | 525,000 | 575,000 |
| | 5,000,000 | 4,610,000 | | 2,275,000 | | 2,700,000 | 2,335,000 |

Table 4.6). Additionally, once the current (new) estimate of the total project cost is calculated, it can be subtracted from the project contract price to obtain a current estimated profit/loss at completion (col. 3, Table 4.6). By multiplying the estimated profit at completion by the percentage complete, profit to date is obtained (col. 6, Table 4.6). In the case of the warehouse project undertaken by It Will Cost You a Million Construction Co., a loss is projected at project completion, as seen in Table 4.6.

Table 4.7 presents the complete status report with revenue and over/underbillings recognized. Project managers are expected to analyze and explain the progress of their projects. If the project manager fails to forecast a cost overrun, the report will show an underbilling. This can be particularly troublesome if the cause of the cost overrun cannot be identified.

> An expected loss must be recognized in the period when the loss is identified—the total expected loss.

## *FINANCIAL STATEMENT ANALYSIS*

In and of themselves, very few of the individual figures in a financial statement are exceptionally significant. The important indicators of a company's condition are the relationships of the individual figures to other quantities, or the magnitude and direction of change with time (since a previous statement).

> Analysis of financial statements is a matter of studying relationships, changes, and trends.

The three commonly used analysis techniques are (1) dollar and percentage change, (2) component percentages, and (3) ratios.

Financial statement/report analysis looks at the relationships and changes between items and groups of items. A **comparative statement** is an analysis report showing changes in statement items by presenting item amounts for two or more successive accounting periods placed side by side in a column format on a single statement (Fig. 4.3). As an aid to controlling operations, a comparative income statement is usually more valuable than a comparative balance sheet.

**comparative statement**
*A financial statement with data for two or more successive accounting periods placed in columns side by side to illustrate changes.*

TABLE 4.7 Contract Status Report for It Will Cost You a Million Construction Co.

| Contract | Contract price | Current estimated total cost | Estimated gross profit at completion | Cost to date | Percent complete | Total gross profit (loss) recognized to date |
|---|---|---|---|---|---|---|
| Column | (1) | (2) | (3) | (4) | (5) | (6) |
| Formula | | 4 + 11 | 1 − 2 | | 4 ÷ 2 | 7 − 4 or 5 × 3 |
| School 1 | $1,000,000 | $850,000 | $150,000 | $450,000 | 53% | $79,412 |
| School 2 | $1,000,000 | $850,000 | $150,000 | $450,000 | 53% | $79,412 |
| Store 1 | $1,000,000 | $950,000 | $50,000 | $450,000 | 47% | $23,684 |
| Store 2 | $1,000,000 | $910,000 | $90,000 | $450,000 | 49% | $44,505 |
| Warehouse | $1,000,000 | $1,050,000 | −$50,000 | $475,000 | 45% | −$22,619 |
| | $5,000,000 | | | $2,275,000 | | |

| Contract | Amount earned to date (revenue) | Total amount billed to date | Overbillings | Underbillings | Re-estimated cost to complete | Contract balance |
|---|---|---|---|---|---|---|
| Column | (7) | (8) | (9) | (10) | (11) | (12) |
| Formula | 5 × 3 or 4 + 6 | | 8 − 7 | | | 1 − 8 |
| School 1 | $529,412 | $700,000 | $170,588 | 0 | $400,000 | $300,000 |
| School 2 | $529,412 | $400,000 | | $129,412 | $400,000 | $600,000 |
| Store 1 | $473,684 | $600,000 | $126,316 | | $500,000 | $400,000 |
| Store 2 | $494,505 | $475,000 | | $19,505 | $460,000 | $525,000 |
| Warehouse | $452,381 | $525,000 | $72,619 | | $575,000 | $475,000 |
| | | $2,700,000 | | | $2,335,000 | |

TABLE 4.8 Analysis of Low Bid Construction Corporation's Financial Situation

| Year | 2019 | 2018 | 2017 |
|---|---|---|---|
| Revenue | $174,063,148 | $163,573,258 | $210,002,272 |
| Cost of revenue | 169,404,787 | 158,935,103 | 200,070,826 |
| Dollar change in revenue | $10,489,890 | −$46,429,014 | |
| Percentage change in revenue | 6.4% | −22.1% | |
| Dollar change cost of revenue | $10,469,684 | −$41,135,723 | |
| Percentage change cost of revenue | 6.6% | −20.6% | |

## Dollar and Percentage Change

The dollar magnitude (amount) of change from one reporting period to another is significant and expressing the change in percentage terms adds perspective. The *dollar magnitude of change* is the difference between the amount of a comparison statement (year) and a base statement (year). The *percentage change* is computed by dividing the amount of the change between the statements by the magnitude for the base statement. Table 4.8 presents the dollar magnitude and percentage change of Low Bid Construction Corporation's revenue and cost of revenue for the 3-year period 2017 through 2019.

**TABLE 4.9** Component Percentage Analysis of Current Assets (Thousands)

| Current assets | 2019 | 2018 | 2017 |
|---|---|---|---|
| Cash | $67,054 15.7% | $53,215 13.1% | $48,310 10.5% |
| Accounts receivable | $175,513 41.0% | $187,311 46.3% | $222,341 48.1% |
| Unbilled receivables | $74,552 17.4% | $74,514 18.4% | $101,564 22.1% |
| Total current assets | $427,722 | $405,014 | $459,249 |

Though the company cannot be happy about the decrease in revenue shown by the comparison, one possible cause can be eliminated. Because the percentage changes are approximately the same (e.g., revenue decreases 22.1% and likewise the cost of revenue decreases 20.6%), it seems an increase in cost was not the cause of the revenue decrease experienced in 2018.

## Component Percentages

*Component percentages* indicate the relative size of each item included within a total. Each item on a balance sheet can be expressed as a percentage of total current assets, total assets, total current liabilities, or total liabilities. Such a display of total assets illustrates the relative importance of current and noncurrent assets as well as the relative amount of financing obtained from current creditors, by long-term borrowing, and from the owners.

Table 4.9 presents a component percentage of current assets analysis. In this case the analysis demonstrates how the company has been increasing its cash position and decreasing accounts receivable and unbilled receivables (underbillings).

## Ratios

A *ratio* is the relationship expression of two quantities (items) expressed as the quotient of one divided by the other. To compute a meaningful ratio, there must be a significant relationship between the two items considered. The current ratio [discussed before, Eq. (4.3)] summarizes the relationship between current assets and current liabilities. Besides the current ratio, other important financial ratios for construction companies are

$$\text{Quick ratio} = \frac{\text{Cash} + \text{Receivables}}{\text{Current liabilities}} \tag{4.9}$$

The *quick ratio* measures the short-term liquidity of a company.

$$\text{Debt to worth} = \frac{\text{Total liabilities}}{\text{Total shareholders' equity}} \tag{4.10}$$

The *debt ratio* measures the firm's leverage. The margin of protection to creditors is higher when the ratio is lower because the owners are contributing the bulk of the funds for the business. Properly leveraged contractors have debt-to-worth ratios of 2:1 or less.

$$\text{Receivables to payables} = \frac{\text{Receivables}}{\text{Payables}} \tag{4.11}$$

The suggested range for the *receivables-to-payables ratio* varies with the type of construction activity undertaken by the firm.

Based on the type of industry being studied, there are rules of thumb for evaluating financial ratios. In the case of construction companies:

- Current ratio should be above 1.3:1.
- Quick ratio should be above 1.1:1.
- Debt-to-worth should be below 2.0:1.
- Receivables to payables (commercial general contractors) should be around 1.5:1.
- Receivables to payables (trade/heavy contractors) should be around 2.0:1.
- Receivables to payables (labor-intensive trade contractors) should be around 3.0:1.
- One further rule of thumb concerns a construction company's capacity to do work. A firm's backlog of work should not be more than approximately 10 times the company's working capital amount. This is an important factor a surety will consider before granting a bond for new work.

# CHAPTER 5
# CONSTRUCTION ESTIMATES

*Estimating is the process of determining how to construct work in the most economical manner and within the time specified by the contract. All estimates should be prepared using a consistent format. A project work breakdown structure supports estimate development. When complete project plans and specifications are available, it is possible to prepare a detailed task-by-task estimate. Conversely, during the early stages of project development when construction details are not available, parametric estimating techniques are used. Direct costs are the costs attributable to a single construction task. Indirect costs are the costs of project management; these cannot be attributed to a single work task.*

## INTRODUCTION

One of the most consequential elements governing the success of a construction company is the ability of its staff, estimators, and experienced project supervisory personnel to prepare fully detailed and accurate cost estimates for performing work. For a typical facility, cost estimates formulated during project development vary in accuracy as project scope and features become more detailed (Fig. 5.1). Typically four different types of estimates are prepared during project development:

1. Conceptual; by the project designer or the owners project team.
2. Feasibility; by the project designer for Design–Bid–Build (DBB) project or by the contractor for Design–Build (DB).
3. Engineering/design; by the project designer for DBB project or by the contractor for DB. There can be multiple estimates as design progresses.
4. Construction; by the contractor.

### Conceptual Estimates

An estimate at this stage in a project's development is, by necessity, based on very general parameters of facility size, anticipated quality of construction, and use of the facility. A **conceptual estimate** should be prepared with considerable input from the project owner. The owner must approve the scope document as it serves as the basis for the estimate. The accuracy expected is in the order of plus or minus 25%. These estimates are prepared for early budgets to determine project feasibility and for securing project financing. They must be prepared by individuals having considerable experience.

**conceptual estimate** *An estimate prepared from a conceptual description of a project. This is done before plans, specifications, and other project details have been developed.*

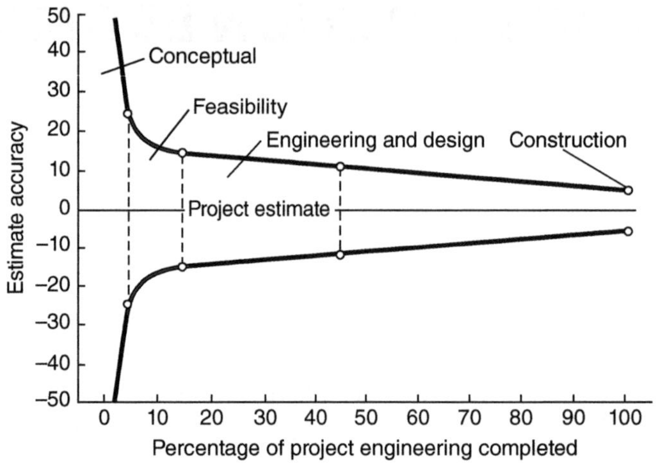


**FIGURE 5.1** Estimate accuracy versus percentage of engineering complete.

## Feasibility Estimates

**feasibility estimate** *A cost estimate prepared before complete construction documents are available.*

Using preliminary design information and the project scope, a broad budget **feasibility estimate** can be prepared. Major items of equipment can be priced and their costs incorporated into the estimate. With more detailed scope definition, the accuracy expected is in the order of plus or minus 10–15%.

## Engineering and Design Estimates

**value engineering** *An evaluation of a project's objective functions against expected cost for the purpose of bettering the objectives in terms of cost and functional parameters.*

Based on schematic-level design documents, major quantities can be quantified (tons of steel) and types of construction (steel vs. concrete) identified. Typically, architectural/engineer organizations update an estimate at specified milestone points (15%, 30%, and 60% of design) or when there is a scope change. Consequently, as design progresses accuracy of an estimate improves into a range of plus or minus 5–10%. The designer and project owner may use these early estimates to review expected total project costs and to identify design cost drivers. **Value engineering** (VE) analyses are often performed based on the design estimate projection of project cost.

## Estimates for Construction

These are itemized cost computations based on a complete set of contract documents. Construction estimates can be prepared using historical average cost data or by creating crews and considering productivity. The method used depends on the type of work. Building work contractors make greater use of historical data whereas heavy/highway contractors typically estimate using crew productivity. In either case the expected accuracy is plus or minus 5%.

In the past, construction estimates by contractors were prepared based on a complete project design supported by a detailed set of specifications prepared by or for the project owner. The estimate would include all direct costs to complete the individual construction tasks plus field supervision, general overhead, and desired profit. The proliferation of Alternate Project Delivery methods has changed the traditional roles, processes, and responsibilities of owners, designers, and contractors.

With the DB project procurement method contractors must prepare project cost estimates prior to completion of project design. When the Construction Management-At-Risk delivery method is used, the construction manager commits to deliver the project at a guaranteed maximum price which is based on the construction documents and specifications at the time of the guarantee.

## Estimating Methodology

When adequate design information is provided, detailed estimating methods can be employed. When construction details cannot be reasonably assumed, historical unit prices can be used to price work based on gross parameters, square foot of building space. The methodologies and procedures described here are applicable whether the cost estimate is prepared manually or by using computer-estimating software systems. There are database-driven estimating software based on the universally accepted Construction Specifications Institute **(CSI)** MasterFormat divisions. The CSI system is used to format project construction specifications. It provides a master list with subject titles based on work results or construction practice. Use of the CSI system of organizing project requirements/specifications supports a standard system of presentation.

**CSI**
*The Construction Specifications Institute provides technical information to enhance communication among all disciplines of nonresidential building design and construction.*

The CSI MasterFormat has two major grouping of information: (1) Procurement and Contracting Requirements and (2) Specifications. Within the Specifications group there are 50 numbered divisions (Table 5.1). This format provides the building construction industry with a uniform system for organizing project information.

When analyzing construction project work for an estimate, the estimator should identify those tasks driving the major portion of the project costs. Table 5.2 is such an analysis by one contractor for high-tech facilities work. These tasks can be identified by applying the 80/20 rule—approximately 80% of the project cost is contained in 20% of the tasks. Because these substantial tasks account for most of the project cost, they should receive special attention and effort during both estimate preparation and review.

Based on the information shown in Table 5.2, it is clear that the estimator should carefully develop the mechanical and electrical parts of the estimates for high-tech facilities. In addition, **indirect costs** (overhead expense) are a significant portion of the total project cost, so indirect cost must be prudently considered. Indirect costs include expenses such as job supervision, office staff, project office facilities, and temporary utilities. The two tasks, mechanical and electrical, plus indirect costs typically account for between 61 and 84% of the project cost when constructing these facilities.

**indirect costs**
*Overhead or auxiliary costs incurred in completing a project but not attributable to any specific tasks.*

## Work Breakdown Structure

All estimates should be prepared using a consistent format. A work breakdown structure (WBS) of the tasks needed to complete an item of work should be developed to support project pricing. The WBS provides an ordered hierarchy framework for summarizing information and for quantitative reporting to customers (the client/owner) and management. *A Guide to the Project Management Body of Knowledge* (*PMBOK® Guide*) published by the Project Management Institute, Inc., defines the WBS as a "deliverable oriented hierarchical decomposition of the work to be executed by the project team." The individual WBS for each task provides a hierarchical outline of the work required to deliver the project (Fig. 5.2).

**TABLE 5.1** CSI MasterFormat

| **MasterFormat 2018** | |
|---|---|
| **Procurement and Contracting Requirements Group** | |
| Division 00 Requirements: | |
| **Specifications Group** | |
| ***General Requirements Subgroup:***<br>Division 01 General Requirements | |
| *Facility Construction Subgroup* | *Site and Infrastructure Subgroup* |
| 02 Existing conditions | 30 Reserved |
| 03 Concrete | 31 Earthwork |
| 04 Masonry | 32 Exterior Improvements |
| 05 Metals | 33 Utilities |
| 06 Wood, Plastics, and Composites | 34 Transportation |
| 07 Thermal and Moisture Protection | 35 Waterway and Marine Construction |
| 08 Openings | 36–39 Reserved |
| 09 Finishes | *Process Equipment Subgroup* |
| 10 Specialties | 40 Process Integration |
| 11 Equipment | 41 Materials Processing and Handling Equipment |
| 12 Furnishings | 42 Process Heating, Cooling, and Drying Equipment |
| 13 Special Construction | 43 Process Gas and Liquid Handling, Purification |
| 14 Conveying Systems | and Storage Equipment |
| 15–19 Reserved | 44 Pollution Control Equipment |
| *Facility Services Subgroup* | 45 Industry-Specific Manufacturing Equipment |
| 20 Reserved | 46 Water and Wastewater Equipment |
| 21 Fire Suppression | 47 Reserved |
| 22 Plumbing | 48 Electrical Power Generation |
| 23 Heating, Ventilation and Air Conditioning (HVAC) | 49 Reserved |
| 24 Reserved | |
| 25 Integrated Automation | |
| 26 Electrical | |
| 27 Communications | |
| 28 Electronic Safety and Security | |
| 29 Reserved | |

The WBS provides

1. An organized system for collecting project cost data in a standard format and later use of the data to support estimating, cost reporting, and cost tracking
2. A checklist for categorizing costs
3. A means to maintain historical cost data in a standard format

**TABLE 5.2** One Company's Cost Experience for High-Tech Facilities

| Task types | Cost range (%) | Cost percent average |
|---|---|---|
| Site work | 1–3 | 2.0 |
| Concrete | 5–10 | 6.5 |
| Structural | 5–8 | 6.0 |
| Architectural finishes | 6–9 | 6.0 |
| Tech-specific rooms | 6–13 | 8.0 |
| Mechanical | 36–48 | 42.0 |
| Electrical | 11–14 | 12.0 |
| Indirect costs | 14–22 | 17.5 |
| | | 100.0 |

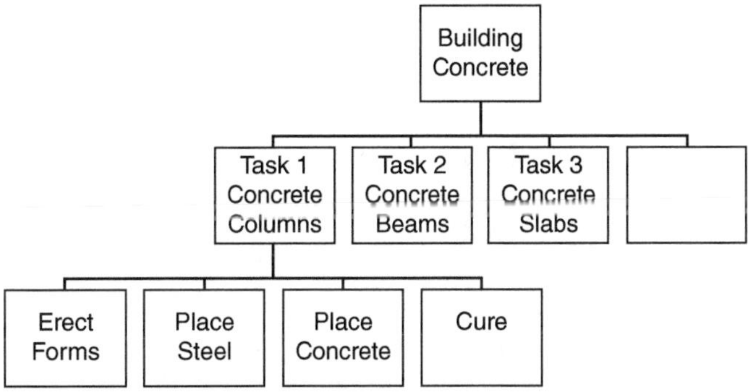


**FIGURE 5.2** Hierarchical WBS example for a concrete building.

Separate unique WBSs can be developed for different types of building construction work. The partial estimate shown in Fig. 5.3 follows the WBS approach. For the Approach Forebay Wall Concrete Task, there are subtasks for forming, reinforcing steel, and concrete.

The use of a WBS product-oriented format identifies all project-required work. It provides a common framework for preparing cost estimates and collecting construction project cost data. A WBS model consists of stepped levels of detail and each level can have many parts.

Level 1 Facility (building)

Level 2 (System) 02—Superstructure (columns, beams, floors)

Level 3 (Subsystem) 01—construction steps

The WBS elements must be

- **Definable**—can be described and easily understood by all project participants.
- **Manageable**—a meaningful unit of work where specific responsibility and authority can be assigned to a responsible individual.
- **Estimable**—time duration and cost can be estimated.
- **Independent**—minimum interface with or dependence on other ongoing elements (i.e., assignable to a single control account and is clearly distinguishable from other work packages).
- **Integratable**—integrates with other project work elements and with higher level cost estimates and schedules to deliver the entire project.

| Bid item | Description | Units | Bid quantity | Labor | Burden | Permanent material | Construction material | Direct total | $/unit |
|---|---|---|---|---|---|---|---|---|---|
| 270000 | APPROACH FOREBAY WALLS | LS | 1 | $3,855,323 | $5,697,439 | $9,379,140 | $2,481,735 | $38,799,408 | |
| 272000 | CONCRETE WORKS | $yd^3$ | 79,238 | $1,970,629 | $2,858,369 | $7,679,610 | $2,057,170 | $14,565,777 | $183.82 |
| 272200 | WALL TOE | $yd^3$ | 39,883 | $523,942 | $749,520 | $3,668,550 | $152,206 | $5,094,219 | |
| 272300 | WALL & COUNTERFORTS | $yd^3$ | 29,435 | $1,109,069 | $1,617,520 | $3,002,580 | $1,484,961 | $7,214,130 | |
| 272400 | CABLE DUCT | $yd^3$ | $9,920 | $337,618 | $491,327 | $1,008,480 | $420,003 | $2,257,428 | |
| | | | 79,238 | $1,970,629 | $2,858,367 | $7,679,610 | $2,057,170 | $14,565,777 | $183.82 |
| 272210 | FORMS | $yd^2$ | 3,124 | $55,711 | $81,792 | - | $116,780 | $254,283 | |
| 272310 | FORMS | $yd^2$ | 38,520 | $686,930 | $1,008,516 | - | $1,439,934 | $3,135,380 | |
| 272410 | FORMS | $yd^2$ | 10,811 | $192,793 | $283,050 | - | $404,131 | $879,974 | |
| | | | 52,455 | $935,434 | $1,373,357 | - | $1,960,845 | $4,469,637 | $53.88 |
| 272220 | REINFORCEMENT STEEL | TONS | 2,990 | $374,307 | $539,246 | $1,674,400 | $16,130 | $2,604,083 | |
| 272320 | REINFORCEMENT STEEL | TONS | 2,208 | $372,713 | $540,143 | $1,236,480 | $35,735 | $2,185,071 | |
| 272420 | REINFORCEMENT STEEL | TONS | 738 | $118,621 | $172,363 | $413,280 | $10,531 | $714,795 | |
| | | | 5,936 | $865,642 | $1,251,752 | $3,324,160 | $62,396 | $5,503,949 | $69.46 |
| 272230 | CONCRETE | $yd^3$ | 39,883 | $93,924 | $128,483 | $1,994,150 | $19,296 | $2,235,853 | |
| 272330 | CONCRETE | $yd^3$ | 29,435 | $49,426 | $68,861 | $1,766,100 | $9,292 | $1,893,679 | |
| 272430 | CONCRETE | $yd^3$ | 9,920 | $26,204 | $35,914 | $595,200 | $5,341 | $662,659 | |
| | | | 79,238 | $169,553 | $233,258 | $4,355,450 | $33,929 | $4,792,191 | $60.48 |

**FIGURE 5.3** Hierarchical WBS estimate example for a concrete structure.

- **Measurable**—work progress can be measured; has start and completion date and measurable interim milestones.
- **Adaptable**—sufficiently flexible so the addition/elimination of work scope can be readily accommodated in the WBS framework.

## Degree of Detail

All cost estimates should be prepared on the basis of calculated quantities and the production rates used must reflect the project work environment. This is accomplished by separating construction work into incremental parts. These parts are commonly referred to as construction tasks and are the line-by-line listings of the estimate. Each task is defined and priced as accurately as possible.

> Tasks are seldom spelled out in the contract documents, but these are necessary for evaluating the requirements and developing task cost.

At the most detailed level, each task is typically related to and performed by a crew. The estimator develops the task description by defining the type of effort required to construct an item of work. In Fig. 5.3, the tasks are Formwork, Reinforcement, and Concrete. To complete a work item, multiple tasks are usually required. Again these three tasks are needed to complete the Wall Toe, Wall and Counterforts, and Cable Duct for the Approach Forebay Walls.

For the estimate to be credible, the task descriptions should be as complete and accurate as possible. This will aid in later review and analysis. Whenever a significant number of design assumptions are necessary, such as with a DB project, the estimator may have to use historical cost data from similar projects constructed in the recent past and/or use parametric-estimating techniques.

## Parametric Estimates

An alternative method to detailed task-by-task estimate preparation, especially in the early stages of project development when design details are not available, is parametric estimating. This approach bases the project estimate on historical cost data adjusted for the circumstance of the present project. The parameters used typically include the physical measures of project work (e.g., size, square feet; building type, hospital; foundation type, piling; exterior closure material, concrete block; roof type and material, composition; and number of floors). The appropriateness of the parametric method depends on the extent of project definition available, the similarity between the project and the historical data, the ability to calculate details, and known parameters or project factors.

**Parametric Estimate Example:** Estimate the cost of a 250,000-sf warehouse building. Similar buildings have been constructed in the past several years. One, built 6 years ago, cost $2,250,000 and had 150,000 sf of floor space. Two years ago, a small 80,000-sf warehouse was completed for $1,680,000, and last year a 200,000-sf structure was completed for $3,400,000. Using the historical data, a sf cost can be deduced and applied to the proposed warehouse to arrive at a parametric estimate of the cost.

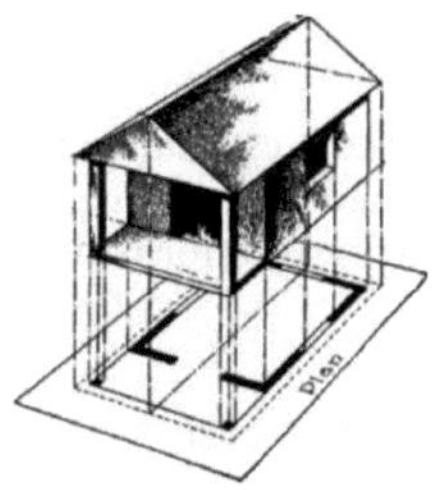


$$\text{6 years ago } \frac{\$2{,}250{,}000}{150{,}000\text{ sf}} = \$15.00 \text{ per sf}$$

$$\text{2 years ago } \frac{\$1{,}680{,}000}{80{,}000\text{ sf}} = \$21.00 \text{ per sf}$$

$$\text{1 year ago } \frac{\$3{,}400{,}000}{200{,}000\text{ sf}} = \$17.00 \text{ per sf}$$

Therefore, the average cost per sf was \$17.666 per sf [(\$15 + 21 + 17)/3]. Using a straight average, the estimated cost for the new warehouse would be \$4,416,500 (250,000 sf × \$17.666 per sf).

However, when the dates of the past construction are reviewed it appears inflation has impacted the square foot cost between the first warehouse constructed 6 years ago and the latter two. Moreover, when examining the two more recently constructed warehouses, size does seem to lead to lower cost per square foot. Therefore, a \$17.00 per sf price should be prudently considered. At \$17.00 per sf the estimated cost would be \$4,250,000 or \$166,666 less than when an average of the three buildings. Parametric estimating requires careful consideration of all the historical data.

There are multiple mathematical methods to arrive at specific indices. In the Parametric Estimate Example, a simple average of the individual costs was used first. Three common methods of deriving an index value are

$$\text{Index value (average)} = \frac{\sum_{i=1}^{n} c_i}{n} \tag{5.1}$$

$$\text{Index value (inverse)} = \frac{n}{\sum_{i=1}^{n}\left(\frac{1}{C_i}\right)} \tag{5.2}$$

$$\text{Index value (root of the product)} = \left(C_1 \times C_s \times C_3 \times \cdots \times C_n\right)^{\frac{1}{n}} \tag{5.3}$$

where $C$ = the individual costs elements
$n$ = the numbers of cost elements

**Index Value Example:** Considering the cost elements derived in the Parametric Estimate Example, calculate the index values using the three different methods. The cost elements were \$15.00, \$17.00, and \$21.00 per sf.

$$\text{Using Eq. (5.1): } \frac{\$15.00 + \$17.00 + \$21.00}{3} = \$17.666 \text{ per sf}$$

$$\text{Using Eq. (5.2): } \frac{3}{\left(\frac{1}{\$15.00} + \frac{1}{\$17.00} + \frac{1}{\$21.00}\right)} = \$17.330 \text{ per sf}$$

$$\text{Using Eq. (5.3): } (\$15.00 \times \$17.00 \times \$21.00)^{1/3} = \$17.495 \text{ per sf}$$

From the results shown in the Index Value Example, it is clear the different methods yield similar results and the choice of which method to use is more a matter of situational consideration.

## Cost-Estimating Relationships

It is often desirable to refine parametric estimates by the use of cost-estimating relationships (CERs). If the cost data presented in the Parametric Estimate Example had come from projects in several different locations, index values published by construction industry cost information service companies could be used to adjust the cost to a particular base city.

$$\frac{\text{Index for City A}}{\text{Index for City B}} \times \text{City B cost} = \text{City A cost}$$

Similarly, indices can be applied to adjust cost data for

**Inflation**—Unit cost can be adjusted for the date difference between the historical projects and the estimated project.

**Project size**—Project size can affect cost; therefore, in developing a CER, size of the historical projects in relation to the estimated project must be factored into the cost relationship.

**Unit cost**—The costs of certain items are independent of project size; therefore, the estimator must have a clear understanding of the proposed project scope.

### Accuracy and Completeness

Accuracy and completeness are critical factors in all cost estimates. Accuracy implies both correctness and exactness while completeness means all elements have been given appropriate consideration.

An accurate and complete estimate establishes accountability and creates management confidence in the estimated project cost. It is always prudent to make a deliberate examination of the contract documents for confusion about project scope, conflicting requirements within the specifications, and to identify the primary risk factors.

### Review

All construction cost estimates must be given an independent review. In the case of a construction company, the estimate is usually reviewed by or with other estimators in the firm and by company management. Engineering firms may have their estimates reviewed and checked by an independent estimating service. Estimates are reviewed to confirm the validity of assumptions and the logic used in estimating the cost of construction tasks. The review should include a check of the quantities, prices, the basis for production values, and arithmetic. It is important for the reviewer to develop and use a uniform checklist procedure in the assessment process.

## *ESTIMATE PREPARATION BASICS*

Construction cost estimates consist of

1. Descriptions of work elements to be accomplished (tasks)—prime and paint wall, two coats
2. Defining the quantity of work required for each task (700 sf of wall)
3. The cost for each task quantity based on an assumed productivity (87 sf/hr single coat)
4. A time to complete each task [(700 sf/87 sf per hr) × 2 coats = 16.1 hours]

A **unit cost** is developed for each task to increase the accuracy of the estimating procedure. Unit cost provides the data for comparisons to historic experience. Lump-sum estimating, when used at the task level, must be documented.

**unit cost**
*The cost of a single unit of work.*

**construction documents**
*The drawings and specifications setting forth the requirements for the project work.*

**general conditions**
*The contract part used to define the rights, responsibilities, and relationships of all parties to the contract.*

### Review the Contract Documents

The first step in preparing a construction estimate is to thoroughly examine the **construction documents** starting with the general and supplemental conditions of the contract. Clauses in the **general conditions** of the specifications can affect project indirect cost and methods of construction.

In addition, the estimator should carefully review the supplemental (special) conditions as these describe the unique characteristics of the project and alter the general conditions. The estimator should also review any environmental impact profile and mitigation commitment statements if the work must be built under such agreements. During this examination, it is appropriate to make notes about things affecting construction duration, the cost to perform the work, or adding indirect and overhead cost to the work.

If the owner conducts a **prebid conference**, the estimator should definitely attend. Before attending such meetings it is necessary to thoroughly review the project plans and specifications as the meeting will offer an opportunity to resolve questions.

**prebid conference**
*A meeting prior to bid opening for the purpose of explaining the project and answering bidder's questions with respect to the contract documents and the work.*

## Plan the Work

The estimator must thoroughly understand the scope of work and the project environment. To do this, it is necessary to review the drawings, specifications, and other references. It is always a good practice to visit the project site. Do not just drive by. Walk the ground slowly if possible. Take time to observe the surroundings and features possibly impacting project work and the delivery of materials.

> The detailed review enables the estimator to formulate a construction sequence and duration. A site visit is **strongly** recommended to enable the estimator to relate the physical site characteristics to the design parameters and to develop the construction plan based on the project environment.

This is particularly important in the case of projects with difficult site conditions, major maintenance and repair projects, alteration/addition projects, and environmental projects. The construction sequence must be developed as soon as possible and should be used to provide a checklist of construction requirements throughout the cost-estimating process.

## Quantities

A good quantity take-off is an essential part of the cost estimate. It must be as accurate as possible, and should be based on all available engineering and design data. After the project scope has been analyzed and broken down into the construction tasks, tasks must be quantified prior to pricing. Quantities should be shown in standard units of measure and must be consistent with design units (English or metric). The detail to which the quantities are prepared for each task is dependent on the level of design detail.

> Project notes should be added at the appropriate level in the estimate to explain the basis for the quantity calculations and to explicitly explain contingencies.

## Types of Cost

Various types of cost elements must be evaluated in detail.

***Direct costs.*** The costs attributed to a single task of construction work are known as **direct costs**. These costs are usually associated with a construction crew performing a task using specific materials and equipment. The costs of labor to erect forms or placement of concrete in the forms are direct task costs. Subcontracted costs should be considered direct costs to the prime contractor in estimates.

**direct costs**
*All cost elements associated with a specific item of project work.*

***Indirect costs.*** Costs not attributed to a single task of construction work are classified as indirect costs. These costs include overhead and required bonds.

Estimates based on a detailed design will be developed based on separate direct cost pricing of labor, materials, construction equipment, **supplies**, and subcontractors. Applicable indirect costs will be added to reflect the total construction cost. Other pricing considerations include escalation and construction contingencies. These will be added to the construction costs to determine the total project direct cost without the risk and profit assigned by management.

**supplies**
*Materials used during the construction process but not permanently incorporated into the completed facility.*

## Production

It is helpful to have a common method of recording and reporting production (productivity). The most convenient and useful are: *unit of work done* and *unit of time*. These are used in calculating productivity for a particular activity and are a function of the specific work task being analyzed. To make accurate and meaningful comparisons and conclusions about productivity, it is best to use standardized terms. *Production rate* is the relationship of work done and the time required to accomplish the work [Eq. (5.4)]. It can be cubic yards per hour, square feet of wall per shift (also indicate the duration of the shift), or installed piping per hour.

$$\text{Production rate} = \frac{\text{Work done}}{\text{Unit of time}} \tag{5.4}$$

- *Work done.* This denotes the unit of production accomplished. It can be the volume or weight of the material moved, the number of windows installed, the quantity of insulation, or any similar measurement of production.
- *Unit of time.* This denotes a designated time unit such as a minute, an hour, a 10-hr shift, a day, or any other duration for the unit of accomplished work.

**Time-Required Formula**—The inverse of the production-rate formula [Eq. (5.4)] is sometimes useful when scheduling a project because it defines the time required to accomplish a measure of work.

The most convenient and useful *units of time* and *work done* for a particular task or piece of equipment are a function of the specific work task. To be able to make accurate and meaningful comparisons and conclusions, it is best to standardize the production data terms.

## Productivity and Price Resources

Various productivity and pricing resources should be available to the estimator. When using the data from such resources, the estimator's experience and ability to relate the data to a specific circumstance is important. To do this, the estimator must visualize the processes needed to accomplish the tasks. *Historical databases*, commercial publications or company records, are the most commonly used productivity and pricing resources. A company's historical production and costs data from similar past work are excellent resources when adequate details have been recorded and adjustments to project specifics can be defined. Portions of other estimates having similar work tasks can be retrieved and repriced to the current production rates.

## Development of Specific Tasks

Descriptions of task activities must adequately define the scope and material requirements for the work. The unit cost for each task is developed as a direct cost with separate costing for the labor, equipment, permanent material, and construction material (forms, temporary bracing) components. Notes to explain key factors in the pricing methodology should accompany the task development. Comparison with published pricing guides is recommended.

## Labor Unit Cost

This cost is based on using a defined crew to perform the task at an assigned production rate. Hourly rates for each craft member of a crew are multiplied by the labor hours to arrive at the crew **labor cost** for a task. The total crew labor cost is divided by the expected production to derive the labor cost/unit.

**labor cost**
*Wages paid directly to on-site personnel and fringes.*

## Equipment Unit Cost

The only reason for purchasing equipment is to perform work, which will generate a profit for the company. The expense associated with productive machine work is commonly referred to as ownership and operating cost (O&O). O&O cost is stated on an hourly basis (i.e., $90/hr for a dozer) because it is used in calculating the cost per unit of machine production. If a dozer can push 300 cy per hour and it has a $90/hr O&O cost, production cost is $0.300/cy ($90/hr ÷ 300 cy/hr). The estimator/planner can use the cost per cubic yard figure directly.

***Ownership cost*** is the cumulative result of those cash flows an owner experiences whether or not the machine is productively employed on a job. Most of these cash flows are expenses (purchase price), but a few are cash inflows (salvage value).

***Operating cost*** is the sum of those expenses an owner experiences by working a machine on a project. Typical expenses include (1) fuel, (2) lubricants, filters, and grease, (3) repairs, (4) tires, and (5) replacement of high-wear items.

***Operator wages*** are sometimes included under operating costs, but because of wage variance between jobs, the general practice is to keep operator wages as a separate cost category.

## Material Unit Cost

This cost is developed using vendor quotes, historical costs, commercial pricing sources, or component calculations. The price should include taxes, delivery, unloading, storage, and in some cases a factor for wastage.

## Unit Cost Reference Data

There are, through subscription or purchase, estimating cost data books. The basis of the cost data is typically explained along with adjustment methodology. These publications can provide a valuable "second opinion" to the estimator.

## Costs and Pricing

The cost for each task is developed by summing the direct cost elements for labor, equipment, permanent materials, construction materials, and subcontractors. The indirect costs and other markups associated with each task or work item should be identified but are considered separately.

| CREWS COMPOSITION | | | | | | |
|---|---|---|---|---|---|---|
| Overburden | | | | | | |
| | | PRODUCTIVITY = | 475 | M3/HR @ 80% | Crew E01110 | |
| | 1 | HYD EXC 385BL OF | 6 | M3 BUCKET | 1 | Foreman |
| | 1 | LOADER 988FX OF | 0 | M3 BUCKET | 1 | Operator Cl 1 |
| | 4 | CAT 773B CAPACITY | 55 | TONS | 5 | Operator |
| | 1 | DOZER D8 TO HELP THE MOVE THE MATERIAL | | | 3 | Labors |
| Rock | | | | | | |
| | | PRODUCTIVITY = | 795 | M3/HR @ 80% | Crew E01114 | |
| | 1 | HYD EXC 385BL OF | 6 | M3 BUCKET | 1 | Foreman |
| | 1 | LOADER 988FX OF | 9 | M3 BUCKET | 2 | Operator Cl 1 |
| | 8 | CAT 773B CAPACITY | 55 | TONS | 12 | Operator Cl 3 |
| | 3 | DOZER D8 WITH SINGLE SHANK | | | 4 | Labors |
| | 1 | DOZER D8 TO HELP THE MOVE THE MATERIAL | | | | |
| Rock - Structural | | | | | | |
| | | PRODUCTIVITY = | 288 | M3/HR @ 80% | Crew E01121 | |
| | 1 | HYD EXC 345BL OF | 4 | M3 BUCKET | 1 | Foreman |
| | 1 | LOADER 988FX OF | 0 | M3 BUCKET | 1 | Operator Cl 2 |
| | 4 | CAT 769D CAPACITY | 36 | TONS | 6 | Operator Cl 3 |
| | 1 | DOZER D8 WITH SINGLE SHANK | | | 3 | Labors |
| | 1 | DOZER D8 TO HELP THE MOVE THE MATERIAL | | | | |

**FIGURE 5.4** Crew documentation for an excavation project.

When using historical pricing, adjustments must be made for project location, work methodology, quantity of work, and any other dissimilarities affecting prices.

For cost estimates prepared during preliminary or planning phases, when a detailed design is not available, predetermined unit prices for key systems adjusted to current pricing levels can be used. The estimator must use extreme care and sound judgment when using predetermined unit costs. The basis for the unit costs should be well documented and included in the supporting data of the estimate.

## Supporting Documentation

The estimator should always remain mindful of the documentation necessary to support the estimate. Support documentation includes project work narratives (Fig. 5.4) and schedule, backup data, and drawings and sketches.

**Narrative of Contract Costs** This part of the *construction cost estimate* consists primarily of those sheets, with notes, which describe the scope tasks and costing. It contains discussions, considerations, and the development of the construction plan. The types of items normally included are:

- ***Project narrative.*** The project narrative provides general details of the project. The narrative defines the assumptions made during the preparation of the cost estimate. It describes the project requirements in sufficient detail to provide a clear understanding of the scope of work. It provides a summary of project details including: (1) length, width, height, and shape of primary features; (2) special problems to be expected in performing the work; (3) site conditions affecting the work; (4) time for mobilization and demobilization of all equipment; and (5) the reasons for unusually high or low unit prices.

Other factors to be considered in the project narrative are the use of overtime, shift work, limited working hours, phasing, and subcontracting. Project-related details include site access, borrow areas, construction methodology, unusual conditions (soil, water, or weather), unique construction techniques, equipment/labor availability, and environmental concerns if appropriate.

**Narrative Example:** CSI Division 6 Wood, Plastics, and Composites

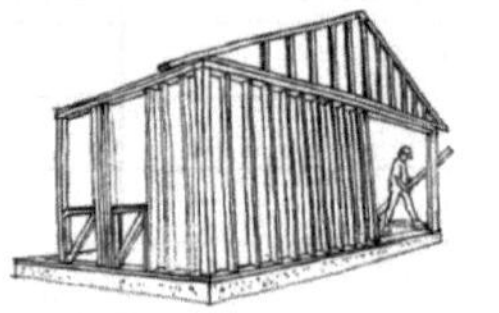

Rough carpentry includes wood blocking at the roof and interior blocking.

Finish carpentry includes vanity tops in public toilets, dressing room counters and cabinets, concession counters, ticket counters, projection room stands, wardrobe units, storage shelving, and suite cabinets and counters.

- ***Construction schedule.*** The estimator will prepare a construction schedule (Fig. 5.5) to support the cost estimate. The schedule must be consistent with the contract specified completion date. It may be in the form of a bar chart or network analysis system, but it must identify the sequence and duration of the major tasks on which the cost estimate is developed.

The **preliminary schedule** must be prepared in sufficient *detail to adequately develop the required labor, equipment, crew sizes, and production rates* required for each of the major construction tasks.

**preliminary schedule** *A schedule prepared to support development of the cost estimate.*

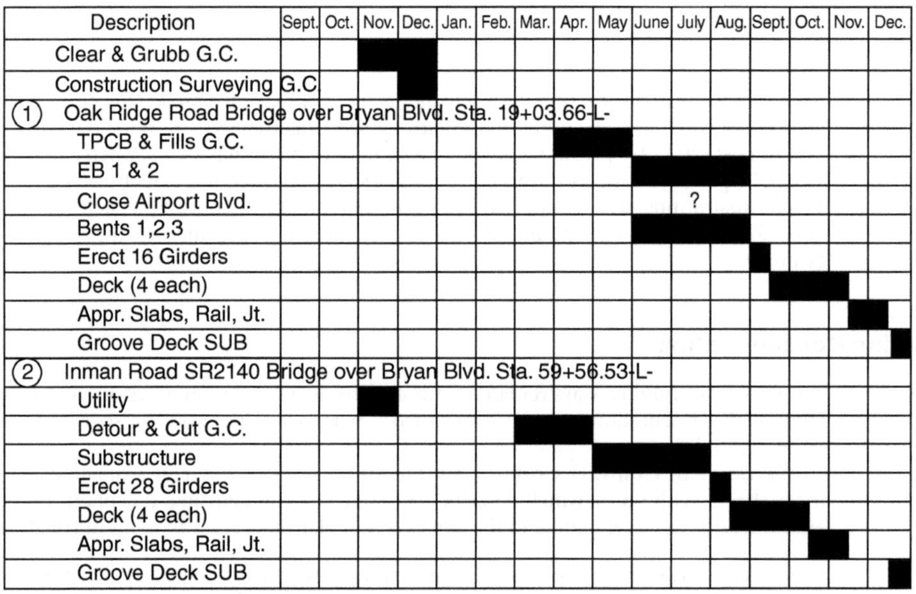


**FIGURE 5.5** Schedule prepared by a bridge estimator. Note clearing and grubbing item belongs to the general contractor. The bridge contractor is bidding as a subcontractor.

- ***Equipment and materials utilization.*** On projects requiring large construction equipment, it is necessary to sufficiently plan the equipment usage against the work schedule to identify the actual number of cranes and dozers, and allow for proper mobilization to ensure that the demand for the equipment is not over- or understated. Required long lead times materials can become critical to the construction schedule. These should be noted, and adequately considered.
- ***Labor discussion and utilization.*** The estimate should clearly state the sources for the various labor classifications and rates. This would include the composition of skills in each crew and what those skills are doing, or can do, according to union rules if applicable. When extensive overtime beyond the normal workday is used in the estimate, an explanation should be provided.

**Estimate Backup Data** This part of the estimate consists of support and backup documentation.

- ***Quantity computations.*** The quantity take-off computations for the estimated tasks should be organized by tasks for each bid item and retained as backup.
- ***Crew, labor, and equipment rates.*** The details used to prepare and express the crew composition and associated rates for labor and equipment costs must be shown. The information contained on these sheets provides the backup support for the task unit labor and equipment costs.
- ***Production rates.*** The information/sources used to develop and analyze crew production rates must be documented.
- ***Mobilization, preparatory work, and demobilization.*** These costs should be itemized and totaled separately. They may be combined at the summary level with overhead if they are not paid as separate bid items.
- ***Overhead costs.*** The itemization and calculations of overhead costs, for both the job site and the home office, should be presented.
- ***Bond and Insurance costs.*** Bond costs should be calculated.

> The quantity take-off should reference the drawing and clearly explain the computations for each item.

**Quotations** Quotations should be collected and compiled by task or bid item into an organized reference. When quotations were not obtained for significant material and supply items, the basis for the cost used in the estimate must be described.

> Quotations should be considered proprietary information and must be kept confidential to protect the information entrusted to the estimator and company.

**Drawings and Sketches** Include all other information pertinent to the estimate such as drawings and sketches used as the basis of the costing. Drawings may include a project map showing the location of the work with respect to principal cities, roads, railways, and waterways; a site map showing the location of the work, storage areas, existing access roads, any existing usable facilities, and a work layout. Supporting documents such as sketches, soil boring, and material classifications are also important.

## LABOR

The cost of labor is probably the most variable and difficult to estimate. The local labor market and conditions should be investigated to determine the available supply of all classes of labor and their competence. Local work practices must also be studied to ascertain their effect on productivity.

***Direct labor costs*** are defined as base wages plus labor cost fringes (additives) including payroll taxes, fringe benefits, travel, and overtime allowances paid by the contractor for personnel who perform a specific construction task. The various union crafts usually negotiate their own wage rates and working conditions. Therefore, if union labor is used, these must individually be examined and understood. In addition to the actual workers, there are generally working crew foremen who receive an hourly wage and are considered part of the direct labor costs.

***Indirect labor costs*** are wages and labor cost fringes *paid to contractor personnel whose effort cannot be attributed to a specific construction task.* Personnel such as superintendents, engineers, clerks, and site cleanup laborers are usually included as indirect labor costs (project overhead).

## Crews

Direct labor cost sums are developed by work tasks. A labor crew, including equipment, usually performs each work task. Consequently, the crew must be defined, its cost determined, and a production rate established for the task. Crews may vary in size and mix of skills. The number and size of each crew should be based on such considerations as having sufficient workers to perform a task, within the construction schedule, and workspace limitations. Once the crew compositions have been developed, the task labor costs can be determined based on the production rate of the crew and the labor wage rates.

## Wage Rates

A wage rate must be determined for each labor craft. The rate will have two parts: (1) the base wage rate and (2) payroll taxes, insurance, and fringe benefits. The base wage rate is subject to an overtime adjustment. An overtime premium is not applied to taxes. Social Security has a wage base limit. The Social Security tax rate is currently 12.4%, with half of the tax paid by the employer and the other half by the employee. After the maximum income amount is reached ($132,900 in 2019) only the Medicare tax is applied to access income.

Wage rates on government-funded work are generally well defined. The Davis-Bacon Act, PL 74-403, requires a contractor performing construction in the United States for the federal government to pay not less than the prevailing rates established by the Department of Labor. A schedule of minimum rates is included in the project specifications. Figure 5.6 presents a sample Davis-Bacon wage determination. Determinations can be found online.

Wherever labor is in short supply for certain crafts, the work is in a remote area, or when local rates are higher than the set rate scale, higher wage rates should be used instead of the minimum wage since they will be required to attract labor to the job. The wage rate should be adjusted to include travel time or night differential when these are a customary requirement.

## Overtime and Shift Differential

Based on the construction schedule, the estimator should carefully consider the available working time to accomplish each task. The efficiency of a second or third shift should be adjusted to account for the productivity difference. Productivity, for most types of construction operations, will not be as high during second or third shifts compared to the day shift. A three-shift operation is normally avoided because of lower labor efficiency and equipment maintenance requirements.

Overtime should be included in the labor cost computation when work in excess of 40 hr is required to meet the construction schedule or is the custom of labor in the project area. Overtime is normally calculated as a percentage of the base wage rate. The Fair Labor Standard Act (FLSA) specifies overtime pay rate of one and one-half (1.5) times the regular hourly rate. But the rate may be double time, depending on the existing labor agreements or local custom. Taxes and insurance costs are applied to overtime, but **fringe benefits** and travel and/or subsistence costs are not.

**fringe benefits**
*Employee benefits paid by an employer in addition to an employee's base pay. These may be paid directly to the employee or to various agencies on behalf of the employee.*

| | Rates | Fringes |
|---|---|---|
| *PAIN0051M 06/16/2000 | | |
| PAINTERS: | | |
| Brush, Roller, Spray and Drywall Finishers | 20.23 | 5.47 |
| *PLAS0891A 05/01/2000 | Rates | Fringes |
| CEMENT MASONS | 19.80 | 3.895 |
| PLUM0005E 09/01/1999 | Rates | Fringes |
| PLUMBERS: | | |
| A-partment Buildings over 4 stories (except hotels), schools, colleges, and speculative office buildings, strip shopping centers, churches, water coolers, room air conditioning units, appliances, packaged ice machines, and light commercial refrigeration and/or air conditioning systems serving a single business in a single story building and not to exceed 5 h.p. or tons, self-contained package unit up to and including 5 h.p. or tons. | 16.54 | 4.835 |
| All other work | 24.85 | 7.735 |
| *PLUM0602A 08/01/2000 | Rates | Fringes |
| STEAMFITTERS, REFRIGERATION AND AIR CONDITIONING MECHANICS: | | |
| Light commercial refrigeration and/or air conditioning systems serving a single business; the air conditioning systems shall not total more than 15 tons and the refrigeration system shall not total more than 7 1/2 tons; apartment buildings over 4 stories with individual units not to exceed 5 tons (excluding spilt units) | 13.75 | 7.505 |
| All other work | 25.71 | 7.505 |

**FIGURE 5.6** A Department of Labor wage decision.

**Overtime Calculation Example:** Calculate the percentage adjustment to be used for adjusting labor wages on a project when the crews will work 8 hr of overtime per week.

| Work hours | Equivalent straight time hours |
|---|---|
| 40 hr at straight time = | 40.0 |
| 8 hr at 1½ times = | 12.0 |
| | 52.0 hr paid equivalent straight time |
| 52 hr paid/48 hr worked = | 1.0833 |
| | (1.0833 – 1) × 100% = 8.33% |

## Taxes and Insurance

Wages paid to employees are subject to payroll taxes such as the Federal Insurance Contributions Act (FICA), Federal Unemployment Tax Act (FUTA), and State Unemployment taxes. The FICA tax is for the federal Social Security and Medicare programs, while FUTA is the federal employer tax used to fund state workforce agencies. Rates for all taxes and insurance should be verified prior to computation for these and any additional components.

***Workman's compensation*** and *employer's liability insurance* costs applicable for the state in which the work is performed should be included in the composite wage rate. Texas is the only state not requiring companies to carry Workers' Compensation Insurance for their employees. These two coverages should be included in the composite wage rate. The project compensation rate is based on the classification of the major construction work and applies to all crafts employed.

***Unemployment compensation taxes*** are composed of both state and federal taxes. Unemployment compensation tax will vary with the state, while the federal unemployment tax will be constant for all projects. Insurance rates can be obtained from the state unemployment office.

***Social Security tax*** is paid by both the employee and the company. Half is paid out of the employee's base pay, and this portion does not represent an extra cost to the employer, but the other half is paid by the employer and is an additional labor cost. The rates and the income ceilings (wage base limit) on which social security taxes must be paid vary from year to year. Therefore, the estimator must verify the rate to be used in the cost estimate. Rates can be obtained from the Social Security Administration.

The total percentage of the applicable taxes and insurance is summed and then applied to the basic hourly wage rate plus overtime for the various crafts. The Overtime Calculation Example illustrates a method for deriving the total tax and insurance percentage. Since rates are subject to change and in some cases vary by region, the calculations shown are presented as an example only. Actual values must be determined by the estimator for the specific project location.

**Overtime Calculation Example:** Calculate the total tax and overhead percentage to be used for adjusting the basic labor rates.

| | |
|---|---|
| Workman's compensation and employer's liability (this varies according to project location, state, and contractor) | 7.60% |
| State unemployment compensation (varies with each state) | 3.20% |
| Federal unemployment compensation (FUTA) | 0.80% |
| Social Security and Medicaid (FICA) | 7.65% |
| Total taxes and insurance | 19.25% |

The rates for indirect and overhead labor must also include these applicable costs.

The calculation, however, is not as simple as shown in the Overtime Calculation Example because the percentages only apply to certain limited amounts of the employee's wages. As an example, the FUTA tax rate is 6.2% of taxable wages but the taxable wage base is only the first $132,900 (2019) in wages during a calendar year.

## Fringe Benefits

Fringe benefits may include health and welfare, pension, and apprentice training depending on the craft and the location of the work. These summed costs are usually expressed as an hourly cost with the possible exception of vacation, which may be easily converted to an hourly cost. On federally funded projects, the type of fringe benefits and the amount for the various crafts can usually be found with the Department of Labor wage determination in the specifications. Nonunion contractors pay comparable fringe benefits directly to their employees.

The Fringe Benefit Example illustrates the calculations for fringe benefits. Since the values change and vary by region and contractor, the calculations shown are presented as an example only. The estimator must determine actual company values.

**Fringe Benefit Example:** Calculate the total fringe benefit costs added to the basic labor rates.

| | |
|---|---|
| Health and welfare | $0.70/hr |
| Pension | 0.75/hr |
| Apprentice training | 0.00/hr (N/A in this case) |
| Total fringe benefits | $1.45/hr |

Some fringe benefits and travel/subsistence costs are subject to payroll taxes. For example, vacation benefits are taxable and should be added to the basic wage rate.

## *LABOR PRODUCTIVITY*

The unit of work one tradesman or one crew can accomplish per hour or per 8-hr shift is the labor productivity for a task. Labor productivity is subject to many diverse and unpredictable factors. A primary factor often overlooked is *leadership*. The estimator must always inquire into who will manage the project. Most project managers, superintendents, and project engineers are very ambitious. The project leaders who are ambitious first and foremost for the company, and not for themselves personally, usually achieve better results—their labor force is more productive. Leaders who help everyone succeed set up their projects for success.

There is no substitution for estimator knowledge and experience when estimating labor productivity. For some types of work, crew productivity is determined by the skill of the crane operator. For some labor-based crews, the task productivity of craftsmen such as carpenters, steelworkers, and masons may be based on the average of historical data tempered with the experience of the estimator.

### Productivity Adjustment

The complexity of the variables affecting productivity makes it very difficult to estimate. Production rates are usually based on averaging past production rates for the same or similar work. The estimator must incorporate particular job factors and conditions to adjust historical data to the project being estimated. Other sources for production rates include reference manuals, field office reports, construction logbooks, and observation of ongoing construction.

> Labor productivity is the critical risk factor affecting profitability. Take time to observe your crews at work.

The labor effort needed to perform a particular task varies with many factors, such as

- Relative experience, capability
- Morale of the workers

- Size and complexity of the job
- Availability of materials, tools, and equipment
- Climatic and topographic conditions
- Degree of mechanization
- Quality of job supervision to include planning and sequencing of work
- Amount of task repetition
- Existing labor-management agreements and/or trade practices

The effect of these labor efficiency factors and local work practices must be considered when selecting productivity rates.

## CONSTRUCTION EQUIPMENT AND PLANT

Construction equipment and plant refers to the tools, instruments, machinery, and other mechanical implements required in the performance of construction work. Construction plant is defined as concrete batch plants, aggregate-processing plants, conveying systems, and any other processing plants erected in place at the job site and are essentially stationary or fixed in place. Equipment includes portable or mobile machines or tools ranging from small hand tools through tractors, cranes, and trucks. For estimating purposes, plant and equipment are grouped together as equipment costs.

### Selection of Equipment

An important consideration in the preparation of an estimate is the selection of the proper equipment to perform the required tasks. The estimator should carefully consider number, size, and function of equipment to arrive at optimum equipment usage. Some factors to consider during the selection process are

- Job progress schedule (production rate)
- Availability of space, mobility, and availability of equipment
- Equipment capabilities
- Distances material must be moved
- Steepness and direction of grades
- Weather conditions
- Hauling restrictions
- Mobilization and demobilization costs

The estimator must be familiar with construction equipment and job-site conditions. The equipment selected should conform to contract requirements and be suitable for the materials to be handled and the project conditions.

### Machine Production Rates

After determining the type of machines to be employed, the estimator should select the specific machine size providing a production rate suited to the efficient and economical performance of the work. The size and number of units required will be influenced by the equipment production rate, job size, availability of space for equipment operations, and the project construction schedule for the various work tasks.

### Mobilization and Demobilization

Mobilization costs for equipment include the cost of loading at the contractor's yard; transportation to the construction site, including permits; unloading at the site; necessary assembly and testing; and standby costs during mobilization and demobilization. All labor, equipment, and supply costs required to mobilize the equipment should be included in the mobilization cost. Demobilization costs should be based on the portion of the equipment returned to the contractor's storage yard.

Mobilization and demobilization costs for plant should be based on the delivered cost of the item, plus erection, taxes, and dismantling costs at the end of the project. Maintenance and repair are operating costs and should be distributed throughout work accomplishment.

### Small Tools

The costs of small power and hand tools and miscellaneous noncapitalized equipment and supplies are usually estimated as a percentage of the labor cost. The allowance must be determined by the estimator in each case, based on experience for the type of work involved. Unit prices based on historical data already include a small tools allowance. The small tool cost will be considered as part of equipment cost. Such allowance can range as high as 12% of direct labor cost but is usually much lower. Small tool expense must not be duplicated in the overhead rate percentages.

## PERMANENT MATERIALS AND SUPPLIES

**Permanent materials** are those items physically incorporated into the finished structure. Supplies are those items used in construction but these do not become physically incorporated into the project (such as concrete formwork). For the purpose of estimating, both can be considered materials unless they need to be separated because of different tax rates.

**permanent materials**
*Materials installed permanently in the completed project.*

### Sources of Pricing Data

Prices for materials and supplies may be acquired by obtaining quotations or from catalogs and historical data. The estimator should review the pricing and assess the reasonableness prior to use. Care should be taken to make proper allowances for quantity discounts, inflation, waste, and other factors affecting cost.

**Quotes** from Manufacturers and Suppliers Quotes should be obtained for all significant materials and installed equipment, and for specialized or items not readily available. Quotations may be received either in writing, telephonically, or by fax. It is preferable to obtain quotes for each project to ensure the estimate has the current price and meets current project's specifications. If possible, more than one quote should be obtained to be reasonably sure the prices are competitive. The estimator should attempt to determine and ensure all discounts are applied to the cost included in the estimate. *Quotes should be kept proprietary to preserve the confidentiality entrusted.* A telephone quotation data sheet similar to the one shown in Fig. 5.7 should be used for recording telephonic quotations.

**quotes**
*To make an offer at a guaranteed price.*

Waste and loss considerations may be included in material unit price computations. This methodology results in a quantity take-off of work placement not altered to reflect material losses. However, the alternative methodology of increasing the measured quantity by waste and loss quantity is acceptable if the excess quantity will not be used for any other purpose. The methodology used by the estimator should consider the impact of charging labor on the excess quantity. In either case, a statement is required in the estimate explaining the methodology used.

TELEPHONE QUOTATION SHEET CSI Number:

Firm quoting: Project:

Address:

Phone number:

and FAX numbers:

Person quoting: Estimator:

Item quoted: Date: Total quantity quoted:

Amount:

Description:

Includes sales tax: YES:____ NO:____

Per plans & specs: YES:____ NO:____

EXCEPTIONS:

FREIGHT INCLUDED TO:

____ FAS PORT ____ FOB FACTORY ____OTHER ____FOB JOB

| | | |
|---|---|---|
| | TOTAL QUOTATION | $____________ |
| ALLOWANCE FOR WASTE AND LOSSS | | $____________ |
| Weight | Export packing | $____________ |
| Volume | U.S. inland freight | $____________ |
| Quote valid for___days | TOTAL MATERIAL COST | $____________ |
| REMARKS: | | |

**FIGURE 5.7** Example of a telephone quote sheet.

## Forward Pricing

Sometimes quotes are requested in advance of the expected purchase date. However, suppliers are reluctant to guarantee future prices and often will only quote current prices. Therefore, it may be necessary to adjust current prices to reflect the cost expected at the actual future purchase date. This cost adjustment, if required, should not be included as a contingency, but should be clearly and separately defined in each estimate. Adjust current pricing to future pricing using specific escalation factors. Computations of adjustment should be clear and should be maintained as cost estimate backup support.

## Freight

The estimator should check the basis for the price quotes to determine if they include delivery. If they do not include delivery, freight costs to the project site must be determined and added to the pricing. The supplier can usually furnish an approximate delivery cost. For delivery charge, free on board (FOB) refers to the point to which the seller will deliver goods without additional charge to the buyer. If the materials or supplies are FOB factory or warehouse, freight costs to the construction site should be added to the cost of the materials or supplies.

If the cost of materials or supplies includes partial delivery, FOB to the nearest rail station, the cost of unloading and transporting the materials or supplies should be included in the estimate. If the materials or supplies are a large quantity in bulk, which would require extensive equipment for unloading and hauling, it may be desirable to prepare a labor and equipment estimate for the material handling and delivery.

### Handling and Storage

The contractor is usually required to off-load, handle and stockpile, or warehouse materials on site. These costs should be included in the estimate. An item of electronic equipment requiring special low-humidity storage might have this special cost added to the direct cost of the equipment. For common items, such as construction materials or equipment needing secure storage, the cost for the security fencing, temporary building, and material handling should be considered as an indirect cost and be included in the job-site overhead cost.

### Taxes

When applicable, state and local sales taxes should be added to the cost of materials or supplies. Care should be taken to understand how local sales taxes might be applicable and to properly include in the estimate. The estimator should verify the tax rates and the applicability of these rates for the project location. Sales tax is considered to be a direct cost of the materials and supplies, and is included in the estimated material costs.

## SUBCONTRACTED WORK

In construction, specialty items such as plumbing, heating, electrical, roofing, and tile work are usually more effectively performed by subcontractors. With so many specialties being performed, subcontract work becomes a very significant portion of the total cost of a building project. Since each estimate should be prepared as practically and as realistically as possible, subcontract costs become an important consideration.

The estimator must first determine the parts of the work the contractor will subcontract. When the work to be subcontracted has been determined, those items will be identified in the estimate.

The cost of subcontracted work is the total cost to the prime contractor for the work performed. Subcontractor's costs include direct labor, materials and supplies, equipment, second-tier subcontracts, mobilization and demobilization, transportation, setup, and charges for overhead and profit. The total subcontract cost is considered a direct cost to the prime contractor.

### Use of Quotations

The estimator may use quotes for the expected subcontracted work when reviewed and verified as reasonable. In lieu of a quotation, each subcontract task should be estimated independently by the contractor as a direct cost with an appropriate rate of subcontractor's overhead and profit added.

## OVERHEAD COSTS

Overhead costs are those costs not attributed to a single construction task. Costs applicable to a particular item of work should be considered a direct cost to the item and not included in overhead costs. The overhead costs are customarily divided into two categories:

- Job office overhead, also referred to as general conditions or field office overhead
- General home office overhead, commonly referred to as **general and administrative overhead** (G&A)

**general and administrative overhead**
*The contractor's general operating expenses not related to a specific project but incurred in operating the business.*

The estimator must be sure the applied overhead costs are not duplicated between the two categories. Because of the nature of overhead costs, it is not practical to discuss all overhead items. Specific considerations must be carefully evaluated for each project. The estimator must use considerable care and judgment in estimating overhead costs. In the case of building construction, many indirect cost items are frequently described in the General Requirements Section (CSI Division 01) of the contract specifications. If not related to a specific work task, these costs must be identified and appropriately assigned as overhead costs.

Overhead will vary from project to project and may even vary from month to month over the duration of a project. Job overhead items should be estimated in detail for all projects. Detailing of overhead costs for subcontract work is recommended when the impact of these costs is significant.

## Job Office Overhead

Job overhead costs are expenses occurring at the project site as a result of managing the work. Examples of job overhead costs are

- Job supervision and office personnel
- Engineering and shop drawings/surveys
- Site security
- Temporary facilities, project office
- Temporary material storage
- Temporary utilities, electricity, and water
- Preparatory work and laboratory testing
- Telephone and communications
- Permits and licenses
- Insurance (project coverage)
- Quality control
- Operation and maintenance of temporary job-site facilities

The costs of mobilization and preparatory work, including the setup and removal of construction facilities and equipment are part of overhead costs, unless there is a specific bid item. For large projects, the cost for each part of this initial work should be estimated on a labor, materials, and equipment basis.

## Duration of Overhead Items

After the overhead items have been listed, a cost must be determined for each. Each item should be evaluated separately. Some items, such as erection of the project office may occur only once during the project. The estimator should use the developed project schedule in estimating overhead duration requirements. Costs reflective of each particular item during the scheduled period should then be applied. The product of duration and unit cost is the overhead cost for the item.

> If a job is not finished on time, the contractor can lose money because project overhead costs exceed the amount included in the estimate.

## General and Administrative Overhead

Home office overhead expenses are those incurred by the contractor in the overall management of the business. Since they are not incurred for any one specific project, they must be apportioned to all projects.

Each contracting company is organized differently. Each incurs costs differently from varying sources and manages operations of the home office by its own methodology. Home office costs are not standard or fixed. Even though the cost for a specific contractor varies from period to period, a rate is normally averaged as a computation of total home office costs over a defined period (usually a year) divided by the total volume of business during the period. This rate computation methodology allows distribution and projection to future project estimates. When more specific data are not available, the cost estimator may include empirical rates. Empirical G&A rates typically range from 3% for large contractors to 10% for small contractors. Home office costs are usually included in the estimate of overhead as the product of an average experienced percentage rate times the expected contract amount. Typical categories of home office overhead are

- Main office building, furniture, equipment
- Management and office staff, salary, and expenses
- Home office utilities
- General communications and travel
- Supplies
- Corporate vehicles
- General business insurance
- Taxes

## Distribution of Overhead

The contractor's overhead costs should be summed and distributed to the various bid items. A proportionate distribution is commonly made by percentage ratio of total direct costs to the direct cost of each item. Regardless of the method of distribution, the estimates should clearly demonstrate the procedures and cost principles applied.

# CHAPTER 6
# ESTIMATING BUILDING PROJECTS

*The architects and engineers who design buildings use drawings and symbolic language to communicate their ideas to contractors. The calculation of material quantities can be simplified by an understanding of a few principles of measurement. In the case of building estimates, uniformity and consistency are achieved by using the Construction Specification Institute (CSI) item-coding format. Estimates are organized to both calculate expected cost and identify risk.*

## *SYMBOLS*

The success of a construction company depends on its ability to produce an accurate project cost estimate and schedule. The estimating process is also the first step in the identification of project risks. Do not bid a project if the estimate is suspect.

> Accurate cost estimates allow the contractor to win profitable jobs.

A detailed and accurate quantity take-off is the foundation of all project estimates. The *units* of the take-off should reflect each work task. The architects and engineers who design commercial buildings communicate their ideas to constructors by means of the symbolic language of drawings and through the written specifications. The drawings and their accompanying notes show and explain how the building is arranged and how it is to be constructed.

Because of the amount and type of information shown on the drawings, and because much of the information is repeated from project to project, symbols (Fig. 6.1) have been developed to communicate information. The lines, notes, and symbols are a shorthand language. This shorthand language is an indispensable tool for those who design and construct buildings. The drawing or "plans" are part of the contract between the owner and contractor. To estimate the cost of the work, the contractor must understand the information communicated by the symbols.

### Computer-Aided Design

The plans for many projects are developed using computer-aided design (CAD) or computer-aided design and drafting (CADD) tools. CAD technology has replaced the manual drafting of past decades with an automated process. When plans are transmitted to the contractor in CAD formats it is possible to extract construction quantities electronically. The estimator can automatically import quantities directly from CAD files and perform adjustments as necessary.

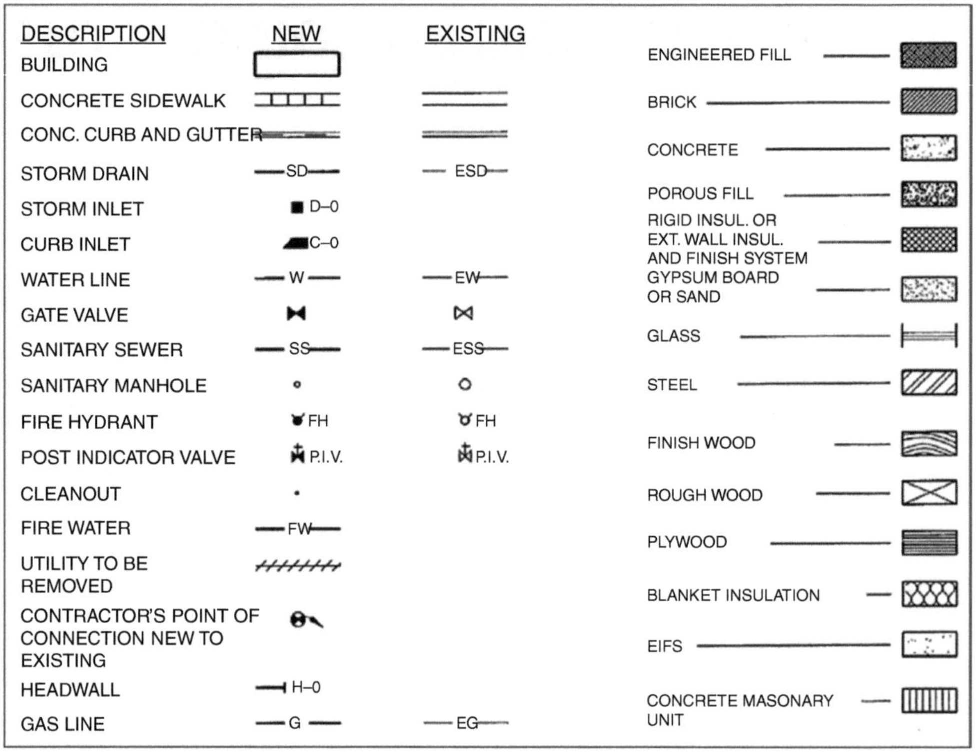


**FIGURE 6.1** Some of the common symbols used in communicating construction information.

## UNDERSTANDING PROJECT DRAWINGS

The information provided to a contractor in a set of building plans is usually organized into categories:

1. Title sheet (including site plan)
2. Floor plans
3. Elevations and sections
4. Special construction details
5. Footing and foundation
6. Structural
7. Plumbing; electrical; and heating, ventilating, and air-conditioning (HVAC)

### General Information

The first page in a set of drawings is the title sheet. Depending on the preference of the architect/engineer (A/E), other information such as an index of drawings and a plot plan may be included on the title sheet. Sometimes because of the size of the project, it takes more than one sheet to present all of this information. A plot plan, sometimes called a *site plan,* is drawn to scale and indicates the location of the proposed building on the site (Fig. 6.2).

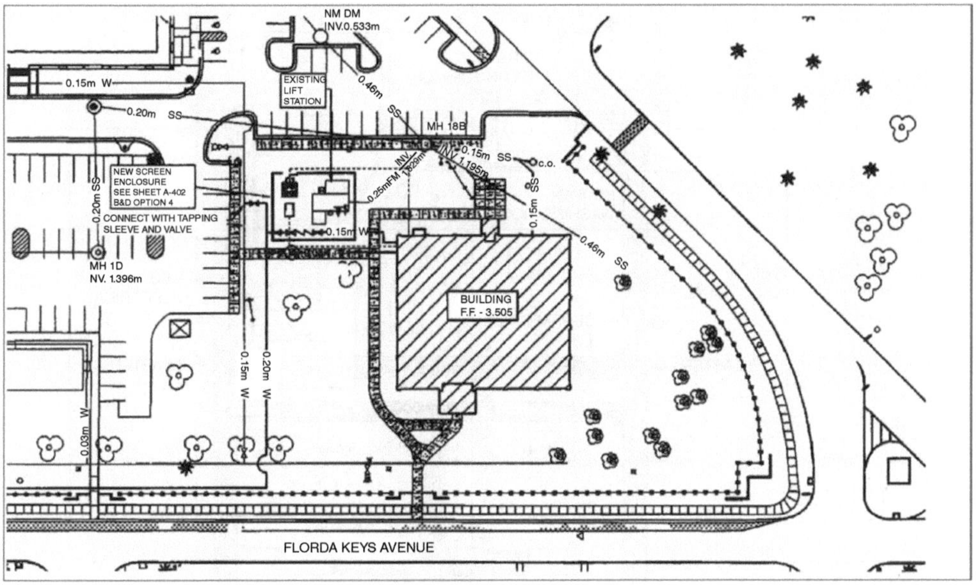


**FIGURE 6.2** Part of a plot plan for a small building project.

On the plot plan, dashed lines represent existing grades, while solid lines indicate finish grades. The elevation figures on the plot are based on a "**benchmark**." A benchmark is a point whose elevation is used to reference all other elevations on the plans. Many times, the point's exact elevation in feet above sea level is used for this purpose, but it could also be a random elevation such as 100 ft. Existing buildings, roadways, and utilities are also displayed on the plot plan.

**benchmark**
*Mark on some permanent object fixed to the ground from which elevations are taken.*

## Floor Plans

The floor plan represents a horizontal slice through a building showing, to scale, a view looking downward, a plan view. Think in terms of taking the roof and ceiling off a building and studying the layout of the floor. A separate floor plan drawing is made for each floor of the building.

These sheets indicate the location and arrangement of **load-bearing walls,** doors, windows, and vertical transportation. Dimensions are provided on the floor plans for determining exact locations in the horizontal plane. These sheets also provide information concerning materials to be used. A sample floor plan for only a portion of a building is shown in Fig. 6.3.

**load-bearing wall**
*A wall supporting any vertical load other than its own weight.*

Figure 6.3 plan of the toilet and shower areas between two rooms has reference numbers inside square box symbols (□); these are not a part of the drawing but are used to reference the following discussion presented here of information shown on the drawing.

1. The title block references the project and drawing.
2. A plan key is used if the whole building plan is not shown on one drawing. The shaded area indicates the shown section.
3. A scale reference is found on all drawings. There may be different scales used on the same drawing sheet.

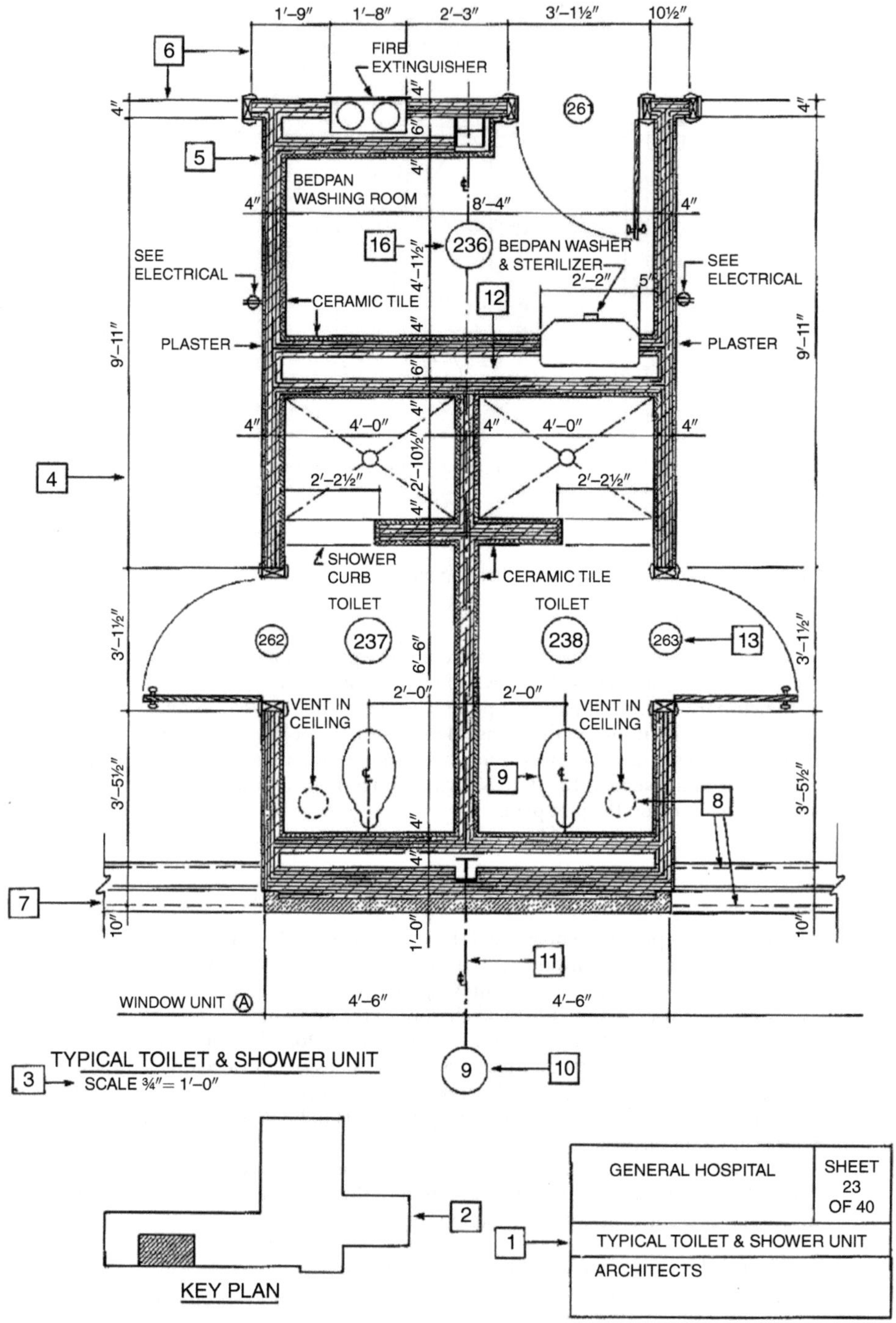


**FIGURE 6.3** Portion of a building floor plan.
*(Drawing reprinted courtesy of F. W. Dodge, McGraw-Hill Construction Information Group.)*

4. Dimension lines show distances between points.

> Always use dimension values, as drawings may not be exactly to scale.

5. Main object lines indicate outlines of walls and structural objects, which the dimensions reference.
6. Extension of lines beyond the main object is for the purpose of easy dimensioning.
7. Broken lines indicate a part continues but is not shown on the drawing.
8. Dashed lines indicate an object or area not visible from the vantage point of the cut in a plan, elevation, or section drawing. The objects are indicated to show their relationship to other parts of the building. In Fig. 6.3, the dashed lines indicate the exterior wall lines. These are hidden from view because of the window frame and sill. The ceiling vent is above the cut level.
9. Equipment lines show fixtures and other equipment.
10. Column line identifiers are provided to aid in explicitly describing locations or points within the building; numbers are usually used for the left to right identifiers (usually increasing left to right) and letters for the bottom to top identifiers on the sheet (usually increasing top to bottom).
11. Centerlines are used to determine the location of equipment, fixtures, and structural elements such as columns. The symbol L superimposed on a C identifies the centerline (℄).
12. Pipe chases are void spaces between walls. They are typically for the purpose of accommodating plumbing, HVAC ducts, and wiring.
13. Small circled numbers at door openings identify specific doors. Larger circled numbers in the center of a space identify the room or area. Using these numbers, additional information can be found on the door, window, and room schedules (Fig. 6.4).

## Elevations and Sections

An **elevation** illustrates the vertical elements of a building, either exterior or interior. Section views reveal the arrangement of materials and systems (Fig. 6.5). Section views enable a builder to both estimate and perform the work properly.

**elevation**
*A drawing showing exterior or interior faces of building walls.*

## Special Details

Floor plans or the elevations and sections cannot show all construction details adequately. Therefore, special large-scale detail drawings are found throughout a set of plans. These special drawings explain construction and design features. Figure 6.4 is a special construction detail illustrating door details. The small circles enclosing numbers indicate the areas shown in details. The "Interior Door Schedule" on the bottom of Fig. 6.4 lists door dimensions and other pertinent information. Plans will contain schedules for windows, and wall, ceiling, and floor finishes of each room.

## Footing and Foundation

Footings and foundation walls transfer the building loads to the ground below. A second function of a foundation wall is to keep moisture out of the underground parts of the building. Usually, a waterproofing compound is applied to the exterior surface of foundation walls. Figure 6.6 shows a footing and foundation wall drawing. On the drawing, the foundation walls are shown by two solid lines (=)

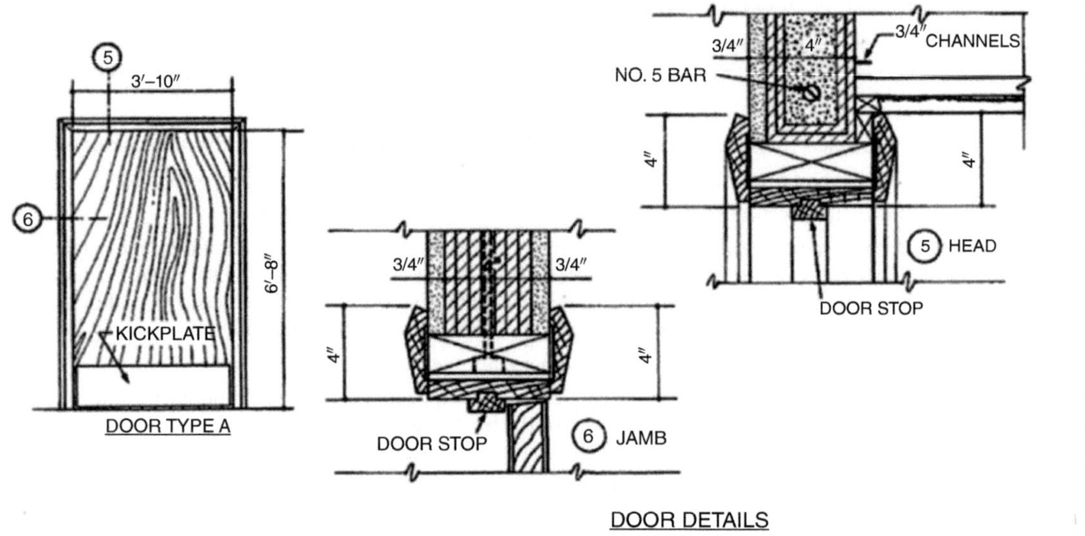


| INTERIOR DOOR SCHEDULE | | | | |
|---|---|---|---|---|
| DOOR NO. | DOOR TYPE | DOOR SIZE | HEAD DETAIL | JAMB DETAIL |
| 259 | A | 3′–10″ × 6′–8″ × 13/4″ | 5 | 6 |
| 260 | A | 3′–10″ × 6′–8″ × 13/4″ | 5 | 6 |
| 261 | D | 2′–8″ × 6′–8″ × 13/4″ | 9 | 10 |
| 262 | D | 2′–8″ × 6′–8″ × 13/4″ | 9 | 10 |
| 263 | D | 2′–8″ × 6′–8″ × 13/4″ | 9 | 10 |
| 264 | A | 3′–10″ × 6′–8″ × 13/4″ | 5 | 6 |
| 265 | A | 3′–10″ × 6′–8″ × 13/4″ | 5 | 6 |

**FIGURE 6.4** Detail drawing of a door head and jamb, and a door schedule.
*(Drawing reprinted courtesy of F. W. Dodge, McGraw-Hill Construction Information Group.)*

while the dashed lines (– – –) on either side of the solid lines delineate the footings on which the foundation wall is constructed. The circled letters and numbers are the column line identifiers (discussed in the floor plans section under item 10). They aid in identifying column locations.

## Structural Plans

The structural drawings contained in a set of plans indicate the size and location of major structural elements and give specific dimensions and details. Steel, concrete, and timber are widely used structural framing materials. The steel supplier takes the designers plans and prepares detailed drawings for structural steel framing work. These drawings are referred to as *shop drawings.* A part of a steel roof-framing plan is shown in Fig. 6.7. The W12 × 26 terminology indicates a *wide flange beam,* nominally 12 in. deep, weighing 26 lb per linear foot. The ℄ ½″ × 11 ½″ indicates a steel plate attached to the beam to support masonry above (section S-S, Fig. 6.7). The OW JOIST 12H4 indicates an *open*

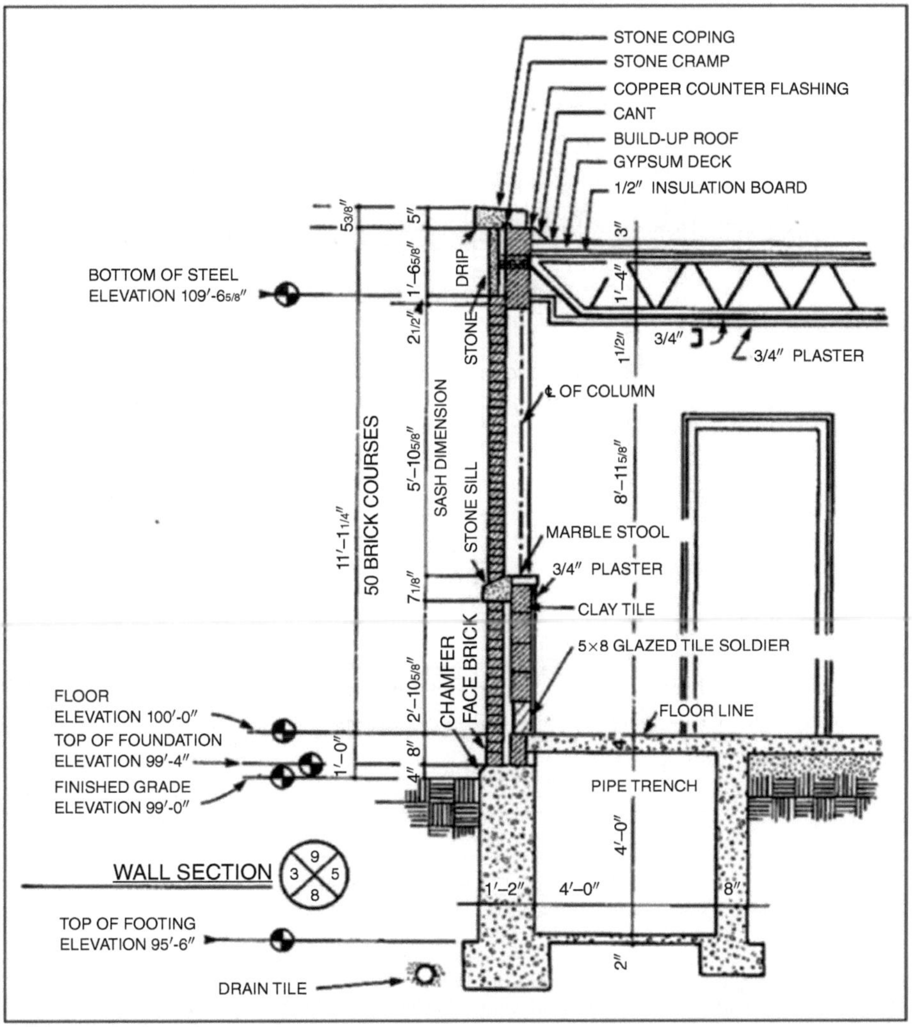


**FIGURE 6.5** Section view of a building wall system.
*(Drawing reprinted courtesy of F. W. Dodge, McGraw-Hill Construction Information Group.)*

*web steel joist* 12 in. deep, sometimes referred to as a bar joist. The last number, "4" in the 12H4, identifies the size of the top and bottom cords (section S-S, Fig. 6.7).

## Mechanical, Plumbing, and Electrical

The mechanical, plumbing, and electrical drawings are used to locate pipes, fixtures, ducts, outlets, lights, and wiring runs. The detailed descriptions of these and the other equipment to be installed, and the work required, are described in the written specifications (Fig. 6.8). A portion of the HVAC ductwork for a section of building office space is shown in Fig. 6.9.

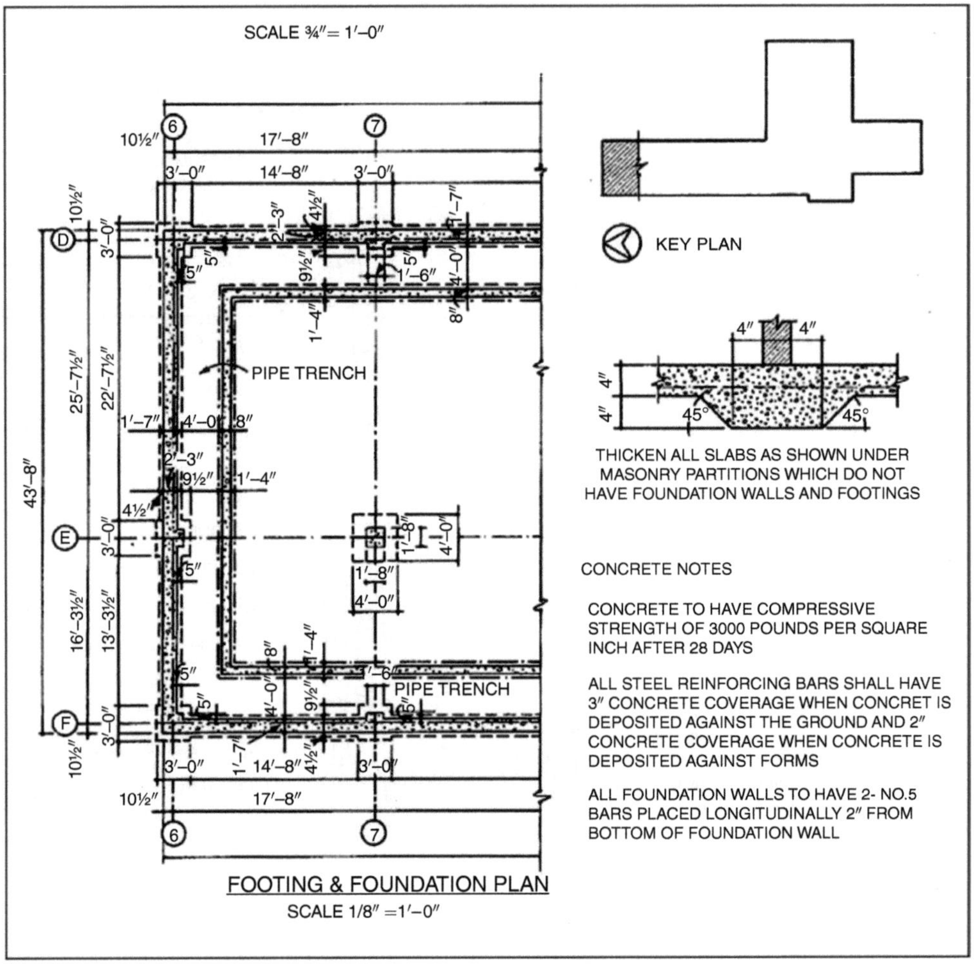


**FIGURE 6.6** Footing and foundation plan for a building.
*(Drawing reprinted courtesy of F. W. Dodge, McGraw-Hill Construction Information Group.)*

## SPECIFICATIONS

The project specification (specs) provides detail information concerning the materials to be used, scope of work, installation processes, and required work or performance quality. The three primary specification types are:

- Prescriptive
- Performance
- Proprietary

Prescriptive or detail specifications describe exactly how work should be accomplished. Performance specifications state requirements in terms of product performance. When performance specifications are used the contractor has flexibility to provide innovative, technologically advanced, and

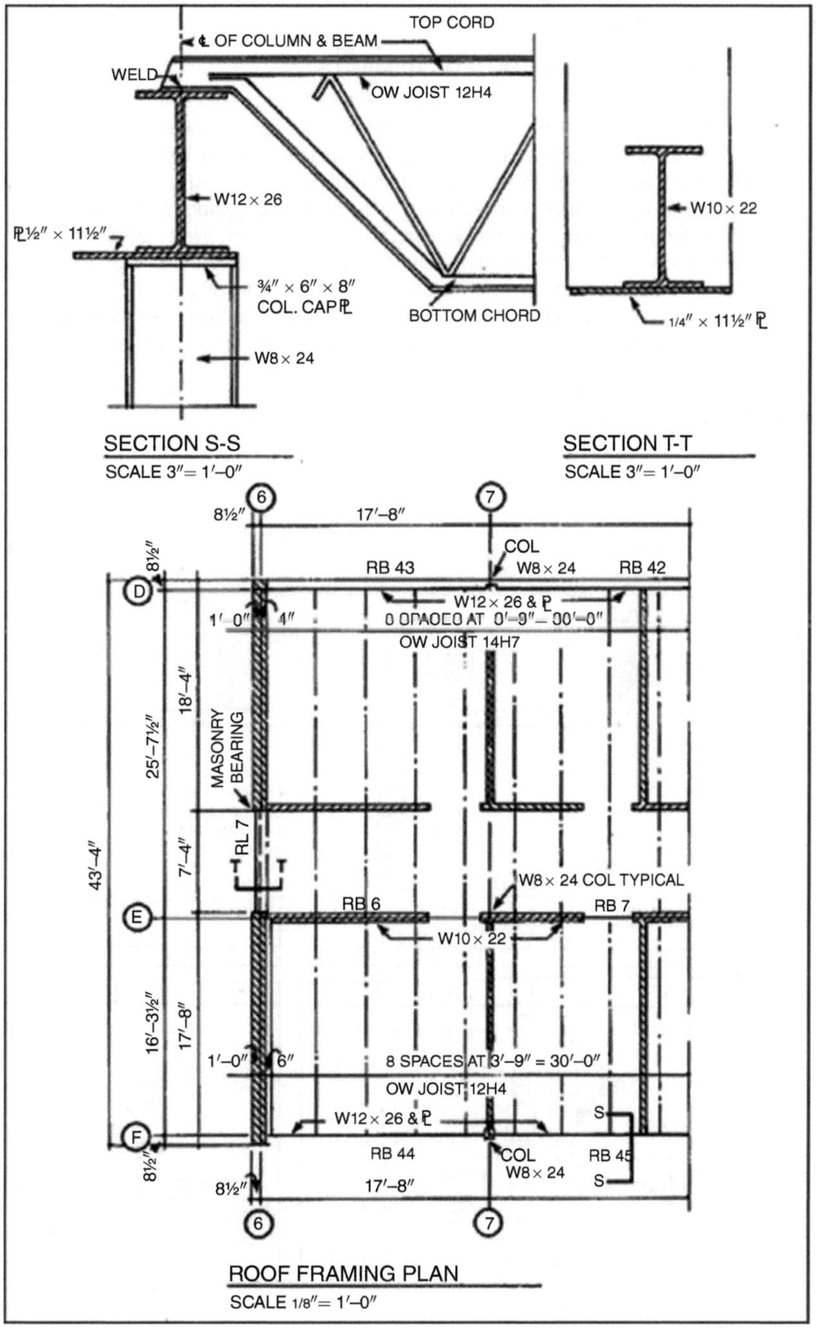


**FIGURE 6.7** Roof framing plan.
*(Drawing Reprinted courtesy of F. W. Dodge, McGraw-Hill Construction Information Group.)*

**PART 3 EXECUTION**

**3.1 GENERAL INSTALLATION REQUIREMENTS**

Hubless cast-iron pipe shall not be installed under concrete floor slabs. Piping located in air plenums shall conform to NFPA 90A requirements. Unprotected plastic pipe shall not be installed in air plenum. Piping located in shafts that constitute air ducts or that enclose air ducts shall be noncombustible in accordance with NFPA 90A. Installation of plastic pipe where in compliance with NFPA may be installed in accordance with PPFA-01. The plumbing system shall be installed complete with necessary fixtures, fittings, traps, valves, and accessories. Water and drainage piping shall be extended 5 feet outside the building, unless otherwise indicated. A gate valve and drain shall be installed on the water service line inside the building approximately 6 inches above the floor from point of entry. Piping shall be connected to the exterior service lines or capped or plugged if the exterior service is not in place. Sewer and water pipes shall be laid in separate trenches, except when otherwise shown.

**FIGURE 6.8** Execution portion of a plumbing specification.

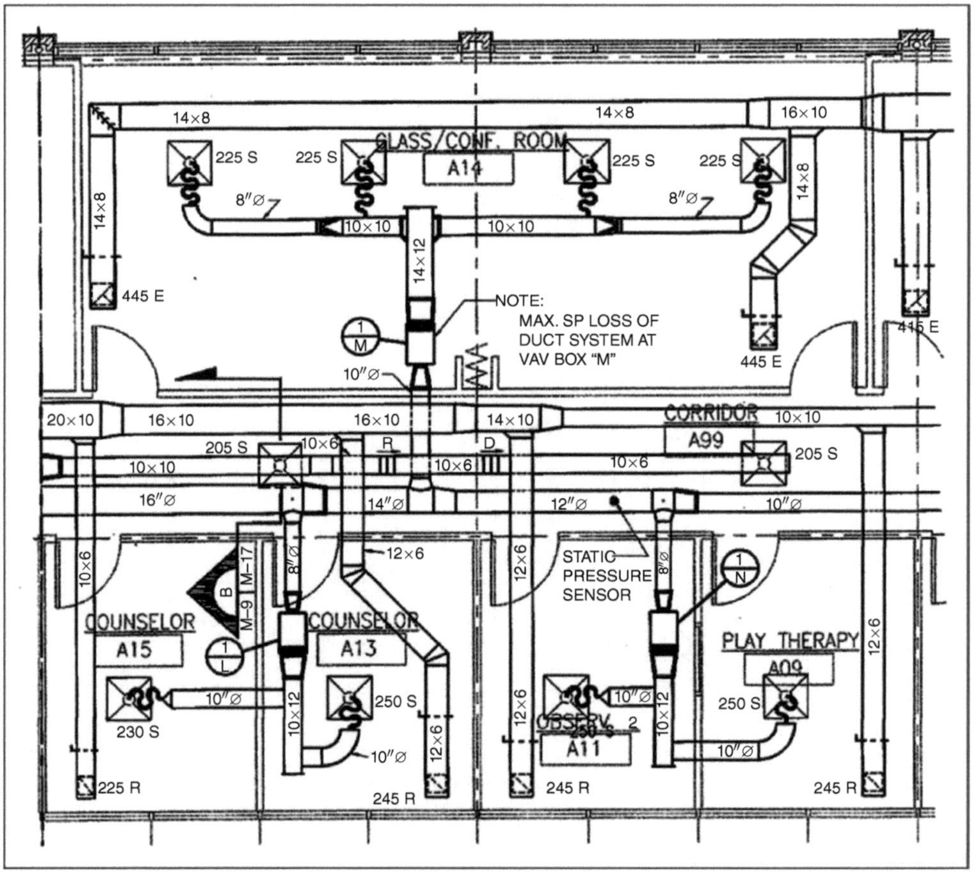


**FIGURE 6.9** Layout of office space HVAC ductwork.

**RESILIENT FLOORING**

**PART 1 GENERAL**

**1.1 REFERENCES**

The publications listed below form a part of this specification to the extent referenced. The publications are referred to in the text by basic designation only.

AMERICAN SOCIETY FOR TESTING AND MATERIALS (ASTM)

| | |
|---|---|
| ASTM D 4078 | (1992; R 1996) Water Emulsion Floor Polish |
| ASTM E 648 | (1997) Critical Radiant Flux of Floor-Covering Systems Using a Radiant Heat Energy Source |

**FIGURE 6.10** General portion of a flooring specification.

best-value solutions; however, quality risk is transferred to the contractor. Proprietary specifications are a special type of specification used for products or pieces of equipment. When included in a contract they force the builder to use only the specified product or equipment of a specific manufacturer.

The Construction Specifications Institute (CSI) system of specification organization and format is commonly used for building construction work. The basic building construction CSI divisions under the 2018 MasterFormat are shown in Table 5.1.

In 2009 CSI published a GreenFormat. It provides a standard organizational structure for information addressing sustainable elements associated with materials, products, systems, and technologies used in a project. The GreenFormat has five information categories: (1) product general information, (2) product properties, (3) product life cycle, (4) manufacturer sustainability policies, and (5) manufacturer support documentation.

## Organization of a CSI Technical Specifications

CSI provides a consistent organization structure for the information presented in a technical specification. Each subsection within a division will have three major sections: (1) General, (2) Products, and (3) Execution. As the names imply, each provides a particular part of the information the builder needs to properly perform the work.

The general section discusses the scope of work and general requirements under which the work must be executed (Fig. 6.10). The order of subsections for the general sections is: (1) references; (2) scope; (3) related work; (4) system description; (5) quality assurance; (6) submittals; (7) delivery, storage, and handling; (8) project site conditions; (9) sequencing/scheduling; (10) alternatives/alterations; (11) allowances; (12) unit prices; and (13) warranty.

The products section details the materials to be incorporated into the work (Fig. 6.11). The order of subsections for the products sections is: (1) acceptable manufacturers, (2) materials, (3) equipment, (4) mixes, and (5) fabrication.

The execution section explains the workmanship, quality, and installation requirements (Fig. 6.8). The order of subsections for the execution sections is presented here with typical verbiage of a flooring specification.

1. *Inspection*—The contractor shall examine and verify that site conditions are in agreement with the design package and shall report all conditions preventing proper installation.
2. *Preparation*—Flooring shall be in a smooth, true, level plane, except where indicated as sloped.
3. *Installation/Application/Erection*—Wall base shall be installed with adhesive in accordance with the manufacturer's written instructions. Base joints shall be tight and base shall be even with adjacent resilient flooring. Voids along the top edge of base at masonry walls shall be filled with caulk.

**RESILIENT FLOORING**

**PART 2 PRODUCTS**

**2.1 VINYL-COMPOSITION TILE TYPE VCT-1**

Vinyl-composition tile shall conform to ASTM F 1066, Class 2 (through pattern tile), Composition 1, asbestos-free, and shall be 12 inches square and 1/8 inch thick. Tile shall have the color and pattern uniformly distributed throughout the thickness of the tile. Flooring in any one continuous area shall be from the same lot and shall have the same shade and pattern.

**2.4 ADHESIVE**

Adhesive for flooring and wall base shall be as recommended by the flooring manufacturer.

**FIGURE 6.11** Parts of the products portion of a flooring specification.

4. *Field quality control, adjusting, and cleaning*—Immediately on completion of installation of tile in a room or an area, flooring and adjacent surfaces shall be cleaned to remove all surplus adhesive.
5. *Protection*—From the time of laying until acceptance, flooring shall be protected from damage as recommended by the flooring manufacturer.
6. *Extra stock/Spare parts*—Provide extra stock materials from original dye lots, for use in facility operations and maintenance (approximately 2% of the total floor surface for each color, surface texture, and format of Manufactured Product).
7. *Schedules* (this refers to reference schedules, see Fig. 6.4, not the time schedule for the work)—The following painting schedules identify the surfaces to be painted and prescribe the paint to be used and the number of coats of paint to be applied.

The contractor is bound to produce the quality specified by the contract specifications; therefore, it is necessary to carefully study information presented in the specifications before attempting to price the work.

## MEASUREMENT

The project plans allow an estimator to calculate or count the material quantities required to construct the work. The calculating process can be simplified by understanding a few principles of measurement. Consider only the bottom footer (Fig. 6.12) of the foundation wall in Fig. 6.5. If it were necessary to compute the volume of concrete, one procedure would be to multiply the cross-sectional area by the total length of the wall centerline. But usually it is also necessary to compute the formwork contact area. The length of the inside and outside formwork will differ from the centerline length.

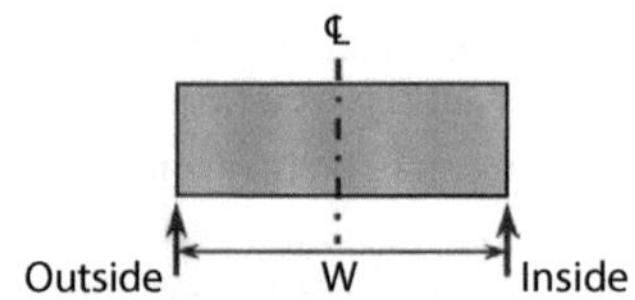


**FIGURE 6.12** Measurement of a rectangular section.

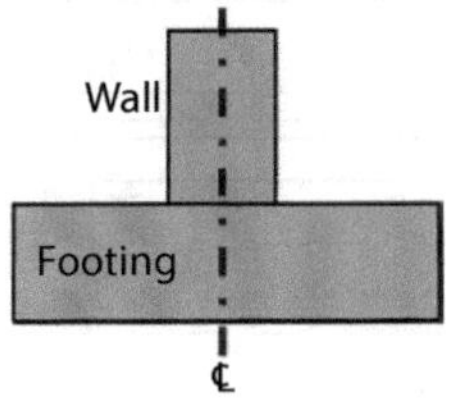


**FIGURE 6.13** Two features symmetrical about the same centerline.

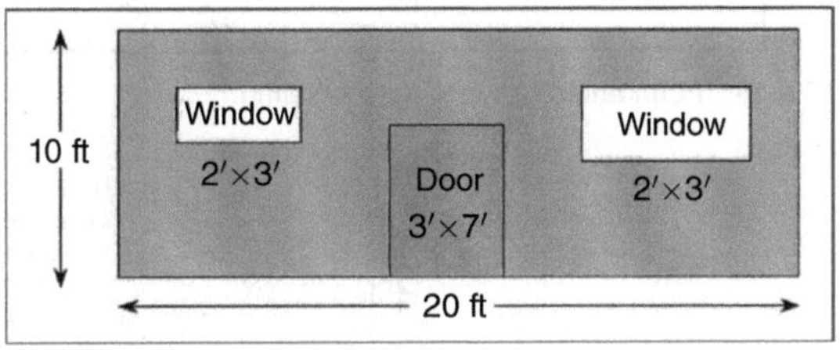


**FIGURE 6.14** Computing wall surface area.

## Perimeter

Three perimeters can be defined for the solid sections circumscribing a building (Fig. 6.12).

1. *Mean perimeter*—The true measure of an object's length. It is the length of the line about which the object is symmetrical, the center line. If the object was stretched out in a straight line this would be the resulting length.
2. *Inside perimeter*
3. *Outside perimeter*

The mean perimeter can be used to compute volumes but the inside and outside perimeters are useful in computing surface areas. If a wall and footing are symmetrical about the same centerline, their mean perimeters will be the same (Fig. 6.13).

All descriptions in an estimate are assumed to describe a positive item, but it is necessary to enter negative items when developing a take-off. It is easier to calculate the total area of a wall and to then deduct the door and window space when computing wall surface area (Fig. 6.14). A deduction space totally within a measured area is termed a *void*. A deduction not totally within a measured area is termed a *want*, and a deduction enclosed on three sides by a measured area is a *recess*. The empty spaces in the upper right and lower left of Fig. 6.15 are wants. The cavity in the lower center is a recess.

Because the dimensioning for the foundation drawing shown in Fig. 6.15 has distance between exterior points, it is easy to calculate the building's outside perimeter (OP).

$$\text{Outside perimeter} = \sum \text{outside dimensions} \tag{6.1}$$

**Perimeter Calculation Example:** Calculation of outside perimeter for the Fig. 6.15 wall.

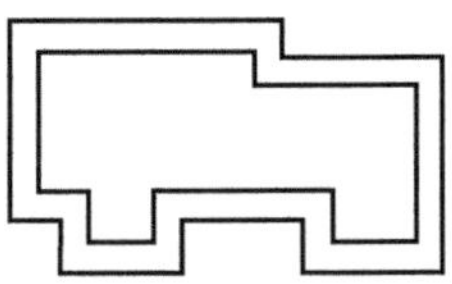

$$\text{Outside perimeter} = 20 + 4 + 10 + 16 + 8 + 4 + 9 + 4 + 9 + 4 + 4 + 16 \Rightarrow 108 \text{ ft}$$

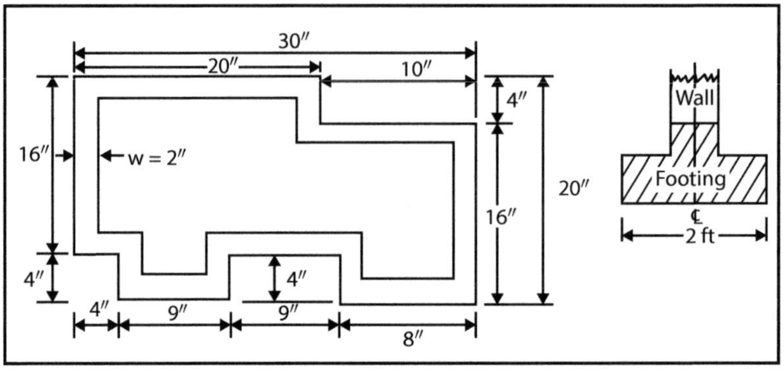


**FIGURE 6.15** Foundation drawing for a building.

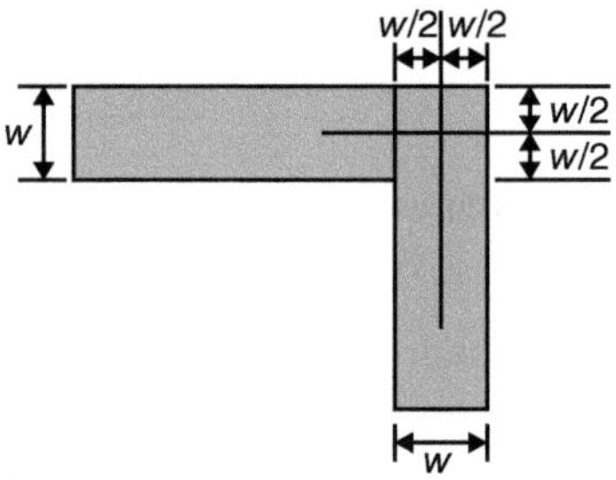


**FIGURE 6.16** Measurement around corners.

The total length of the building is 30 ft. The OP calculation used length values of 20, 10, 8, 9, 9, and 4 for a sum of 60, equaling twice the length. The width of the building is 20 ft and in the OP calculation the width values used were 4, 16, 4, and 16 for a sum of 40, equaling twice the width. The depth of the recess is 4 ft and in the OP calculation the recess values were 4 and 4 for a sum of 8, equaling twice the depth of the recess. From this example a simple measurement rule can be stated.

$$\text{Outside perimeter} = 2 \times (\text{length} + \text{width} + \text{recess}) \tag{6.2}$$

The outside perimeter can be calculated easily using Eq. (6.2) and the information in Fig. 6.15.

$$\text{Outside perimeter} = 2 \times (30 + 20 + 4) \Rightarrow 108 \text{ ft}$$

## Relationship between Perimeters

Examination of Fig. 6.16 provides an understanding of the relationship between perimeters. The term "$W$" in the figure stands for the wall thickness. If the total length of the outside perimeter is known, the length of the mean perimeter can be calculated using Eq. (6.3):

$$\text{Mean perimeter} = \text{OP} - 4\left[2 \times \left(\frac{w}{2}\right)\right] \tag{6.3}$$

where $W$ = thickness of the section.

The equation deducts one half the thickness twice to account for what happens at each corner. At a "want" or "recess," the deductions are canceled out by an addition at the next corner. Equation (6.4) is a general formula for calculating perimeter distance once one perimeter is known.

$$\text{Perimeter}_1 = \text{Perimeter}_2 + 4 \times \left[2 \times \left(\pm\text{horizontal distance between } P_1 \text{ and } P_2\right)\right] \tag{6.4}$$

A positive sign is employed when the perimeter distance is increasing, moving from an inside perimeter to an outside perimeter. A negative sign is used when the perimeter is decreasing, moving from the outside to the inside of the space.

**Volume Example:** The thickness of the foundation wall shown in Fig. 6.15 is 2 ft. Calculate the volume of concrete required to build this foundation wall if its cross-sectional area is 4 sf. If the wall is 2 ft in height, what is the required formwork in square feet contact area (sfca)?

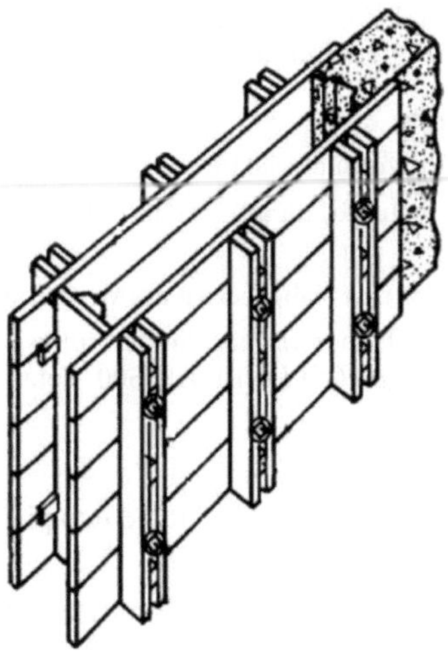

Outside perimeter = 2 × (30 + 20 + 4) ⇒ 108 ft

$$\text{Mean perimeter} = 108 - 4\left[2 \times \left(\frac{2}{2}\right)\right] = 100 \text{ ft}$$

Volume concrete = 100 ft × 4 sf ⇒ 400 cf or 14.8 cy
Inside perimeter = OP + 4 × [2 × (−2)]
Inside perimeter = 108 + 4 × [−4] ⇒ 92 ft
Inside forms = 92 x 2 ⇒ 1846 sfca
Outside forms = 108 × 2 ⇒ 216 sfca
Total forms required = 216 + 184 ⇒ 400 sfca

## CONCRETE CONSTRUCTION (CSI DIVISION 3)

Concrete is a universal construction material because of its unique properties of strength, durability, and workability. Estimates are organized to serve the needs of the contracting organization, with a critical need being identification and assessment of risk. Therefore, as discussed in Chapter 5, columns in the estimate format segregate project risk. The typical column headings are labor, equipment, material, other, and subcontracts. It is also necessary to organize the estimate for uniformity and consistency. In the case of building estimates, uniformity and consistency are achieved by using the Construction Specification Institute (CSI) item-coding format. Concrete is Division 3 in the CSI format. Concrete forms and accessories are in the 03 10 00 subsection, concrete reinforcement is in subsection 03 20 00, cast-in-place concrete is in subsection 03 30 00, precast concrete is in subsection 03 40 00, and cast decks and underlayment is in subsection 03 50 00.

**FIGURE 6.17** Assembled reinforcing steel ready to be lifted into place.

Placement of concrete is in units of volume, while forming is in units of form contact area, usually square feet contact area (sfca). The basic concrete operations are:

1. Preparation for placement of the liquid concrete:
   - Prepare surface—fine grading and compaction of material below footings and on-grade slabs
   - Fabricate forms—job constructed or manufactured, quantity required to satisfy the placement schedule, number of reuses
   - Erect forms—type of form (wall, column, beam, or slab), height above working surface
   - Embedded items—angle iron ledgers, channel door frames, ladder rungs, and miscellaneous fabrications may be furnished and installed under CSI Division 05500 Metal Fabrications
2. Reinforcing steel:
   - Fabrication—prefabricated or job fabricated
   - Placement—hand placed in the forms, or cages assembled and lifted into place (Fig. 6.17)
3. Placing concrete—size of individual placement:
   - Delivery to the site—from ready mix plant, onsite plant, haul distance affects job site handling time
   - Transport to point of placement—truck, chute, bucket, pump
   - Placement—floor, wall, column, beam, or slab
   - Consolidation
4. Finishing—refer to the specifications, rubbing, point and patch
5. Curing and protection of concrete—exposed surfaces, ponding, compound, hot and cold weather protection

The concrete material cost will be influenced by the specified item strength requirement, commonly a 28-day compressive strength. Strength requirements, which are usually expressed in pounds per square inch (psi), will be found in the project specifications. The concrete quantity take-off should separately quantify concrete volumes by mix design: strength, admixtures, and color and/or any other special features.

When performing the quantity take-off, the estimator should also create a set of supporting informational notes. These notes identify the specific sections and details found in the plans and used to support the quantity calculations—*All reinforcing is to be fabricated in accordance with CRSI standards*; *Proposed forming system does not match the rebar spacing.*

## Formwork

Building layouts normally have repeating structural bays. The use of repeating bays introduces construction economy. Such repetition allows for a formwork cost savings and dictates the quantity of formwork material. The formwork material must be able to withstand repeated usage if the sequence of work is to be maintained. The architecture floor plans typically provide schedules specifying finish quality. Smoother finishes require higher quality form-facing materials. The structural plans will provide information concerning expansion joints, placement joints or strips, reinforcing steel configuration, and penetrations.

**Pilaster**
*Rectangular pier attached to a wall for the purpose of strengthening the wall; also a decorative column attached to a wall.*

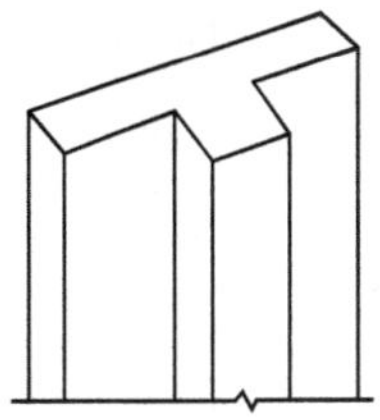

**Walls** The cost of wall forming depends on wall height, the number and size of openings, intersections, **pilasters** (Fig. 6.18), haunches, and blockouts. If there are openings in the wall, the opening areas are usually not deducted from the form quantity. However, there would be a "deduct" when calculating concrete volume. Good construction practice is to make the form continuous over the opening so as not to destroy the forming material for future reuse. This practice minimizes the need to cut form material. But the additional blockout form material needed for placement around the opening must be calculated. If ledges are present, it is necessary to add forming material for ledge forms. Ledges are formed by adding a box to the inside of a wall form. This means the contact area is created twice.

In the case of many features, it is easiest to measure through the feature and then add back the extra forming material. Consider the pilaster wall shown in Fig. 6.18. When calculating the amount of form material, measure the length and height of the wall, and then add the required side material for the pilasters. The extra material for the pilaster form is calculated by the formula:

$$\text{Extra material per pilaster} = 2 \times w \times h \tag{6.5}$$

where $w$ = pilaster projection width
$h$ = pilaster projection height

It is sometimes required to fasten a *chamfer* strip to the inside of the forms where the concrete would make a right angle corner. Chamfering keeps the concrete edges from chipping.

When a wall is being constructed in segments, it will be necessary to calculate the extra forming material required to form the bulkheads at the end of each segment. The quantity of bulkhead is often kept as a separate quantity because these forms require drilled holes for the protruding reinforcing bars.

**Columns** The form quantity for a column is the perimeter multiplied by the column height. For circular columns, the formulas are:

$$\textbf{Volume}\text{, circular column} = \pi \times r^2 \times h \tag{6.6}$$

where $r$ = radius of the column
$h$ = column height

$$\textbf{Surface area}\text{, circular column} = \pi \times d \times h \tag{6.7}$$

where $d$ = diameter of the column
$H$ = column height

**Beams** The formwork for a beam consists of the **beam** bottom and sides. The quantity of forming material for one side of an internal beam is the depth of the beam below the slab, times the length ($D \times L$). The forming for the inside of an exterior beam conforms with an interior beam (depth to the bottom of the slab), but the outside form extends to the top of the slab. The minor amount of unnecessary

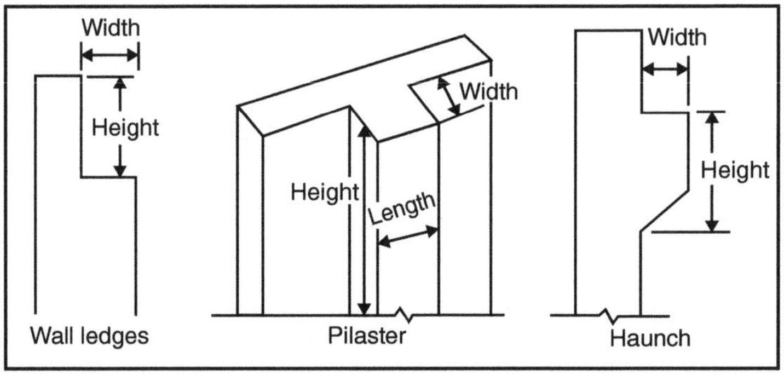


**FIGURE 6.18** Wall variations.

beam bottom forming material for the case when the beam crosses a column is not usually deducted from the total. When two beams intersect at a column the quantities of the deepest beam should be calculated based on the beam running through the column. The shallow beam should be figured as running only to the edge of the deeper beam. Beam quantities should be accumulated by height of required shoring. Accumulation increments of 5 ft are common because this dimension matches shoring tower heights.

**beam**
*A load carrying horizontal structural member.*

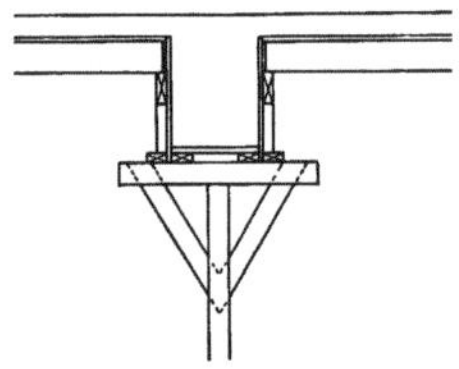

**Slabs** A slab-on-grade rests directly on a prepared ground surface. Under the slab, it is common to have a stone base and a vapor barrier. The slab may have reinforcing steel bars, but typically the reinforcing is wire mesh. Edge forms may constitute the only formwork required, but many times even these are not needed as previously constructed walls serve as the edge forms.

Elevated flat slabs require a bottom form supported by shoring (Fig. 6.19). The outside edges of the beam forms provide the perimeter forms of the slab.

**FIGURE 6.19** Floor slab formwork and shoring.

## Reinforcing

The quantity of reinforcing steel is calculated by summing bar lengths according to size and then converting (Table 6.1) the total length into weight, either pounds (lb), hundred weight (CWT), or tons (tn). Bar laps must be accounted for when calculating reinforcing steel quantities. Lap lengths are usually specified. The cost of reinforcing steel is affected by:

1. Bar grade and tensile strength specification
2. Whether they are deformed, have ridges of indentations on their surface, or if they can be smooth
3. Extent to which they must be fabricated, cut, and bent
4. Whether they can be plain, or must be galvanized or epoxy coated

**on center**
*The dimensional distance between the center of one member and the center of the next.*

The designation #5 dowels @ 12"OC (upper right, Fig. 6.20) identifies the use of #5 "L"- (90° angle) shaped dowels placed every 12 in. **on centers** along the length of the footing. Dowels are short smooth steel bars used to provide a mechanical connection between slabs or other elements. In this case, the element is probably a wall. The dowel will provide reinforcing continuity between the wall and the footing. The designation #5 @ 6" OC CONT means the #5 bars are spaced 6 in. apart on centers across the footing width and placed lengthwise continuously. The #4 @ 12" OC means the #4 bars are placed perpendicularly across the longitudinal #5 bars and they must

**TABLE 6.1** Reinforcing Steel Data

| Bar size | Weight (pounds per linear ft) | Diameter (in.) | Cross-sectional area (sq. in.) | Perimeter (in.) |
|---|---|---|---|---|
| #3 | 0.376 | 0.375 | 0.11 | 1.178 |
| #4 | 0.668 | 0.500 | 0.20 | 1.571 |
| #5 | 1.043 | 0.625 | 0.31 | 1.963 |
| #6 | 1.502 | 0.750 | 0.44 | 2.356 |
| #7 | 2.044 | 0.875 | 0.60 | 2.470 |
| #8 | 2.670 | 1.000 | 0.79 | 3.142 |
| #9 | 3.400 | 1.125 | 1.00 | 3.544 |
| #10 | 4.303 | 1.270 | 1.27 | 3.990 |
| #11 | 5.313 | 1.410 | 1.56 | 4.430 |
| #14 | 7.650 | 1.693 | 2.25 | 5.320 |
| #18 | 13.600 | 2.257 | 4.00 | 7.090 |

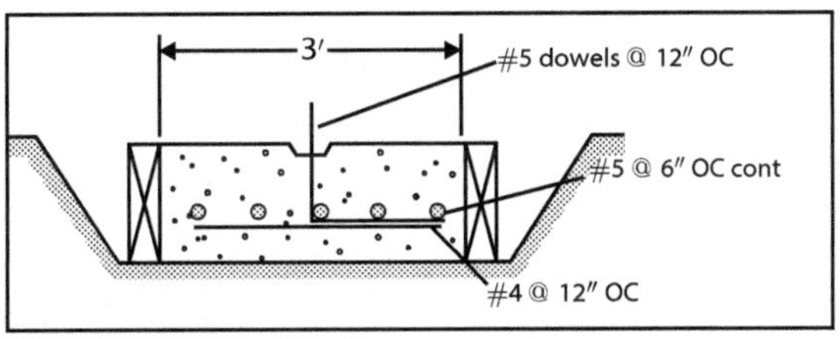


**FIGURE 6.20** Placement of reinforcing steel in a concrete footing.

have a 12-in. on-center spacing. The #4 transverse steel serves to hold the longitudinal steel in the proper position.

## Curing

To control the rate of concrete hydration, the fresh concrete must be cured. Proper curing produces concrete capable of excellent long-performing service. Curing serves two purposes:

- It retains moisture in the placement so the concrete can continue to gain strength.
- It delays drying shrinkage until the concrete is strong enough to resist shrinkage cracking.

Curing methods may be stated in the specifications for the project or selection of a method may be at the contractor's discretion. Common concrete curing methods include:

- Leave the forms in place—They provide protection against moisture loss for the formed concrete surfaces.
- Water cure—The concrete is flooded, ponded, or continually mist sprayed.
- Water-retaining methods—Use coverings such as sand, canvas, burlap, or straw and keep the covering wet.

**Reinforcing Steel Example:** Calculate the reinforcing steel required for the footing shown in Fig. 6.20. The footing is 36 ft in length. The L dowel has a height of 2 ft. The longitudinal bars require 6 in. of cover at their ends and a transverse bar is placed at the end of the longitudinal bars.

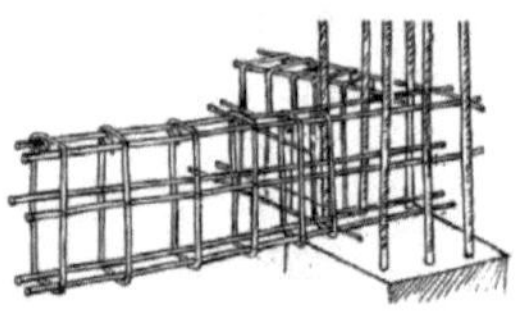

Length of longitudinal bars: 36 ft − 0.5 ft − 0.5 ft = 35 ft

Number of longitudinal bars: 5

Total length of longitudinal bars: 35 ft × 5 = 175 ft of #5 bars

Length of transverse bars: 2 ft

Number of transverse bars: 35 + 1 = 36

Total length of transverse bars: 36 × 2 ft = 72 ft of #4 bars

Length of dowel bars: 1 ft + 2 ft = 3 ft

Number of dowel bars: 35 + 1 = 36

Total length of dowel bars: 36 × 3 ft = 108 ft of #5 bars

Weight of #4 bars: 72 ft total length x 0.668 lb per lf = 48.1 lb

Weight of #5 bars: 283 ft total length (175 ft + 108 ft) x 1.043 lb per lf = 295.2 lb

Total weight reinforcing steel: 48.1 lb + 295.2 lb = 343.3 lb

- Waterproof paper or plastic film seal—These are applied as soon as the concrete is hard enough to resist surface damage.
- Chemical membranes—The chemical application should be made as soon as the concrete is finished.

The surface areas to receive a curing treatment should be totaled based on the types of treatment. The calculations would be performed during the formwork take-off. Historical production rates are used to calculate the labor cost to accomplish the curing. Material cost and application rates for a curing compound are obtained from the manufacturers.

## Pricing Concrete

Estimating the cost of concrete work is normally accomplished by applying historical production rates against the work task quantities. The historical production rates are based on a specific crew, to include equipment. Companies should develop historical unit costs for the types of work they perform. These costs are developed from actual experience on previous projects. Therefore, once the formwork quantity (by type, slab, column, or beam) is known, the work can be priced. If the company does not have experience with a particular work item, typical crews and production rate information can be found in published reference books. The reference books will state the assumed production rate; the labor, equipment, material, and other prices; and crew composition and pay rates (Table 6.2), equipment rates, and the overhead and profit percentages used in their cost calculations. The price of the delivered concrete will be a quote from the ready mix company. Always check to supplier's delivery rate capability.

**Concrete Pricing Example:** Calculate the bare cost (without overhead and profit) to place 75 cy of 3,000 psi concrete. The concrete is to be placed in 8-in.-wide wall forms by use of a concrete pump truck. This data was found in an estimating cost publication. The publication did not provide a concrete price but a local supplier quoted $90 per cubic yard delivered.

| | Production rate | Material cost/cy | Labor cost/cy | Equipment cost/cy | Total cost/cy |
|---|---|---|---|---|---|
| Walls, 8-**in.** width pumped | 12.5 cy/hr | | $24.00 | $7.70 | $31.70 |

| | Production rate | Material cost/cy | Labor cost/cy | Equipment cost/cy | Total cost/cy |
|---|---|---|---|---|---|
| Walls, 8-**in.** width pumped | 12.5 cy/hr | $90.00 | $24 | $7.70 | $31.70 |
| 75 cy | 6 hours | $6,750.00 | $1,800.00 | $577.50 | $9,127.50 |

It will take 6 hr to complete the placement. Therefore, the total bare cost is $9,127.50 (this is just to provide the concrete and to place it in the forms).

**Concrete Pricing Example 2:** Estimate the cost to place 123 cy of concrete using a pump. The calculated placing time is 9.84 hours; therefore, use 10 hours to calculate placement labor and equipment costs. The vibratos will cost $6.24 per hour and the combined owning and operating (O&O) cost for the pump is $110 per hour. This is placement only, no pricing for the concrete.

| Task | 123 cy, placed by concrete pump | | | | | | |
|---|---|---|---|---|---|---|---|
| Production rate | 12.5 cy per hr | | Total hours | 10 | | | |
| | No. | Labor rate | Labor | Equipment | Total | Unit cost ($ per cy) | |
| Forman | 1 | $52.30 | $523.00 | | $523.00 | $4.252 | |
| Labor | 3 | 30.75 | 922.50 | | 922.50 | 7.500 | |
| Cement mason | 1 | 46.85 | 468.50 | | 468.50 | 3.809 | |
| Equipment operator | 1 | 44.50 | 445.00 | | 445.00 | 3.618 | |
| Concrete pump | 1 | | | $1,100.00 | 1,100.00 | 8.943 | |
| Vibrator | 2 | | | 125.00 | 125.00 | 1.016 | |
| | | | $3,584.00 | $1,225.00 | $3,584.00 | $29.138 | |

**TABLE 6.2** Concrete Crew Composition and Pay Rate Data from a Reference Book

| Concrete crew | Bare costs |
|---|---|
| 1 Carpenter foreman (outside) | $55.30/hr |
| 2 Carpenters | 52.90/hr |
| 2 Labors | 42.12/hr |
| 1 Cement Finisher | 48.30/hr |
| 1 Gas engine vibrator | 30.40/day |

## *MASONRY CONSTRUCTION (CSI DIVISION 4)*

Brick, subsection 04 21 clay unit masonry, and concrete masonry units, subsection 04 22, are the two most common types of masonry, but Division 4—Masonry, also includes glass block, subsection 04 23; and stone, subsection 04 43. Many varieties of texture and appearance can exist under the classification of brick or concrete masonry and the estimator should always understand local custom and terms used to describe masonry units. The term "concrete masonry unit" (CMU) refers to concrete block. Concrete block is usually a relatively large unit with hollow cores. The common size is a nominal 8-by-8-by-16 in. An accurate estimate of masonry cost requires knowledge of various factors, including the type of structural element, thickness of the section, kind of mortar, style of mortar joint, class of workmanship, and the skill of the workers. There are several categories of masonry walls:

1. Conventional bonded walls
2. Drainage-type walls, including masonry veneer, cavity type, and masonry-bonded hollow walls; the success of this type of wall depends on the care taken to provide continuous means for water to escape
3. Barrier-type walls—metal tied and reinforced masonry

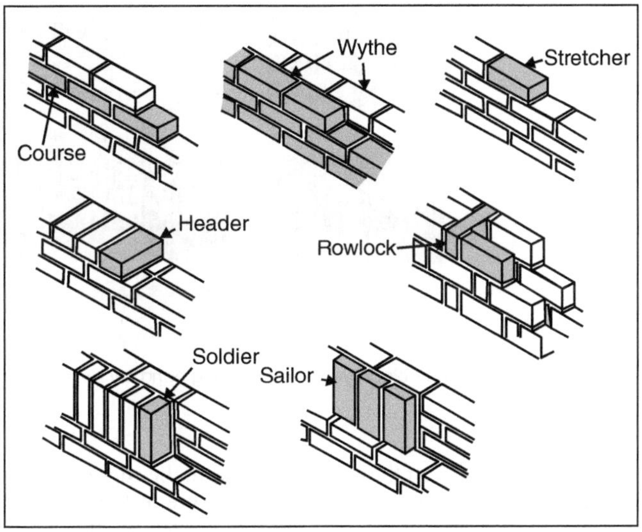


**FIGURE 6.21** Brick positions in a wall to form patterns.

## Position of the Brick

Most brick is laid as *stretchers* with the long narrow face dimension horizontal. The terms applied to brick surface positions are (1) stretcher, (2) header, (3) rowlock, (4) rowlock stretcher, (5) soldier, and (6) sailor (Fig. 6.21).

## Bond

The term "bond," in reference to masonry, can have three different meanings. *Structural bond* refers to the method by which individual masonry units are interlocked or tied together to cause the entire assembly to act as a single structural unit. The adhesion of mortar to the masonry units is termed *mortar bond*. The term used to describe how the masonry units are presented in the face of a structure is *pattern bond*. The pattern formed by the units and the mortar joints on the face of a wall make up the pattern bond. The pattern may be dictated for structural reasons or it may be for purely decorative design. Three of the basic structural bonds are:

- *Running bond*. The simplest of the basic pattern bonds, the running bond consists of all stretchers (Fig. 6.22). Because there are no headers in this bond, metal ties are usually used. Running bond is used largely in cavity wall construction.
- *Common* or *American bond*. This is a variation of the running bond with a **course** of full-length headers at regular intervals.
- *English bond*. This is made up of alternating courses of stretchers and headers.

**course**
*A continuous uniform height row of stone or brick.*

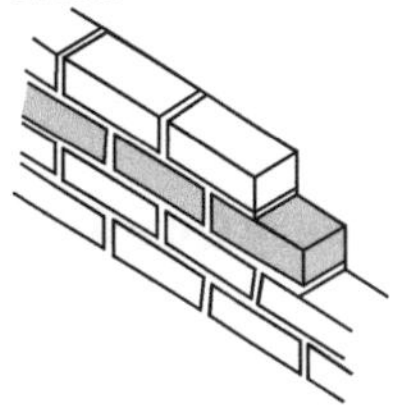

## Joints

The mortar between masonry units must be struck to form the joint and the specifications may require a specific type of joint. A *flush* joint (sometimes referred to as a rough cut) is created when the

**FIGURE 6.22** Concrete block wall with a running bond lay pattern.

mason simply strikes off the excess mortar with the edge of a trowel. *Concave* and *V-shaped* joints are formed using a steel-jointing tool. The pressure of the tool compacts the mortar. The compacted mortar and the shape of the joint together creates an effective seal to resist rain water penetration. Concave and V joints are recommended for use in areas subjected to heavy rains.

## Factors Impacting Masonry Cost

Many factors influence the labor cost of building masonry. There is no set number of masonry units a mason will lay per hour, but some unions do have rules concerning the number of units a mason can place per day. Solid brick walls are used less frequently than in the past. Today most brick is a facing backed up with concrete blocks or other types of wall systems. In some areas prefabricated brick panels are available. The panels are hung onto the building eliminating or minimizing on-site scaffolding needs. However, the size of masonry panels is limited by transportation and erection limitations.

Masonry production will vary with the required grade of workmanship, the style of bond in which the units are laid, the width and kind of mortar joint, and the type of building—straight walls or walls with multiple door and window openings. Common bond is the most economical method of placing face masonry units. Other styles, such as English, require more labor because of the various patterns and the required additional cutting, splitting of the bricks.

The cost of cleaning face brick varies with the kind of brick. In the case of smooth face brick, it may only be necessary to wash the wall, but other masonry cleaning requirements might include (1) brush hand cleaning, (2) high-pressure water, or (3) chemical cleaning.

The height of the wall will affect production by adding **scaffolding** requirements (Fig. 6.23) and the need for lifting equipment (Fig. 6.24).

**scaffolding**
*A temporary structure, designed and built by the contractor, to provide elevated working space.*

## Masonry Quantities

The quantity take-off steps for a masonry estimate are (1) determine the surface area of masonry and (2) convert the area to the number of masonry units based on the type of unit and the required placement pattern. All openings should be deducted in full regardless of their size. From the elevation view plan sheets (Fig. 6.25), it is usually easy to identify masonry surfaces, but to calculate the area of those surfaces it is generally necessary to consult the floor plans, which give longitudinal dimensions, and the section drawing (Fig. 6.26), which provides height information.

**FIGURE 6.23** Climbing scaffolding for blocking a building wall.

**FIGURE 6.24** Using a forklift to handle brick.

The section drawings will be referenced on the elevation views. The two dark left-pointing arrows at the top of Fig. 6.25 let the viewer know it is necessary to consult sheet A8-1 for TYP. 1 and 2 details. The A7-1 indicates the sheet of this elevation view. TYP stands for typical. TYP. 1 will show a view through the window area and TYP. 2 will show a view of the solid wall. The typicals in this particular case do not by themselves provide sufficient information to calculate the vertical dimension of the brick. It is necessary to also consult several special details.

Above the window in WALL SECTION @ WINDOW view 1, Fig. 6.26 special detail 1 on sheet A9-1 (Fig. 6.27) is referenced. From studying the elevation drawing and the two section views in Fig. 6.26 it is clear the majority of the brick must be laid in a running bond pattern. Because there are no headers in a running bond pattern, metal ties are ordinarily used to tie the brick exterior wall to the inner wall of the structure.

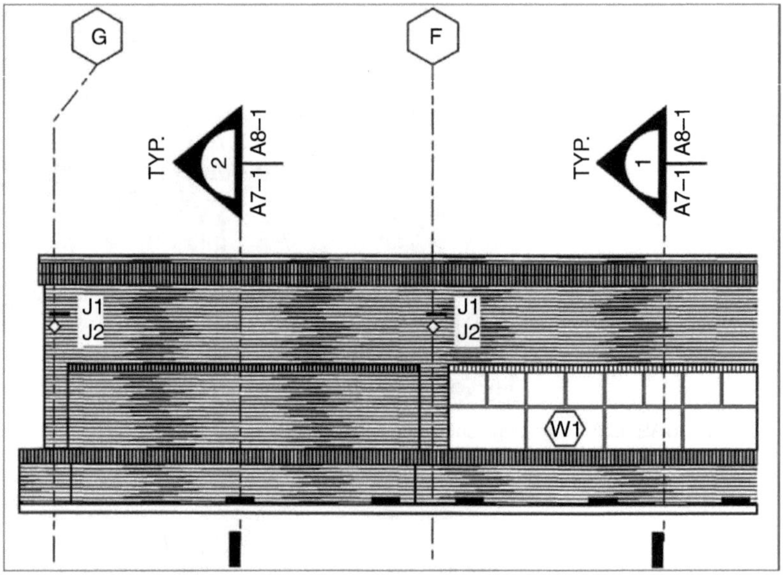


**FIGURE 6.25** Elevation view of a building having facing brick.

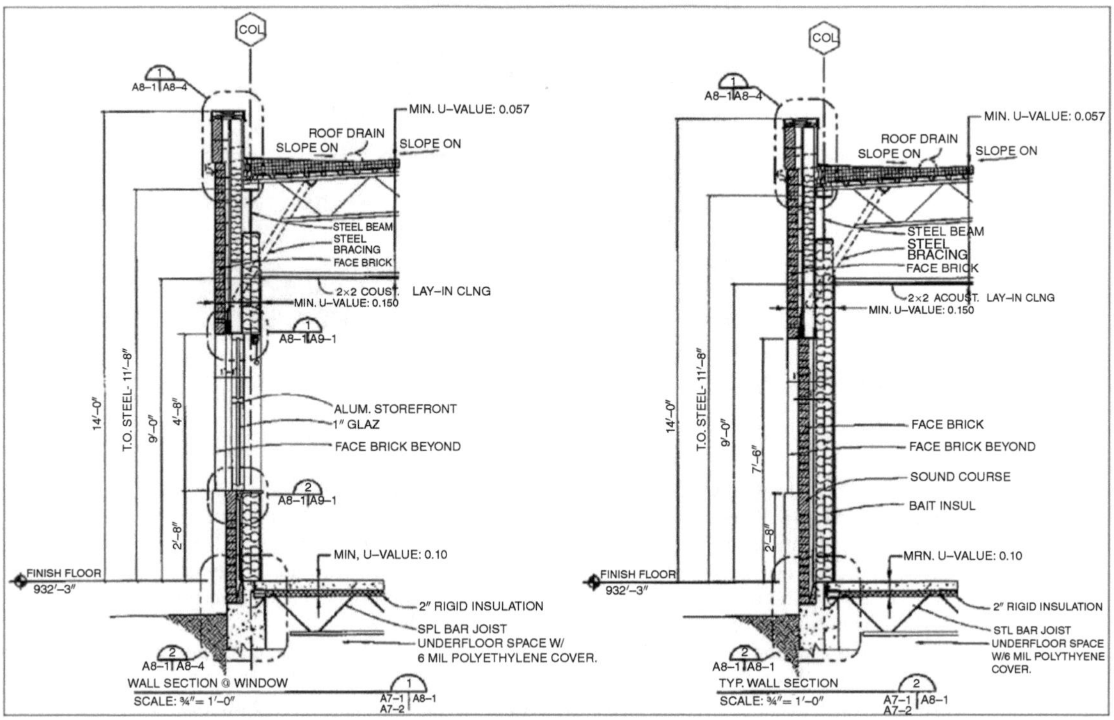


**FIGURE 6.26** Section views of the wall shown in Fig. 6.25.

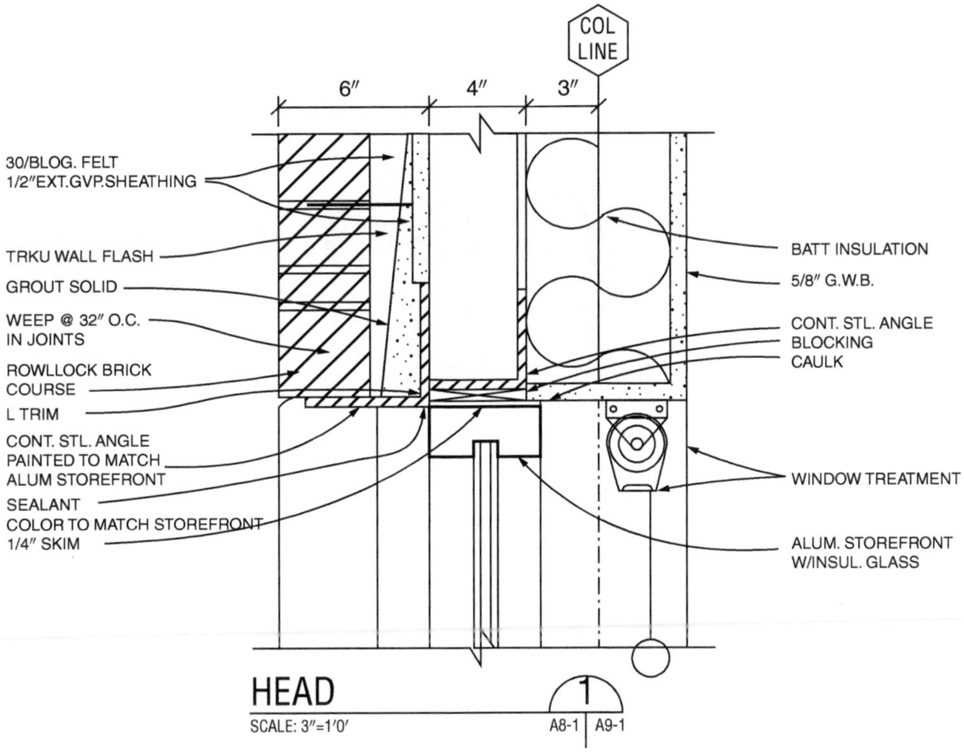


**FIGURE 6.27** Detail of window head, referenced from left drawing in Fig. 6.26.

The plans show a variation in the brick pattern (Fig. 6.25). At the level directly under the window, there is a soldier course and at the level directly above the window there is a rowlock course (also see Fig. 6.25).

> Before starting a masonry take-off, the estimator must carefully study many different drawings in the plan set.

Brick walls are usually designated on the plans as 4 in. or 4½ in., or 8 in. or 9 in. The estimator need not consider this variation in thickness because a 4-in. or 4½-in. wall is one brick thick; and an 8-in. or 9-in. wall is two bricks thick.

After the surface area of the wall is obtained, it is necessary to determine the number of bricks required per square foot. The nominal dimensions for a standard nonmodular brick is 8 in. × 2¼ in. × 3¾ in. The nominal dimensions for a standard modular brick are 8 in. × 2⅔ in. × 4 in. Sizes will vary somewhat depending on the brick's location in the kiln during burning, the brick in the center of the kiln usually being more thoroughly burned, but the nominal dimensions yield satisfactory estimating results.

The thickness of the mortar joint must be added to the brick dimensions. Mortar thickness will vary but 3/8 in. is the normal thickness. Consequently, when a standard nonmodular brick is laid as a stretcher and the mortar thickness is included, the dimensions are 8⅜ in. long by 2⅝ in. in height and

**FIGURE 6.28** Pallet of bricks stacked at the job site.

it has a surface area of 21.984 sq in. (8⅜ in. × 2⅝ in.). To obtain the number of bricks required per square foot of wall, divide 144 sq in./sf by 21.984 sq in. For a 4-in. (one brick thick) wall, 6.55 bricks (144 sq in./sf ÷ 21.984 sq in.) are required per sf of wall surface.

The waste factor for brick wall construction should not exceed 1½ to 2% unless irregular or poor quality bricks are used. However, bricks and CMUs are ordered by the pallet (Fig. 6.28) so the order must be in pallet multiples.

The nominal dimensions for standard modular concrete blocks are 7⅝ in. × 7⅝ in. × 15⅝ in. When laid up with ⅜-in. mortar joints, the unit is 8 in. high and 16 in. long, requiring 112.5 units per 100 sf, not including an allowance for waste.

**Brick Quantity Calculation Example:** The outside perimeter of a brick wall is 30 ft. Calculate the bricks required to build this wall if its height is 12 ft and the bricks will all be laid as soldiers. Standard modular bricks have been specified for the project. Assume a 1.5% waste factor.

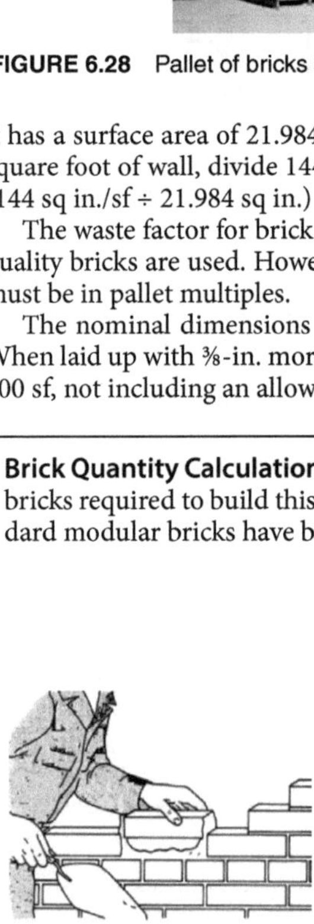

Area of wall = 30 × 12 ⇒ 360 sq ft

$$\text{Area of one soldier brick} = \left(2\frac{2}{3} + \frac{3}{8}\text{ in.}\right) \times \left(8 + \frac{3}{8}\text{ in.}\right) \Rightarrow 25.471\text{ sq in}$$

$$\text{Bricks per sf} = \left(\frac{144\text{ sq in.} / \text{ft}}{25.474\text{ sq in.}}\right) \Rightarrow 5.653\text{ bricks per sf}$$

Number of bricks without waste

= 360 sq ft × 5.653 bricks per sf ⇒ 2,035 bricks

Number of bricks with waste

= 2,035 bricks × 1.015 ⇒ 2,066 bricks

## Pricing Masonry

Many companies use historical unit costs to price masonry work. If the company does not have experience with a particular type of masonry work, average pricing information can be found in published reference books. Table 6.3 presents information on the composition of masonry crews.

**CMU Calculation Example:** Calculate the bare cost (no overhead and profit) to place 360 sf of concrete block. The blocks are being used to construct an exterior wall. Do not include the cost of scaffolding. The wall is to be 8 in. thick. The following data was found in an estimating cost data publication.

| | Production rate | Material cost/sf | Labor cost/sf | Equipment cost/sf | Total cost/sf |
|---|---|---|---|---|---|
| CMU walls 8″ thick, exterior | 45.6 sf/hr | $3.39 | $4.65 | | $8.04 |

| | Production rate | Material cost/sf | Labor cost/sf | Total cost/sf |
|---|---|---|---|---|
| CMU walls 8″ thick, exterior | 45.6 sf/hr | $3.39 | $4.65 | $8.04 |
| 360 sf | 7.9 hours | $1,220.40 | $1,674.00 | $2,894.40 |

Therefore, with a standard crew the total bare cost is $2,894.40 and it will take 8 hr to complete the placement.

**TABLE 6.3** Typical Masonry Crew Composition

| Masonry crew |
|---|
| 1 Bricklayer (or block) foreman |
| 3 Bricklayers |
| 3 Bricklayer helpers |

## *METALS (CSI DIVISION 5)*

CSI Division 5 includes structural steel framing, subsection 05 12 (Fig. 6.29); metal joists, subsection 05 21 (Fig. 6.30); steel decking, subsection 05 31; structural metal stud framing, subsection 05 41 (Fig. 6.31), and formed metal fabrications, subsection 05 58. Fabricated steel can be purchased by the general contractor and erected by company forces. It is more common, however, to subcontract both the fabrication and erection.

A structural steel estimate involves a large amount of detail work. The quantities of structural members are determined from the plans and then their weights computed. The cost of shop fabrication (purchase cost) is requested, to which is added freight; and the cost of erecting the steel on the job (crane, areal lifts, temporary support). Material prices depend on where material is purchased. Mill prices apply to large orders. Warehouse prices apply to small orders from local suppliers. Prices may be quoted as of *time of delivery*; therefore, the estimator should be very cognizant of escalation clauses.

### Steel Shapes and Nomenclature Information

When estimating the quantity of structural steel required for a project, each class of work (column bases, columns, girders, beams, trusses) should be estimated separately because each involves

**FIGURE 6.29** Steel frame building under construction.

**FIGURE 6.30** Steel framing with metal joist used to support second-story floor.

different labor operations in fabrication and erection. The steel member structural shapes are shown in Figs. 6.32 and 6.33 with a nomenclature explanation provided beside each shape in the figures.

## Sequence of Take-Off

When performing a steel take-off it is best to maintain a sequence of execution. One strategy is to work by structural types, columns, beams, and then flooring, and to work across the structure by column line identifiers. It is also good practice to work up the structure one floor at a time. This is illustrated in the following Steel Weight Calculation example.

**FIGURE 6.31** Metal stud framing used in an office building.

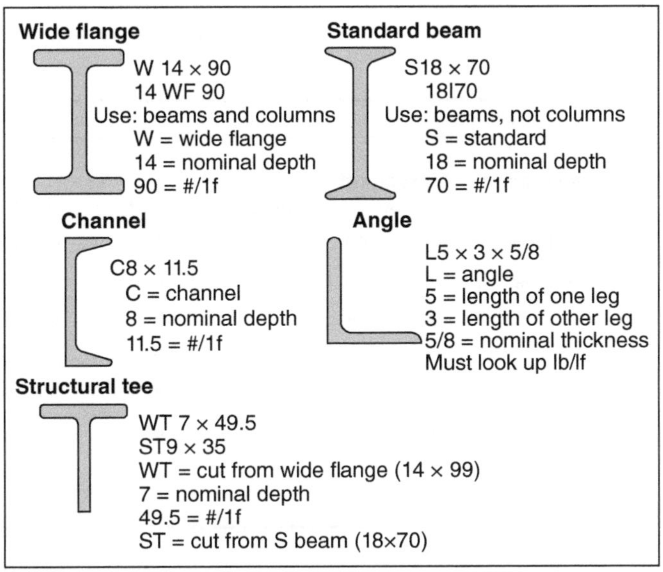


**FIGURE 6.32** Common steel member structural shapes.

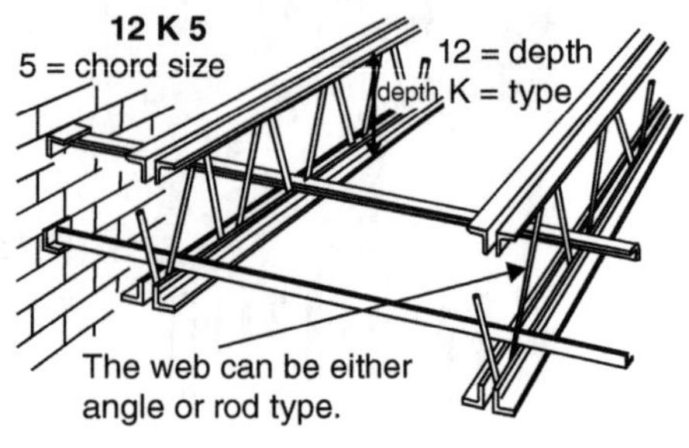


**FIGURE 6.33** Steel joist.

**Steel Weight Calculation Example:** Calculate the total weight of steel (tons) required to frame one floor of the building shown in the figure. The 10J3 steel joists (4.83 lb/ft) are over all four bays, but for clarity are shown only once.

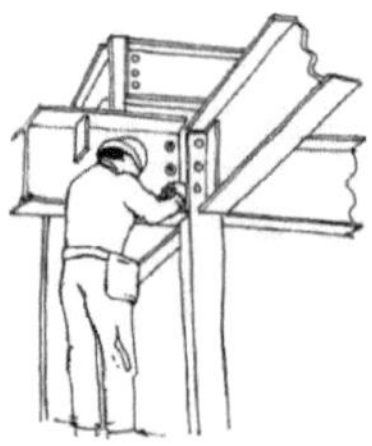

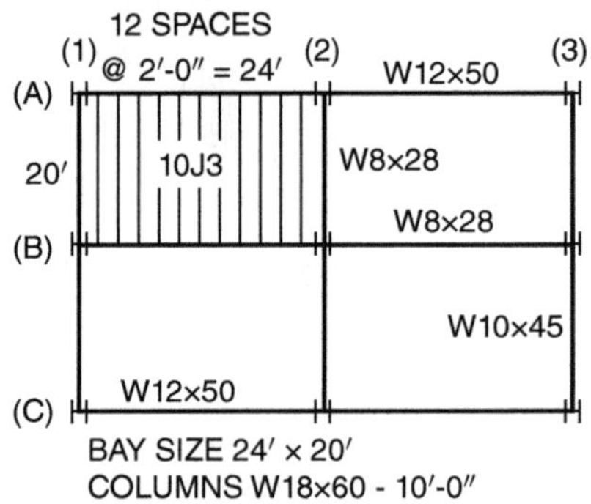


| Steel take-off | | | 1st Floor | | |
|---|---|---|---|---|---|
| Type of section or member | Structural shape | Number | Length | Weight/ft | Total pounds |
| Columns | W18 × 60 | 9 | 10 ft | 60 #/ft | 5,400.0 |
| Beams | W12 × 50 | 4 | 24 ft | 50 #/ft | 4,800.0 |
| | W10 × 45 | 4 | 20 ft | 45 #/ft | 3,600.0 |
| | W8 × 28 | 2 | 24 ft | 28 #/ft | 1,344.0 |
| | W8 × 28 | 2 | 20 ft | 28 #/ft | 1,120.0 |
| Joist | 10J3 | (4 × 11) | 20 ft | 4.83 #/ft | 4,250.4 |
| | | | | Total weight lb. | 20,514.4 |
| | | | | Total weight tons | 10.2572 |

$$\text{Total weight} = \frac{20{,}514\ \text{lb}}{2{,}000\ \text{lb / ton}} \Rightarrow 10.26\ \text{tons}$$

### Pricing Steel

Contractors typically receive a material price from a fabricator for the cost of the required steel members. On a small job, it may be possible to buy standard members from a warehouse, but for most projects the estimator will submit a set of project plans to a manufacturer or fabricator so they can prepare their own quantity take-off. The manufacturer or fabricator will then supply a material quotation. Steel joists are usually sold directly from the manufacturer to the contractor (or subcontractor).

Few contractors have the trained personnel and equipment (cranes or hoists) to erect structural steel. Consequently, most general contractors rely on experienced specialty subcontractors to handle this task. Based on historical project data, contractors relate labor hours to tons of steel by building type. Such information can also be found in estimating reference books. An example would be: industrial buildings, one story, beam and girder construction, 12.90 tons per 8-hr day with a specific crew. The labor cost to erect the steel can be calculated by applying the project labor rates to the labor crew and multiplying by the crew time. Crew time is calculated by dividing the project tonnage by the crew production, in tons per hour or per day.

**Steel Pricing Example:** The total weight of the framing steel for a building is 38.7 tons. A steel erection crew can complete 12.9 tons in an 8-hr day. The daily cost for the crew is $2,673.60. Calculate the number of days needed to erect the steel and the erection cost on a per ton basis.

$$\text{Time required} = \frac{38.7 \text{ tons}}{12.9 \text{ tons / day}} \Rightarrow 3 \text{ days}$$

$$\text{Cost} = 3 \text{ days} \times \$2{,}673.60 \Rightarrow \$8{,}020.80$$

$$\text{or } \frac{\$8{,}020.80}{38.7 \text{ ton}} = \$207.26 \text{ per ton}$$

## *WOOD, PLASTIC, AND COMPOSITES (CSI DIVISION 6)*

Lumber is a commodity sensitive to supply and demand. Thus, it is important to always check current market prices when estimating a project. To accurately perform a carpentry take-off, the estimator should have an understanding of timber design and construction procedures.

While it may not be clear from the plans, joists and **studs** should be doubled in many locations. Wood construction can be divided into several specific types of work:

- Heavy framing and planking to include glued and laminated pieces
- Rough carpentry—framing, sheathing, roofing, and door and window installation
- Finish carpentry
  - Interior—flooring, stairs, trim
  - Exterior—siding, porches, corners, doors, and windows
- Custom woodwork—millwork, factory built, cabinets

**studs**
*Vertical framing members in a wall normally spaced 16″ or 24″ on center.*

### Lumber

The dimensions used to designate lumber sizes denote the cross-sectional dimensions of the board, dimensioned lumber, or timber, as shown in Fig. 6.34. Boards are thin pieces 1 in. in thickness. **Dimensional lumber** refers to the intermediate-sized pieces used for rough carpentry work.

**dimensional lumber**
*Framing lumber cut to standardized width and depth, specified in inches.*

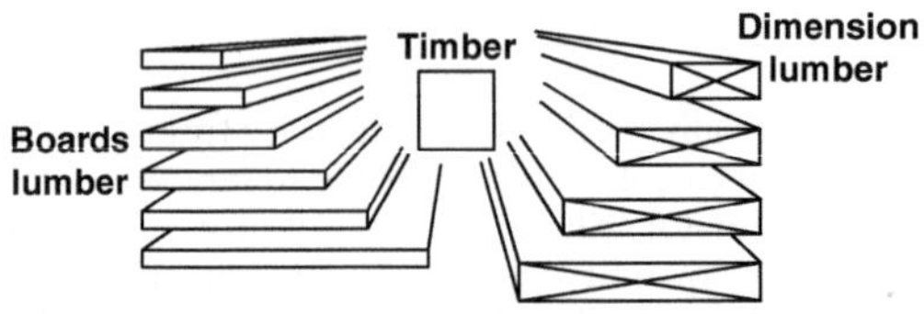


| Boards | | Dimension lumber | | Timbers | |
|---|---|---|---|---|---|
| Nominal | Actual | Nominal | Actual | Nominal | Actual |
| 1 × 2 | $\frac{3}{4}'' \times 1\frac{1}{2}''$ | 2 × 2 | $1\frac{1}{2}'' \times 1\frac{1}{2}''$ | 6 × 6 | $5\frac{1}{2}'' \times 5\frac{1}{2}''$ |
| 1 × 3 | $\frac{3}{4}'' \times 2\frac{1}{2}''$ | 2 × 4 | $1\frac{1}{2}'' \times 3\frac{1}{2}''$ | 6 × 8 | $5\frac{1}{2}'' \times 7\frac{1}{4}''$ |
| 1 × 4 | $\frac{3}{4}'' \times 3\frac{1}{2}''$ | 2 × 6 | $1\frac{1}{2}'' \times 5\frac{1}{2}''$ | 6 × 10 | $5\frac{1}{2}'' \times 9\frac{1}{4}''$ |
| 1 × 6 | $\frac{3}{4}'' \times 5\frac{1}{2}''$ | 2 × 8 | $1\frac{1}{2}'' \times 7\frac{1}{2}''$ | 8 × 8 | $7\frac{1}{4}'' \times 7\frac{1}{4}''$ |
| 1 × 8 | $\frac{3}{4}'' \times 7\frac{1}{2}''$ | 2 × 10 | $1\frac{1}{2}'' \times 9\frac{1}{4}''$ | 8 × 10 | $7\frac{1}{4}'' \times 9\frac{1}{4}''$ |
| 1 × 10 | $\frac{3}{4}'' \times 9\frac{1}{2}''$ | 2 × 12 | $1\frac{1}{2}'' \times 11\frac{1}{4}''$ | | |
| 1 × 12 | $\frac{3}{4}'' \times 11\frac{1}{2}''$ | 4 × 4 | $3\frac{1}{2}'' \times 3\frac{1}{2}''$ | | |
| | | 4 × 6 | $3\frac{1}{2}'' \times 5\frac{1}{2}''$ | | |

**FIGURE 6.34** Types and sizes of lumber.

Timbers are the large size pieces used for heavy framing. Boards and dimensioned pieces are usually dressed (finished) on all four sides, the designation being S4S, surfaced four sides. The numbers used to designate lumber size refer to the **nominal size.**

The actual or **dressed lumber** dimensions will be less than nominal as can be seen by the data in Fig. 6.34. A 2 by 4 is actually 1½ in. thick by 3½ in. in width after surfacing. Framing lumber can usually only be purchased in lengths which are multiples of 2 ft., i.e., 8, 10, 12, 14 ft. The one common exception is precut studs, which are 7.5 ft in length.

The common system used to measure lumber is known as **board measure.** Many types of lumber are sawed in random widths and lengths. This is particularly the case for valuable hardwoods. To provide a unit of measure for such lumber the board foot (BF) was developed. It is a volume measure and is the common quantity unit for lumber. A board foot is 1 in. × 1 ft × 1 ft or 144 cubic inches. The formula used to calculate BF is given by Eq. (6.8). Nominal material dimensions are used in the equation.

**nominal size**
*Size of lumber before dressing, rather than its actual or finished size.*

**dressed lumber**
*Lumber machined and smoothed at the mill, to usually 1/2 in. less than nominal (rough) size.*

**board measure**
*System of lumber measurement. The unit is 1 BF, which is 1 ft square by approximately 1 in. thick.*

$$\text{BF} = \text{length (ft)} \times \frac{\text{dressed size(inches} \times \text{inches)}}{12 \text{ inches per foot}} \tag{6.8}$$

**Lumber Quantity Example:** A building requires 256 studs for wall framing. A stud is a precut piece of lumber 2″ × 4″ by 7.5 ft in length. How many board feet of studs should be ordered?

BF = 256 × 7.5 ft × (2 in. × 4 in.)/(12 in. per foot) ⇒ 1,280

## Framing Take-Off

The term "timber frame" refers to a variety of different construction methods used to create a load-bearing structure from timber (Fig. 6.35). Most timber-framed buildings today have a **platform framing** design. This consists of a number of modular panel-like components. A platform-framed structure is assembled one-story level at a time. Although the components are made from relatively lightweight softwood timber, when they are fastened together they form a very strong, rigid structure. The principal parts of a platform-framed structure are identified in Fig. 6.35.

**platform framing** *Framing in which each story is built upon the other.*

A second type of wood-frame construction is the **balloon framing** system. With balloon framing the studs are continuous from the foundation sill to the top wall plate. Floor structures (one, two, or more) are hung from the studs (Fig. 6.35). Another timber construction method is "post and beam." Post and beam structures are characterized by the concentration of structural loads on fewer and larger sized pieces of timber.

**balloon framing** *Building-frame construction in which each of the exterior studs are one piece from the foundation to the roof of the structure.*

**Joist** To count the number of joists required for a floor system, divide the length of the floor perpendicular to the joists, by the spacing distance of the joists and add 1 to allow for the extra joist required at the end of the span (Fig. 6.36).

**Sill Plates** The length of sill plate lumber is equal to the perimeter of the structure.

**Studs** To obtain the number of studs take the length of each wall or partition and divide by the stud spacing distance, then add to the calculated number:

- Two studs per wall opening—doors and windows
- Two studs per corner—interior and exterior (Fig. 6.37)
- One stud per wall lead

Studs are 7.5 ft in length for a structure having an 8-ft ceiling. This is because there will be a sole and two top plates to complete the framing system.

**Wall Plates** Use the length of wall multiplied by the number of plates, normally two on top and one sole.

**Headers** The length of header (end of joist) lumber is twice the length of the structure, because there is a header at each end of the joists.

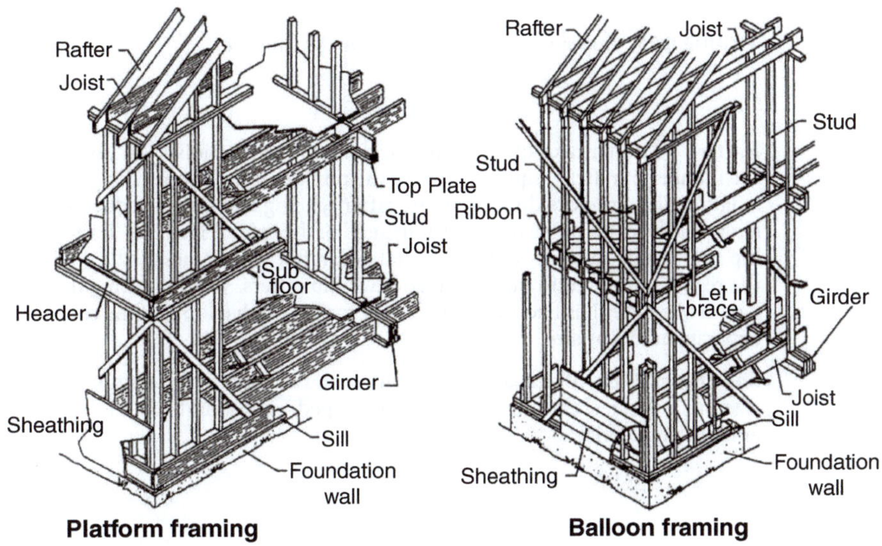


**FIGURE 6.35** Wood-frame construction.

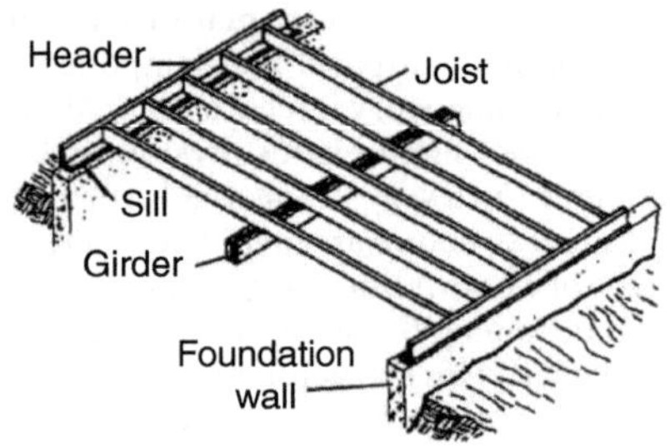


**FIGURE 6.36** Joists floor framing.

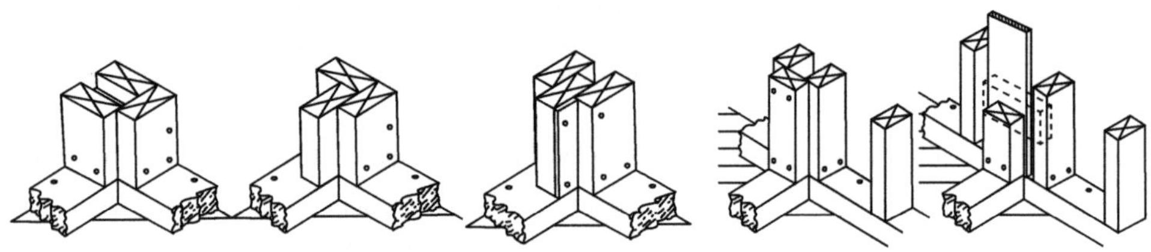

**FIGURE 6.37** Examples of corner stud assemblies.

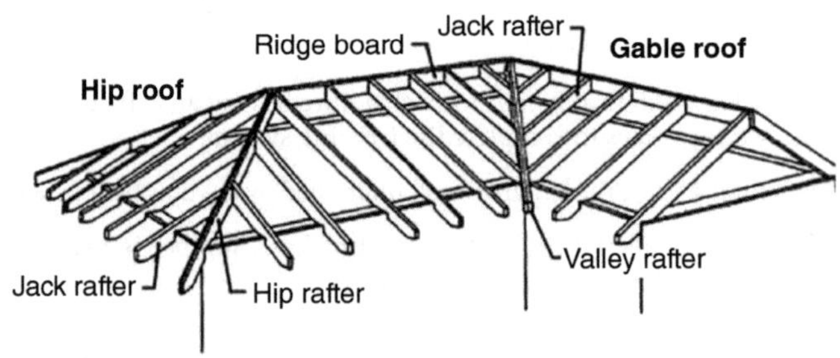


**FIGURE 6.38** Roofing system for a timber-framed structure.

**pitch**
*Slope of a roof usually expressed as a ratio.*

**Rafters** A rafter take-off is similar to the one used for joists. To obtain the number of rafters required, double the number found from taking the length of the structure divided by the rafter spacing and the one added to allow for the extra rafter required at the end of the span (Fig. 6.38). Treat a hip roof just like a gable roof as the total jack rafter requirement will equal the regular rafter requirement, but add two hip rafters. When figuring the length of a rafter it is necessary to account for the **pitch** (slope) of the roof. Separate rafters by size.

**Sheathing** Take-off the area of sheathing and convert to the number of panels based on the panel size used. Some sheathing and siding is lapped, so account for the lap when calculating the gross amount of material.

**Lumber Take-off Example:** Take-off the number of studs required for the house shown in the figure. The studs are to be spaced 16 in. on centers.

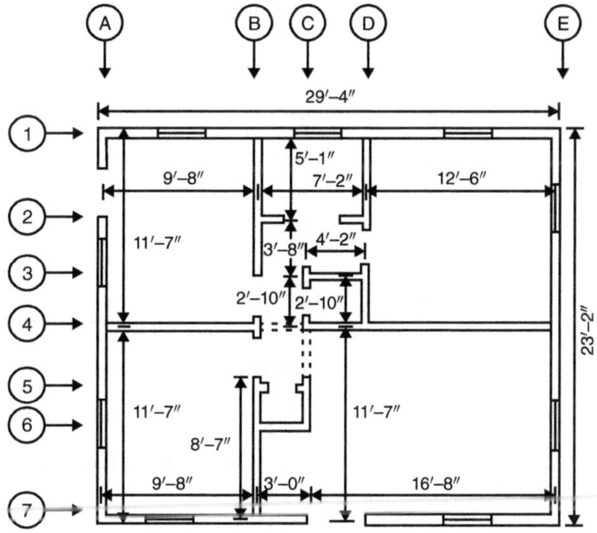


The wall lengths are:

| Column line | Length | Column line | Length |
|---|---|---|---|
| 1 | 29 ft-4 in. | A | 23 ft-2 in. |
| 2 | 7 ft-2 in. | B | 23 ft-2 in. |
| 3 | 4 ft-2 in. | C | 14 ft-5 in. [(2 ft-10 in.) + (11 ft-7 in.)] |
| 4 | 29 ft-4 in. | D | 11 ft-7 in. |
| 5 | 3 ft-0 in. | E | 23 ft-2 in. |
| 6 | 3 ft-0 in. | | |
| 7 | 29 ft-4 in. | | |
| Total | 105 ft-4 in. | | 95 ft-6 in. |

Total length of wall = (105 ft-4 in.) + (95 ft-6 in.)

= 200 ft-10 in. or 2,410 in.

$$\text{Number of studs} = \frac{2410\text{ in.}}{16\text{ in.}} \Rightarrow 151$$

Add 2 studs per wall opening or corner:

| Column line | Number of openings | Number of corners | Column line | Number of openings | Number of corners |
|---|---|---|---|---|---|
| 1 | 3 | 2 | A | 3 | 2 |
| 2 | 1 | 2 | B | 2 | 2 |
| 3 | 0 | 2 | C | 2 | 1 |
| 4 | 1 | 2 | D | 1 | 2 |
| 5 | 1 | 2 | E | 2 | 2 |
| 6 | 0 | 2 | | | |
| 7 | 3 | | | | |
| Total | 9 | 14 | | 10 | 9 |

Total number of wall openings or corners: 9 + 14 + 10 + 9 = 42

Total number of studs to add: 42 × 2 = 84

Total number of studs: 151 + 84 = 235

## THERMAL AND MOISTURE PROTECTION (CSI DIVISION 7)

Thermal and Moisture Protection specifications address all the materials and processes to seal the building against moisture, damp proofing, waterproofing, roofing; thermal; and air penetration. This includes associated insulation and accessories. Roofing can be subdivided into various types:

- Built-Up Roofing (BUR) Membrane
- Metal Roofing—Galvanized steel, aluminum, copper, tin
- Modified Bitumen Roofing—reinforce roof fabrics as bitumen carriers
- Thermoset (EPDM) Roof Membrane—a single-ply rubber material
- Thermoplastic Polyolefin (TPO) and Poly Vinyl Chloride (PVC) Roof Membranes; for oil and/or fat emission facilities such as restaurants
- Vegetative "Green" Roofing System; manages water, drainage, and improves energy efficiency

Flat or low slope commercial roofs are often used to maintain the watertight integrity of big and small box buildings, apartment buildings, and large public buildings. Such roofing systems are used on about 40% of the buildings constructed for commercial purposes.

Other common types of roof coverings include asphalt shingles, wood shingles or shakes, and built-up. Roofing comes under CSI Division 7, Thermal and Moisture Protection. Asphalt shingles, part of subsection 07 22, are manufactured with a thermally activated asphaltic sealant to bond the shingles together once applied to the roof and exposed to sufficient heat from sunlight.

A wood shingle is tapered and smooth sawn on both surfaces for a uniform appearance. A wood shake, part of subsection 07 22, is split to reveal the uneven natural grain of the wood on the face. Wood shakes are usually much thicker than wood shingles.

A built-up bituminous roof, subsection 07 51, is made of alternating layers of fiberglass or asphalt-saturated felts, standard tar-and-gravel roof. They have multiple layers (piles). The number of layers is matched to durability requirements. Built-up roofs are common in commercial construction. There are many other roof types such as sheet metal, subsection 07 61, and tile, subsection 07 32.

In preparing a detailed man-hour estimate for roofing, it is necessary to include allowances for unloading, storing materials, and hoisting at the project site. Important factors affecting the cost of a roof system are:

- Size of the roof (roof area)—The first step in estimating roof cost is to calculate the size of the roof (roof area). Obviously this is the most important factor affecting quantity of materials and labor.
- Slope of the roof—A flat or low slope roof for a commercial building will usually have a minimum **slope** of ¼ in. per foot. This is because ¼ in. is a common building code requirement.

**slope**
*The slope of a roof is defined as rise over run. It is expressed in inches of rise per foot (i.e., 4 in 12″).*

It is important to note the slope of the roof. The steeper the roof the more expensive the installation. OSHA defines a low-slope roof as a roof having a slope of less than or equal to 4 in. of vertical rise for every 12 in. horizontal length. The OSHA definition is used as a basis for implementing low-slope fall-protection measures, such as warning-line systems and safety monitors.

| Falls are the leading cause of construction deaths. |
|---|

Each worker on a steep-slope roof with unprotected sides and edges 6 ft or more above a lower level must be protected by either guardrail systems with toe boards, a safety-net system, or a personal fall arrest (PFA) system.

- Number of penetrations in the roof—Either a hole has to be cut into the membrane or the penetration must be covered.
- Type of flashing system—Flashing may be plastic, copper, asphalt, rubber, or other material. Usually it is taken off in linear feet.
- New or retrofit roof.
- Geographic location of project/weather conditions.

Many roofing and siding products are bought by the square. One square is equal to an area of 100 sq. ft.

## OPENINGS (CSI DIVISION 8)

Door and windows are included in CSI Division 8, Openings. There are separate subsections for metal doors, 08 13; wood doors, 08 14; composite doors, 08 16; and different window frame types metal, wood, and plastic. Some of the items in this division (CSI Division 8) are furnished by material suppliers and installed by the general contractor, but it is also possible subcontractors may bid these as "in-place" items. Care must be exercised to ensure the estimate covers all costs, but the estimator must be careful items or tasks are counted only once as several different installation trades may be involved in completing the work.

Hollow metal frames could be installed by the general contractor, or by subcontractors involved in masonry or metal stud and **drywall construction**. Metal windows could be supplied by one company, installed by the general contractor, and glazed, glass glazing is subsection 08 81, by a subcontractor. In the estimate, it is important to include labor cost under the proper trade.

**drywall construction**
*Interior wall covering other than plaster, usually referred to as "gypsum board" or "wallboard."*

**TABLE 6.4** Door Checklist

| Metal doors | Wood doors | Glass |
|---|---|---|
| Type of metal | Hollow or solid core | Type of glass |
| Metal gauge | Fire rating | Thickness |
| Door core material | Flush or raised panel | Type of frame |
| Fire rating | Finish | Size—measure to the next even inch, list width first and then height |
| Finish | | |

## Door and Window Schedules

Doors and windows can look alike, but still have significant differences. In the case of doors the estimator must always check the items in Table 6.4.

Information concerning these requirements will be found in the specifications or on the door schedule (Fig. 6.39). The door hardware cost will add considerably to the price of the work. The hardware requirements will be delineated in the door schedule; note the "HDW SET" column in Fig. 6.39. The estimator should, therefore, become very familiar with the schedules the A/E includes in the plans. Door schedules normally list the door openings by "mark" numbers or letters, e.g., #1, #2, #3 or A, B, C. After the mark designation there will be the door size, material, frame, fire rating, remarks, and possibly other important information. The "mark" designation refers to the particular type and has nothing to do with the total quantity required.

**Door Schedule Example:** From the door schedule shown in Fig. 6.39 identify the fire rating for door 113A. Door 101 is listed as Type A. What is a Type A door? What kind of door is door 114B?

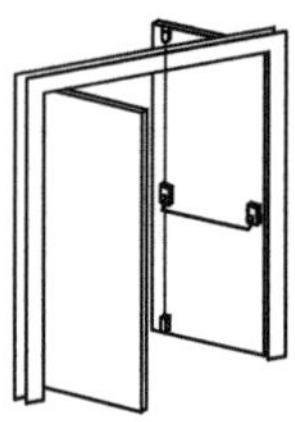

Door 113A must have a fire rating of 3/4 hr.
Door 101 is flush solid core wood.
Door 114B is an overhead door.

## Pricing Doors and Windows

The labor time required to install a door or window unit is dependent on the quality of workmanship specified, the type of tools and equipment used, and whether the work involves only a few units or multiple units in a production line-type operation. Estimating tables in reference texts can be consulted if a company does not have a historical record of labor time for this type of work. When using such tables it is important to carefully select the proper classification of work. As a case in point, one reference has the time for a carpenter, to fit and hang, a single wood door at 1 hour, but reading further there is a note about hanging a large number of doors, a production line job. If it is a production job the carpenter would only require 0.6 hr per door. Some of this time difference is setup time for hanging doors.

In the case of glass, it is important to note if the glazing is from the inside or outside of the building. If from the outside, ladders, scaffolding, temporary structures, or aerial lifts may be necessary for

DOOR SCHEDULE

| DOOR NO. | WIDTH | HEIGHT | THICK | TYPE | FRAME | FIRE LABEL (HRS) | WEATHER STRIP | HDW. SET | JAMB NO. | HEAD NO. | THRESHOLD | REMARKS |
|---|---|---|---|---|---|---|---|---|---|---|---|---|
| 101 | 914 | 2134 | 44 | A | STL-1 | — | — | 1 | J1 | H1 | — | — |
| 102 | PR. 914 | 2134 | 44 | C | STL-4 | — | X | 100 | J7, J7A | H2A/A5-3 | — | — |
| 103 | 914 | 2134 | 44 | B | ALUM. | — | X | 101 | 3/A9-3 | 2/A9-3 | — | SEE⟨W⟩A9-1 |
| 104 | 914 | 2134 | 44 | A | STL-1 | — | — | 2 | J1 | H1 | — | — |
| 105 | 914 | 2134 | 44 | C | STL-2 | 3/4 | — | 4 | J3 | H3 | — | — |
| 106 | 914 | 2134 | 44 | A | STL-1 | — | — | 2 | J1 | H1 | — | — |
| 107 | 914 | 2134 | 44 | A | STL-1 | — | — | 1 | J1 | H1 | — | — |
| 108 | 914 | 2134 | 44 | A | STL-1 | — | — | 3 | J1 | H1 | — | — |
| 109 | 914 | 2134 | 44 | C | STL-2 | — | — | 5 | J2 | H2 | — | — |
| 110 | 914 | 2134 | 44 | A | STL-1 | — | — | 1 | J1 | H1 | — | — |
| 111 | 914 | 2134 | 44 | A | STL-1 | — | — | 1 | J1 | H1 | — | — |
| 112 | 914 | 2134 | 44 | C | STL-2 | — | — | 5 | J2 | H2 | — | — |
| 113 | 914 | 2134 | 44 | C | STL-2 | — | X | 102 | J5, J5A | H5 | — | — |
| 113A | 914 | 2134 | 44 | C | STL-2 | 3/4 | — | 4 | J3 | H3 | — | — |
| 114 | 1219 | 2134 | 44 | C | STL-2 | — | X | 104 | J4 | H4 | — | — |
| 114A | PR. 914 | 2134 | 44 | C | STL-2 | — | X | 103 | J6, J6A | H6 | — | — |
| 114B | 3658 | 4267 | — | D | — | — | X | — | H3/A5-3SM | H3/A5-3SM | — | OVERHEAD DOOR |
| 114C | 3658 | 4267 | — | D | — | — | X | — | J3/J5-3 | H3/A5-3 | — | OVERHEAD DOOR |
| 116 | 914 | 2134 | 44 | C | STL-2 | — | — | 5 | Jt | H1 | — | — |
| 117 | 914 | 2134 | 44 | C | STL-3 | — | X | 105 | J4 | H1/A5-3 | — | — |
| 118 | 914 | 2134 | 44 | E | — | — | — | — | — | — | — | VAULT |
| 119 | 914 | 2134 | 44 | C | STL-2 | — | — | 1 | Jt | H1 | — | — |
| 121 | PR. 914 | 2134 | 44 | D | STL-2 | — | X | 100 | J4 | H4 | — | — |

FLUSH SOLID CORE WOOD

TYPE A

DOOR TYPE

SCALE: 1:50

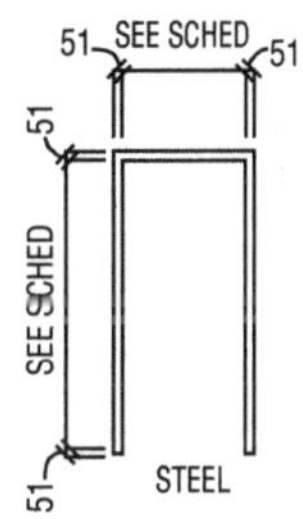


TYPE STL-

FRAME TYPE

SCALE: 1:50

**FIGURE 6.39** Door schedule for a small project. This project was dimensioned in metric units so width, height, and thickness are given in mm.

performing the work safely. As the height above the ground increases, additional labor and equipment are also needed for delivering materials.

## *FINISHES (CSI DIVISION 9)*

Before commencing an estimate of the cost to accomplish finish work (CSI Division 9), the estimator should carefully review the entire set of project plans for additional work related to the primary task and coordination with other work. Structural plans are examined to determine the type of exterior walls and other specifics bearing on finish work. The architectural plans and details provide wall (CSI 09 70), ceiling (CSI 09 50), and floor (CSI 09 60) dimension information.

## Finish Schedules

Room finish schedules should be included in the plans (Figs. 6.40, 6.41, and 6.42). These delineate by room number and name: (1) floor material, (2) base material, (3) wall material (each wall), and (4) ceiling materials and ceiling height. In addition, there will be notes with special instructions. If the plans do not contain a finish schedule the estimator should generate one.

## Finish Work Quantity Take-Off

Measure the actual surfaces as accurately as possible from the plans. Each room should be an independent take-off. The quantity survey should reflect the different labor requirements. The labor productivity when painting a flat surface is different compared to painting a small-diameter pipe, so in the first case, square footage of surface area is a good measure of work, but in the second case, linear foot measure is more appropriate. Surfaces are grouped based on type of surface, finish, application method, and expected labor production rates. Application method and labor production are directly related, so the estimator must make assumptions about how the work will be accomplished. Some general measurement rules are:

1. No object is considered to be less than one linear foot.
2. Small openings in a continuous surface are not deducted (it is just as difficult to work around the opening).

| | | FLOOR | | BASE | | WALLS | | | | | | | | CEILING | | | |
|---|---|---|---|---|---|---|---|---|---|---|---|---|---|---|---|---|---|
| | | | | | | NORTH | | EAST | | SOUTH | | WEST | | | | | |
| ROOM NO. | ROOM NAME | MAT. | COLOR | MAT. | COLOR | MAT. | COLOR | MAT. | COLOR | MAT. | COLOR | MAT. | COLOR | MAT. | COLOR | HEIGHT | REMARKS |
| A01 | SOCIAL WORK | CP1 | CPT-1 | RB | RB-1 | GWP | PT-1 | GWP | PT-1 | GWP | PT-1 | GWP | AL | ACT* | ACT-1 | 9′-0″ | *NOTE 1 |
| A02 | COUNSELOR | CP1 | CPT-1 | RB | RB-1 | GWP | PT-1 | GWP | PT-1 | GWP | PT-1 | GWP | PT-1 | ACT* | ACT-1 | 9′-0″ | *NOTE 1 |
| A03 | SOCIAL WORK | CP1 | CPT-1 | RB | RB-1 | GWP | PT-1 | GWP | PT-1 | GWP | PT-1 | AL | | ACT* | ACT-1 | 9′-0″ | *NOTE 1 |
| A04 | COUNSELOR | CP1 | CPT-1 | RB | RB-1 | GWP | PT-1 | GWP | PT-1 | GWP | PT-1 | GWP | PT-1 | ACT* | ACT-1 | 9′-0″ | *NOTE 1 |
| A05 | SOCIAL WORK | CP1 | CPT-1 | RB | RB-1 | GWP | PT-1 | GWP | PT-1 | AL | | GWP | PT-1 | ACT* | ACT-1 | 9′-0″ | *NOTE 1 |
| A06 | COUNSELOR | CP1 | CPT-1 | RB | RB-1 | GWP | PT-1 | GWP | PT-1 | GWP | PT-1 | GWP | PT-1 | ACT* | ACT-1 | 9′-0″ | *NOTE 1 |
| A07 | OBSERV. 1 | CP1 | CPT-1 | RB | RB-1 | GWP | PT-1 | GWP | PT-1 | AL | | GWP | PT-1 | ACT* | ACT-1 | 9′-0″ | *NOTE 1 |
| A09 | PLAY THERAPY | CP1 | CPT-1 | RB | RB-1 | GWP | PT-1 | GWP | PT-1 | AL | | GWP | PT-1 | ACT* | ACT-1 | 9′-0″ | *NOTE 1 |
| A10 | GROUP THER. 1 | CP1 | CPT-1 | RB | RB-1 | GWP | PT-1 | GWP | PT-1 | GWP | PT-1 | GWP | PT-2 | ACT* | ACT-1 | 9′-0″ | *NOTE 1 |
| A11 | OBSERV. 2 | CP1 | CPT-1 | RB | RB-1 | GWP | PT-1 | GWP | PT-1 | AL | | GWP | PT-1 | ACT* | ACT-1 | 9′-0″ | *NOTE 1, **NOTE 5 |
| A12 | GROUP THER. 2 | CP1 | CPT-1 | RB | RB-1 | GWP | PT-1 | GWP | PT-2 | GWP | PT-1 | GWP | PT-1 | ACT* | ACT-1 | 9′-0″ | *NOTE 1 |
| A13 | COUNSELOR | CP1 | CPT-1 | RB | RB-1 | GWP | PT-1 | GWP | PT-1 | AL | | GWP | PT-1 | ACT* | ACT-1 | 9′-0″ | *NOTE 1, **NOTE 5 |
| A14 | CLASS/CONF. ROOM | CP1 | CPT-1 | RB | RB-1 | GWP | PT-1 | GWP | PT-3 | GWP | PT-1 | GWP | PT-3 | ACT* | ACT-1 | 9′-0″ | *NOTE 1 |
| A15 | COUNSELOR | CP1 | CPT-1 | RB | RB-1 | GWP | PT-1 | GWP | PT-1 | AL | | GWP | PT-1 | ACT* | ACT-1 | 9′-0″ | *NOTE 1, **NOTE 5 |
| A99 | CORRIDOR | CP1 | CPT-1 | RB | RB-1 | GWP | PT-1 | GWP | PT-1 | GWP | PT-1 | GWP | PT-1 | ACT* | ACT-1 | 9′-0″ | |
| A99E | VESTIBULE | CP1 | CPT-1 | RB | RB-1 | GWP | PT-1 | AL | | GWP | PT-1 | AL | | ACT* | ACT-1 | 9′-0″ | *NOTE 1 |

**FIGURE 6.40** Finish schedule for a small project.

**Finish Schedule Legend**

| | | | |
|---|---|---|---|
| ACT | 2″×2″ Acoustical Ceiling Tile | EFS | Exterior Wall Insulation and Finish System |
| AL | Aluminium Storefront | EXP | Exposed |
| BRK | Exposed Brick Front | GWP | Gypsum Wall Board—Painted |
| CMU | Concrete Masonry Unit | NONE | None |
| CON | Concrete | RB | Rubber Base |
| CPT | Carpet | VCT | Vinyl Composition Tile |
| CS | Concrete Sealed | VWC | Vinyl Wall Covering |
| CT | Ceramic Tile | | |

**FIGURE 6.41** Abbreviation legend for the Fig. 6.40 finish schedule.

**Color Legend**

**Floor**

| | |
|---|---|
| CPT-1 | Mannington Carpets<br>Pattern: The Cambrian Collection.<br>"Caldera"<br>Color # Villarrica (VIRI) |
| CPT-2 | Mannington Carpets<br>Pattern: Nepenthe II<br>Border Carpet—Broadloom<br>Color # Forest Shade (FOSH) |
| CT-1 | American Olean<br>Color # A50 Bone<br>Size 2″×2″ |
| CT-2 | American Olean<br>Color # R22 Woodland |
| VCT-1 | Armstrong<br>Pattern: Imperial Texture<br>Color # 51911 Classic White |
| VCT-2 | Armstrong<br>Pattern: Imperial Texture<br>Color # 51901 Taupe |
| VCT-3 | Armstrong<br>Pattern: Imperial Texture<br>Color # 51915 Charcoal |
| CS-1 | Concrete Sealer<br>Natural Grey |

**Base**

| | |
|---|---|
| RB-1 | Roppe Rubber<br>Color # 40 Fawn<br>Size 4″ Cove Base |
| CT-3 | American Olean<br>Color # S13 Classic Bone<br>Size 4″×4″ |

**Walls**

| | |
|---|---|
| VWC-1 | Tower Wallcovering<br>Pattern: Corfu<br>Color # T2-CF-02 Bisque |
| PT-1 | Devoe Raynolds<br>Color # 2H42G Ice Beige<br>Latex Semigloss |
| PT-2 | Devoe Raynolds<br>Color # 1M501D Chocola Tint<br>Latex Semigloss |
| PT-3 | Devoe Raynolds<br>Color # 1U57B Deacon<br>Latex Semigloss |

**Ceilings**

| | |
|---|---|
| ACT-1 | Armstrong World Industries<br>Pattern: Minaboard # 770,<br>Natural fissured<br>Size 2″×2″ Lay-in-with white suspension grid |
| ACT-2 | Armstrong World Industries<br>Pattern: Scored beaded cirrus #624<br>Size 2″×2″ Tegular edge lay-in with white suspension grid |
| PT-1 | Devoe Raynolds<br>Color # 2H42G Ice Beige |

**FIGURE 6.42** Color legend for Fig. 6.40 finish schedule.

Most types of flooring are measured on a square foot basis; however, sheet flooring is measured by the square yard. **Baseboard** is measured by the linear foot. Drywall (gypsum board or wallboard) is measured in square feet, but it may be calculated by the number of sheets. With drywall it is also necessary to know the linear feet of joints to be taped. The estimator must consult the specifications concerning the type of board required, such as water resistant or low-sag. There may be a thickness requirement stated either on the drawings or in the specifications. In the case of painting, it is necessary to consult the specifications to determine the number of coats. With painting it is necessary to ascertain the amount of protection and masking required. Irregular spaces require complex geometry calculations, so it is good to measure slightly beyond the area to allow for slower production rates.

**baseboard**
*Finish board covering where the wall and floor meet.*

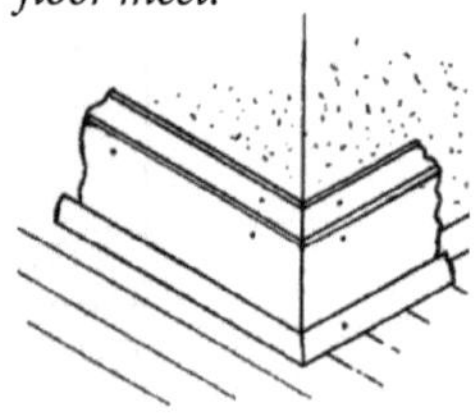

There may be unique items and these are counted (each). Features like doors, door frames, exposed metal, or other single items should be individually counted for painting.

## Estimating Finishes

Labor production rates for finish work can be found in estimating reference manuals. Some general considerations based on the type of work are presented here. An important factor to consider is climate, high levels of moisture or low temperatures can restrict finish work activities.

> There is better productivity for finishing larger areas and similar spaces. There may also be discounted pricing for larger quantities of materials.

> Finishing small or irregular spaces can be tedious.

Estimating data publications such as the *Building Construction Costs with RSMeans data* (from Gordian) provide insight into project factors affecting cost. In the case of floors these could include underlayment or subflooring, stringent leveling, staining of concrete, and for wood floors sanding and finishing requirements. A company's historical date from recent projects can be used to produce accurate estimates. When a unit cost, including labor, material, and equipment, for finish work is calculated, approximately half of the cost will be material and half installation. This holds true unless an expensive type of material is specified. Other issues to factor into the costing are:

- Are the specifications sole source or open?
- An unstated sole source specification is created when there is the need to match existing finishes or use matching products.
- On a new project there can be coordination issues with other trades.
- New concrete floors must cure properly before flooring can be installed.
- With renovation work there can be significant cutting and patching.
- Hazardous materials (asbestos and lead paint) can be issues in renovations. Request a copy of inspections from the client.
- Is there a work period requirement to complete renovation work outside the traditional day shift?

**Flooring Estimate** There are many types of flooring. In the case of tile flooring, it is necessary to estimate both the time and materials required to prepare and prime the receiving surface, and the time to actually lay the flooring. The productivity of placing tile flooring varies with the size and shape of the room or space. Carpeting is manufactured in rolls of standard width, 12 ft, 15 ft, and 13 ft 6 in. Because of this, the estimator must carefully determine how the carpet will be placed in the spaces. Irregular shapes or unusually sized rooms can result in a high waste factor.

**Painting Estimate** Factors to consider when estimating painting include (1) covering capacity, (2) thickness of coat, (3) roughness of surface, and (4) absorption of material—prime, second, or third coat. Do not overlook the cost of protection, scaffolding, and other equipment. Labor costs vary extensively with the grade of workmanship specified. The production rates and required quality of paint to cover a fixed area will vary with the coat and surface texture. Paint manufacturers may say one gallon of paint covers 250–400 square feet. But the high square footage would be for a smooth surface. If the area is rough and previously unpainted a gallon will probably only cover 250 square feet.

**Ceiling Estimate** Acoustical tile ceiling treatments are used to absorb and deaden sound in buildings. They may be either cemented directly to dry plaster, concrete, or gypsum board, or installed into a suspended grid system. Suspended grid drop ceilings are common to retail stores, office buildings, and school classrooms. The suspended metal grid system is filled with individual tiles. Tiles come factory cut and prefinished. Labor production rates can be found in estimating reference manuals, but always remember to adjust rates based on working conditions (Fig. 6.43).

## Finish Work Analysis

Two assessments can be used to validate a finish estimate. The first considers the area of the take-off.

- Is the total area commensurate with the size (square footage) of the building?
- Have all spaces, rooms, been included?
- Have floors only receiving a sealer been included?
- Have painted ceiling been included?
- Total ceilings area and total floor area should be within 10% of total building area.

**FIGURE 6.43** Workman using an aerial-lift while installing a ceiling.

The second assessment investigates cost. Do the flooring, ceiling, and wall painting unit prices fall within historical averages. An exposed ceiling is common in mechanical rooms but can be specified for the entire building. While there are no ceiling tiles, is the ceiling area painted to hide the mechanical, electrical, and plumbing services. A project owner may be seeking a higher level of green construction. The specifications will state such requirements and they can substantially increase the cost of materials.

## PLUMBING (CSI DIVISION 22) AND HEATING, VENTILATION, AND AIR CONDITIONING (CSI DIVISION 23)

Building piping, plumbing fixtures, and heating/ventilation/air-conditioning equipment are all found under CSI Divisions 22 and 23. Piping (Fig. 6.44) may be related to plumbing, fire Suppression (Division 21), or heating, ventilation, and air-conditioning systems (HVAC). To locate all requirements impacting the cost of mechanical items, it is necessary to carefully review the structural and architectural plans and specifications. Mounting requirements, wall and ceiling penetrations, and the depth of excavations should be checked. Specialty subcontractors perform most mechanical work but the general contractor may be responsible for furnishing support. It is important to have a clear understanding of which contractor is performing each work item. A general contractor can arrive at a budget estimate for plumbing installations using a percentage of total project cost. The average range is between 3 and 12%, with stores and warehouses at the lower end and apartments and motels at the high end. Commercial buildings may have a slightly higher upper range.

Plumbing is a system of piping, apparatus, and fixtures for water distribution, subsection 22 11, and waste disposal, subsection 22 13 sanitary sewerage, within a building. Pipes and fittings for plumbing systems are classified into four basic groups: (1) cast-iron soil pipe, (2) galvanized-steel/iron pipe, (3) copper tubing, and (4) plastic pipe. When preparing a detailed plumbing estimate it will be necessary to calculate the length of runs for each pipe size, as piping is priced by the linear foot. The quantity and type of joints and fittings must be determined. Fittings can add another 25–50% to material costs. Suspended pipe hangers are usually spaced every 10 ft; however, for cast iron pipe the hanger spacing is 5 ft, and for plastic pipe three hangers are needed every 10 ft. The number of each type of fixture must also be determined. Cost for many plumbing items reflects three separate costs: (1) the fixture cost, (2) the rough-in cost, and (3) final setting of the fixture.

Ductwork, subsection 23 31 (Fig. 6.45) for HVAC systems, is taken-off by linear footage for rectangular ducts. The linear number is converted to pounds as cost is usually computed based on the poundage required. It is also necessary to itemize fittings and diffusers, registers, and intakes. In addition, it is often necessary to have a crane available to set rooftop HVAC units.

**FIGURE 6.44** Piping in a commercial building.

**FIGURE 6.45** Ductwork for a building's HVAC system.

## *ELECTRICAL (CSI DIVISION 26)*

Power distribution, branch power, and branch lighting are parts of CSI Division 26, Electrical. As in the case of mechanical work, specialty subcontractors perform most electrical work, but the general contractor may be responsible for furnishing support or coordinating the electrical and mechanical subcontractors. Electrical work can include construction of electrical distribution lines, outdoor lighting, and interior electrical work. On average, electrical work is about 8–11% of office building construction cost.

Before beginning a quantity take-off the estimator should study the:

- Plot plan—for obstruction to underground cable work
- Floor plans—wall construction types (wood, block, concrete, metal stud)
- Ceiling plans—show lighting fixtures and other electrical features keyed to a legend
- Roof framing plan—connections to rooftop units

The take-off itself will involve counting fixtures, counting devices, and measuring conduit and wire. On the electrical drawings there should be a fixture schedule, an electrical panel schedule (Fig. 6.46), and a legend describing the different symbols used by the designer. The number of fixtures shown on the schedule should be checked against a room-by-room count. Smaller conduit and wire is detailed only on the plans, whereas larger runs for feeders are found on the riser diagram. The riser diagram must be superimposed on the plans to measure run distance.

Electrical work is a two-step process. Once the building is framed, but before the walls and ceiling are closed in, the electricians perform rough-in work. This work includes installing service mains, switches, panels, conduits, and outlet boxes and pulling cable through conduit and splicing in

**FIGURE 6.46** Panel boxes in a commercial building.

electrical boxes. After the walls and ceilings are finished the electrician returns to perform the electrical finish work of installing and connecting receptacles, switches, light fixtures, controls, and appliances.

## *IMPORTANT POINTS*

The drawings and their accompanying notes show and explain how the building is laid out, how it is constructed, what materials must be used, and how systems operate. Estimates are organized to serve the needs of the contracting organization, with a critical need being identification and assessment of risk.

| A detailed and accurate take-off is the basis of an accurate estimate. |
|---|

The *units* of the take-off should reflect how the work tasks are measured. Placement of concrete is in units of volume, while forming is in units (sf) of form contact area. A structural steel estimate involves a large amount of detail work. The quantities of structural members are found on the plans and weights are computed; the cost of shop fabrication (purchase cost) is by quotations, to which is added freight cost; and finally, the cost of erecting the steel on the job.

Some of the items in CSI Division 8, for example, door and windows, are furnished by material suppliers and installed by the general contractor, but it is also possible these may be bid by subcontractors as "in-place" items. Before beginning an estimate of the cost to accomplish finish work, the estimator should carefully review the entire set of project plans for (1) additional work requirements, (2) related work, and (3) coordination with other work. As in the case of mechanical work, specialty subcontractors perform most electrical work but the general contractor may be responsible for furnishing support or coordinating the electrical and mechanical subcontractors.

## DESIGN–BUILD ESTIMATING

With the proliferation of DB contracting and building information modeling (BIM), estimating is becoming a continuous process. It is relatively easy to change and refine design during DB project development and each change requires a new cost estimate. Each proposed change must be evaluated in terms of cost and schedule effects. As the project design continues to be refined still another estimate iteration is necessary. As the refinement process continues in BIM with automated extraction of project quantities, the estimator is challenged with an exponentially larger mass of data. As a result, estimators must find ways to organize, manage, and control the mountain of project data. While there can be data management issues, there is also the opportunity to automatically validate estimated cost against historical averages and to automatically perform a first validation of the estimate.

Progressive companies are providing estimators the software to work with 3D models. With such software the project plans and estimating systems are integrated allowing estimators to drag and drop, automatically count, and move information from the drawing to the estimate. The visualization model provides the estimator information concerning what has been extracted (taken-off) from the plans and incorporated into the estimate. Estimating is becoming a visual process. If the proposal is successful, the estimator passes visual models to the execution team

BIM support for project estimating is rapidly becoming the norm. Still there are issues, designers rarely provide complete BIM models of a project. So the need remains to have and use 2D drawing and the specifications for resolution of questions concerning the designer's intent. Still BIM is a productivity enhancer for estimating and particularly important in the alternate contracting environment. There are tools for combining 2D and 3D input into one set of information. The array of new software will allow smart contractors to efficiently create more accurate project estimates.

## RESOURCES

*Formwork for Concrete* (SP-004), 8th ed. (2014). American Concrete Institute, Farmington Hills, MI.

*Concrete Masonry Handbook*, 6th ed. (2018). Portland Cement Association, Skokie, IL.

*PDCA Cost & Estimating Guide*, Vol. II: Rates & Tables, 4th ed. (2013). Painting and Decorating Contractors of America, Maryland Heights, MO.

*Manual of Standard Practice*, 29th ed. (2018). Concrete Reinforcing Steel Institute (CRSI), Schaumburg, IL.

*Standard Estimating Practice*, 10th ed. (2017). American Society of Professional Estimators, Nashville, TN.

# CHAPTER 7
# SCHEDULING PROJECTS

*The critical path method (CPM) is a planning and control technique used to provide an accurate, timely, and easily understood graphic of project task sequencing. It can also be used to allocate resources over time. One of the most important features of the CPM is the logic diagram. This diagram portrays the relationships between project activities. The critical path through the network is the longest time duration path through the diagram and defines the minimum project duration. Any activity not on the critical path will contain float. Float is additional time available to complete an activity beyond the activity's assigned work duration. To communicate schedule information in a meaningful manner, the CPM network needs to be expressed as a calendar schedule with points in time, such as an activity start expressed as a specific date.*

## PROJECT PLANNING

Planning is the process of considering task sequencing alternatives and appropriate ordering of the project work. Planning creates an orderly sequence of work tasks, explains the principles to be followed in carrying forth the plan, and defines the duration of a project. It serves the manager by identifying what needs to be done (tasks and actions), the sequence of the tasks, how long each task should take to be completed, and the person or job title of the person responsible for a specific task or action.

The goal of planning is to minimize resource expenditures while satisfactorily completing a given task. Planning aims at producing an efficient use of equipment, materials, and labor, together with ensuring a coordinated effort. Effective planning requires continual monitoring of task accomplishment—progress. A comparison of actual progress to scheduled progress helps a manager identify problems early and permits development of revised plans to maintain the proper course toward the objective.

Scheduling and planning tools to keep up with the advances in technology surrounding us daily are continuing to develop quickly. Project teams have gone from marking up a copy of a schedule in a PDF format and making schedule updates, to using an app on their phones to communicate with the schedule in real time. Field teams can now actualize activities and update activity completion percentage without returning to the office, opening a computer, and trying to recall what happened days ago. This is just one example of how technology is helping field and management teams create a fluid real-time schedule document, not a piece of wallpaper gracing the wall of a jobsite trailer.

## HISTORICAL PERSPECTIVE

Construction managers employ scheduling practices, principles, and techniques derived from earlier concepts and experiences. Frederick Taylor (1856–1915) conducted studies aimed at improving productivity at the Bethlehem Steel Company in Pittsburgh and in 1911 he published *Principles of Scientific Management*. Today we utilize Taylor's concept of breaking work into elementary parts: the work breakdown structure used for scheduling and cost control. Henry L. Gantt, an associate of Taylor's, developed the **Gantt chart**, or **bar chart**. This is a visual tool for scheduling and planning work tasks. Gantt did his work in response to the need for better industrial logistical planning during World War I. The Gantt chart is used for scheduling multiple overlapping tasks.

**bar chart**
*A graphical representation of planned activities (vertical axis), the estimated activity durations (horizontal axis), and the planned sequence of activities.*

In the 1950s two major scheduling methods were developed. The first was the program evaluation and review technique (PERT) by Willard Frazar. It came out of the Navy's need in managing the Polaris Missile Program. The second was the critical path method (CPM) by Morgan R. Walker of E. I. du Pont de Nemours and James E. Kelly of Rand Remington Univac.

## THE PLANNING PROCESS

When planning construction work, the manager must carefully assess all factors impacting task performance. The goal is to (1) safely deliver the project, (2) at the desired quality level, (3) on budget, and (4) by the required date—safety, quality, cost, and schedule.

### Assumptions Based on Facts

Many of the factors affecting task accomplishment relate to the environment within which the work will be undertaken. Climate can significantly impact time and cost. Always consider the consequences of weather on operations. A hot environment can reduce productivity while cold weather can add operation costs for providing heating. In such situations, the evaluation of this impactor alone can be more important than all other factors. Contractors try to have buildings "closed-in" before winter weather brings cold temperatures and moisture. When this is not possible, special measures may be necessary to keep the work progressing (Fig. 7.1).

**FIGURE 7.1** Building structural frame covered with blankets with the interior heated so construction can proceed during winter weather.

### Select the Course of Action

Various courses of action are compared in terms of personnel, material, equipment, and time. This can be difficult because planners must consider multiple sources of uncertainty, and in addition, the plan must address all aspects of the project including administration and logistics. It should provide a roadmap for successful completion of the work. Important aspects requiring attention are:

- Evaluating availability of labor, equipment, and materials
- Executing mobilization, including moving onto the job site, office location, and laydown areas
- Securing appropriate site access for delivering supplies and equipment
- Planning for inclement weather
- Providing site drainage

## *ACTIVITIES/TASKS*

A common technique used to understand and organize a complex and multidimensional undertaking is to break the project into smaller pieces (divide into subparts). In construction, this technique is applied in both planning and estimating. To create a construction plan, the first step is to identify all of the work tasks necessary to accomplish the project. In the terminology of scheduling, tasks are referred to as activities. Each activity is a discrete task. It is also necessary to think about critical success factors and key milestone dates for each activity.

The project manager must carefully study all of the contract documents, construct the project mentally, and break it down into its component activities. At the strategic level, the activities to build the steel frame building, in Fig. 7.2, are:

- Foundation
- Floor slab
- Steel frame
- Roofing
- Exterior walls
- Interior finish work

Activities should be stated consistent with the level of management control. When developing an activity list, always consider the purpose of the schedule and who will use it. The number and detail of the listed activities will vary from job to job and will depend on the intended level of control.

Planning involves the scheduling of activities. But management also includes monitoring and controlling the execution of the work; therefore, the activities provide the building blocks for focusing management attention.

> Activity descriptions should be concise and unambiguous. The description communicates the scope and location of the activity. A verb should be used in the description of production-type activities.

The Associated General Contractors of America (AGC) has defined five activity characteristics:

1. *Activities consume time.* To be classified as an activity, a project task must consume time.
2. *Activities usually consume physical resources.* A project activity usually consumes labor, material, or equipment resources. However, a few construction activities do not consume physical resources, such as curing time for concrete.

**FIGURE 7.2** An activity is a time-consuming, definable task, effecting steel frame erection.

3. *Activities have definable start and finish points.* An activity represents a definable scope of work. A crew begins work—a starting point in time—and when all of the work is completed, there is an ending point in time.
4. *Activities are assignable.* Each activity is to be accomplished by a particular member or crew of the construction team. This characteristic of an activity facilitates management of the work.
5. *Activities are measurable.* The progress toward completion of an activity's scope of work must be measurable.

An activity is, therefore, a task, function, or decision with *a time duration.*

## Activity Duration

One of the most important steps in planning a project is estimating the time required to complete each activity. The duration of an activity is a function of the magnitude of work to be accomplished and the work production rate. Task production rates are based on the planned composition of labor and equipment used to perform the activity. Carelessly made estimates of production rates may cause uneconomical use of personnel, materials, equipment, and time.

When a contractor prepares a project estimate for bidding, the estimator calculates the quantities of material to be put in place and assumes a production rate to use in arriving at the cost for each work task. In determining production rates, the estimator considers construction methods and techniques. The selected method establishes the task duration. The bid preparation estimate serves as the base for calculating an activity duration.

**Activity Duration Example:** In estimating a three-story building project, the plans call for 480 fluorescent light fixtures on each floor. In developing the bid estimate, the estimator used a production rate of two fixtures per man-hour. This production rate was based on a crew size of five electricians. The normal workday will be 8 hr. What is the duration (in days) for the activity: Install fluorescent-light fixtures on the first-floor?

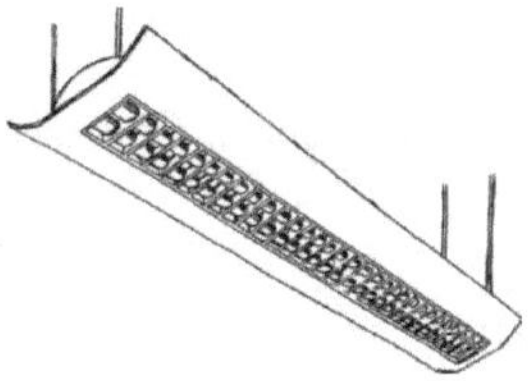

$$\text{Production rate } \frac{2 \text{ fixtures}}{1 \text{ man-hour}} \times 5 \text{ electricians} = 10 \text{ fixtures per hour}$$

$$\text{Per-floor activity duration hours } \frac{480 \text{ fixtures}}{\text{floor}} \times \frac{1 \text{ h}}{10 \text{ fixtures}} = 48 \text{ hr per floor}$$

$$\text{Per-floor activity duration days } \frac{48 \text{ hr per floor}}{8 \text{ hr per day}} = 6 \text{ days per floor}$$

The time unit used to express an activity's duration should match the objective of the scheduling effort. In the early stages of project development, a milestone schedule based on activity durations expressed in months might be appropriate. During the construction stage of a project, it is more common to use weeks, days, or hours as the time unit for activities. In the case of a long-duration project, the use of a time unit expressed in weeks might be appropriate. For the majority of projects, the activity time unit is days. This use of days usually matches the terms in the construction contract, as it is common practice to express contractual time in either work or calendar days. In the case of a process plant shutdown or short-duration complex projects, the use of an hour time unit is often necessary.

> All schedule activities must have durations expressed in the same unit of time.

## BAR CHARTS

In 1917, Henry Gantt conceived a chart scheduling method. A Gantt chart presents planned activities as stacked horizontal bands against a background of dates (along the horizontal axis). To visualize activity relationships and durations, the Gantt or bar chart is a commonly used project planning and control tool. Almost all of today's commercially available scheduling software presents information in bar chart formats (Fig. 7.3). It is a simple and concise graphical picture for managing a project. It is also easy with a bar chart to compare planned production against actual production.

The bar chart is widely used as a construction scheduling tool because of its simplicity, ease of preparation, and graphical format. Normally the activities are listed in chronological order according to their start date. Discontinuous bars are sometimes used on hand-drawn bar charts to represent interruptions to activity work. The better practice is to have individual activities with their own bars.

**Steel frame building**

| **Weeks** | 1 | 2 | 3 | 4 | 5 | 6 | 7 | 8 | 9 | 10 | 11 | 12 | 13 | 14 |
|---|---|---|---|---|---|---|---|---|---|---|---|---|---|---|
| Foundation | ■ | ■ | ■ | | | | | | | | | | | |
| Floor slab | | | | ■ | ■ | | | | | | | | | |
| Steel frame | | | | | | ■ | ■ | | | | | | | |
| Roofing | | | | | | | | ■ | ■ | | | | | |
| Exterior walls | | | | | | | | | | ■ | ■ | | | |
| Interior finish work | | | | | | | | | | | | ■ | ■ | ■ |

**FIGURE 7.3** Bar chart of the Steel Building activities in Fig. 7.2.

Field personnel can easily understand the information presented on the bar chart. This is an important advantage. In addition, it is an especially useful tool for preliminary planning and scheduling. If cost, equipment, or personnel requirements are superimposed on the activities, the total resources required at any time can be computed.

Disadvantages of a bar chart are:

- It does not clearly show the detailed sequence of the activities.
- It does not show which activities are *critical* to the successful and timely completion of the project.

## CRITICAL PATH METHOD

The **critical path method** focuses management's attention on the relationships between the critical sequences of activities. It is an activity relationship representation of the project. The evaluation of critical tasks, those controlling project duration, allows for the determination of project duration. Because of the size and complexity of major construction projects (Fig. 7.4), CPM scheduling is usually applied using computer software to make the calculations.

**critical path method**
*A scheduling method where activities are arranged based on interrelationships and the longest time path through the network (the critical path) can be determined.*

> A CPM schedule calls attention to the activities that must be completed before other activities can begin.

**FIGURE 7.4** Major building project in downtown Los Angeles.

The critical path method overcomes the disadvantages of a bar chart and provides an accurate, timely, and easily understood graphic of the project. One of the most important features of the CPM is the logic diagram. The logic activity network or logic diagram graphically portrays the relationships between project activities. With this additional information, it is easier to plan, schedule, and control the project.

CPM calculations define a time window within which an activity can be performed without delaying the project.

Using the CPM to develop a schedule requires detailed investigation into all identifiable project tasks. This means the manager must visualize the project from start to finish and must estimate time and resource requirements for each task. It is good practice to also obtain task input from superintendents and subcontractors. The advantages of using a CPM logic network include:

- Reduces the risk of overlooking essential tasks and provides a map for long-range planning and coordination of the project
- Provides a clear graphic of the interrelationships between project activities
- Focuses management's attention by identifying the critical activities controlling project duration
- Generates information about the project so the manager can make rational and timely decisions if complications develop during the progression of the work
- Enables the project manager to easily determine what resources are needed to accomplish the project and when these resources need to be made available
- Allows the project manager to quickly determine what additional resources will be needed if the project must be completed earlier than originally planned
- Provides completed project feedback for improving techniques on future projects

A CPM is not a substitute for appropriate construction knowledge and planning. It does not make decisions for the project manager, nor can it contribute anything tangible to the actual construction.

**The great danger:** Anyone with operating system proficiency can, with scheduling software, construct what appears to be a reasonable schedule.

It is easy to enter various project activities into a scheduling program and create relationships so a schedule of activities seems to proceed in a sensible manner. Unfortunately, there is no way to tell by looking at the plotted schedule whether this is a true picture of physical construction relationships or simply a fine-looking diagram. Very often, there is no internal schedule logic and, therefore, the schedule is essentially useless from the standpoint of monitoring and analyzing progress. The creation of such schedules is usually the result of a lack of construction knowledge or of inexperience.

Schedules are produced to support management of tasks and the information contained in the schedule must be conveyed to many project participants (e.g., subcontractors, suppliers, and the leaders of trade crews). In the case of a large complex project, this is a huge information dump. For a participant receiving only part of the information, it is virtually impossible to check the supplied information. An example of this would be to only supply an electrical contractor with the electrical activities of a schedule; in such a case it would not be possible to determine how other trade activities impact the electrical work.

## Activity Logic Network

The activity logic network provides a graphical picture of the sequence of construction tasks. Before the diagram can be developed, the project must first be constructed mentally to determine activity relationships. The manager does this by asking the following questions for each activity on the activity list:

- Can this activity be started at the beginning of the project? There are no predecessor activities. These are called **start activities**, and there may be several. (Start)
- Which activities must be finished before this activity begins? These are called **predecessor activities**. (Precedence)
- Which activities may either start or finish at the same time as this activity? (Concurrence)
- Which activities cannot begin until this activity is finished? These are called **successor activities**. (Succession)

**start activities**
*The first activities in a network or a start date constraint shown by an artificial activity.*

**predecessor activities**
*Activities that must be completed before another activity can start.*

**successor activities**
*Activities that cannot start until another activity is completed.*

In the diagram, the sequencing of activities is clear; in some cases this can be dictated by a spatial sequence—foundation completed before the floor slab can be placed (Fig. 7.2). Conversely, in other cases, the sequencing may be dictated by structural reasons and these are not always clear by *only* referring to the project plans. In many cases, the scheduler and project manager must carefully study other project documents, such as the shop drawings submitted by fabricators or subcontractors. Sometimes when schedules are developed, the sequencing of activities is based on an assumed equipment constraint, such as using only one crane. Equipment and crew resource constraints are discussed later in the scheduling activities section of this chapter.

One way to determine these relationships is to list all activities in a columnar format with a second column to the right of the activities list titled "Preceded Immediately By" (PIB). For each activity, list all other *necessary immediately* preceding activity numbers (identifier codes) in the second column. If the activity can begin at the beginning of the entire project, write "None."

A network has only *one start* and *one finish* activity for the CPM algorithm to work. Therefore, in the case of a network with multiple activities having no predecessor, a dummy "start" activity can be inserted at the beginning of the network. Similarly, a dummy "finish" activity is placed at the end of a network and all activities with no successor activity are tied to this final activity.

**Preceded Immediately By Example:** Our schedule to build a concrete slab on grade has seven activities: excavate, build forms, procure reinforcing steel, fine grade, set forms, place reinforcing steel, and place and finish concrete. Use the preceded immediately by (PIB) method to organize these activities for a CPM schedule.

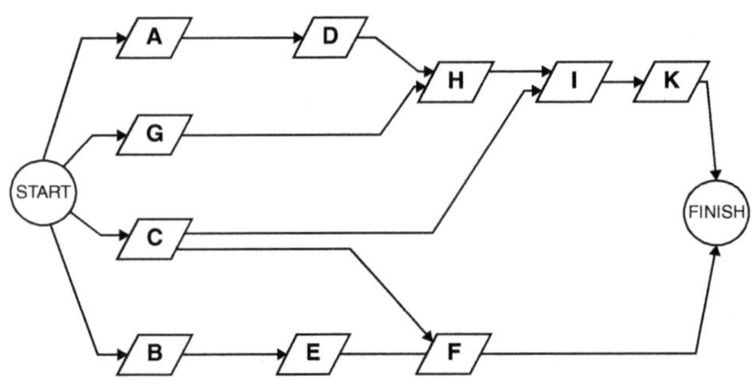


| Activity | PIB |
|---|---|
| A. Excavate | None |
| B. Build forms | None |
| C. Procure reinforcing steel | None |
| D. Fine grade | A |
| E. Set forms | A and B |
| F. Place reinforcing steel | C, D, and E |
| G. Place and finish concrete | F |

There are two CPM logic-diagramming formats: (1) **activity-on-the-arrow**, and (2) **activity-on-the-node**. With the activity-on-the-arrow (AOA) format (Fig 7.5), the arrows of the diagram represent the activities of the logic diagram. Unless the diagram is time scaled, the length of the arrows has no relationship to activity time durations. With activity-on-the node, the nodes serve as the activities. When using the AOA diagramming format, it is often necessary to use **dummy activities** (arrows) to denote dependencies. Dummy activities consume no resources. AOA diagramming is sometimes referred to as I–J diagramming, with the I and J referencing to the beginning and ending nodes, respectively, of an activity arrow. By the logic of the diagram activities (arrows) leaving a node cannot begin until all of the activities heading into the node have been completed.

**activity-on-the-arrow** *A diagram where an arrow represents the activities of the schedule.*

**activity-on-the-node** *A diagram where the nodes represent the activities of the schedule.*

**dummy activities** *A pseudo activity with duration of zero.*

Activity-on-the-arrow logic diagramming used to be the most popular CPM method. Today most CPM users employ the activity-on-the-node (AON) format, where each node represents an activity. This is sometimes referred to as "precedence diagramming." The two basic logic symbols on the precedence diagram are the *node* and precedence *arrow*, but what they represent is reversed from the AOA case.

**Nodes** A node is simply a box used to represent an activity. Each activity on the activities list is represented by a node on the logic diagram. The node is of a standard shape and format and contains activity information. Notations on each node indicate the activity's alphanumeric identifier (ID code) and duration. Sometimes additional information, such as early and late start times, early and late finish times, and required resources is also included (Fig. 7.6). Each activity in a network should have a unique identifier.

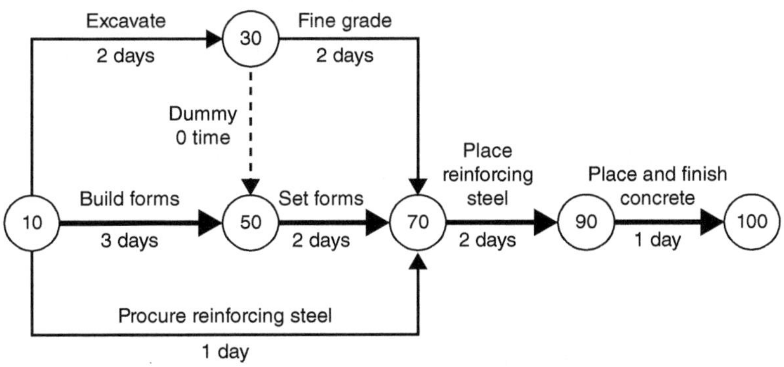


**FIGURE 7.5** Example of an activity-on-the-arrow network diagram.

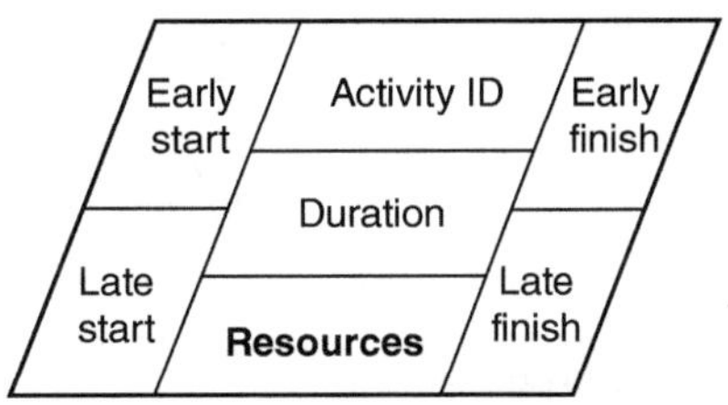


**FIGURE 7.6** Example of activity node information, the specific arrangement will depend on the scheduling software used.

Codes are essential for sorting and categorizing activities when using computer-based scheduling. Activities are identified by user-defined codes; these aid in describing activities. Each activity can have one or more activity codes. Common code categories are:

- Responsibility
- Area of work
- Phase of project
- Subphase of the project
- Type of work
- Specification section
- Bid item

When creating activity identifiers, the practice is to use a numbering system. The numbers should increase in magnitude when moving from project start to finish. Reading project status reports generated from the schedule is much easier when directional numbering is used. It is also the practice when first numbering activities to leave gaps in the numbering, so initially multiples of 5 might be used initially: 5, 10, 15, etc. This permits easy inclusion of additional activities as planning progresses or when change orders are issued.

**Precedence Arrow** The precedence arrows show the order sequence and relationship between activities (such as which activities must precede and follow another activity) when using AON scheduling. The configuration of the diagram's nodes and arrows is the result of the PIB list (or the answers to the five previous questions concerning each activity). Following the diagram logic, an activity cannot begin until all preceding activities are complete. Table 7.1 presents some common logic relationships.

The logic network is constructed without regard to how long an activity will last or whether all required resources are available. It only displays the relationships among activities, provides project understanding, and improves communications. Once the network has been developed, the project manager assigns activity duration and resource requirements to each activity.

**TABLE 7.1** Activity Logic Relationships

| Logic relationships | Diagram example |
|---|---|
| SEQUENTIAL LOGIC<br>Activity 20 cannot start until activity 10 is completed. | 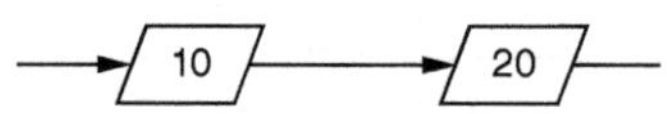 |
| CONCURRENT LOGIC<br>Activity 5 and 10 can proceed concurrently.<br>**Multiple Predecessor Logic**<br>Activity 20 cannot start until both activities 5 and 10 are completed. | 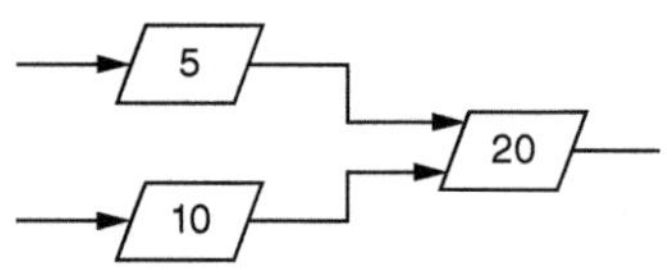 |
| Activity 30 and 40 cannot start until both activities 10 and 20 are completed. | 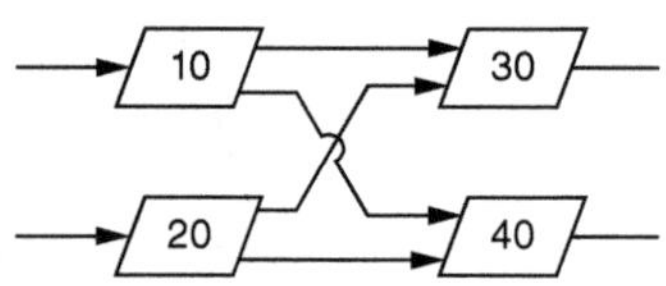 |
| **Multiple Successor Logic**<br>Activity 20 must be completed before either 30 or 40 can start, 30 can start only after 10 and 20 are completed, 40 can start immediately after 20 is completed. | 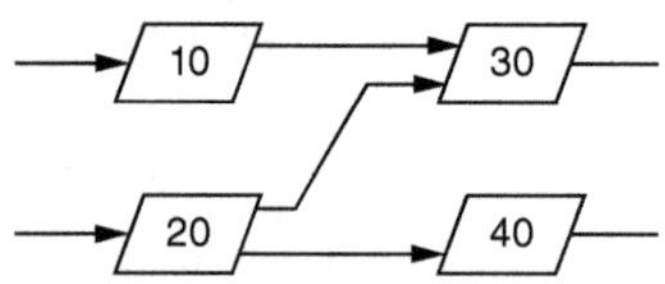 |

## Schedule Calculations

The next step in the CPM process is to calculate the earliest and latest times at which the activities can transpire without violating the network logic or increasing the project's overall duration. This provides the manager with a time frame for each activity. Within this time frame, the activity must be completed, or other activities will be delayed. When an activity is delayed the ripple effects may cause project delay. Once the activity times are calculated, the critical path activities can be identified. By carefully managing the critical activities, it is possible to reduce the time to complete the project. An activity cannot begin until all activities previous to it (arrows leading to it in the logic node) are completed.

## Forward Pass

A forward computational sequence through the logic network will yield:

1. The earliest time each activity in the network can start and finish.
2. The minimum overall duration of the project.

In performing the **forward pass** calculations, all successor activities are started as early as possible. The calculations maintain the rigor of the network logic; therefore, the earliest an activity can start is when all predecessors to the activity are completed. Furthermore, each activity is postulated to finish as soon as possible. This finish requirement yields Eq. (7.1) (Fig. 7.7).

**forward pass**
*The calculations to determine the earliest start and finish time of the activities and the minimum project duration.*

$$\text{Early finish}_n\ (\text{EF}) = \text{Early start}_n\ (\text{ES}) + \text{Duration}_n \qquad (7.1)$$

where $n$ denotes the $n$th activity.

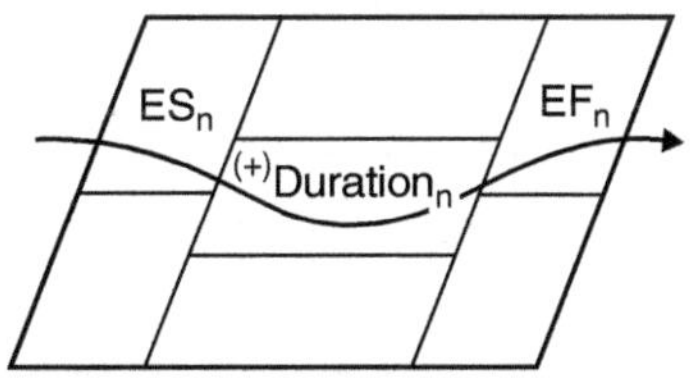


**FIGURE 7.7** Calculation of an activity's early finish.

**Early Start/Early Finish** Taking into account the network logic, the early start time (ES) of an activity is the earliest point in time when the activity may start. In this text, the early start time is positioned in the upper left corner of the activity node box. Some scheduling software may position or display this information in another location.

The starting point for performing a forward pass is the first activity in the network. In the case of the first activity of a project, the earliest time it may start is zero (can be thought of as the end of day 0 or the beginning of day 1).

> The event-time numbers calculated for an activity represent the end of the time period. Thus, a start or finish time of day 5 would mean the end of the fifth day (or the beginning of the sixth day).

If only one precedence arrow leads into an activity, then the activity's early start time is the same as the previous activity's (at the tail of the arrow) early finish time. To determine an activity's early start time when more than one arrow leads into its node, select the *largest* early finish time of all activities at the tail of the arrows [Eq. (7.2)]. Logically, an activity cannot begin until *all* preceding activities are complete (Fig. 7.8).

$$\text{Early start}_n = \text{Maximum early finish of all predecessor activities} \qquad (7.2)$$

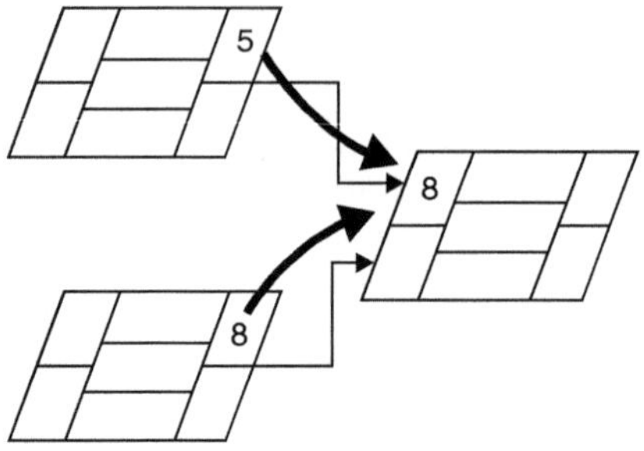


**FIGURE 7.8** Calculation of early start, use largest preceding early finish.

Add the duration of each activity (center of the node in our nomenclature) to the early start time to compute the early finish time (EF) [Eq. (7.1)] and position the EF in the upper right corner of the activity node. The early finish time is the earliest time the activity may finish.

Using this systematic process, work through the entire logic diagram computing all early start and early finish times from the beginning activities to the completion (finish) of the project. This computational sequence through the logic diagram completes the forward pass. The overall duration for the project will be the EF of the last activity in the network. For the network shown in Fig. 7.9, the forward pass yields a project duration of 22 days, as determined by the sequence of construction and the time duration assigned to each activity.

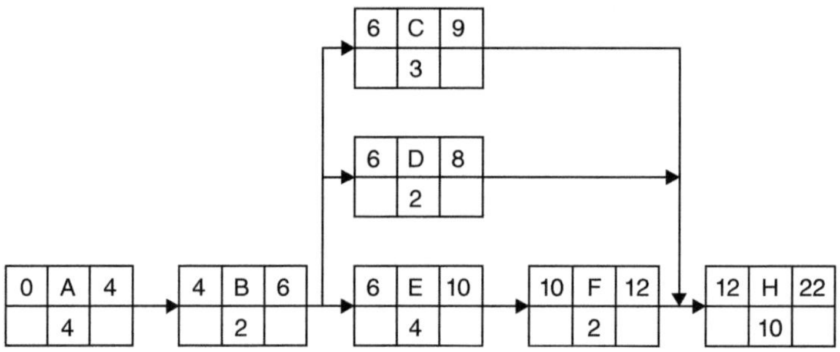


**FIGURE 7.9** Forward pass calculations.

## Backward Pass

A backward computational sequence through the logic network will produce the latest point in time for the start and finish of each activity with the provision of still maintaining the minimum overall project duration. The **backward pass** calculation starts with the last activity in the network. This last activity is assigned a late finish time equal to its early finish time as calculated by the forward pass.

**backward pass**
*A calculation to determine the late start and late finish times of the project activities under the condition of maintaining the project's minimum duration.*

**Late Finish/Late Start** To calculate late finish time and late start time of an activity, follow the precedence arrows backward through the logic diagram (right to left). Taking into account the network logic, the late finish time (LF) of an activity is the latest time; it may finish without delaying the entire project.

For this text, the late finish time is positioned in the lower right corner of the activity node box. Since the last activity in the Fig. 7.9 network (activity H) had an early finish of 22 days, the latest time for this activity to finish and not lengthen the project duration is the end of the 22nd day. The number 22 should, therefore, be assigned as the LF for this network and placed in the lower right corner of the H node.

The preceding activity's late finish time (at the tail of the precedent arrow) is the succeeding activity's late start time (at the head of the precedent arrow). To determine an activity's late finish time when more than one arrow tail leads away from its node, choose the *smallest* late start time of all activities at the arrows' heads (Fig. 7.10). Logically, an activity must finish before *all* follow-on activities may begin.

To compute the late start time (LS) of an activity use Eq. (7.3), which states: subtract the activity's duration (center of the node) from its late finish time (Fig. 7.11). Position the LS in the lower left corner of the activity node. The late start time is the latest time the activity may start without delaying the entire project.

$$\text{Late start}_n\ (\text{LS}) = \text{Late finish}_n\ (\text{LF}) - \text{Duration}_n \qquad (7.3)$$

Using this backward systematic process, work through the entire logic diagram (against the arrows), computing all late finish and late start times. This computational movement back through the logic diagram is known as the backward pass. The late start times of the first activity must be zero. For the network shown in Fig. 7.12, the late start time of activity A is zero.

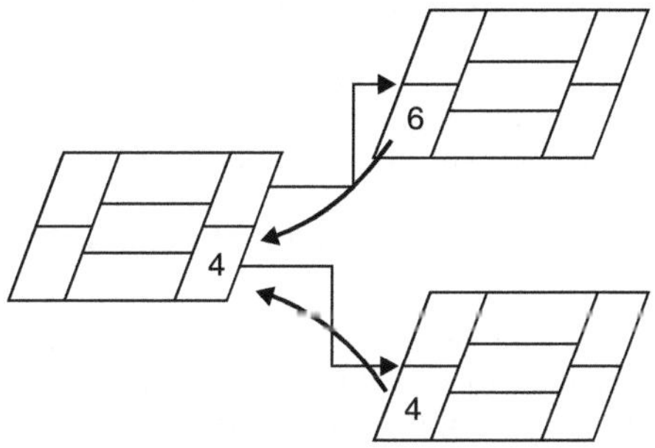


**FIGURE 7.10** Calculation of an activity's late finish, use the smallest late start time of all succeeding activities.

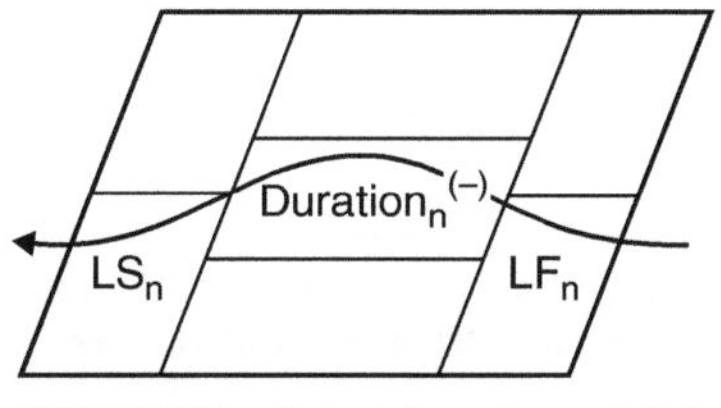


**FIGURE 7.11** Calculation of an activity's late start.

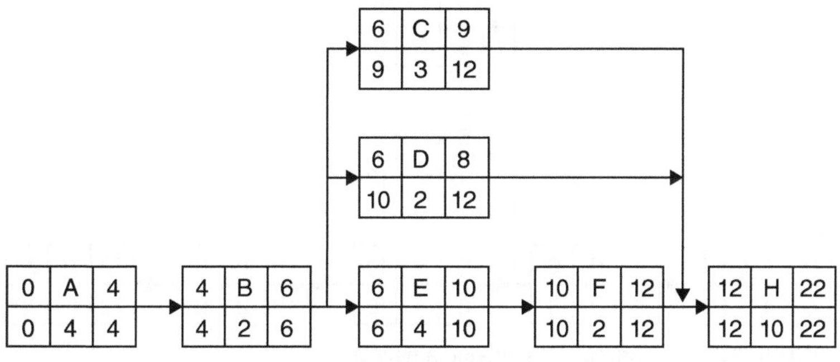


**FIGURE 7.12** Backward pass calculations.

## Critical Path and Critical Activities

The critical path of a schedule network is the longest time duration path through the network. It establishes the minimum overall project time duration. All activities on the critical path are by definition critical. A critical activity can be determined from the logic network by applying either of these rules:

1. The early start and late start times for a particular activity are the same.
2. The early finish and late finish times for a particular activity are the same.

For the network shown in Fig. 7.12, activities A, B, E, F, and H meet the listed rules, thus making them critical activities. A critical activity, if delayed by any amount of time, will delay the entire project's completion by the same amount of time. Critical activities are linked together forming a path from the start activity to the finish activity called a critical path. There can be more than one critical path through the network and the critical path may branch out or come back together at any point. All critical paths must be continuous; any critical path not starting at the start node and ending at the finish node indicates a logic mistake.

Critical paths are indicated on the logic diagram by methods such as double lines, bold lines, or color highlighted lines (Fig. 7.13). Any activity node not on the critical path will contain float. **Float** is additional time available to complete an activity beyond the activity's work duration, such as having 6 days available to do 4 days' worth of work. Activities on the critical path have no float.

**float**
*Activity scheduling flexibility range while still not delaying a following activity or the project.*

**total float**
*The duration an activity can be delayed without delaying project completion.*

**Total float** (**TF**) is the amount of time an activity can be delayed without delaying project completion. Total float is calculated assuming all preceding activities are finished as early as possible and all succeeding activities are started as late as possible. Total float for an activity can be determined by either Eq. (7.4) or (7.5).

$$\text{Total float activity}_n = \text{Late start activity}_n - \text{Early start activity}_n \tag{7.4}$$

$$\text{Total float activity}_n = \text{Late finish activity}_n - \text{Early finish activity}_n \tag{7.5}$$

where $n$ denotes the $n$th activity.

Both equations yield the same result.

> Total float is calculated and reported individually for each activity in a network. However, total float is a ***network path attribute*** and not associated with any one specific activity along the path.

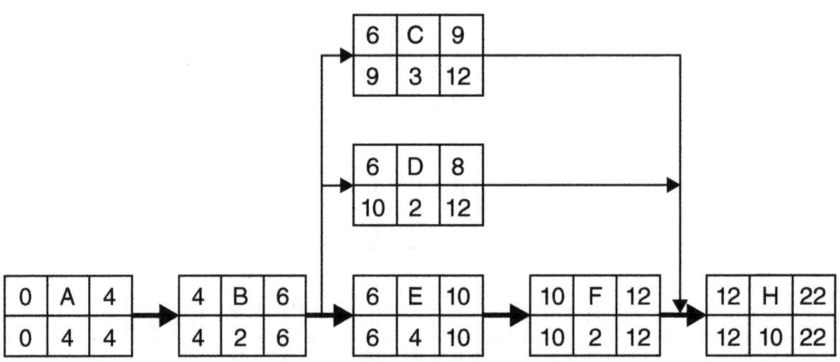


**FIGURE 7.13** The critical path through a network.

**Free float** (**FF**) is the time duration an activity can be delayed without delaying the project's completion and *without delaying the start of any succeeding activity.* Free float is the property of an activity, not a network path. Free float for an activity is determined by

$$\text{Free float}_n = \begin{array}{c}\text{Minimum early start of all}\\ \text{successor activities}\end{array} - \text{Early finish}_n \tag{7.6}$$

where $n$ denotes the $n$th activity.

Activity C of the network shown in Fig. 7.13 is not on the critical path. Therefore, the activity has float. Free float would be 3 (Minimum early start of successor, 12 – EF of 9).

An activity's **interfering float** (**IF**) will delay the start of one or more following noncritical activities but will not delay the project's completion. Interfering float for an activity is determined by

$$\text{Interfering float}_n = \text{Late finish}_n - \begin{array}{c}\text{Smallest early start of}\\ \text{succeeding activity (ies)}\end{array} \tag{7.7}$$

where $n$ denotes the $n$th activity.

If by the network logic, the activity in question has more than one successor, use the smallest successor ES in Eq. (7.7). The aggregate of free float and interfering float equals the total float, as seen in Eq. (7.8).

**interfering float**
*The time available to delay an activity without delaying the project's completion. Using interfering float causes one or more following noncritical activities to be delayed.*

$$\text{Total float}_n = \text{Free float}_n + \text{Interfering float}_n \tag{7.8}$$

**Float Calculation Example:** The project manager is reviewing the group of activities on the Fig. 7.14 project schedule (activities RPB, TIB, RPW, FIB, and CWW). The time durations are in days.

a. What is the total float for each of the activities shown?

Using Eq. (7.4), (Total float$_n$ = Late start$_n$ – Early start$_n$), the total float for each activity is:

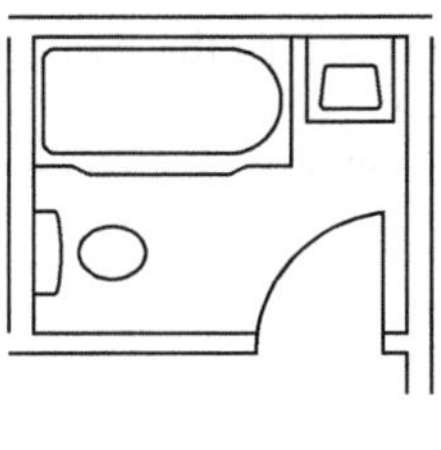

| Activity | Activity ID | Total float |
|---|---|---|
| Rough plumbing bathrooms | RPB | 3 – 0 = 3 |
| Tile bathrooms | TIB | 25 – 13 = 12 |
| Rough plumbing washrooms | RPW | 16 – 13 = 3 |
| Finish bathrooms | FIB | 32 – 20 = 12 |
| Cabinet work washrooms | CWW | 31 – 31 = 0 |

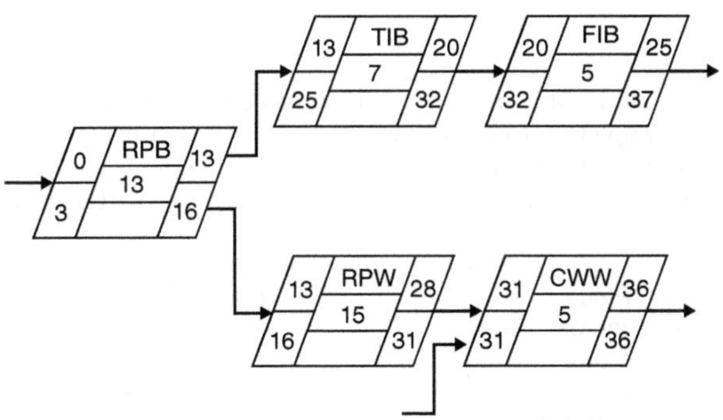


FIGURE 7.14 Portion of a CPM diagram.

b. By inspection can you determine the free float for any of these activities?

Recall that free float is the time an activity can be delayed without delaying the project's estimated completion time and without delaying the start of any succeeding activity.

| Activity | Free float | |
|---|---|---|
| RPB | 0 | If we would delay RPB, the result would be to also delay TIB and RPW, as RPB is the only predecessor of each and therefore controls their start time. |
| TIB | 0 | If we would delay TIB, the result would be to also delay FIB, as TIB is the only predecessor of FIB and therefore controls FIB's start time. |
| RPW | 3 | We can delay RPW for 3 days and not delay the start of CWW, since the other predecessor of CWW sets RPW's late start date. |
| FIB | Cannot determine | We do not have the necessary information about downstream activities for which FIB is a predecessor. |
| CWW | 0 | ES and LS are equal; this is a critical activity, so no float of any type. |

## PRECEDENCE LOGIC DIAGRAMS

**precedence diagramming** *Nodes represent activities and the connecting arrows indicate the relationship between activities.*

**Precedence diagramming** is an extension of the activity on the node scheduling method. The precedence method allows the overlapping of concurrent activities. Most scheduling software allows the use of the precedence method. The advantages of precedence modeling are:

- Concurrent activities can be easily modeled without dividing the work task into a number of discrete activities.
- The workflow of continuous operations is better represented.

The terms *lag* and *lead* are used to describe the amount of time between the start or finish of an activity and the start or finish of the activity's predecessor or successor. Lag is the time an activity must be delayed from the start or finish of a predecessor. Lead is the time an activity precedes the start or finish of a successor. The four precedence relationships are:

- Finish-to-start (FS)
- Start-to-start (SS)
- Finish-to-finish (FF)
- Start-to-finish (SF)

### Finish-to-Start

The finish-to-start relationship is used when the successor activity cannot start until the predecessor activity has been completely accomplished (Fig. 7.15). When using precedence diagramming a lag notation can be added to the finish-to-start relationship to denote a required delay between the finish of the predecessor activity and the start of the succeeding activity (Fig. 7.16).

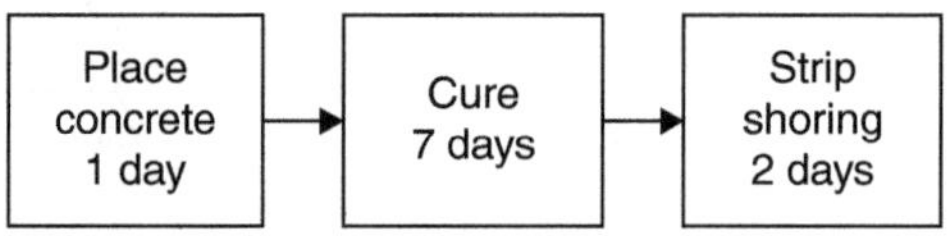


**FIGURE 7.15** AON network finish-to-start relationship without lag.

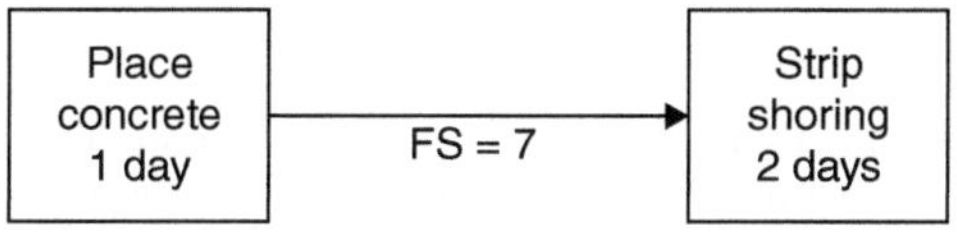


**FIGURE 7.16** AON network finish-to-start relationship with lag.

Assuming activities are performed in a continuous manner, the early start of the succeeding activity and late finish of the preceding activity can be calculated using Eqs. (7.9) and (7.10).

$$\text{Early start}_{n+1} = \text{Early finish}_n + \text{Lag} \tag{7.9}$$

$$\text{Late finish}_n = \text{Late start}_{n+1} - \text{Lag} \tag{7.10}$$

## Start-to-Start

Many construction activities can start after a certain amount of work has been accomplished on a preceding activity. If only finish-to-start diagramming notation is used, the preceding activity would have to be broken into incremental units of work to represent this condition (Fig. 7.17). Using start-to-start notation with a lag, the same relationship can be accounted for using fewer activities (Fig. 7.18).

The early start of the succeeding activity and late finish of preceding activities can be calculated by Eqs. (7.11) and (7.12). The equations are valid, however, only if the activities are performed in a continuous manner.

$$\text{Early start}_{n+1} = \text{Early start}_n + \text{Lag} \tag{7.11}$$

$$\text{Late finish}_n = \text{Late start}_{n+1} - \text{Lag} \tag{7.12}$$

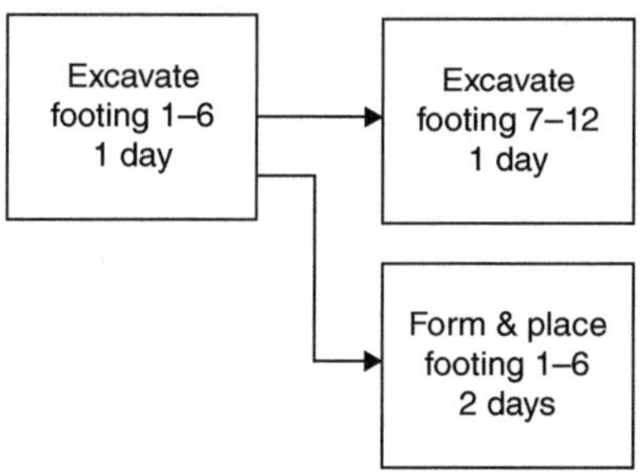


**FIGURE 7.17** AON network finish-to-start relationship.

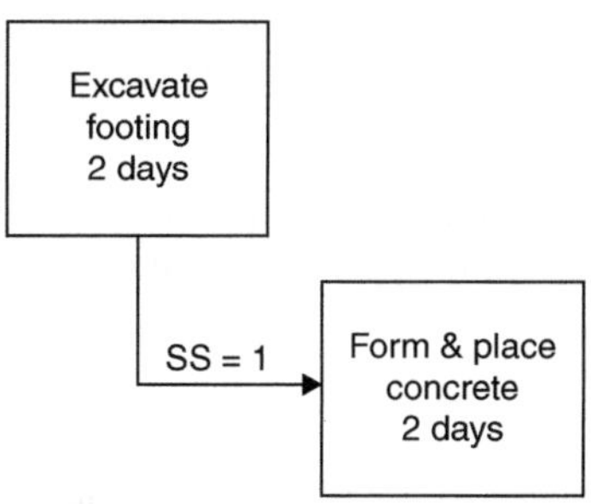


**FIGURE 7.18** AON network start-to-start relationship with lag.

### Finish-to-Finish

Many construction activities can take place concurrently but with the succeeding activity lagging some time interval behind the accomplishment of a certain amount of work on a preceding activity. When finish-to-start diagramming notation is used, the trailing activity must be broken into incremental units of work to represent this condition (Fig. 7.19). By use of finish-to-finish notation with a lag, the same relationship can be described using fewer activities (Fig. 7.20).

The early start of the succeeding activity and late finish of preceding activities can be calculated by Eqs. (7.13) and (7.14). Again, it is assumed the activities are performed in a continuous manner.

$$\text{Early start}_{n+1} = \text{Early finish}_n + \text{Lag} - \text{Duration}_{n+1} \tag{7.13}$$

$$\text{Late finish}_n = \text{Late start}_{n+1} - \text{Lag} \tag{7.14}$$

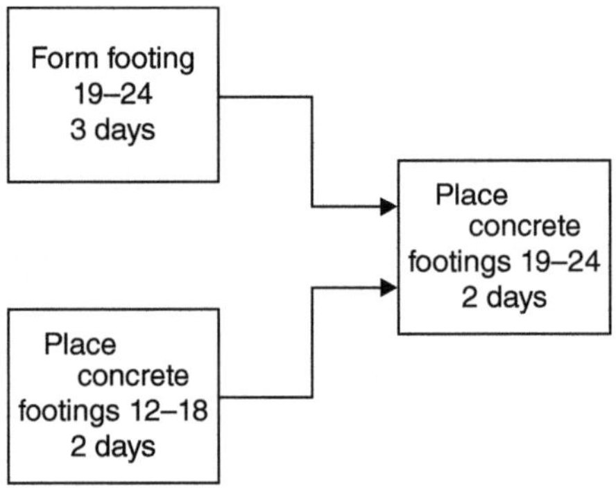


**FIGURE 7.19** AON network finish-to-start.

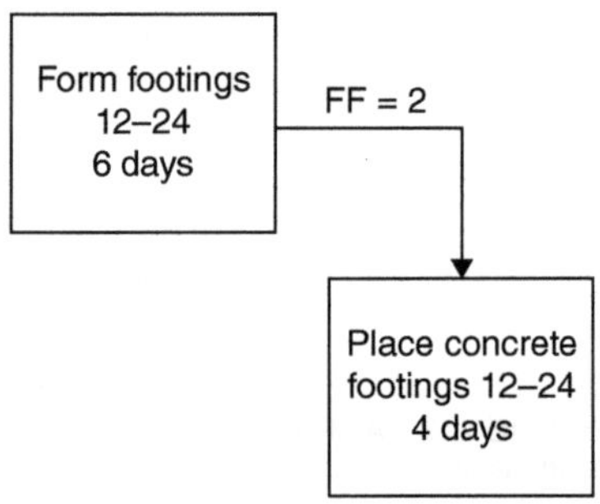


**FIGURE 7.20** AON network finish-to-finish relationship with lag.

### Start-to-Finish

Most scheduling software will not support the use of a start-to-finish relationship. This is probably because other relationships allow appropriate modeling of construction processes.

## CALENDAR DATE SCHEDULE

If all activities are scheduled to begin on their ES dates, the schedule is known as an early start schedule. This logic diagram schedule provides the manager with considerable planning information. But everything, to this point, was done based on workday time durations. To communicate schedule information in a meaningful manner, the workday schedule must be converted to a calendar schedule with activity starts expressed as a specific date. If done manually, this is a simple counting procedure, and if the schedule was created using a software package, the conversion can be done automatically.

The project start date is usually specified in the contract documents. Other decisions or contract requirements possibly affecting the calendar conversion are:

1. *Workdays per week.* This may be specified in the contract or it can be a contractor decision.
2. *Holidays to be observed.* Again, this may be specified in the contract or it can be a contractor decision.
3. *Weather days.* These are a function of specific work activities, the climatic conditions at the project location, and at what point during the seasons the project begins.

## SCHEDULING ACTIVITIES

It is usually assumed the activities will commence on their early start dates and typically the first CPM network is constructed under this assumption. But this does not have to be the case, and what the CPM really does is provide management with the information needed to schedule the project activities in a manner satisfying multiple work constraints. If the resource information is included with each activity, it is possible to determine the magnitude of resource requirements for each day of the project.

An activity can take place at any time within the interval defined by its early start and late finish dates, and still not delay the project. This provides flexibility in scheduling when an activity is actually accomplished. Using this flexibility, a manager can exercise control over the magnitude of resource requirements needed on a given day. A simple scheduling process can be graphically illustrated using the activities in the network shown in Fig. 7.21. The critical path for this network is emphasized by heavier activity connecting arrows.

The first step is to create a bar chart of the network with all activities listed in identifier order. Ordering the activities by ID sequence makes it easier to track successor activities. After each activity, place a reminder note in parentheses of all immediately dependent activities (see the activity column of Fig. 7.22). For example, because activities P and Q cannot begin until activity O is completed, annotate activity O in the schedule like this: O (P, Q). The next step is to mark on the bar chart the time frame for each activity during which it may be performed without delaying the project or violating any of the diagram logic (sequence) relationships (Fig. 7.22).

Consider activity J in Fig. 7.21; the ES shows the early start of this activity as the end of day 3 (or the beginning of day 4). Thus, the beginning of day 4 to the end of day 6 (as determined from the LF) is the available time span in which to complete this activity (Fig. 4.22). Because of the nature of the logic diagram, this activity cannot be scheduled earlier, as activity I must be completed first. It cannot be scheduled later, as then it would delay the entire project. Considering the end of day convention, where a start date means the end of the day or beginning of the following day, the left bracket is placed at the beginning (morning) of the following day, e.g., "ES+1."

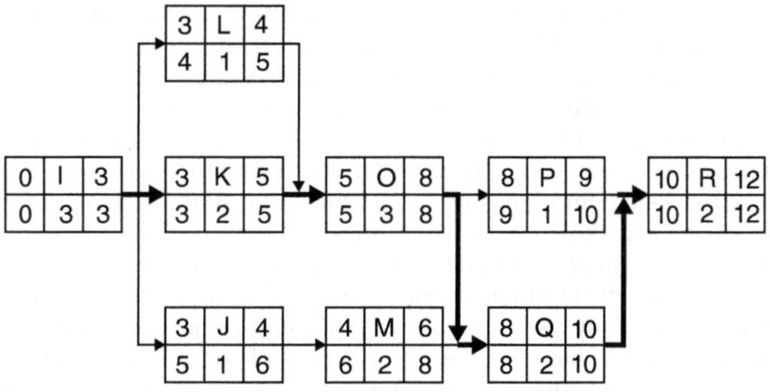


**FIGURE 7.21** Scheduling network activities.

| | Workdays | | | | | | | | | | | |
|---|---|---|---|---|---|---|---|---|---|---|---|---|
| **Activity** | **1** | **2** | **3** | **4** | **5** | **6** | **7** | **8** | **9** | **10** | **11** | **12** |
| I (j, K, L) | 2 | 2 | 2 | | | | | | | | | |
| J (M) | | | | 1 | × | × | | | | | | |
| K (O) | | | | 3 | 3 | | | | | | | |
| L (O) | | | | 2 | | | | | | | | |
| M (Q) | | | | | 1 | 1 | | | | | | |
| O (P, Q) | | | | | | 4 | 4 | 4 | | | | |
| P (R) | | | | | | | | | 2 | | | |
| Q (R) | | | | | | | | | 3 | 3 | | |
| R | | | | | | | | | | | 2 | 2 |
| Sum | 2 | 2 | 2 | 6 | 4 | 5 | 4 | 4 | 5 | 3 | 2 | 2 |

**FIGURE 7.22** Bar chart for scheduling network activities.

Once the brackets are placed correctly, the next step is to formulate a trial schedule, with each activity starting as soon as possible within the time frame, or flush with the left bracket. To schedule a particular activity, place the number of resources required inside each box along the activity line. Do not exceed the activity's duration; stop at the end of the early finish day. The remaining boxes within the brackets are left blank for now and will become either free or interfering float. In Fig. 4.22, the boxes of the critical activities have a bold outline.

The diagonal lines forming an × on the last 2 days of the activity J bracketed area denote days of interfering float. Some activities have free float (activities L, M, and P), and some have all interfering float (activity J). The placement of the ×s serves as a ripple effect reminder. When an activity is moved into this time frame, succeeding activities are impacted.

## Crew and Equipment Resources

If carpenter crews are identified as a controlling resource, the numbers within the activity data boxes indicate how many carpenter crews are required for each of the activities. It is necessary to consider only major controlling resources when developing these schedules. Depending on the nature of the work, each project has its own controlling resources. On a building job, the controlling resource might be the cranes. On a mass concrete project, the controlling resources might be forming crews and placement crews. There are usually several controlling resources, and all would have to be considered when scheduling the work. In cases where many different kinds of resources are necessary for an activity, managers may choose to use several lines contained within one set of brackets and dedicate each line to a different type of resource. The resource requirement for any particular day is the sum of those required for all the individual activities taking place on that day. The last row of Fig. 7.22 shows the resource requirement sum for each day of the project. There are computer-scheduling algorithms available to aid in scheduling resources on large projects.

The resource required to complete activity M is one carpenter crew for 2 days. To show this activity scheduled as soon as possible, place the number 1 (number of carpenter crews) in only the first two boxes of the bracketed duration (Fig. 7.22). Scheduling all the activities as soon as possible yields the early start schedule as shown in Fig. 7.22. All activities are scheduled to begin at their ES times.

The number of *available* resources is often less than the needed requirement for some days, and activities must be delayed (into float whenever possible) to spread resource use across the time frame of the project. When daily resource requirements exceed what is available, the manager must resource-constrain the project to have as little effect as possible on project duration. The terminology is not completely universal but constraining usually means the project will have to be extended because of a resource deficiency. However, sometimes resources can be shifted in such a way as to avoid delaying the overall project duration. The term *leveling* or *leveling project resources* implies no extension of project duration.

If the objective, considering the network in Fig. 7.21, is to limit the number of crews to five or less, and to avoid a situation where there are peaks and valleys in the number of crews required over the duration of the project, Fig. 7.23 illustrates one solution. In Fig. 7.23 activity J has been shifted 2 days into its interfering float. As a result of shifting activity J, activity M also had to be shifted 2 days, as it is dependent on J. In addition, activity L was shifted 1 day but this was into free float, so no other activity was affected. If any of the critical activities are shifted to achieve the desired leveling of resources, the result will be to lengthen the project duration.

Planning the work according to the schedule in Fig. 7.23 results in a uniform increase in the number of required crews, a level requirement for five crews from day 5 through day 9, and then a uniform decrease of crew requirements at the end of the project. This is a desirable situation because it is difficult to manage the work when the required number of resources is constantly changing. The problem with this schedule is that, with only the exception of activity P, we have now made every activity critical. If anything should delay a critical activity, the project will also be delayed.

Another possible schedule that holds the maximum number of crews to five or less is presented in Fig. 7.24. With this schedule, some float has been retained for activities J, M, and P. However, on day 8

| | Workdays | | | | | | | | | | | |
|---|---|---|---|---|---|---|---|---|---|---|---|---|
| Activity | 1 | 2 | 3 | 4 | 5 | 6 | 7 | 8 | 9 | 10 | 11 | 12 |
| I (J, K, L) | 2 | 2 | 2 | | | | | | | | | |
| J (M) | | | | | X | X 1 | | | | | | |
| K (O) | | | | 3 | 3 | | | | | | | |
| L (O) | | | | | 2 | | | | | | | |
| M (Q) | | | | | | | 1 | 1 | | | | |
| O (P, Q) | | | | | | 4 | 4 | 4 | | | | |
| P (R) | | | | | | | | | 2 | | | |
| Q (R) | | | | | | | | | 3 | 3 | | |
| R | | | | | | | | | | | 2 | 2 |
| Sum | 2 | 2 | 2 | 3 | 5 | 5 | 5 | 5 | 5 | 3 | 2 | 2 |

**FIGURE 7.23** A smooth resource-constrained schedule, maximum five crews.

| | Workdays | | | | | | | | | | | |
|---|---|---|---|---|---|---|---|---|---|---|---|---|
| Activity | 1 | 2 | 3 | 4 | 5 | 6 | 7 | 8 | 9 | 10 | 11 | 12 |
| I (J, K, L) | 2 | 2 | 2 | | | | | | | | | |
| J (M) | | | | 1 | X | X | | | | | | |
| K (O) | | | | 3 | 3 | | | | | | | |
| L (O) | | | | | 2 | | | | | | | |
| M (Q) | | | | | | 1 | 1 | | | | | |
| O (P, Q) | | | | | | 4 | 4 | 4 | | | | |
| P (R) | | | | | | | | | 2 | | | |
| Q (R) | | | | | | | | | 3 | 3 | | |
| R | | | | | | | | | | | 2 | 2 |
| Sum | 2 | 2 | 2 | 4 | 5 | 5 | 5 | 4 | 5 | 3 | 2 | 2 |

**FIGURE 7.24** Resource-constrained schedule, maximum five crews.

it is necessary to lay off one crew and then on day 9 one more crew is needed for only 1 day. If the resource were a crane instead of a carpenter crew, there would be the added consideration of the cost to mobilize and demobilize the machine. In such a case, the manager may decide to pay the extra day of machine rent and retain schedule flexibility. Leveling of crew resources is always a challenge when trying to keep productive crews together.

### Crashing

If the duration of the project schedule exceeds the desired completion date, the manager should examine the logic diagram's critical path to find activity durations, which may be shortened. This is known as *expediting, compressing,* or *crashing* the project. The only way to reduce the duration of the project is to shorten the duration of the critical activities.

> Shortening a noncritical activity will not shorten the project duration, but it will increase cost.

However, increasing the allocation of resources to critical path activities may reduce the duration of the project. Additional equipment and personnel can be committed, or the same equipment and personnel can be used for longer hours. Normally, a moderately extended workday is the most economical and productive solution.

> Extended workdays minimize the effects of nonproductive setup time on task accomplishment.

Managers may also choose to work double shifts or weekends. When expediting activities, always consider the long-term effects on safety, morale, equipment, and consequential decreases in efficiency.

## *PULL DRIVEN SCHEDULES*

All of the previous discussion has been based on moving in sequence from the first task to a completed project, otherwise known as push scheduling. A better approach to short-term planning is pull scheduling. Using the CPM schedule hanging on the project office wall, the manager tells the team, "The schedule shows we should be placing concrete three weeks from now. Can we instead place concrete next week? What do we have to do to make it happen?"

Now the team has a target and starts to think about needed tasks, crews, equipment, and materials to achieve a result. The focus is on the desired execution of a task by a specified date. This is pull scheduling. Attention is on streamlining and reducing project duration. To succeed, it is usually necessary to pull the entire team together and let everyone express their ideas (Fig. 7.25).

> Pull scheduling is greatly beneficial in reducing project duration and in building the team.

## *COMPUTER SUPPORT*

Plan drawings and paper printouts have been replaced with 3-Dimensional (3D), 4-Dimensional (4D), and 5-Dimensional (5D) planning tools. A 4D planning tool includes time or schedule, paired

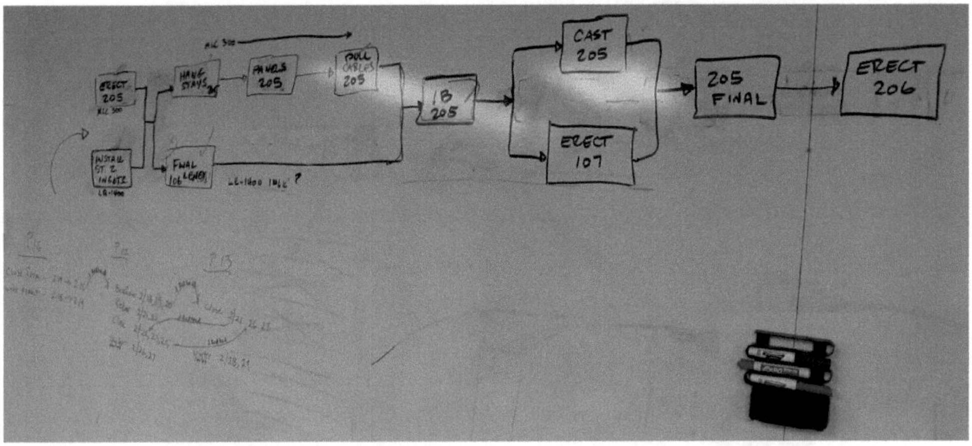


**FIGURE 7.25** Trying out Pull Schedule ideas.

with the 3D model of the project. When viewing a project in 4D, a digital twin of the project is built in real-time with "pieces" or elements being inserted into the model on the date the schedule shows them to be constructed.

- 1-Dimensional: Only one dimension needed to describe, length or a line.
- 2-Dimensional: 2D presents only horizontal and vertical (X and Y) dimensions.
- 3-Dimensional: The use of geometric data to create a three-dimensional (X, Y, and Z) representation.
- 4-Dimensional: A presentation of how three physical dimensions change with time.
- 5-Dimensional: The addition of either cost or revenue to a 4D model.

Figure 7.26 shows a 3D view of a project completed based on the construction schedule work tasks. Visualizing the plan drawings in a 3D view and then making the view come to life on the screen is extremely beneficial in creating a consistent plan, communicating the plan to all stakeholders, and overall execution of the project.

A 5D planning tool adds the revenue (or cost) of the project schedule elements to the 4D model. If a crew is installing reinforcing bar, the software equates the dollar amount the project team will receive upon completion of the activity. Utilizing schedule software with the capability to assigned revenue to activities will create a graphical representation of the revenue or cost for the project team. This permits the analysis of the project cash flow during construction based on schedule elements. In Fig. 7.27, the total project cost graph (5D) can be seen with the calendar date (4D), while the scheduled activity or construction element is being constructed. Adding revenue or cost to the 4D model allows the project team to forecast the projected revenue or cost at any given time throughout the project and, at the same time, correlate revenue to a project visual of the work.

Artificial Intelligence (AI) is starting to make a robust appearance in the construction industry. AI is being utilized at the project level as a support tool when forecasting activity durations/cost for similar activities based on actualized or past productions. The program will advise the project team if similar activities had a different duration than what is being proposed for the schedule. AI will present suggested durations or cost and suggest adjustments to the proposed work schedule. This allows the team to recognize the schedule effects of a potentially longer or more costly activity item and adjust the work plan to the work tasks.

AI can also be utilized to analyze multiple iterations of constructing a project. For example, the user will set the parameters of the basic logic for constructing the project. Then using AI, the program will run hundreds of scenarios. These yield different schedule times and revenue options. The project team is then able to analyze and determine if the project plan proposed is in fact the fastest and most

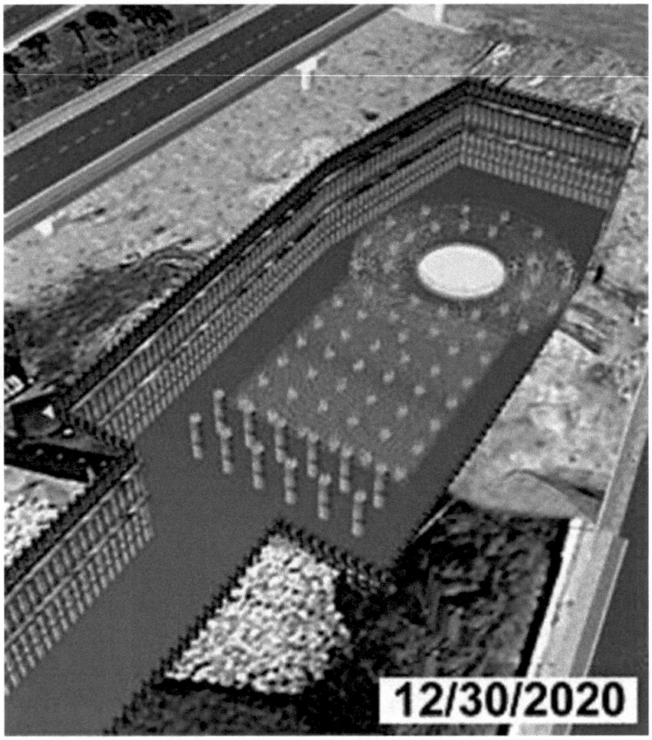


**FIGURE 7.26** 3-Dimensional view being built according to the project schedule.

cost-effective plan. Although this technology is new, it is making waves in the construction industry because it offers a way to improve project planning and cost projections. It helps project teams understand how to execute a project in a safer, quicker, and more efficient manner.

Virtual Reality (VR) and Augmented Reality (AR) are still other trending construction technologies. VR allows a user to experience a real-life view of a construction site and overlay the 3D model of what the project will look like during or after construction. This provides project stakeholders a "live step" into a building before it is constructed. The viewer can look around, walk around, turn lights on and off, and see the proposed final product. This technology not only allows for design changes to be made prior to construction, but also ultimately creates safe work sites and identifies potential hazards or changes before crews place the first concrete or drive the first nail.

Computerized scheduling programs range in price from a few hundred dollars to thousands of dollars, based in part on their flexibility, reporting capabilities, graphics, and resource allocation characteristics. The 3D and 4D CAD systems are more expensive than traditional scheduling software but their cost can easily be justified by the benefits provided.

With the advent of project websites and the pervasive nature of e-mail, project data can be shared with as many or as few people as desired around the world. Online project collaboration has enabled remote schedule review via PDAs (personal data assistants) and likely soon via cell phone, in addition to all typical computer platforms.

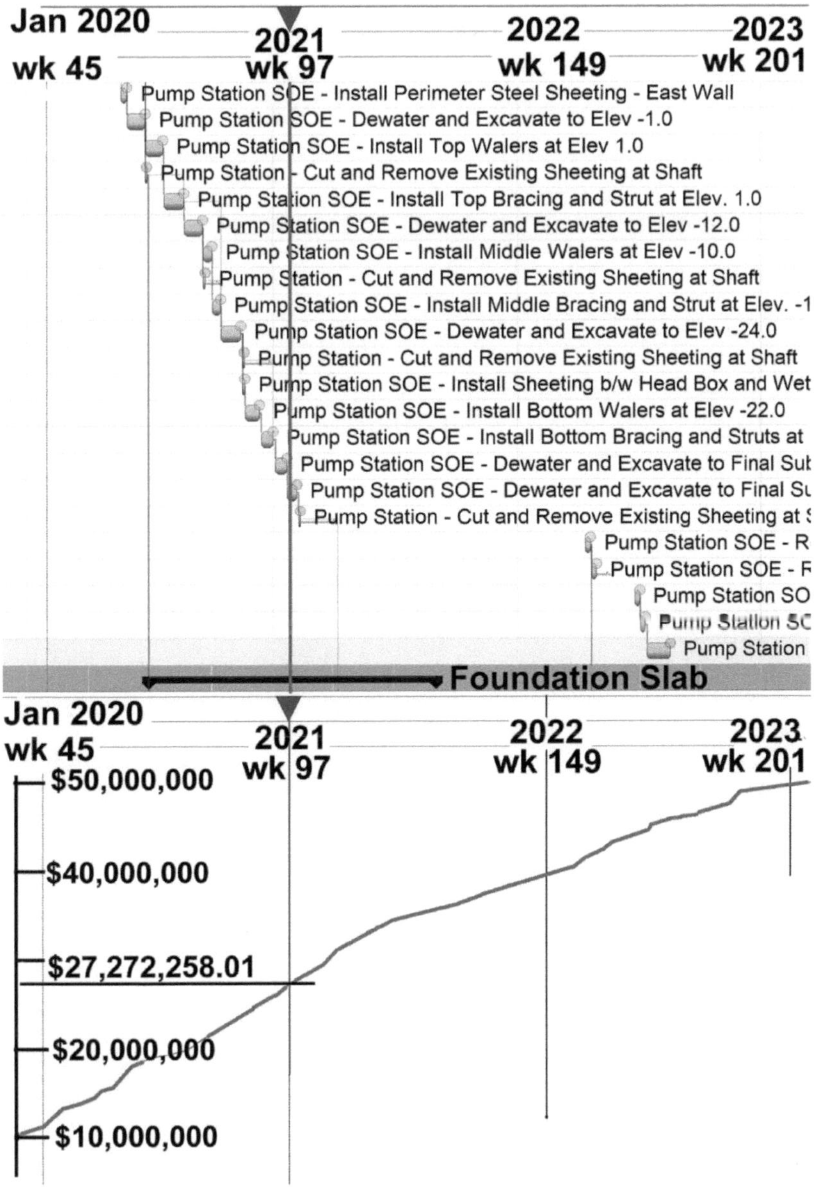


**FIGURE 7.27** 5-Dimensional planning tool adds the revenue (or cost) of the 3D view in Fig. 7.25.

## RESOURCE

*Construction Planning & Scheduling*, 2nd ed. (2004). The Associated General Contractors of America, Arlington, VA.

# CHAPTER 8
# RISK MANAGEMENT

*Why would you not want to know what could be preventing you from achieving your project goals and objects? Why would you not want to know what you could do to improve safety, cost, schedule, and management reputation? These questions are the essence of a formal risk management process. Risk Management is a key knowledge area for good project management practice; it is a systematic process of planning, identifying, evaluating, responding, and monitoring a project's threats and opportunities.*

## OVERVIEW

Risks are threats and opportunities to achieving project goals. Identifying and assessing these risks, through a risk management process, will greatly improve the chances for delivering a successful project.

Formal risk assessment or risk analysis may be the most underutilized or misunderstood project management process. The programming, the environmental process, the project selection process, the project delivery method selection, and the procurement method selection decisions and the project execution plan are all risk allocation decisions. Risks should be allocated to the party in the best position to control each specific risk; shifting risks to a party not equipped to address the risk will result in higher project delivery cost, delays, decreased quality, or disputes.

As they develop and deliver their projects, organizations can benefit tremendously from a formal risk management process. A formal risk management process should be implemented throughout the life of all projects. Risk assessments do not only identify threats to achieving project goals, but also identify opportunities for achieving the goals.

The risk management process should be simple and the effort tailored to the scope, duration, and size of the project. Basic steps in a formal risk management process (Fig. 8.1) are as follows:

1. Develop a risk management plan.
2. Identify risks (threats and opportunities).
3. Evaluate the identified risks (qualitatively and quantitatively, if necessary).
4. Develop risk-response strategies.
5. Monitor the risks and risk management plan.

The best practice is to regularly update the initial risk assessment. Updates serve to identify additional risks (threats and opportunities) and allow for adjustment to the risks previously identified based on current information as the project progresses.

| The process is iterative. |
|---|

There are numerous sources dedicated to detailed discussions on the risk management process. This chapter is intended to provide an overview and sufficient guidance to conduct a successful risk management process, which will lead to a more successful project.

One of the significant products resulting from a formal risk management process is the risk register. This is typically a spreadsheet that lists all identified risks, their rankings, response strategies, and responsible parties for implementing the response/mitigation strategies. The risk register serves as an excellent communication tool for the team.

The benefits of conducting risk assessment include

- Improving communications among team members
- Improving communications with stakeholders
- Better utilization of resources.
- Increasing what goes right (opportunity identification and responses)
- Decreasing what goes wrong (threat identification and responses)
- Minimizing contingency (cost and schedule)
- Increasing likelihood of achieving project goals and objectives
- Improving project management
- Improving decision-making
- Reducing bias of team members and others
- Prioritizing actions

Common reasons for not doing formal risk management and responses to each are outlined in Table 8.1. Risk literature presents these in much greater detail but the heart of the arguments is contained in these short statements.

The risk management process (Fig. 8.1) provides a comprehensive approach to addressing uncertainty from a variety of sources. The risk management approach outlined in a Risk Management Plan (RMP) is intended to reduce the probability and impacts of adverse risk events on a project. The RMP defines and describes the strategies, actions, and processes the project management team will implement during each phase of the project in order to provide the maximum opportunity for success. Taken together, these elements comprise a risk management approach integral to the structure of the project and inherent within each discipline of the delivery process.

**FIGURE 8.1** The risk management process.

**TABLE 8.1** Why a Formal Risk Management Process?

| Arguments against | Responses |
|---|---|
| • Risk management processes waste project team time and cost money. | • Proper application of risk management processes saves time and money. |
| • Responding to identified risks costs money. | • Risk responses are an investment in the future, spend now to gain later. |
| • Risk management does not work. | • It has proven its value in many industries and is a key knowledge area for successful project management. |
| • Risk management processes are scaremongering. | • Real risks are project uncertainties needing careful management attention. They can also be opportunities. |
| • The benefits are not quantifiable. | • The benefits can be measured. |
| • We are too busy doing day-to-day work. | • The process will save valuable project management time. |
| • We do it every day; it is what we do. | • A structured approach is better than a false sense control. |

## Risk Management Plan

The purpose of a Risk Management Plan is to provide the project team a guide for the successful implementation of a risk management program. The plan is intended to assure that all project personnel can develop a working knowledge of the risk management process and thereby contribute to an effective program. The plan identifies the objectives, processes, tools, and responsible parties necessary to successfully manage project risk. The RMP is typically an appendix to the project's Project Management Plan (PMP). As each project is unique, the RMP becomes a key communication tool for risk management.

> The Risk Management Plan should be provided to all project participants. This ensures a project-wide understanding of the risk management process.

As a project evolves, its risk profile will change and the focus of risk management practice will change. Lessons learned from previous risk assessments can be applied to future work. Therefore, an RMP should be considered a "living document" and should be updated to reflect changes in managerial methodology or project structure.

The RMP highlights specific areas of management focus as identified through the risk review process. As this plan is implemented, it provides a means for managing the project's risk. The principal topics covered within an RMP include the following:

- Introduction—Provides an overview of risk management and the RMP
- Goals and Objectives of Risk Management Process—Establishes measures to complete the project within budget and on schedule through the implementation and utilization of cost and schedule contingency procedures, risk management, and the development of available risk management capacity
- Risk Management Scope—Identifies the scope of work for the Risk Management effort
- Risk Management Organization—A description of key Project personnel roles for implementing the Risk Management Plan and describes the Risk Assessment Committee
- Risk Management Process—A description of the risk management process the project will follow (the five steps shown in Fig. 8.1)

## Goals and Objectives of the Risk Management Process

The RMP provides guidance for the identification of key risks (threats and opportunities) and respective response plans by defining a recommended risk management methodology.

By implementing the plan, the project team can prioritize and define response actions, estimate the effect of accepted and/or retained risks, and assist in developing the Project's contingency budget. The resulting information can then be used to run "what-if" scenarios when variable field conditions or other project challenges are encountered. This becomes a valuable on-going project decision-making tool.

Program management utilizes a number of "good management practice" tools to assist in meeting project objectives. Each tool has a unique way of helping to achieve project success. Risk management seeks to achieve project success by identifying and effectively managing identified opportunities and threats impacting the project objectives. For example, the plan could define a successful project with the following statements:

- Deliver the project safely, on schedule, and within budget.
- Ensure the sustainability of the project.
- Maintain an effective outreach and involvement program in coordination with the Design-Builder.
- Provide an equal opportunity and non-discriminatory workplace and meet goals for disadvantaged business enterprise and minority and female employment programs.
- Maintain relationships with project stakeholders, including neighbors and government partners.
- Maximize the project owner's investment.
- Minimize adverse effects on vehicular traffic operations during construction.
- Meet contract requirements, particularly in relation to construction, quality, environment, future maintenance and operations, public outreach, and other specifications.
- Maintain an effective contract oversight program.

The objectives of applying risk management to a project are as follows:

- Manage risks that jeopardize project success.
  - Increase the likelihood of the project achieving its objectives and minimize surprises.
  - Prioritize risks for senior management focus.
  - Ensure that contract documents reflect the owner's risk appetite and project objectives.
- Provide the basis for early planning to ensure an effective and affordable project.
  - Generate a proactive risk management culture within the project management team.
  - Minimize reactive and unprepared responses to sudden events.
  - Implement a formal process for the assessment of threats and opportunities leading to effective risk response planning and opportunity exploitation.
  - Increase the project's teams ability to understand and manage stakeholders.
  - Support the project's decision-making process and evaluating alternative options.
- Provide a rational method for managing the cost estimate contingency and schedule float.
  - Establish a rational method for calculating realistic and defendable contingency budgets (cost and schedule) and a control process to monitor their use; provide managers with the means to decide where best to invest the project's time and money.
  - Comply with owner's/funding agency's requirements, as applicable.

| An understanding of project objectives is a risk management prerequisite. |
|---|

The risk identification process identifies all risks, some of which threaten the ability to secure the project objectives and opportunities to enhance meeting the project objectives. Therefore, it is important for the project team to agree and record the project objectives as outlined above.

## RISK MANAGEMENT SCOPE

The risk management scope for a project typically includes details on the following:

- Facilitate the risk assessment, which includes risk identification, analysis, and evaluation.
- Facilitate the identification of risk response actions.
- Maintain a risk management information system, including: risk register, risk assessment documentation, and risk communications.
- Facilitate and document regularly scheduled risk response meetings with project executives.
- Perform quantitative analysis to forecast project cost and schedule exposure.
- Perform quantitative analysis of residual risks.
- Support the cost and schedule contingency management process, including the use of contingency drawdown charts.
- Perform estimate at completion (EAC) forecasting.
- Utilize the risk-based decision-making process to support decisions.
- Document and maintain all discussions and agreements regarding allocation and management of project risks.
- Interface with funding and partners with respect to risk management requirements.
- Maintain and document all funding and partner requirements.
- Disseminate, communicate, and make presentations regarding the project's efforts for managing risks.

## RISK MANAGEMENT ORGANIZATION

The risk management organization functions as part of the project management organization, as described in the PMP. All project personnel are part of the risk management process, but the Risk Assessment Committee (RAC) is responsible for risk management.

The success of the risk program is significantly influenced by the performance of the risk owners and the RAC. The RAC is composed of a select set of key project personnel who have the project's latest risk status readily available to them at all times and the flexibility to manage the risks as they choose. The degree to which the RAC drives the process and the commitment they demonstrate to the project greatly influence the development of a risk culture and, in turn, the success of the project's risk program. The RAC is responsible for the following:

- Identifying and monitoring project risks during the advancement of the project
- Evaluating the risk response details, focusing the project risk efforts, and setting priorities and goals
- Implementing and monitoring risk responses to limit project risk impacts
- Assuring that the identified risks are being addressed using the appropriate response procedures conducted by the assigned risk owners
- Establishing policy and adjusting contingency plans as the project progresses
- Is the only entity that can change risk likelihood values and consequence (impact) values after these values have been agreed to and placed in the risk register as a result of project risk workshops or other meetings.

The RAC member should cover all major project activities and the organizational structure of the project, such as

- Project Director
- Commercial Manager
- Design Compliance Engineer

- Construction Compliance Engineer
- Quality Verification Manager
- Outreach Director
- Environmental Compliance Manager
- Contract Manager
- Project Controls Lead
- Risk Manager
- Diversity Compliance Manager
- Superintendent Lead—Foundations
- Superintendent Lead—Structure
- Superintendent Lead—HVAC
- Superintendent Lead—Claddings
- Superintendent Lead—Facilities and Utilities
- Superintendent Lead—Security and Operations
- Superintendent Lead—Other
- Legal Advisor
- Oversight agency*
- Technical Experts*

*The Project Manager will typically determine the need for Technical Experts and the representatives of other organizations to attend the RAC or other risk meetings depending on the necessity, and specific meeting agenda.

RAC meeting attendees should vote on any request for a risk reduction after responses are implemented. Majority vote is required and results should be available to all participants. It is anticipated that the RAC will meet more frequently in the beginning of a project and then less frequently as the project progresses. It may be necessary to conduct an RAC meeting after a significant event occurs to address the changed risk profile. The RAC meetings may be part of other meetings, such as Progress Meetings.

The RAC members should have clearly defined roles. Table 8.2 is an example communication plan or responsibility assignment matrix, for RAC members showing who is responsible, accountable, to be consulted, and to be informed for the different risk management activities.

**TABLE 8.2** Responsibility Assignment Matrix Example

| | Risk Assessment Committee member | | | | | | |
|---|---|---|---|---|---|---|---|
| **Risk management activity** | **Project Director** | **Risk Management & Project Controls Manager** | **Project Risk Manager** | **Construction Compliance Manager** | **Design Compliance Manager** | **Design Quality Assurance Manager** | **Construction Quality Assurance Manager** |
| Risk Management Plan | Accountable | Responsible | Responsible | Consulted | Consulted | Consulted | Consulted |
| Risk Assessment | Accountable | Participate | Responsible | Participate | Participate | Participate | Participate |
| Risk Assessment Report | Accountable | Informed | Responsible | Informed | Informed | Informed | Informed |
| Risk Response Actions | Accountable | Responsible | Participate | Responsible | Responsible | Consulted | Consulted |
| Risk Response Report | Accountable | Informed | Responsible | Informed | Informed | Informed | Informed |

### Risk Management Process

The risk management process consists of five steps shown in Fig. 8.1, planning for risk management (establishing the project context), identifying risks (threats and opportunities), evaluating the risks (qualitatively or quantitatively), developing risk responses, and monitoring the risks. This process is iterative and, depending on the project duration, will be completed many times. The process follows the ISO 31000: 2009 "Risk Management-Principles and Guidelines" and augmented as appropriate by the "Practice Standard for Project Risk Management" (2009) published by the Project Management Institute (PMI).

### Establishing Context

Establishing the context defines the project parameters to be considered in the risk assessment process. The scope for the risk assessment process should be determined and the criteria should be set against which the risks will be assessed. Understanding the social, economic, and natural environment in which the project is to be built is critical in defining the context for project decisions, activities, and events.

## *IDENTIFICATION OF RISKS (THREATS AND OPPORTUNITIES)*

The risk assessment process covers a complete understanding of the risk profile of either the project or a defined element of the project. Risk assessment is a systematic, formal process to identify, analyze, and evaluate risks whether qualitatively or quantitatively. The Project Director, supported by the risk management team, will identify the elements of the project needing a risk assessment and the timing of such evaluations.

A risk cannot be managed unless it is first identified. Therefore, the purpose of risk identification is to search for and to locate sources of risk exposure, before they are realized. These sources of risk exposure are called risk events. The risk identification process begins with the project team compiling a list of the project's risk events. Identified risks should be in the form of a risk statement that follows a "cause → event → impact" logic. The suggested form of a risk statement is as follows:

"If <RISK EVENT> happens, then <CONSEQUENCE> will <IMPACT> <PROJECT OBJECTIVE(S)>."

Or

"If <RISK EVENT> happens, then <CONSEQUENCE> will happen to <WHO>, causing a/an <IMPACT> of <RESULT>."

**brainstorming**
*An informal method for obtaining a variety of ideas and inputs. It occurs in a group setting where each member provides suggestions for solutions to a particular problem.*

There are numerous tools available to assist in the identification of risks (threats and opportunities). The most common method of risk identification is **brainstorming**. These brainstorming meetings should include all project stakeholders. Ideally this meeting would include external and internal stakeholders. Following are brief descriptions of the more common risk identification methods.

**Delphi Technique:** a structured method for using informed judgment for decision making or long-range forecasting. It involves a panel of experts answering rounds of questions on a topic and receiving opinion feedback from fellow group members.

**Subject matter expert:** a person who possesses bona fide expertise and knowledge about what is needed to accomplish a particular job, task, or process.

**Crawford Slip Method:** a method involving gathering input from people on slips of paper. This is a simple yet effective type of brainstorming. It gives the opinion of all team members' equal weight, however, quiet they might naturally be in group meetings.

**Partnering:** a collaborative process normally set up at the outset of a project. It typically involves facilitated sessions, attended by the major decision-makers, where project goals and issues are discussed. Partnering sessions occur throughout the life of the project.

## RISK REGISTER

The primary document used to capture the risk information is the risk register. The risk register will be managed by the RAC. The Risk Register is a "living document" and it is updated more frequently than the RMP. Table 8.3 summarizes the information typically included in a risk register.

### Risk Breakdown Structure

The risks in the register can be further categorized according to a predetermined risk breakdown structure (RBS). An example is shown in Table 8.4. An RBS is a tool for achieving more complete risk identification, better recognition of key risk areas, more effective reporting, and to facilitate comparison of risk profiles between projects. The RBS serves as a template to ensure all areas of the project are reviewed for the identification of risks and a comprehensive list of risks is developed. It also serves to group and capture risks in major categories.

**TABLE 8.3** Typically Risk Register Data Elements

| Section | Description |
|---|---|
| **Risk ID** | A unique number assigned to an individual risk to differentiate it from others. |
| **Risk Category** | A classification based on source and affect. |
| **Risk Description** | A description of the risk, including an overall risk event description, as well as cause and consequence. |
| **Risk Rating** | An overall qualitative rating for each risk. Could be color coded (green, yellow, orange, or red) based on likelihood of cost and schedule impacts. |
| **Probability of Occurrence** | Risk occurrence evaluated on a 0–100% scale. |
| **Expected Impact** | The range (minimum, most likely, and maximum) of the cost and schedule delay if the risk was to occur. It could also include quality, reputation, and environment impact. |
| **Risk Owner** | Individual responsible for managing an assigned risk. |
| **Risk Response** | Risk Response Plan includes specific actions to implement the mitigation plan, and where possible, specify the deliverable, person responsible for the deliverable, and date the deliverable is due for each action. |

**TABLE 8.4** Example Risk Breakdown Structure

| Reference | Category |
|---|---|
| 1 | Management/Organization Function |
| 2 | Compliance (Environmental/Permitting) |
| 3 | ROW/Real Estate |
| 4 | Utilities/3rd Parties |
| 5 | Design Changes |
| 6 | Construction (Site Constraints, Field Conditions, Weather) |
| 7 | Procurement/Material/Equipment |
| 8 | Quality/Safety/Security |
| 9 | Commissioning |
| 10 | Other Risks |

## *EVALUATING RISK (RISK RATING)*

The objective of a risk rating is to establish an overall qualitative term for the probability and consequence for each risk. The qualitative terms are typically: Low (acceptable), Medium (tolerable), High (substantial), Very High (significant) for each identified risk (Table 8.5). This risk rating may then be utilized for the purposes of risk prioritization. The rating of each risk is recorded and updated in the risk register based upon input from RAC members using the rating system described in your RMP.

All risks are evaluated based on two aspects: (1) probability of occurrence (likelihood) and (2) consequence (impact). Probability of occurrence is in terms of a consensus percentage from the RAC members. The consequence or impact are not limited to cost and schedule but should also include other categories relevant to the projects goals.

- Consequence/Impact (I)
- Monetary impact range (minimum and maximum) in dollars (C)
- Schedule impact range (minimum and maximum) usually in months (S)
- Quality (Q)
- Reputation (R)
- Environment (E)

The risk rating (RR) for each risk is simply the probability (P) times the impact (I):

$$RR = P \times I$$

A qualitative risk rating can then be determined for each "very high" and "high" risks (are for all risks if desired, but this is not usual because of the effort involved). If sufficient information is available at the outset, the team may go directly to a quantitative risk assessment. The calculation is the same, $RR = P \times I$. Table 8.6 is an example probability matrix. Table 8.7 is an example impact matrix.

The individual risks are often plotted on a chart, known as a heat-map, to graphically display where each risk is ranked and to display the overall risk profile of the project. Figure 8.2 is an example heat map showing qualitative scoring.

It should be noted that risk analysis involves consideration of the causes and sources of risk, their positive and negative consequences, and the likelihood of consequence occurrence. Risk analysis can be undertaken with varying degrees of detail, depending on the risk, the purpose of the analysis, and the information, data, and resources available. Analysis can be qualitative, semiquantitative or quantitative, or a combination of these, depending on the circumstances.

**TABLE 8.5** Risk-Rating Action Matrix

| Rating | Risk Response Action |
|---|---|
| Very High (significant) | Unacceptable—requires priority response(s) |
| High (substantial) | Unacceptable—requires response(s) |
| Medium (tolerable) | Evaluate response(s) |
| Low (acceptable) | Acceptable/Monitor |

**TABLE 8.6** Example Probability Matrix

| | Very unlikely | Unlikely | Possible | Likely | Very likely |
|---|---|---|---|---|---|
| **Probability** | 1 | 2 | 3 | 4 | 5 |
| | <5% | 5–20% | 21–50% | 51–75% | 76–100% |

**TABLE 8.7** Sample Impact Matrix

| Impact | Insignificant 1 | Minor 2 | Moderate 3 | Significant 4 | Severe 5 |
|---|---|---|---|---|---|
| **Cost** | <$10 M | $11–25 M | $26–50 M | $51–100 M | >$100 M |
| **Time** | <1 week | 1 week–1 month | 1–2 months | 3–6 months | >6 months |
| **Quality** | No risk | Acceptable, minor repairs | Reject and rework | Significant effort to repair or replace | Practically irreversible structural compromise and/or lack of final project acceptance |
| **Reputation** | No press coverage; no stakeholder reaction | Press coverage is neutral; stakeholders' reaction is neutral | Press coverage and stakeholder reaction is generally negative; local or regional coverage | Press coverage and stakeholder reaction is negative; local, regional, or statewide coverage | Press coverage is negative; stakeholder reaction is negative; regional, statewide, or national coverage |
| **Health & Safety** | No injuries or illness | Injury or occupational illness not resulting in a lost work day | Injury or occupational illness resulting in 1 or more lost work day(s) | Permanent partial disability, injuries, or occupational illness may result in hospitalization | Could result in death, or permanent total disability |
| **Environment** | No environmental impact | Minimal environmental impact | Reversible moderate environmental impact | Reversible significant environmental impact | Irreversible significant environmental impact |

| | | Impact (I) | | | | |
|---|---|---|---|---|---|---|
| | | 1 Insignificant | 2 Minor | 3 Moderate | 4 Significant | 5 Severe |
| Probability (P) | 5 Very likely | 5 | 10 | 15 | 20 | 25 |
| | 4 Likely | 4 | 8 | 12 | 16 | 20 |
| | 3 Possible | 3 | 6 | 9 | 12 | 15 |
| | 2 Unlikely | 2 | 4 | 6 | 8 | 10 |
| | 1 Very unlikely | 1 | 2 | 3 | 4 | 5 |

**FIGURE 8.2** Qualitative risk ranking heat map.

Qualitative risk analysis aims to prioritize individual risks to be treated. On the other hand, the quantitative risk analysis process provides a numerical estimate of the overall effect of risk on project objectives, based on the current plans and information. It also provides a basis for evaluating the effectiveness of risk management or risk allocation strategies.

Qualitative risk analysis is the process of prioritizing risks for further analysis or action by assessing their likelihood (probability of occurrence) and their impact (magnitude of consequence), using qualitative values.

## COST AND SCHEDULE EVALUATION

In conducting a formal cost and schedule assessment, a project will need to perform a quantitative risk analysis (QRA). The QRA is a natural extension of cost estimating and scheduling. It addresses the risks of a project (cost and schedule risk events) and the uncertainty that exists in development of the project at the time of the assessment (e.g., uncertainties of cost to complete estimates). Uncertainty is associated with almost every cost element and schedule activity. QRA integrates the uncertainty identified in the current cost and schedule with the identified risk events and reports a combined exposure for a true, point in time, snapshot of the project element being studied. The analysis produces an uncertainty in the project's baseline cost estimate and schedule. These considerations are necessary when preparing or updating cost estimates because there is little likelihood the cost estimate will equal the actual project cost or that the original schedule will be maintained.

A cost and schedule risk analysis not only helps approximate the probability of deviation from plans, but also defines the requisite amount of contingency to be added to an estimate or schedule to achieve a desired confidence level. The project team can formulate the risk assessment results from the development of an integrated risk model where cost and schedule are generated simultaneously in response to the potential occurrence of risks (e.g., Monte Carlo simulation). The results will be the combination of all of the simulation runs; however, the actual project execution assumes a single simulation of the cost and schedule models.

A schedule risk assessment indicates which activities appear most frequently on the critical path and may require schedule contingency (float). During a simulation, an activity may or may not appear on the critical path. After the simulation, the number of appearances on the critical path is an indicator of how critical the activity will be to project completion.

The QRA is built on the following inputs:

- Project's latest risk register
- Project's latest schedule
- Cost estimate

The analysis is divided into three parts: (1) cost estimate uncertainty; (2) assessment of risk events resulting in cost risk; and (3) risk events resulting in schedule risk and schedule delay costs (delay days and delay day cost).

The three assessments are integrated in a risk model and the QRA is performed using Monte Carlo simulation (e.g., through @Risk software[1]) to develop cumulative distribution curves relating cost and schedule to confidence levels.

The following provides details of the three assessments.

**Cost Estimate Uncertainty:** The cost estimate serves as the building block for the analysis model. It is necessary to first capture uncertainty within the cost estimate. This is done by evaluating each cost element and then setting cost ranges over which each element is expected to exist.

**Cost Risk Events:** Identified risk events from the project risk register, which result from the risk workshop, are then added to the model. Workshop participants identify risk event impacts—the range in dollars or delays in days they expected the risk event would impact the project.

[1] The software @RISK supplied by the Palisade Corporation is a standard tool for conducting quantitative integrated cost and schedule risk analysis. @RISK is a Windows-based tool, which is an "add-on" to Microsoft Excel®; @RISK links directly to Excel® to add risk analysis capabilities. It allows users to define uncertain cell values in Excel® as probability distributions using functions.

**Schedule Risks:** The schedule component, a project rolled-up schedule with relevant activities, critical path, and flow, is modeled by assigning each schedule risk event to its relevant activity in the schedule component. During the simulation, durations are modified based on the number of risks being simulated to occur and their random consequence. Each iteration generates a number of delay days based on the duration of the project, which are multiplied by a cost per day then added to the cost distribution. The schedule provides a baseline with which to analyze the possible duration of the project. Since the relationships of different risks are already embedded in the predecessor-successor relationships in the schedule, it is not necessary to assign extra causal links among schedule risks.

The resulting statistics from the QRA consist of the output curves forecasting cost and schedule as a function of confidence level as well as the top cost and schedule risks. A 70% confidence level, which is generally regarded as an industry standard, is used to establish cost and schedule exposure. The project can also set contingency levels to match a desired level of confidence from the risk assessment results. The simulation analysis has three outcomes:

- Supports the calculation of the expected distribution of project cost, which is used to estimate the contingency budgets.
- Provides a more accurate shortlist of risks prioritized by importance based on their expected impact in cost and schedule.
- Provides a confidence level to the overall project budget and schedule.

The main goal is to maintain the project cost and schedule within funding limitations, schedule commitments, and minimum contingencies. The project team can maintain or reduce contingency requirements by actively mitigating identified risks. They can accomplish this by developing mitigation actions for each significant risk and implementing these action plans in a timely manner.

## RISK RESPONSES

The objective of the risk response step, as referred to in the ISO 31000:2009 Risk Management Process, is to minimize the impact of risk on the project objectives through the identification of appropriate responses, so the project team will achieve desired goals under a controlled budget as well as meet schedule expectations. Most risks can be minimized through proactive means by either reducing the likelihood of the risk event occurring or minimizing the impacts if the risk does occur.

### Response Options for an Identified Threat

Risk responses fall into one of seven categories as noted in Table 8.8 and described below. One or more of these response methods may be deployed on individual risks.

**Risk transfer:** Transfer the risk to the party best suited to handle, such as an insurance carrier or another project participant. This methodology includes risk allocation, which is a way to transfer risks through contractual mechanisms to the supply chain.

**TABLE 8.8** Risk Response Strategies

| Threats | Opportunities |
|---|---|
| Transfer | Share |
| Avoid | Exploit |
| Mitigate | Enhance |
| Accept (Actively or Passively) | |

**Risk avoidance:** Prohibit, do not use, the process or procedure causing a risk too high to manage or accept.

**Risk mitigation:** Take action to reduce the likelihood or consequence of a risk or some combination of the two.

**Risk acceptance:** Typically taken if the risk is uncontrollable or the cost of risk reduction measures exceeds the potential risk exposure.

### Response Options for Opportunities

**Share opportunity:** an opportunity response strategy involving efforts to increase the probability of an identified opportunity event by shifting some or all of the opportunities to a third party.

**Exploit opportunity:** an opportunity response strategy of ensuring an identified opportunity will definitely occur in the project's favor.

**Enhance opportunity:** an opportunity response strategy making the effort to increase the probability or impact or both of the identified events.

**Accept opportunity:** a decision to accept or retain an opportunity is typically taken if the opportunity is uncontrollable. The acceptance can be either active or passive.

To manage opportunities and continue the implementation of response strategies, the following steps should be followed by expanding the project's risk/opportunity register to include

- Each opportunity response strategy (a risk may have more than one response)
- Identification of the opportunity owner or person responsible for managing the opportunity
- Priority of opportunity response strategies options by considering
  - Response(s) providing most value
  - Response(s) providing value to more than one opportunity
  - Response(s) addressing the opportunity early in the schedule
- Target date for implementation of opportunity response
- Forecast the relative opportunity reductions of each response strategy

## *CONTROL AND MONITOR RISKS*

The success of the project risk management effort is dependent on the effective execution of the risk responses. The risk monitoring and control process is designed to provide oversight of the implementation of the responses, identify the requirement for additional responses, and determine impacts of any changes in the project's risk profile. Risk monitoring and reporting can occur in specific risk meetings and/or as part of other meetings and reports, such as the weekly project management meeting.

The objectives of the risk monitoring and control process are as follows:

- Review on a monthly basis the current risk profile and identify changes in risk probabilities and impacts with project senior management.
- Monitor on a monthly basis the implementation of risk responses and implement any necessary changes.
- Update the risk register at a minimum on a quarterly basis with any new risks and associated responses based on changes in project scope, project progress, and changing risk generators.
- Conduct regular risk mitigation meetings with the project management team to review the top risks and document discussions.

- Issue, at a minimum, monthly risk reports containing
  - A record of risk activities and meetings conducted during the report month
  - Risk management program-plan status
  - Updated Risk Register (may be a separate document)
- Report compliance with estimate and project partners, as applicable.

## COMMUNICATION AND CONSULTATION

Communication and consultation is a continual dialogue between project parties. It involves sharing information about the management of project risk. It is essential for project team to generate a proactive risk management culture within the project team and minimize reactive and unprepared responses to sudden events. Therefore, effective communication among the project team at each step of the risk management process is very important. In order to enhance the communication and risk culture in the project, the following should be shared with all project parties:

- Updated Risk Register
- Risk Reports
- Quantitative Risk Analysis Results and Reviews
- Agenda and Minutes of Risk Meetings
- Risk Communiqués (short report documents)

## COMMON CONSTRUCTION PROJECT RISKS

A study by Dodge Data and Analytics (commissioned by Autodesk, Inc. and published in 2020) surveyed more than 200 US-based contractors in order to identify and analyze current processes for planning and executing projects. The findings suggest that by adopting specific processes for project management, contractors can reduce risk, thus minimizing downstream problems and improving performance. Project managers can use these indicators as a means of identifying project risks. The study identified seven key performance indicators:

1. **Construction documents:** The frequency with which contractors are capturing errors, omissions, and/or constructability issues in the "bid set" of construction documents.
2. **Request for Information:** The frequency with which contractors log Request for Information (RFIs) and responses.
3. **Change orders:** Frequency of which firms are collecting and documenting change orders including turnaround time, root cause evaluation, and schedule impacts.
4. **Schedule:** The frequency of updating schedules and other related activities and outcome, including the impact of slippages.
5. **Safety and Inspections:** If and how firms are using technology to manage safety and/or inspections.
6. **Labor Productivity:** The top factors negatively impacting labor productivity are Coordination and Communication, Contract Document Quality, and Poor Schedule Management.
7. **Quality and Close Out:** If and how firms are using software to manage punch list and close out activities.

Most risk management experts suggest not using potential risk lists, as they have a tendency to inhibit, or restrict thinking on the specific project being analyzed. A risk list may be beneficial after risks have been identified to see if the list triggers an oversight in the project team's risk identification.

Typical construction project risk categories include

- **Safety Hazards:** height/falls, moving objects/being struck, slips/trips, noise, vibration, trenches, electricity, materials, traffic, equipment/backing-up (see OHSA data)
- **Labor:** shortages, productivity, quality
- **Construction Defects:** refer specifically to any defect in the design, workmanship, or in the materials used on a project
- **Contractual**
- **Subcontractors, Suppliers, Fabricators**
- **Geotechnical or subsurface conditions**
- **Site Protection/Security**
- **Permits/Authorizations**
- **Communication/Coordination**
- **Reputation**
- **Documentation**
- **Third Party** (e.g., utilities, railroad)

Involving stakeholders in a formal process to identify project risks will greatly enhance the identification of threats and opportunities to the project's goals and objectives. Good project management practice means risk analysis is not a one-time activity, but a continuous effort throughout the life of the project. Therefore, the initial risk register should be updated regularly and serve as an excellent communication tool as the project or program progresses and is refined.

# CHAPTER 9
# BUILDING MATERIALS

*There are several kinds of buildings, with the most obvious distinction being commercial and residential. Commercial buildings vary greatly in size, type of construction, number of stories, and intended use. There are many kinds of residential buildings, varying greatly in style, grandeur, and number of individual residences in each building. Engineers should appreciate the impact of selecting different materials and different construction methods. They must also understand the impact materials and methods have on the overall cost of a structure, and on its appearance and life expectancy. The major building materials are concrete, reinforcing steel, structural steel, wood, and masonry, with a more recent emphasis on glass.*

## CONCRETE

Concrete consists of Portland cement, water, and aggregates mixed together in the proper proportions and then allowed to cure and gain strength. There are two types of concrete-mixing operations: (1) transit mixed and (2) central mixed. Unless the project is in a remote location or is relatively large, the concrete is batched in a central batch plant and transported to the job site in transit-mix trucks. A variety of equipment and techniques are used to place the concrete, ranging from simple hand tools to pumps or belt conveyors.

## CEMENTITIOUS MATERIALS

Cementitious materials include the many products that are mixed with either water or some other liquid or both to form a cementing paste. The paste may be formed or molded while plastic but will set into a rigid shape. When sand is added to the paste, mortar is formed. A combination of coarse and fine aggregate (sand) added to the paste forms concrete.

Portland cement concrete is one of the most widely used structural materials (Fig. 9.1). Its versatility, economy, adaptability, availability, and especially its low maintenance requirements make it an excellent building material. The term "concrete" is applicable for many products but is generally used with Portland cement concrete. It consists of Portland cement, water, and aggregates mixed together, placed, consolidated, and allowed to cure.

The Portland cement and water form a paste, which acts as the glue or binder holding the aggregates of the mix together. When only fine aggregate, having a size range between a No. 200 mesh sieve and the No. 4 sieve, is added to the paste, the resulting mixture is termed "mortar." When coarse aggregate is also included, having aggregate sizes larger than the No. 4 sieve but less than 3 in., concrete is produced. Normally, by volume, concrete consists of about three-fourths aggregate and one-fourth paste. The paste typically consists of water–cement ratios between 0.4 and 0.6 by weight. Admixtures are sometimes added for specific performance purposes, such as

**FIGURE 9.1** Building with tilt-up concrete walls.

- Improve workability
- Impart color
- Retard the initial set of the concrete (e.g., when long haul distances are needed)
- Gain rapid hardening and high initial strength (e.g., for post-tensioned elements)
- Improve flowability (e.g., for self-consolidating concrete)
- Waterproof the concrete

Limes were formerly widely used as the sole cementitious material, but their slow setting and hardening are not compatible with modern requirements. Hence, their principal function today is to plasticize the otherwise harsh cements and add resilience to mortars and stuccoes. Use of limes is beneficial because their slow setting promotes healing, the recementing of hairline cracks.

Another class of cements is composed of calcined gypsum and its related products. The gypsum cements are widely used in interior plaster and for fabrication of boards and blocks.

| The solubility of gypsum prevents its use in construction elements exposed to any but extremely dry climates. |
|---|

Oxychloride cements constitute a class of specialty cements of unusual properties. Their cost prohibits their general use in competition with the cheaper cements; but for special uses, such as the production of spark proof floors, they cannot be equaled.

Masonry cements or mortar cements are widely used. While they are, in general, mixtures of one of more of the above-mentioned cements with some admixtures, they deserve special consideration because of their economies.

Other cementitious materials, such as polymers, fly ash, and silica fume, may be used as a cement replacement in concrete. Polymers are plastics with long-chain molecules. Concretes made with them have many qualities superior to those of ordinary concrete.

Silica fume, also known as microsilica, is a waste product of electric-arc furnaces. The silica reacts with the limes in concrete to form a cementitious material. A fume particle has a diameter only 1% of a cement particle.

## Portland Cements

Manufactured cement consists of crushed limestone (CaO) and clay ($SiO_2$ and $Al_2O_3$) burned at temperatures in excess of 2,700°F and then ground into the fine powder. The calcareous materials typically come from limestone but calcite, marl, or shale can be used. So the location of the plant determines the material used.

When only a small amount of cement is needed it is purchased in paper bags, each containing 1 cubic foot (cf) loose measure and weighing 94 lb net.

> Bagged cement must be stored in a dry place on pallets and should be left in the original moisture-proof bags until used.

Concrete ready-mix plants receive their cement in bulk quantities by transport truck or railroad cars. Bulk cement is usually unloaded by low air pressure from special truck trailers or railcars and stored in overhead silos or bins.

ASTM C150 defines requirements for eight types of Portland cement with each having unique properties to control set rate. But the five primary types of cement usually encountered in building construction are

1. Type I is a general-purpose cement intended for use where concrete is not subject to specific exposures, such as sulfate attack from soil or water. It is commonly used for building structures.
2. Type II is a cement modified to resist moderate sulfate attack. It gains strength slightly more slowly than Type I. Because of conditions in California, it is commonly specified for structures in the state.
3. Type III is a high early-strength cement. It will produce design strengths at 7 days or less. The cement granules are ground to a smaller size producing more surface area and allowing more water to hydrate the cement. Because of this it produces higher heat evolution. As a result it should not be used in mass concrete placements. Type III is not always available from dealer's stock and ready-mix sources. Therefore, it must be ordered in advance of need.
4. Type IV is a special cement with low heat of hydration used for mass concrete placements. Its strength develops at a slower rate.
5. Type V is sulfate resistant; it reaches its design strength in 60 days. At location where the concrete will be subject by sulfate liquors (wastewater treatment plants, seawaters, and some natural waters), Type V cement is used.

Type IV and Type V cements are specialty cements and are not normally available from dealer's stock. They must be ordered in advance. See ASTM C150/C150M-17 "Standard Specification for Portland Cement" for additional details concerning each type.

## Mortars

Mortars are composed of a cementitious material, fine aggregate, sand, and water. They are used for bedding unit masonry (Fig. 9.2), for plasters and stuccoes, and with the addition of coarse aggregate, for concrete. Mortars are usually proportioned by volume. A common specification is that not more than 3 $ft^3$ of sand be used with 1 $ft^3$ of cementitious material.

Properties of mortars vary greatly, being dependent on the properties of the cementitious material used, ratio of cementitious material to sand, characteristics and grading of the sand, and ratio of water to solids.

**Workability** is an important property of mortars, particularly of those used in conjunction with unit masonry of high absorption. Workability is controlled by the character of the cement and amount of sand. For example, a mortar made from three parts sand and one part slaked lime putty

**Workability** (mortar) *The ease with which mortar can be handled.*

**FIGURE 9.2** Mortar for bedding Brick Unit Masonry.

will be more workable than one made from two parts sand and one part Portland cement. But the 3:1 mortar has lower strength. By proper selection or mixing of cementitious materials, a satisfactory compromise may usually be obtained, producing a mortar of adequate strength and workability.

**Water retention** is the ratio of the flow after 1-min standard suction to the flow before suction, and is used as an index of the workability of mortars. A high value of water retention is considered desirable for most purposes. There is, however, a wide variation in water retention of mortars made with varying proportions of cement and lime and with varying limes. ASTM C270 "Standard Specification for Mortar for Unit Masonry" requires mortar mixed to an initial flow of 100–115, as determined by the test method of ASTM C109, to have a flow after suction of at least 75%.

**Strength** of mortar is frequently used as a specification requirement, even though it has little relation to the strength of masonry. The strength of mortar is affected primarily by the amount of cement in the matrix. Other factors of importance are the ratio of sand to cementing material, curing conditions, and age when tested.

**Volume change** of mortars constitutes another important property. Normal volume change (as distinguished from unsoundness) may be considered as the shrinkage during early hardening, shrinkage on drying, expansion on wetting, and changes due to temperature. After drying, mortars expand again when wetted. Alternate wetting and drying produces alternate expansion and contraction, which apparently continues indefinitely with Portland-cement mortars.

## Lime Products

Hydrated limes are manufactured by adding water to powdered quicklime. Hydrated limes, Type S (Special) and Type SA (Special Air-Entrained), are fine dry factory-made powder packaged in 50-lb sacks. These are strictly American products because of the more stringent masonry performance requirements than those of other countries (ASTM C206–14 "Standard Specification for Finishing Hydrated Lime").

When mixed with sand and/or cement in the correct proportions, hydrated limes create a lime-based mortar. These products are also suitable for use in base-coat plasters and concrete. For brick, concrete block, and stone construction, Type S hydrated lime increases the mortar bond to masonry units. Bond strength is critical to the prevention of cracking, water leakage, and efflorescence.[1] Type S and SA hydrated limes are also used in exterior plasters.

[1] A white or a grayish tint material, a crystalline deposit of salts, left of the surface of masonry work.

## Plasters

Plasters are used to turn rough wall and ceiling surfaces into smooth decorative areas and to provide surface protection. There are a variety of plastering materials available with more common being gypsum, lime, and cement. The selection of a particular plaster depends on desired effect but all work in a similar manner.

**Gypsum plaster** (plaster of Paris) is a fine powder, usually white. Dry plaster of Paris powder mixed with water re-forms into gypsum. It can be used as either a molding—ornamental work—or gaging plaster. As a gaging plaster it will form a smooth white-coat finish on walls. However, it will disintegrate continually exposure water (moisture). When mixed with perlite or vermiculite aggregate, a lightweight plaster is produced. Gypsum plasters have a strong set, when dry will gain full strength, do not have abnormal volume changes, and have excellent fire-resistance characteristics. The material is also used to manufacture gypsum block, tile, and gypsum board.

**Oxychloride cements** are a mixture of magnesium chloride and magnesium oxide. The French chemist M. Sorel patented the mixture in 1867, so it is also referred to as Sorel cement. The mixture is used to manufacture floor tiles. Floors made of oxychloride cement are spark proof and more resilient than concrete floors.

Oxychloride cement will bond with inorganic and organic aggregates. It is used in making partition block or tile with wood shavings or sawdust as aggregate. It is moderately resistant to water but should not be used under continually wet conditions. This disadvantage, poor water resistance, has limited its use.

## Masonry Cements

Masonry cements are specially formulated for binding brick, block, and stone when mixed with sand (Fig. 9.3). These are cements to which plasticizing materials, lime, and other materials are added to enhance setting time, workability, water retention, and durability.

Masonry cements are available in Type N, Type S, and Type M strengths and used to prepare ASTM Specification C270 Type N, S, or M mortars without addition of additional cements. They are available in a variety of colors. Table 9.1 provides guidance for selecting the appropriate type of mortar.

Masonry cements are sold in bags containing from 70 to 80 lb; each bag nominally contains one cubic foot by volume.

**FIGURE 9.3** Mixing masonry cement.

**TABLE 9.1** Selection of Mortar Type

| Building segment | Mortar type |
|---|---|
| Exterior, above grade | |
| Load-bearing | N, S, or M |
| Non-load-bearing | N |
| Parapet wall | N or S |
| Exterior, at or below grade | S or M |
| Interior | |
| Load-bearing | N or S |
| Non-load-bearing | N |

## ADMIXTURES FOR CONCRETE

Admixtures are ingredients in the concrete mix other than Portland cement, water, and aggregate; these will affect the workability, durability, and economics of the concrete. The primary functions of admixtures are air-entraining, water-reducing, retarding, accelerating, and plasticizing the concrete mixture. These are detailed under ASTM C494 "Standard Specification for Chemical Admixtures for Concrete."

### Air-entraining

Air-entraining admixtures create microscopic air bubbles for relieving internal pressure when freezing water expands within the hardened concrete. ASTM C260 "Standard Specification for Air-Entraining Admixtures for Concrete" covers materials for use of air-entraining admixtures to be added to concrete in the field. Air entrainment may also be achieved by use of Types IIA and IIIA Portland cements.

### Water-reducing

Water-reducing admixtures lower the required water content by about 5–10% to increase strength and create a more consistent setting rate.

### Retarding

Retarding admixtures slow the setting rate of concrete, a very effective admixture for counteracting the accelerating effect of hot weather. Initial set can be retarded from several hours to several days.

### Accelerating

Accelerating admixtures increase the rate of early strength development and reduce the required time for proper curing and protection. Calcium chloride is the most commonly used set-accelerating admixture. Where reinforcing or prestressing steel is present it can contribute to corrosion of the steel.

## Plasticizers

Plasticizers are added to concrete mix with a low-to-normal slump and water–cement ratio to produce high-slump flowing concrete.

## Pozzolans

Pozzolans are a mineral admixture used to replace a percentage of Portland cement in the mix. They enhance the concrete properties and improve mix economics. Commonly used pozzolans include fly ash from coal-fired power plants, silica fume from silicon metal production, and blast furnace slag. Fly ash has been found to be an excellent mineral admixture in Portland cement concrete, improving almost all concrete properties. Up to 30% of Portland cement in the mix can be replaced with fly ash, resulting in stronger, less permeable, more durable, and more economical concrete. Silica fume and blast furnace slag can dramatically increase concrete strength and durability. Slag has the added benefits of disposing a manufacturing byproduct in an environmentally friendly manner, and chemically whitening the concrete color. These pozzolanic replacements to Portland cement are also broadly referred to as Supplementary Cementitious Materials, or SCMs.

## Coloring Admixtures

Colors can be added to concrete for aesthetic purposes. There are mineral oxides or manufactured pigments for this purpose. It is even possible to closely replicate the variegated, multitonal appearance of natural stone. Coloring pigments should comply with ASTM C979 "Standard Specification for Pigments for Integrally Colored Concrete."

While adding less the 6% pigment will have very minor or no effect on the physical properties of the concrete, ACI 301-05 "Specification for Structural Concrete" does warn about using unmodified carbon black. Its use will considerably increase the air-entraining admixture needed to provide resistance to freezing and thawing. Proposed colored mixes should be tested prior to their use in a project. To maintain a consistent color, when coloring agents are used, it is necessary to careful control of all ingredients, batching, and addition of water.

## Fibers for Concrete Mixes

Fibrous materials are used in concrete mixes for secondary reinforcement as an alternative to welded wire fabric. These are micro- or macrofibers depending on the relative size of the fiber. Microfibers have diameters less than 0.3 mm (0.012 in.) while macro-fibers have diameters equal to or greater than 0.3 mm (0.012 in.). The four primary categories of fiber-reinforced concrete (FRC) are:

- Type I: **Steel Fiber**—Reinforced concrete, stainless steel, alloy steel, or carbon steel fibers conforming to ASTM A820.
- Type II: **Glass Fiber**—Reinforced concrete, alkali-resistant glass fibers in accordance with ASTM C1666.
- Type III: **Synthetic Fiber**—Reinforced concrete, man-made fibers such as carbon, nylon, polyester, and polyolefin adhering to ASTM C1116.
- Type IV: **Natural Fiber**—Reinforced concrete, cellulose fibers from a variety of plants addressed in ASTM D7357.

In all cases their greatest use is for crack control of floor slabs, elevated metal pan decks, and other flat work.

Because the fibers are dispersed throughout a mix, they are three-dimensional reinforcement. Orientation of the millions of fibers in concrete is random. Steel fibers will increase impact strength and durability of the concrete. Synthetic fibers used in place of welded-wire fabric provide crack control in flatwork. FRC is blended in paddle mixers. These mixes lack large aggregate and only contain fine sands, cement, and fibers.

Glass-fiber-reinforced concrete (GFRC or GRC) is used primarily to cast building elements such as thin architectural cladding panels, and ornamental domes, statues, planters, and fountains. GFRC cladding is relatively thin, about ½″–1″ thick, and has high architectural value. GFRC is typically mixed with alkali-resistant glass fibers to resist break down in an alkaline environment.

### Miscellaneous Admixtures

Pumping economically is enhanced by a variety of admixtures. When concrete will be in contact with water or damp earth, dampproofing admixtures are used to reduce the moisture transmission.

**Pumping admixtures** decrease the viscosity of harsh or hard-to-pump mixes. Organic and synthetic polymers, fly ash, bentonite, or hydrated lime may be used to increase mix viscosity while under pressure from the pump. Admixtures such as air-entraining agents and some water-reducing and retarding admixtures used primarily for other purposes will also improve the pumpability of the mix.

**Dampproofing admixtures** are intended to reduce passage of water and water vapor through concrete structures. The common dampproofing agents are certain soaps, stearates, and petroleum products. But these will not reduce the permeability of concretes having low cement contents, high water–cement ratios, or a deficiency of fines. These additives will not be effective when the concrete is in contact with water under pressure. The permeability of concrete is lessened by the use of fly ash and silica fume. A mix with a water–cement ratio less than 0.50 will greatly reduce permeability.

**Gas-forming admixtures**, hydrogen peroxide, aluminum powder, or activated carbon are used to create lightweight concrete. These will also serve to maintain placements' initial volume—counteracting settlement and bleeding—by generating or liberating bubbles in the mix. These agents are beneficial when attempting to grout a confined space. The temperature of the fresh mix affects the amount of expansion. When the volume of expansion is critical, careful control temperatures are critical.

## *AGGREGATES*

Aggregates' inert mineral crystals vary in size from boulders to fine sands. These include crushed stone, natural gravels, and blast furnace slag. Aggregates may be further described by their respective sizes. The following size limits represent standards set by the American Society for Testing and Materials (ASTM):

*Gravel* is composed of rounded or semiround particles of rock that will pass a 3 in. and be retained on a 2.0-mm No. 10 sieve. Sizes larger than 10 in. are usually called *boulders*.

*Sand* is disintegrated rock with particles that vary in size from the lower limit of gravel (2.0 mm) down to 0.075 mm (No. 200 sieve). It will be classified as coarse or fine sand, depending on the grain size. Sand is a granular noncohesive material and its particles have a bulky shape.

*Silt* or dust is a material finer than sand, and thus its particles are smaller than 0.075 mm but larger than 0.005 mm. It is a noncohesive material and has little or no strength. Silt compacts very poorly by itself, but can enhance compaction of larger aggregates by filling void spaces or reorienting particle position.

*Clay* is a cohesive material whose particles are less than 0.005 mm.

The more the aggregate used in a Portland-cement concrete mix, the lower the cost of the concrete. The larger the maximum size of coarse aggregate, the lesser the cement paste needed to coat all the

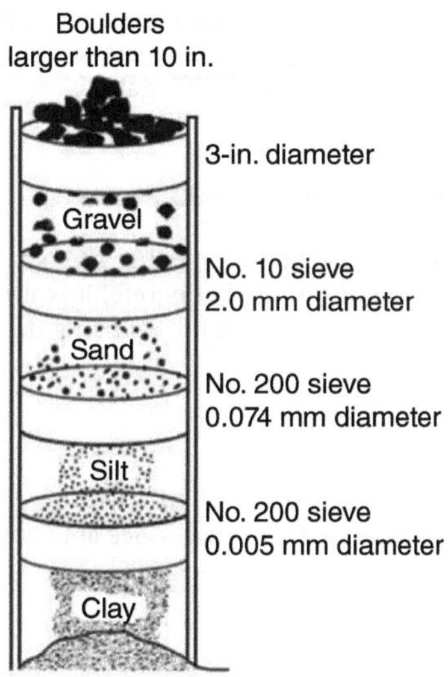


particles and to provide necessary workability. By volume, the majority of a concrete mix is aggregates, usually more than 60%. In the case of asphalt concrete, aggregates are 90–95% of the mix by volume.

The properties of the aggregates used in a concrete mix affect the service life and durability of concrete, asphaltic, mortar, and aggregate base. The surface abrasion resistance of the concrete is almost entirely a function of the properties of the fine aggregate.

The acceptability of the aggregates used in concrete or mortar is a function of

- Gradation
- Amount of fine material passing the No. 200 sieve
- Hardness
- Soundness
- Particle shape (is very important for pavements)
- Alkali reactivity
- Freezing and thawing resistance
- Organic impurities

ASTM C33 "Specification for Concrete Aggregates," C637 "Specification for Aggregates for Radiation-Shielding Concrete," and C330 "Specification for Lightweight Aggregates for Structural Concrete" provide property limits for aggregates used in building construction.

## Gradation of Aggregates

The aggregate size distribution in a concrete mix directly influences the amount of cement required to achieve a specified strength and the **workability** of the mix. Workability affects the economy of placing "liquid" concrete. ASTM C33 provides ranges of fine- and coarse-aggregate grading limits. There is a difference in specifications and performance between using rock fines (manufactured sands) and natural sands with percentages

**Workability** (concrete) *How easily concrete can be mixed, placed, consolidated, and finished.*

of silts and clays. The ideal mix has a dense or well-graded size distribution that optimizes the void content of the combined aggregates. Greater amounts of fine material, passing the No. 200 sieve, will increase the water demand of the aggregate.

## Amount of Fine Material Passing the No. 200 Sieve

Greater amounts of fine material, passing the No. 200 sieve, will increase the water demand of the aggregate. In the case of pumped concrete, it is important for the fine aggregate to have at least 5% passing the No. 100 sieve and about 3% passing the No. 200 sieve.

## Hardness

The Los Angeles Abrasion Test, ASTM C131, or C595 is used to quantify coarse-aggregate hardness. These tests do not directly replicate service abrasion experienced by aggregates in a concrete mix but are empirically related to concrete service life.

## Soundness

Soundness or "soundness and durability" refers to the aggregate's resistance to disintegration by weathering, in particular, freeze-thaw cycles. ASTM C88 "Test Method for Soundness of Aggregates by Use of Sodium Sulfate or Magnesium Sulfate" is used to determine the soundness of an aggregate.

## Particle Shape

Natural sand and gravel are produced by the action of running water so they have round, smooth particle shapes. Manufactured, crushed aggregate (coarse and fine) tend to have bulky angular shapes. A high degree of particle angularity or elongation will increase the amount of cement required to provide desired mix strength and the effort necessary to finish. These features also make pumping the mix difficult. The use of properly graded "angular particles" offset the need to increase the water quantity required to produce quality concrete.

## Alkali Reactivity

Most aggregates are chemically inert. However, aggregates containing forms of silicas or carbonates will react with the alkalis in Portland cement (*alkali-silica reaction,* ASR). This reaction will produce cracks or create pop-outs in the concrete. Some of the tests used to check for ASR potential of an aggregate will give false-positive results, making determination of this problem difficult. Experience records of material for aggregates sources is probably the best predictor of ASR problems.

## Freezing and Thawing Resistance

When the finished concrete is exposed to repeated freeze-and-thaw cycles, the pore structure, absorption, porosity, and permeability of the aggregates used in the mix are particularly important. When these aggregates become saturated and then freeze, they cannot accommodate the expansion of the water they contain. ASTM C666/C666M-15 "Standard Test Method for Resistance of Concrete to Rapid Freezing and Thawing" is used to determine the *resistance of concrete* specimens to *rapidly* repeated cycles of *freezing and thawing.*

### Organic Impurities

Organic impurities in the aggregates of a concrete mix will cause erratic setting times and rates of hardening. ASTM C40 "Test Method for Organic Impurities in Fine Aggregates for Concrete" is used to identify the presence of impurities in fine aggregates. Usually, it is the fine aggregates causing the problems.

## HEAVYWEIGHT AND LIGHTWEIGHT AGGREGATES

Aggregates can be classified on the basis of their bulk specific gravity: heavyweight, normal-weight, and lightweight aggregates.

**Heavyweight aggregates** possess relatively high density and are used to produce high-density concrete. Their bulk specific gravity can range from 3.5 for natural goethites to 5.2 for barites. When such aggregates are used, the mix density can be as much as 60% higher than normal concrete. These aggregates are used for radiation shielding structures, counter weights, and to attenuate vibration and sound.

**Lightweight aggregates** have high absorption values; therefore, a modified approach is needed for concrete proportioning. These aggregates are used for structural and nonstructural concrete. ASTM C330/C330M-17a "Standard Specification for Lightweight Aggregates for Structural Concrete" defines aggregates used to produce concretes generally in the strength range of 3,000–4,000 psi. Vermiculite and perlite are common nonstructural lightweight aggregates (ASTM C332) used in insulating concretes for soundproofing and nonstructural floor toppings.

Lightweight aggregates produce concrete with low thermal conductivities, which equate to good fire protection. When concrete is exposed to extreme heat, the moisture present within the concrete rapidly changes from a liquid to steam having a volume of up to 15 times larger. This creates an unfavorable condition for punch-outs and buckling. The high number and large sizes of pores within lightweight aggregates create pressure-relief regions.

## *PORTLAND-CEMENT CONCRETE*

For successful concrete performance, the mixture must be properly proportioned. The American Concrete Institute (ACI) has a number of recommended practices, including one on proportioning concrete mixtures. Detailed treatment of concrete proportioning is beyond the scope of this text, but some practical considerations include

1. Although it takes water to initiate the hydraulic reaction, as a general rule, the higher the water-cement ratio, the lower the resulting strength and durability of the concrete.
2. The more water used (not to be confused with the water–cement ratio), the higher the slump.
3. Adequate consolidation produces stronger and more durable concrete.
4. The use of properly entrained air enhances almost all concrete properties with little or no decrease in strength if the mix proportions are adjusted for the air.

To the designer, fresh concrete is usually of little importance. To the constructor, fresh concrete is all-important.

> A constructor must mix, transport, place, consolidate, finish, and cure fresh concrete, a semifluid material.

To satisfy both the designer and the constructor, the concrete should

1. Be easily mixed and transported.
2. Have minimal variability throughout, both within a given batch and between batches.
3. Be of proper workability so as to enable proper consolidation, prevent segregation, completely fill the forms, and achieve a proper finish.

Workability is difficult to define, it depends on the situation. One measure of workability is slump, which is a pseudomeasurable value based on an American Society for Testing and Materials standard test (ASTM C143/C143M-15).

The fresh concrete is placed into a hollow frustum of a cone, 4 in. diameter at the top, 8 in. diameter at the bottom, and 12 in. high. After filling and consolidating the concrete in the cone according to a prescribed procedure, the cone is raised from the concrete, allowing the fresh concrete to "slump" down. The amount of slump is measured in inches (or millimeters) from its original height of 12 in., with the stiffest concrete having zero slump and the most fluid concrete having slumps in excess of 8 in. Although the slump measures only one attribute of workability, it is the most widely used measure. Very high slump concrete, often used as backfill under trench pipes or in other difficult to compact areas, is regularly referred to as flowable fill. Table 9.2 gives the recommended slumps for various types of concrete construction.

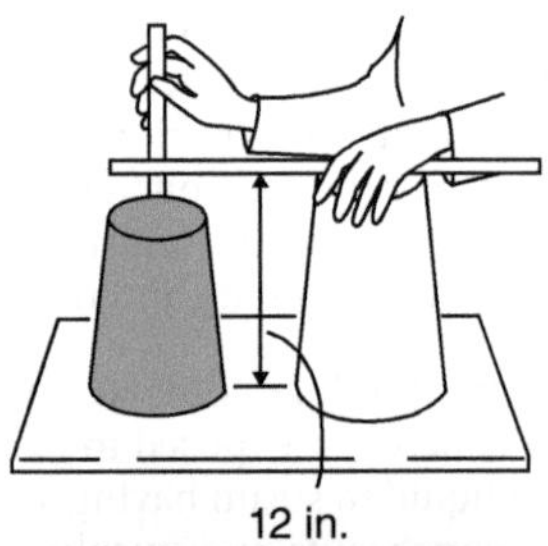

## Normal-Weight Concrete

The nominal weight of concrete is between 140 and 150 lb/ft$^3$.

**Compressive Strength** of normal-weight concrete ranges from 3,000 to 6,000 psi—the lower strength limit for high-strength concrete. Strength increases with age and curing. The

**TABLE 9.2** Selection of Mortar Type

| Type of construction | Slump (in.) | |
|---|---|---|
| | Maximum | Minimum |
| Reinforced foundation walls and footings | 3 | 1 |
| Plain footings, caissons, and substructure walls | 3 | 1 |
| Beams and reinforced walls | 4 | 1 |
| Building columns | 4 | 1 |
| Pavements and slabs | 3 | 1 |
| Mass concrete | 2 | 1 |

**FIGURE 9.4** Standard test cylinders for pigmented concrete.

standard is a 28-day compressive strength. Standard test cylinders 6-in. diameter by 12-in. high (Fig. 9.4) are made and cured in accordance with procedures described for standard-cured specimens in ASTM C31 "Standard Practice for Making and Curing Concrete Test Specimens in the Field." The cylinders are later tested in accordance with ASTM C39 "Standard Test Method for Compressive Strength of Cylindrical Concrete Specimens." The test result strength is the average of at least two cured specimens made from the same concrete sample and tested at the same age.

A standard method for designing a normal-weight concrete mix is given in ACI 211 "Standard Practice for Selecting Proportions for Normal, Heavyweight, and Mass Concrete."

**Volume changes** occur as concrete transitions from a plastic state to become a solid. As the plastic mix dries, shrinkage can produce cracks in the concrete surface. If a mix has an excessive amount of water, shrinkage cracks will likely occur. Excessive water increases the water–cement ratio (W/C), producing a weaker concrete, and all water cannot be absorbed in the hydration process, leading to bleeding to exposed surfaces.

Concrete expands when heated and contracts when cooled. Coefficients of thermal expansion (CTE) range from 3.2 millionths per °F for limestone to 7.0 millionths per °F for quartz. The most notable result of the response of concrete to thermal changes is the movement of slabs, where a 100-foot slab with a 50°F temperature drop and median CTE would contract about 0.3 in. Normal-weight concrete is not designed for fire exposure and will expands on being heated. A side effect is some strength loss and a reduction in the modulus of elasticity.

**Resistance to freezing and thawing** can be accomplished by proper air entrainment in the concrete, use of a mix with a minimum water content, and proper curing of the concrete. The American Concrete Institute (ACI) chapter 318 "Building Code Requirements for Reinforced Concrete" specifies the amount of air to use based upon exposure and maximum aggregate size.

**Chemical attack** may be internal (alkali-silica reaction) or external (sulfate attack or an aggressive service environment). In either case, the basic concerns are the characteristics of the available materials and the environment in which the concrete will be used. Alkali-reactive aggregates should be avoided, but if they must be used, a low-alkali cement complying with ASTM C150 Type II Modified should be selected. If sulfate attack is a concern, a low W/C (0.45 maximum) and air entrainment should be used with either a C150 Type V cement or a C150 Type II cement with C618 Type F fly ash. For protection from attack by other chemicals, a low W/C (0.45 maximum), more concrete cover over the reinforcing steel, a corrosion-protection additive, or a latex-modified concrete should be used. ACI 318 contains requirements for special exposure conditions.

**Acceptable appearance** depends on good workmanship and a supply of consistent materials. The formwork should be watertight and properly oiled before concrete placement. Forms should not be made of wood that will release sugars into the concrete and create a retarded surface finish. During concrete placement, the concrete should have consistent workability. The forms should be uniformly and consistently vibrated to consolidate the concrete. Form liners are an effective means to improve the appearance and texture of the exposed concrete surface. Special workmanship requires accurate placement to yield a desired surface texture indicated in the drawings. They require solid backing and secure attachment to maintain stability during the concrete pour. Sagging and stretching in hot weather must be factored during installation. Joints in form liners are sealed to prevent mortar leaks. The form liner is coated with a release agent similar to other formwork faces.

## Lightweight Concrete

Concrete weighing considerably less than the 144 lb/ft$^3$ of normal-weight concrete may be produced by use of lightweight aggregates or by expanding or foaming the concrete. Lightweight concrete is used principally to reduce the dead load of a structure and lower the cost of foundations. The light weight of the aggregates used for this type of concrete derives from the cellular structure of the particles. Hence, lightweight-aggregate concrete as well as foamed and expanded concretes have excellent fire-protection capabilities because of the internal voids in the aggregates or the concrete itself. When lightweight aggregates are used, they may be both fine and coarse, or lightweight coarse and normal-weight fine (sand), or normal-weight coarse and lightweight fine. The last combination is the least often used. Unit weights range from 90 lb/ft$^3$ (all aggregates lightweight) to 115 lb/ft$^3$ (sand lightweight). Typically, compressive strengths range from 2,500 to 4,000 psi. High-strength lightweight concretes, however, have been produced with maximum unit weights of 125 lb/ft$^3$ and strengths from 6,000 to 9,000 psi. Structural lightweight concretes are defined by the ACI as concretes with a 28-day compressive strength more than 2,500 psi and unit weight of 120 lb/ft$^3$ or less.

The variable amount of water absorbed in the voids of lightweight aggregates makes use of W/C difficult in design of a lightweight-aggregate mix. Air entrainment of 4–6% is desirable to prevent segregation. Maximum size of the coarse aggregate should not exceed half the depth of cover over the reinforcing steel.

Lightweight-aggregate concrete exposed to sulfates should have a compressive strength ranging from 3,750 to 4,750 psi (see ACI 318). For marine structures, the W/C should not exceed 0.40 and at least seven bags of cement should be used per cubic yard of concrete. The modulus of elasticity ($Ec$) of lightweight concrete generally ranges from 1,500,000 to 3,000,000 psi.

**Volume changes** occur in lightweight concrete as in normal-weight concrete, but lightweight concrete is more stable when exposed to heat. Drying shrinkage causes the most undesirable volume changes, because it produces cracks in the surfaces of the concrete. The primary cause of drying-shrinkage cracks is excessive water in the mix. The water has two effects. First, it increases the W/C and weakens the concrete. Second, the additional water beyond that needed for hydration of the cement creates an excessive number of bleed channels to the exposed surfaces. When the cement paste undergoes normal drying shrinkage, these channels cannot provide any resistance to ingress of aggressive chemicals. Creep is an important concern for lightweight concrete, as it is for normal-weight concrete, especially for prestressed concrete ("Standard Practice for Selecting Proportions for Structural Lightweight Concrete," ACI 211.2, and "Guide for Structural Lightweight Aggregate Concrete," ACI 213).

## Heavyweight Concrete

Concretes made with heavyweight aggregates are used for shielding and structural purposes in construction of nuclear reactors and other structures exposed to high-intensity radiation. Heavyweight aggregates are used where heavyweight is needed, such as ship ballast and encasement of underwater pipes, and for making shielding concretes because absorption of such radiation is proportional to density, and consequently, these aggregates have greater capacity for absorption than those ordinarily used for normal concrete. With such aggregates, concrete weighing up to about 385 lb/ft$^3$ can be produced.

Concrete made with limonite or magnetite can develop densities of 210–224 lb/ft$^3$ and compressive strengths of 3,200–5,700 psi. With barite, concrete may weigh 230 lb/ft$^3$ and have a strength of 6,000 psi. With steel punchings and sheared bars as coarse aggregate and steel shot as fine aggregate, densities of 250–288 lb/ft$^3$ and strengths of about 5,600 psi can be attained. Generally, grading of aggregates and mix proportions are similar to those used for normal concrete.

The properties of heavyweight concrete are similar to those of normal-weight concrete. Mixing and placing operations, however, are more difficult than those for normal-weight concrete, because of segregation. Good grading, high cement content, low W/C, and air entrainment should be employed to prevent segregation. Sometimes, heavyweight aggregates are grouted in place to avoid segregation. Heavyweight concretes usually do not have good resistance to weathering or abrasion ("Recommended Practice for Selecting Proportions for Normal, Heavyweight, and Mass Concrete," ACI 211.1).

## High-Performance Concretes

These concretes either have high-design strength (more than 6,000 psi for normal-weight concrete and 5,000 psi for lightweight concrete) or will be subjected to severe service environments. The difference between high-performance concretes and normal-weight concretes is that the former have lower W/C and smaller maximum aggregate size. ACI 318 specifies the W/C and compressive strengths for concrete in severe exposures and the maximum chloride-ion content of concrete. High-performance concrete is defined by either durability or strength-performance characteristics. Durability characteristics are resistance to freeze-thaw, scaling, abrasion, and chloride permeability. The strength characteristics have been defined in four grades as shown in Table 9.3.

High-strength, Portland-cement concretes generally incorporate in the mix fly ash, silica fume, or superplasticizers, or a combination of these admixtures. A retarder is often beneficial in controlling early hydration. The W/C may be as small as 0.25. The maximum size of aggregate should generally be limited to 1/2 in., with 3/8 in. being a commonly chosen size.

With superplasticizers, relatively high strengths can be achieved at early ages, such as 7-day strengths of normal concrete in 3 days and 28-day strengths in 7 days. Compressive strengths exceeding 10,000 psi can be achieved in 90 days.

Aside from reduction in W/C, the use of superplasticizers in the production of high-strength concretes does not require significant changes in mix proportioning. An increase in the range of sand content of about 5%, however, may help avoid a harsh mix. Curing is very important, because strength

**TABLE 9.3** High-Performance Concrete Strength Characteristics

| Characteristic | Grade 1 | Grade 2 | Grade 3 | Grade 4 |
|---|---|---|---|---|
| Compressive strength, ksi | $6 \leq x < 8$ | $8 \leq x < 10$ | $10 \leq x < 14$ | $x \geq 14$ |
| Modulus of elasticity, 103 ksi | $4 \leq x < 6$ | $6 \leq x < 7.5$ | $x \geq 7.5$ | |
| Shrinkage, microstrain, in./in. | $600 \leq x < 800$ | $400 \leq x < 600$ | $x < 400$ | |
| Creep, microstrain, in./in. | $0.41 < x \leq 0.52$ | $0.31 < x \leq 0.41$ | $0.21 < x \leq 0.31$ | $x \leq 0.21$ |

gain halts when water is no longer available for hydration. Also, it is important that proper quantities of air-entraining admixtures be determined by trial. Some air loss may result when melamine or reducers may actually increase air content. Larger amounts of air-entraining agent may be needed for high-strength concretes, especially for low-slump mixes with high cement content and mixes with large amounts of some types of fly ash. Furthermore, some types of superplasticizers and air-entraining admixtures may not be compatible with each other ("State-of-the-Art Report on High-Strength Concrete," ACI 363).

## Nonstructural or Foamed Cellular Concretes

These are formed by the use of admixtures that generate or liberate gas bubbles in concrete in the plastic stage. Aluminum powder, which reacts with the alkalis in cement to release hydrogen, is generally used for this purpose. Hydrogen peroxide, which generates oxygen, or activated carbon, which liberates absorbed air, can also be used. These foaming agents create stable, uniformly dispersed air spaces within the concrete when it sets. Perlite and vermiculite are most frequently used as aggregates. The resulting concrete may weigh 50 lb/ft$^3$ or less and have a compressive strength up to 2,500 psi. Applications of such lightweight concretes include topping and soundproofing barriers over structural concrete slabs.

The effectiveness of the admixture is controlled by the duration of mixing, handling, and placing of the mix relative to the gas-generation rate. The amount of unpolished aluminum powder to be added to a mix may range from 0.005 to 0.02% by weight of cement under normal conditions. Larger quantities, however, may be used to produce lower-strength concretes. More aluminum may be needed at low temperatures to achieve the same amount of concrete expansion, for example, twice as much as 40°F as at 70°F. Furthermore, at low temperatures, to speed up gas generation, it may be necessary to add to the mix alkalis such as sodium hydroxide, hydrated lime, or trisodium phosphate. Also, to prevent the powder from floating on the surface of mixing water, the aluminum may be premixed with sand or combined with other admixtures.

Curing is very important. If good curing practices and jointing are not followed, extensive drying shrinkage may result.

## Polymer Concretes

Plastics with long-chain molecules, called polymers, are used in several ways to enhance concrete properties: replacement of Portland cement, incorporation in a mix as an admixture, and impregnating hardened concrete.

Polymer concretes, such as methyl methacrylate and unsaturated polyester, in which a polymer replaces Portland cement may have more than double the strength and modulus of elasticity of Portland-cement concrete. Creep is reduced and resistance to freezing and thawing cycles is higher with the polymer concretes. After curing for a very short time, for example, overnight at room temperature, polymer concretes are ready for use, whereas ordinary concrete may have to cure for about a week before exposure to service loads.

Monomers and polymers may be used as admixtures for restoring and resurfacing deteriorated concrete surfaces. Latexes of methyl methacrylate, polyester, styrene, epoxy-styrene, furans, styrene-butadiene, and vinylidene chloride have been employed for these purposes. The resulting concrete hardens more rapidly than normal concrete. A polymer admixture may also be used to improve the bonding properties of Portland cement. Inserted in a mix as an emulsion for this purpose, the admixture supplies a significant amount of water to the mix, which becomes available for hydration of the cement. The release of the water also sets the emulsion. Hence, moist curing is not desirable, inasmuch as the emulsion needs to dry to develop the desired strength. A grout or mortar containing the bonding admixture develops a higher bond strength when applied as a thin layer than as a thick one and the bond may be stronger than materials being joined.

Impregnation of concrete with polymers is sometimes used to harden surfaces exposed to heavy traffic. Strength and other properties of the impregnated concrete are similar to those of concrete in which polymers replace Portland cement. Impregnation is achieved by first drying the concrete surface with heat and then soaking the surface with a monomer, such as methyl methacrylate, styrene, acrylonitrile, or tert-butyl styrene. It is subsequently cured with heat.

***Slab Toppings:*** At least partly because of excellent adhesion, epoxies are formulated with sand and other fillers to provide surfacing materials for concrete. Unlike standard concrete topping, epoxy-based surfacing materials can be thin. They are especially useful for smoothing uneven, irregular surfaces. The epoxy cures quickly, allowing use of the surface in a short time.

***Grout:*** Cracked concrete can be repaired with an epoxy grout. The grout is forced into cracks under pressure for deep penetration. Because of its good bonding strength, the epoxy grout can largely restore strength, while, at the same time, sealing the crack against penetration by liquids ("Polymers in Concrete," ACI 548; "Guide for the Use of Polymers in Concrete," ACI 548; and "Polymer Modified Concrete," SP-99, American Concrete Institute).

## CONCRETE MASONRY UNITS

Concrete masonry units (CMUs) are often used for both exterior and interior walls in all types of low-rise buildings and residences (Figs 9.5 and 9.6). They are durable, easy to install, fireproof, and low maintenance. Exposed units provide a durable finish in demanding environments. In factories across the country, CMUs are manufactured in molds from dry, stiff "no-slump" concrete mixtures. The units can be reinforced to meet specific structural applications.

CMUs are modular with the most common nominal hollow (2 core or cell) concrete block size being 8-by-8-by-16-in. (actual dimensions plus width of mortar joint). The order of terms is width, height, and length. The actual block dimensions are 7⅝ × 7⅝ × 15⅝ in. Solid 8-by-8-by-16-in. and 4-by-8-by-16-in. CMUs are also available. In fact, there is a large variety of hollow and solid sizes available.

The typical cement-aggregate ratio for concrete blocks is 1:6 with 60% of the aggregate being fines and 40% coarse. ASTM C-90-91 specifies the compressive strength requirements of CMUs.

**FIGURE 9.5** Concrete Masonry Unit wall construction—low-rise building.

**FIGURE 9.6** Concrete Masonry Unit wall construction—residential basement wall.

ASTM C90-16a "Standard Specification for Loadbearing Concrete Masonry Units" provides detail information on the three classes of CMUs: normal weight, medium weight, and lightweight. Lightweight is less than 105 lb/cf, medium weight is 105 to less than 125 lb/cf, and normal weight is 125 lb/cf or more. ASTM C90 specifies a minimum net area compressive strength of 2,000 lb/in.$^2$ for all three densities, whereas ASTM C129–17 "Standard Specification for Nonloadbearing Concrete Masonry Units" covers hollow and solid non-load-bearing CMUs made from Portland cement, water, and mineral aggregates.

Concrete masonry walls will provide fire ratings of up to 4 hr. Acoustical units are available to dampen sound, and are used in schools and churches to improve internal acoustics. Following ASTM C33, normal-weight aggregates can be blended with lightweight aggregates to provide a tight smooth texture.

There are open-ended units so the CMU can be threaded around reinforcing bars, eliminating the need to lift units over the top of the reinforcing bar, or to thread the reinforcement through the masonry cores after the wall is constructed.

## BRICK MASONRY

Brick, because of its simplicity to place, durability, heat retention capabilities, and fire resistance, is used in many capacities in the built environment.

### Brick Manufacturing

Bricks are formed of natural materials, most commonly clay minerals. Kaolin and shale are typical. Other additives such as manganese are mixed with the clay to produce different shades of brick and improve durability. Modern bricks are fired after being formed by one of three processes:

1. The **molded or soft mud** process uses clays mixed to contain between 20 and 30% water. This mixture is then placed into lubricated molds to form the bricks. This process can be completed by hand or with machines.
2. The **dry press** process works well with clay materials that possess a low plasticity. A small amount of water is mixed with the clay and then pressed into steel molds. Hydraulic or compressed air rams are used to apply the pressure which can range between 500 and 1,500 psi (3.4–10.3 MPa).
3. The **extruded or stiff mud** process mixes between 10 and 15% water into the clay to produce a level of plasticity (Fig. 9.7). Once the clay is mixed with water it enters a de-airing chamber. Under the vacuum in the de-airing chamber, air bubbles and holes are removed from the clay. This gives the clay increased plasticity and workability, which results in a higher strength brick. The clay is then extruded through a die that produces a column of clay. Textures and surface coatings are applied to this column of clay prior to the clay being cut to the size of individual brick. Approximately 90% of the brick in the United States is manufactured using the extrusion process.

**FIGURE 9.7** Extruded brick manufacturing process.

Wet brick contains between 7 and 30% moisture. Prior to firing, bricks are subjected to a drying process where most of the water is evaporated (Fig. 9.8). The drying oven temperature ranges from 100 to 400°F (38–204°C). Heat and humidity must be carefully regulated to avoid cracking the brick.

The process of firing bricks varies depending upon the type of kiln in which the brick is being fired. Tunnel kilns are the most common type of kilns, followed by periodic kilns. Firing in a tunnel kiln requires that the dried brick be loaded onto kiln carts and moved through various temperature zones within the kiln (Fig. 9.9). In a periodic kiln, brick are loaded, fired, and then unloaded.

The firing process takes place in stages, each at different temperatures:

- Final drying where all free water is evaporated occurs at 300–1800°F (149–982°C).
- Dehydration which drives the remaining water out of the clay occurs at 400°F (204°C).
- Oxidation where the clay begins to soften occurs from 1000 to 1800°F (538–982°C).
- Vitrification where the clay is transformed to a ceramic material with a low rate of absorption occurs at temperatures ranging from 1600 to 2400°F (871–1316°C).
- Cooling occurs under controlled conditions to produce the desired color and avoid thermal cracking. The cooling process ranges from 10 hr for tunnels and 5–24 hr for periodic kilns.

**FIGURE 9.8** Brick drying.

**FIGURE 9.9** Brick drying.

## Brick Properties

Brick possesses properties reflective of the raw material and the manufacturing process. Brick **durability** depends upon the degree of firing the brick received. Compressive strength and absorptions values are dependent upon firing temperatures and are used as predictors of durability in brick specifications as well as the saturation coefficient.

Brick **color** depends upon the chemical composition of the clay, the temperature at which it is fired and the method of firing. Iron has the greatest effect on the color of the brick due to the oxidation that occurs during the fire process. The process of flashing creates a reducing atmosphere in the kiln. This process creates a brick with a dark color.

Bricks may have various textures, coatings, and glazes. These are applied during the manufacturing process. These textures may be smooth or sand-finished from the forming mold. Textures can also be applied by rollers or brushes applied to the clay column as it exits the extruder die. Glazes can be applied to the brick surface during the firing process to achieve a variety of colors and reflectance that are impervious to water and water vapor.

During the manufacturing of brick, the raw clay material shrinks. This is accounted for in the forming process to ensure that the desired size of the finish brick is achieved. Shrinkage will vary depending on the type of clay. To produce bricks of uniform size, manufacturers control factors that contribute to shrinkage. However, due to the variation in the raw materials and temperature variations in kilns, complete uniformity is impossible. The ASTM standards for brick allow for size variations within certain tolerances.

The degree of firing, the method of manufacturing, and the raw materials used impact the compressive strength and absorption properties of the brick. Higher firing temperatures are associated with greater values of compressive strength and lower absorption. These properties are a function of the raw materials.

## Brick Sizing

**Specified brick dimensions** are the anticipated manufactured dimensions of brick. These are the dimensions used for design and are indicated in drawings and specifications. The dimensions do not take the size of the mortar joint into account.

**Actual brick dimensions** are the final measurements of the manufactured brick. These dimensions are within certain tolerances that are defined in the ASTM standards for each brick type. The tolerances vary depending on the size and type of the brick.

**Nominal brick dimensions** include the size of the brick plus the width of the mortar joint. They are used in modular construction of masonry elements. Nominal sizes of most bricks fit into a grid of 4″. This grid size is the same as doors, windows, and wood components that are commonly used in buildings (Fig. 9.10).

The predominant mortar joint width used in construction is 3/8″ and is referenced by the International Building Code through TMS 602 *Specification for Masonry Structures*. Always confirm the mortar joint width specified. Table 9.4 illustrates various brick sizes for 3/8″ mortar joints.

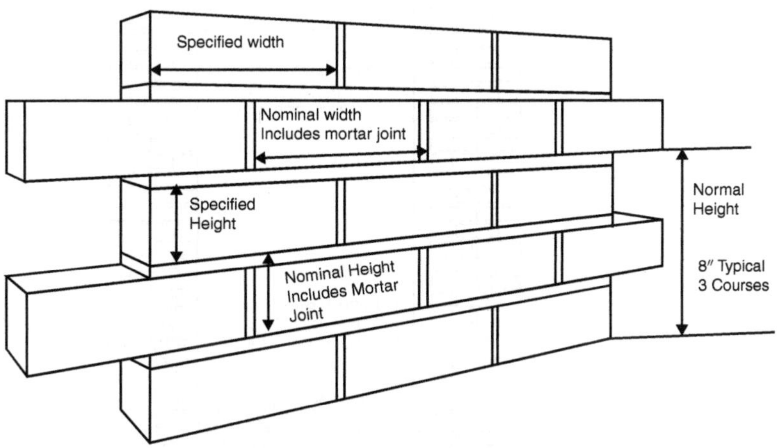


**FIGURE 9.10** Brick dimensions: specified vs. nominal.

**TABLE 9.4** Brick Sizes: 3/8″ Mortar Joint Between Bricks

| Brick type | Specified size D x H x L (in.) | Nominal size D x H x L | Vertical course |
|---|---|---|---|
| Standard | 3 5/8 × 2 ¼ × 8 | Not modular | 3 courses = 8″ |
| Modular | 3 5/8 × 2 ¼ × 7 5/8 | 4 × 2 2/3 × 8 | 3 courses = 8″ |
| Norman | 3 5/8 × 2 ¼ × 11 5/8 | 4 × 2 2/3 × 12 | 3 courses = 8″ |
| Roman | 3 5/8 × 1 5/8 × 11 5/8 | 4 × 2 × 12 | 1 course = 2″ |
| Jumbo | 3 5/8 × 2 3/4 × 8 | 4 × 3 × 8 | 1 course = 3″ |
| Economy | 3 5/8 × 3 5/8 × 7 5/8 | 4 × 4 × 8 | 1 course = 4″ |
| Engineer | 3 5/8 × 2 13/16 × 7 5/8 | 4 × 3 1/5 × 8 | 5 courses = 16″ |
| King | 2 ¾ × 2 5/8 × 9 5/8 | Not modular | 5 courses = 16″ |
| Queen | 2 3/4 × 2 3/4 × 7 5/8 | Not modular | 5 courses = 16″ |
| Utility | 3 5/8 × 3 5/8 × 11 5/8 | 4 × 4 × 12 | 1 course = 4″ |

## Calculating Brick Openings

In brick masonry construction, common building components, such as doors and windows, require openings. Calculating brick openings requires the counting of the courses in the vertical direction or the lengths of bricks in the horizontal direction (Fig. 9.11). This will result in the Masonry Opening. The additional mortar joint must be factored in to determine the actual opening size.

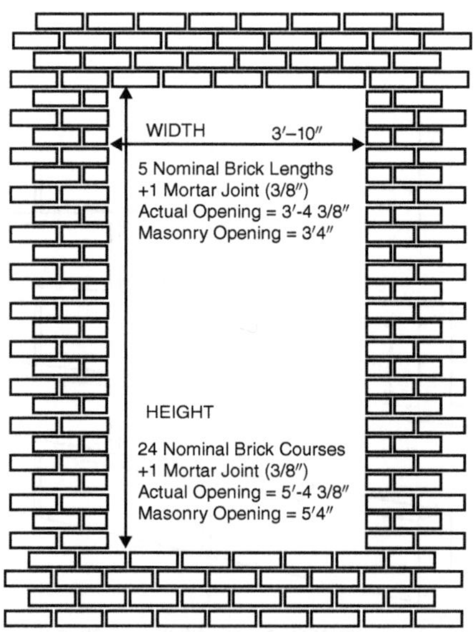


**FIGURE 9.11** Brick openings: nominal masonry opening vs. actual opening size.

## Brick Classifications and Grades

There are many different types of bricks, each with a specific use and application. Selection of bricks is based upon application. Characteristics considered when selecting bricks include their durability, load capacity, and appearance.

Building brick (ASTM C62) is used in both structural and nonstructural applications where appearance is not important. Facing brick (ASTM C216) is used in building applications where appearance is important. Both building and facing bricks are classified as solid bricks. Solid bricks are defined as having holes not greater than 25% of the unit's bed area. The coring of bricks reduces the amount of clay they contain. This assists with the speed of drying and improves the firing of bricks, creating bricks that are lighter and easier to handle. Hollow brick (ASTM C652) is a type of building or facing brick used in structural and veneer construction. This brick is used with anchors or reinforcing. Hollow brick has greater than 25% up to 60% void areas. Brick that is to be applied directly as a veneer is a thin veneer brick (ASTM C1088) and is used for appearance purposes only.

Paving brick is characterized based upon the application. Pedestrian paving brick (ASTM C902) is appropriate for areas of pedestrian or very light vehicle traffic, while heavy vehicular paving brick (ASTM C902) is used where there is standard or heavy vehicle traffic.

Specialized applications of bricks include firebox bricks (ASTM C1261) which are used in residential fireplaces and chemical-resistant bricks (ASTM C279) which are used in situations where the brick may be exposed to chemicals and acids. Bricks are also used in sewer and manhole construction (ASTM C32) and industrial flooring applications (ASTM C410).

Classifications of brick include grade, class, type, application, and use. ASTM criteria associated with each classification may include exposure or use conditions, appearance, physical properties, dimensional tolerances, and void area.

The grade of a brick is an indication of the durability when exposed to freezing temperatures and moisture. Factors that play a role in the durability of a brick include the compressive strength and water absorption. Grades are assigned based upon exposure conditions and resistance to weathering. There are three weathering index regions in the United States that are classified based on the number of freeze-thaw cycles that occur each year. Figure 9.12 shows the brick weathering regions in the United States.

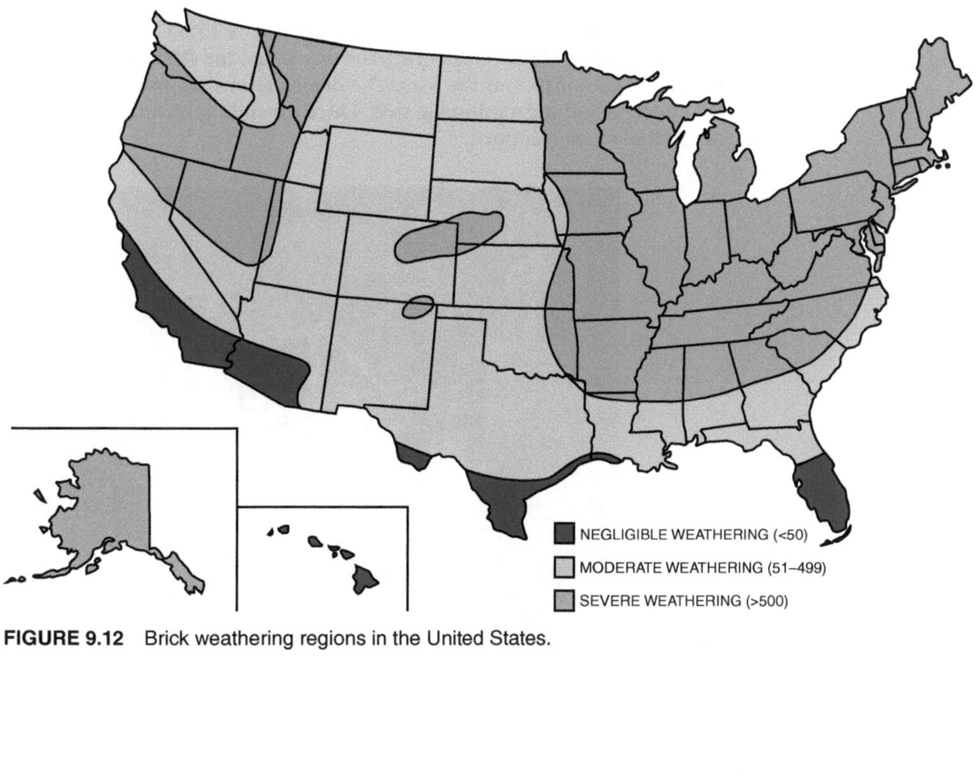


**FIGURE 9.12** Brick weathering regions in the United States.

Bricks graded as Severe Weathering (SW or SX) are the most durable and suitable for use when exposed to all types of freeze-thaw cycles and when the bricks are in contact with the ground.

Bricks graded as Moderate Weathering (MW or MX) are suitable for use where the bricks will not be wet when exposed to freeze-thaw cycles and above grade applications.

Bricks graded as Negligible Weathering (NW or NX) are suitable when the brick will not be exposed to freezing or damp conditions. This brick is also suitable for indoor use.

Brick type classifies bricks based upon their appearance in uniformity of size and shape. There are three different types of facing brick.

1. **FBX**, which allows for the least variation in terms of size of the brick, the least amount of distortion in shape, and a minimum amount of chippage. The dimensional tolerance for this type of brick is +/- 5/32″.
2. **FBS**, which is a general-purpose face brick, with wider variation in shape and size and a greater allowance for chippage. The dimensional tolerance for this type of brick is +/- 1/4″.
3. **FBA** allows for large nonuniformity in shape and size. There is no dimensional tolerance range specified.

Compressive strength is an important material property for structural brick applications. ASTM standards for brick require minimum compressive strengths to ensure durability. These minimums may be as little as one-fifth the actual unit compressive strength of the brick determined in accordance with ASTM C67, Method of Sampling, and Testing Brick and Structural Clay Tile. The specified minimum compressive strengths range between 1,500 and 3,000 psi (10–21 MPa) for facing and building bricks. Structural masonry bricks have higher requirements for compressive strength due to their loading. These high-strength bricks may range between 10,000 and 20,000 psi (69–138 MPa). Brickwork strength depends on the strength of the entire assembly. This includes not only the bricks, but the mortar and any reinforcing included.

## REINFORCING STEEL

Steel, due to its favorable tensile, elastic, and thermal properties, coupled with its ability to develop bonds, is used in construction to reinforce concrete.

Reinforcing bars are produced from molten steel (Fig. 9.13). The molten steel is hot rolled and then run through a series of dies (Fig. 9.14). The dies shape the steel into round bars that have deformations formed in the surface of the bars. The deformations are used to help in transferring loads between the concrete and the reinforcing steel. Once the rolling is completed the bars are cut to standard lengths, bundled, and shipped.

**FIGURE 9.13** Molten steel bars.

**FIGURE 9.14** Reinforcing steel dies.

Reinforcing bars are manufactured in standard diameters. In the United States, rebar numbers are based on the number of 1/8 in. included in the diameter of the bar. With the metric system increasing in use within the U.S. construction industry, a "soft conversion" system of measurement has been developed. Table 9.5 shows the relationships. The bars are exactly the same size but the number corresponds approximately to the diameter of the bar in millimeters. This measure was taken instead of a "hard conversion," which would require changing the sizing system of materials in one measurement system to conform to sizes in the other measurement system and require all the mills to produce different bar sizes.

**TABLE 9.5** ASTM Standard Rebar Sizes

| Bar number | | Risk Assessment Committee Member | | | | | |
|---|---|---|---|---|---|---|---|
| | | Weight | | Bar diameter | | Cross-sectional area | |
| American | SI | lb/ft | kg/m | in. | mm | in.$^2$ | mm$^2$ |
| #3 | #10 | 0.376 | 0.560 | 0.375 | 9.5 | 0.11 | 71 |
| #4 | #13 | 0.668 | 0.944 | 0.500 | 12.7 | 0.20 | 129 |
| #5 | #16 | 1.043 | 1.552 | 0.625 | 15.9 | 0.31 | 199 |
| #6 | #19 | 1.502 | 2.235 | 0.750 | 19.1 | 0.44 | 284 |
| #7 | #22 | 2.044 | 3.042 | 0.875 | 22.2 | 0.60 | 387 |
| #8 | #25 | 2.670 | 3.973 | 1.000 | 25.4 | 0.79 | 510 |
| #9 | #29 | 3.400 | 5.060 | 1.128 | 28.7 | 1.00 | 645 |
| #10 | #32 | 4.303 | 6.404 | 1.270 | 32.3 | 1.27 | 819 |
| #11 | #36 | 5.313 | 7.907 | 1.410 | 35.8 | 1.56 | 1,006 |
| #14 | #43 | 7.650 | 11.38 | 1.693 | 43.0 | 2.25 | 1,452 |
| #18 | #57 | 13.600 | 20.24 | 2.257 | 57.3 | 4.00 | 2,581 |

All deformed steel bars are required to be marked with identifying symbols rolled into one side of each bar (Fig. 9.15). The symbols must consist of the information shown in Table 9.6 and can be oriented to read vertically or horizontally.

Reinforcement bars in the United States are available in various grades of steel and are manufactured according to ASTM Standard A615/A615M "Standard Specification for Deformed and Plain Carbon-Steel Bars for Concrete Reinforcement." The most common grades of steel and their corresponding yield strengths are shown in Table 9.7.

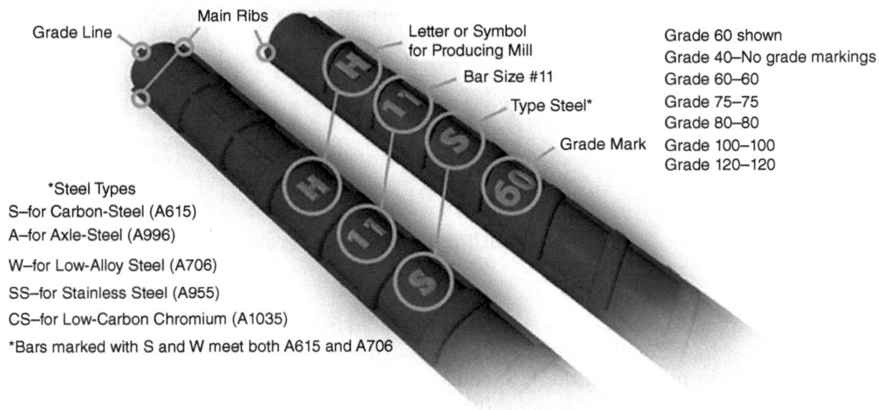


**FIGURE 9.15** Standard bar markings. Retrieved from: Concrete Reinforcing Steel Institute, https://www.crsi.org/images/cmsIT/fckeditor/Image/reinforcing-basics/ASTM-Bar-Marking.jpg

**TABLE 9.6** Reinforcement Bar Marks

| Mark | Meaning |
|---|---|
| Symbol or series of symbols | Designates producer's mill |
| Number | Designates bar size |
| S, W, rail symbol or rail symbol and R, A | Designates the type of steel<br>S—billet-steel<br>W—low-alloy steel<br>Rail symbol or R—rail steel<br>A—axle steel |
| Blank or number | Grade mark (blank for Grade 300/Grade40 and Grade 350/Grade 50, otherwise the grade number) |

**TABLE 9.7** Standard Grades of Reinforcement

| Pound grade | Metric grade | Minimum yield strength | |
|---|---|---|---|
| | | psi | MPA |
| Grade 40 | Grade 280 | 40,000 | 280 |
| Grade 60 | Grade 420 | 60,000 | 420 |
| Grade 75 | Grade 520 | 75,000 | 520 |

Other grades of steel are available depending on the use of the reinforcement. ASTM has several standards that are related to specific applications. ASTM A1035 "Standard Specification for Deformed and Plain, Low-carbon, Chromium, Steel Bars for Concrete Reinforcement" applies to high-strength steels that have yield strengths between 100,000 and 120,000 psi depending on the alloy used. Using this type of bar may improve constructability and resolve congestion issues in tight applications.

ASTM A706 "Standard Specification for Deformed and Plain Low-Alloy Steel Bars for Concrete Reinforcement" is a specification for low-alloy reinforcement bars. These bars are available in numerous grades, including grades 60 and 80. A706 rebar is intended for welding without preheating and is used in applications that require appreciable amounts of welding. A706 rebar can also be used in seismic applications.

Steel is susceptible to corrosion when exposed to air and water. Reinforcing steel in concrete exposed to de-icing salts or is frequently wet must be protected. The most common of these methods is epoxy coating; however, galvanized and stainless steel reinforcements are also used. Epoxy-coated rebar is manufactured according to ASTM A775 or A934 and acts as a barrier, preventing the steel from being exposed to air, water, and oxidizing agents. It is common in bridge decks, parking structures, and structures near salt water. Galvanized rebar is manufactured according to ASTM A767 and involves coating the rebar with zinc. The zinc coating provides a barrier to corrosion and provides a sacrificial layer if the rebar is damaged. Stainless steel rebar is manufactured according to ASTM A995M or with a stainless steel cladding which provides enhanced corrosion resistance.

Welded wire fabric reinforcement is an alternative to standard reinforcement bars. It is a square or rectangular mesh of wires that are factory welded at all intersections. Welded wire fabric is used to resist temperature and shrinkage cracks in concrete slabs and as tie reinforcement in columns. It is manufactured in either plain (ASTM A185) or deformed (ASTM A184M/A) wire according to ASTM standards. It can also be epoxy coated.

## STRUCTURAL STEEL

Steel, due to its strength, durability, speed of installation, ability to be fabricated in different sizes, and cost effectiveness, is a commonly used construction material.

Steel is produced using either the blast furnace or the electric arc furnace. The blast furnace uses coke, iron ore, and limestone to produce pig iron from iron oxides. Pig iron is also referred to as molten steel. The electric arc furnace produces steel by using an electrical current to melt scrap steel and direct reduced iron into molten steel. The carbon-rich molten steel is delivered to a basic oxygen furnace where it is blown with pure oxygen. The process removes the carbon via oxidation, reducing carbon levels from 4% to approximately 0.04%. The molten steel is then poured into the ladle. In the ladle the molten steel undergoes secondary processing and refining. In the refining station, the liquid steel is cleaned of impurities by stirring it with an inert gas such as argon. Impurities such as aluminum oxide and manganese sulfides attach themselves to the argon gas bubbles and float to the steel surface to be absorbed into the slag layer above the steel. The steel chemistry is further fine-tuned by addition of alloys to alter the properties of the steel.

Once the steel reaches the appropriate chemical composition it is then cast. The casting process transforms the liquid steel into a solid predetermined shape and size (Fig. 9.16). The shape is then cooled under controlled conditions. The shape may be used as a semifinished product or sent for further processing by rolling mills or machining. Steel products manufactured by casting include ingots, slabs, billets, and blooms.

To shape steel into a final commercial product, it is either rolled or forged into semifinished cast products or blanks and then processed to achieve its final form and properties (Fig. 9.17). When steel is cast into either ingot, slabs, billets, or blooms, it possesses very poor mechanical and physical properties. The process of shaping and treating the steel involves incrementally reducing the cross-sectional area of the cast product. While the steel is at an elevated temperature, the structure of the cast product is broken down and allowed to cool. During the cooling process, the steel recrystallizes, forming a grain structure, which provides a basis for further processing to establish the final product's

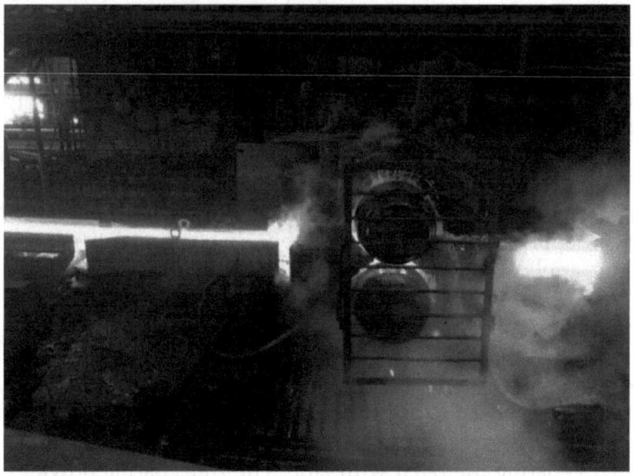

**FIGURE 9.16** Molten steel being cast into billets.

**FIGURE 9.17** Steel billets being moved for shaping.

mechanical properties. Commercial steel products include plates, hot- and cold-rolled coils, structural products, bars and wire rods, rails, seamless tubes, and pipes.

There are many types of structural steel for construction purposes. Structural steel exists in various shapes and grades, depending on the requirements for a specific application. Structural steels are classified by the shape of their cross-sections. These commonly include A, W, I, S, and C. The grade of the steel directly affects the mechanical properties. The requirements of the design will dictate the grade of structural steel used on a project. According to the World Steel Association, there are over 3,500 different grades of steel, encompassing unique physical, chemical, and environmental properties.

**Carbon steels** have properties comprised primarily of the element carbon. This type of steel relies on the carbon content for its structure. Carbon steels are generally categorized by their carbon content as low-carbon (<0.3%), medium-carbon (0.3–0.6%), high-carbon (0.6–1%), and ultrahigh-carbon (1.25–2%) steels. Most of the steel produced in the world is carbon steel. It is commonly used in structural pipe and tubing.

**High-strength low-alloy steels** possess better mechanical properties and are more resistant to corrosion than carbon steels. These improved properties are obtained by adding alloying elements. Common elements include columbium, vanadium, and titanium, used alone or in combination. Weathering steel possesses high resistance against atmospheric corrosion. This steel forms a passive, rustlike layer on the surface that protects the layers underneath from corrosion. High-strength low-alloy steel is commonly used in structural shapes and plates.

**Forged steels** are alloys of carbon and iron that are worked to a predetermined shape. Forging is accomplished by applying mechanical and thermal energy to steel ingots or billets, using specialized machines or hydraulic hammers. This process introduces a uniform grain structure to the material, which improves the structure of the steel matrix by removing voids and gas bubbles. The process increases overall strength of the steel.

**Quenched and tempered alloy steels** are heat-treated steel. The steel is heated below its melting point and then cooled through quenching to improve its physical properties. The quenching process uses water, oil, forced air, or gases to complete the cooling process. This method is used to increase the strength and hardness of steel.

The grading systems applied to steel provide a way to categorize it based on all the different factors that can influence its properties and uses. The ASTM Grading System assigns each metal a letter prefix based on its overall category, as well as a sequentially assigned number that corresponds with that metal's specific properties. "A" is the designation for iron and steel materials.

ASTM A36 steel, with a yield strength of 36,000 psi, is the most common material for structural steel shapes. ASTM A992 has become the most common material for wide flange and I beams in building applications. Steels used for specialized structural applications are low-alloy, high-strength (ASTM A588 and A572), and stainless steels. Steel members are connected to each other using bolts and rivets with plates (Fig. 9.18). These plates attach to the beam webs or they may be welded directly to each other. The American Institute of Steel Construction (AISC) specifies for both framed-beam and heavy framed-beam connections the appropriate number of rows of bolts or rivets based on several standard-diameter fasteners.

Forms of structural steel are standardized by the American Iron and Steel Institute (AISI). In describing the shape of structural steel in construction specifications for beams, tees, and channels, reference is made to flanges and webs. The flange is identified as the flat portion which is normally situated at the top and the bottom of the beam. The web connects the two flanges, and in normal construction is oriented vertically to give the beam its high resistance to bending. The flanges resist bending stresses and the webs resist shear stress. Angles are identified by their leg lengths, which may

**FIGURE 9.18** Bolted steel connections.

be equal or unequal. A distinction is recognized between "bar-size" and "structural-size" shapes. Bar size is used to describe any shape whose major sectional dimension is under 3 in.

The nomenclature for steel shapes follows two standards:

1. For wide flange, bearing pile, S-shapes, channels, and tees: the letter indicates the shape, the first number indicates the nominal height, and the second number indicates the weight per 1 foot of length. For instance, a W12 × 40 is a wide-flange shape that has a nominal height of 12″ and weighs 40 pounds per foot of length.
2. For steel tubes, pipes, plates, and angles: the three numbers indicate the height, width, and thickness of the steel.

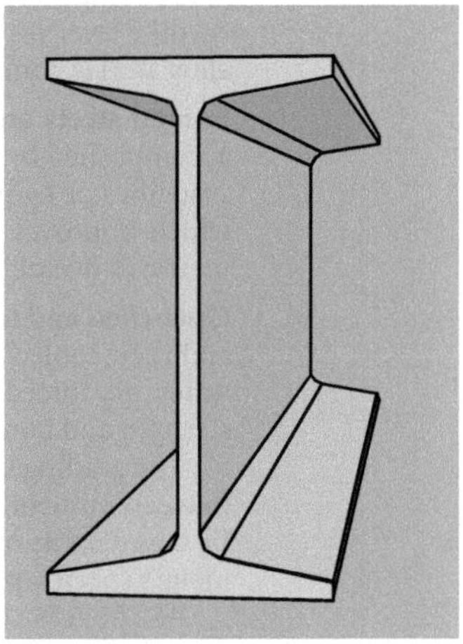

The standard beam has a rolled section with two parallel flanges that are connected by a web. These beams are specified by the letter S, then the nominal web height in inches, then the per-foot weight in pounds. For example an S15 × 50 represents a beam that is 15 in. deep and weighs 50 lb/ft. These beams are commonly lightweight in form. Standard beams have tapered flanges that are thicker than the webs. S-shapes are used less frequently than wide-flange beams because they possess less strength and are less adaptable than wide-flange beams.

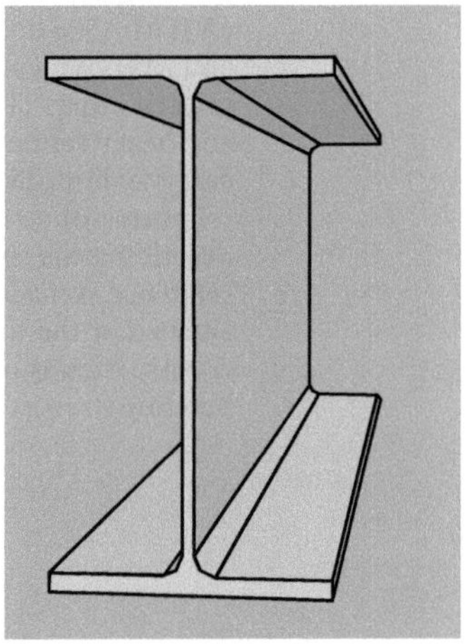

A wide-flange beam is also referred to as an H-beam due to their distinctive shape, similar to a capital letter "H" when viewed in cross-section. Wide-flange beams have flanges that are nearly perpendicular to the web of the material. This shape is very efficient for carrying loads without bending. These beams are used in bridge construction, buildings, and cranes. The thickness of the wide-flange beam will determine the type of applications for which it is best suited. Knowing the depth of section and pounds per foot is essential. Wide-flange beams come in a variety of sizes, lengths, and pounds per foot. The common ASTM grades for these beams are A36 and A992. These beams are specified as W18 × 50 where the W indicates the type of beam, the first number indicates the nominal depth, and the second indicates the weight per foot of beam.

I-beams are commonly used to support trusses and in framing buildings (Fig. 9.19). The beam has two horizontal elements, the flanges, with a vertical element as the web. The web is capable of resisting shear forces, while the horizontal flanges resist most of the beam's bending movement. The I-shape is very effective at carrying shear and bending loads in the web's plane. The difference between an I-beam and an H-beam is all in the shape. H-beams have top and bottom flanges that stick out further from the web than the flanges of I-beams. I-beams are shorter and are not as wide as H-beams. These can also be used as bearing piles where the flange and web have equal thicknesses and the flanges are parallel.

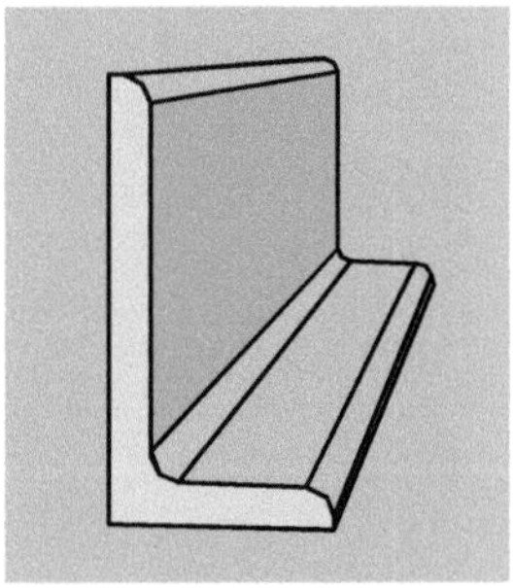

Angle beams are shaped like the letter "L." They are made with two legs that come together at a 90-degree angle. Angle beams are available in shapes that have equal or unequal sizes of each leg. The nomenclature for an angle begins with the letter L and then the size of each leg. It is identified by the dimension and thickness of its legs, for example,

**FIGURE 9.19** Structural steel framing of a building.

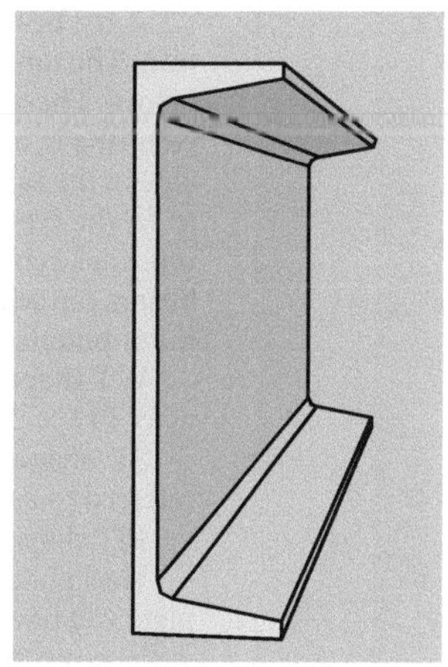

angle 2 in. × 2 in. × 0.5 in. is indicated as L2×2×0.5. When an angle has unequal legs, the dimension of the wider leg is given first. An L2×2×0.5 × L6×3×0.5 indicates an angle with unequal legs. Angles are typically used in floor systems because of the reduced structural depth. They are not commonly used in beam applications. If used as a beam they are doubled up and connected in a manner approximating a tee.

Structural C channels have a C-shaped cross-section. Channels have single webs flanked by upper and lower flanges. These flanges project perpendicularly to the webs, making the channel appear like half a beam with a full web-thickness. The channel flanges have sloped inner flange surfaces and a taper of approximately a 16-2/3% slope. C-shaped beams are cost-effective solutions for short- to medium-span structures. Channels are not ordinarily applied as beams but the flat faces allow them to be bolted up to other flat faces.

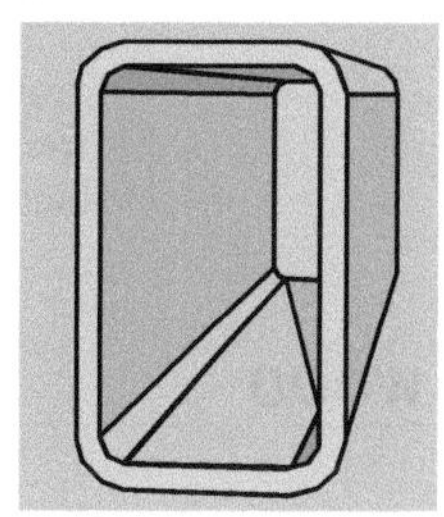

Hollow structural sections (HSS) refer to high-strength welded-steel tubing used as structural elements in buildings. HSSs can be rectangular, square, elliptical, or circular in section. The rectangular shapes are commonly used for welded frames where the loads act upon the structure in several directions. Circular HSSs, sometimes called round structural tubing, are used for load-bearing columns. These shapes all start as welded rounds, with the square and rectangular tubing put through additional forming to create final products with heavy radii at their corners (Fig. 9.20).

Structural steel pipes are important for a variety of construction applications, lending strength and stability. Pipes are hollow, cylindrical tubes that come in a variety of sizes. The pipes are referred to as Pipe 4 STD in construction specification.

**FIGURE 9.20** Hollow structural steel sections awaiting erection.

A tee beam is a load-bearing beam with a T-shaped cross-section. The top of this cross-section is the flange, with the vertical web below. These beams are cut longitudinally through their webs, resulting in a single full-flange and a half-deep web. They are available in the same shapes as full-size beams. These beams are referred to by the type of beam from which the T was split, followed by the nominal web depth in inches and the weight in pounds per foot. Tee beams can withstand large loads but lack the bottom flange, making it not practical for certain applications.

WT shapes are cut from a wide-flange beam and are referred to as WT12 × 38.

ST shapes are cut from American Standard Beam and are referred to as ST12 × 38.

MT shapes are cut from nonstandard I-shapes and are referred to as MT12 × 38.

Steel plate is considered a structural shape whose cross-section is in the form of a flat rectangle. A plate will have a width of greater than 8″ and a thickness of ¼″ or greater. Common uses for plates are with connections between structural members or as component parts of built-up structural members. Plates are referred to by their thickness and width in inches. In practice, a plate may be referred to by its approximate weight per square foot for a specified thickness.

The use of steel in construction has many benefits. Structural steel possesses superior strength and greater durability than other construction materials. It has a high strength-to-weight ratio and tensile strength. Its strength, stiffness, and ductile properties allow it to be fabricated into many different shapes (Figs. 9.21 and 9.22).

## WOOD

Wood is a natural product harvested from trees. How a tree grows determines the strength and properties of wood for engineering and building purposes.

**FIGURE 9.21** Steel being shaped into a curving staircase.

**FIGURE 9.22** Steel truss systems.

## Trees

The bark, whose dead layer protects the growing cells of the tree from insects and whose live layer conducts nutrients to the tree, is useless as a structural material. Growing inside the bark is a thin layer of the tree called the cambium layer. The layer of wood within the cambium is the sapwood. The fibers of the sapwood are active in the life process of the tree. This layer stores nutrients and pumps sap from the roots and distributes to the rest of the tree. As the sapwood dies it becomes heartwood. The heartwood is at the center of the tree and consists of dark dead tree fibers. The darker shade of the heartwood is due to the resins, gums, and minerals it contains. At the center of the tree is a thin vein of soft tissue called pith extending the length of the tree and possessing no strength.

Wood consists of tubular cells whose long axes are parallel to the long axis of the trunk. The cells are made of cellulose and are bound together by a substance called lignin. The direction of the long axes of the cells is referred to as the grain of the wood. The direction of the grain is important as this is what determines physical and aesthetic properties of wood cut from the tree. Properties of wood parallel to the grain are very different than those that are perpendicular to the grain.

Tree growth is rapid in temperate climates during the spring when water is readily available and the air is cool. As water becomes less available in the summer, growth slows. Springwood is lighter and less dense and produces a larger band of growth. Summerwood is heavier, darker, harder, and stronger than springwood, because of dense thicker fiber walls. These alternating growth patterns produce concentric rings of springwood and summerwood. The age of the tree can be determined by the number of these rings. The strength of a piece of wood can be judged by the percentage of summerwood compared to springwood showing in cross-section.

**Softwood** refers to lumber that has been cut from a coniferous or an evergreen tree such as pine, cedar, or spruce. Softwoods are frequently used as building materials.

**Hardwood** refers to lumber cut from broadleaf or angiosperms such as walnut, hickory, or maple.

Softwoods do not have the same cellular structure as hardwoods, giving it an uninteresting grain pattern. Softwoods account for most of the lumber used in construction today, as they are plentiful, inexpensive, and rapidly renewable as they are fast growing. Hardwoods have a much more complex cellular structure that produces an interesting grain that makes them the best choice for furniture or interior finishes.

Lumber is produced in sawmills and how the tree is cut and processed affects the use of the material, the cost, and the material properties. Once lumber is cut, it must be seasoned. The seasoning process involves drying wood to a specific moisture content to avoid additional shrinkage when the wood is put into use. The water content to which the lumber is dried depends upon the specifications for its use. Framing lumber must be dried to a moisture content of 15% to be considered seasoned.

## Lumber Grading

Before the lumber leaves the mill, it is **graded**. The grading of lumber is for either structural purposes or appearance depending upon its use. The structural grading is completed by either visual inspection or by a machine. In machine grading, the structural properties of the wood are assessed and a grade is stamped automatically on the piece of wood. In the machine method, the wood is either flexed to determine the resistance to bending or by scanning the wood electronically to measure its density. Visual grading is accomplished by trained inspectors who inspect the wood, judge it, and then stamp it with a grade. The inspectors are trained in industry-wide grading rules that are based on accepted standards. When grading for appearance, the inspector is visually reviewing the material for imperfections like knots or checks. Figure 9.23 is a sample grading stamp used by the Western Wood Products Association.

**FIGURE 9.23** Western Wood Products Grading Stamp. (Downloaded from Western Wood Products website.)

**WWPA certification mark**—Indicates the lumber bearing this trademark has been graded under the supervision of Western Wood Products Association.

12 **Mill identification**—This is the originating mill identifier, either an assigned mill number or by the firm's name or brand.

STAND **Grade designation**—The grade is shown by the grade name, number, or abbreviation.

D FIR **Species**—Indicates species by individual or combination species name.

S-DRY **Seasoning**—The moisture content or conditioning of the piece at the time of manufacturing.

**MC-15, KD-15**—15% maximum moisture content.
**S-DRY, KD**—19% maximum moisture content.
**S-GRN**—Over 19% moisture content (unseasoned).

The grading criteria is different based upon the application or intended use of the lumber. Tables 9.8, 9.9, and 9.10 detail the grading criteria for dimensional lumber, structural decking, and timber, respectively. Table 9.11 outlines the criteria used in the appearance grading of lumber.

The structural properties of wood vary significantly based upon the grade, the species, and the direction in which the wood is loaded with respect to the grain. Wood is stronger when loaded parallel to the grain compared to loading perpendicular to the grain.

**TABLE 9.8** Dimensional Lumber Grades

| Dimensional lumber grades | | | |
|---|---|---|---|
| **Product** | **Lumber grades** | **WWPA western lumber grading rules section reference** | **Typical uses** |
| Structural light framing (SLF)<br>2″– 4″ thick<br>2″–4″ wide | Select structural<br>No. 1<br>No. 2<br>No. 3 | 42.10<br>42.11<br>42.12<br>42.13 | Structural applications where the highest design values are needed in light of framing sizes |
| Light framing (LF)<br>2″–4″ thick<br>2″–4″ wide | Construction<br>Standard<br>Utility | 40.11<br>40.12<br>40.13 | Where high-strength values are not required, such as wall framing, plates, sills, cripples, blocking, etc. |
| Stud<br>2″–4″ thick<br>2″ and wider | STUD | 41.13 | An optional all-purpose grade designed primarily for stud uses, including bearing walls |
| Structural joists and planks (SJ&P) | Select structural<br>No. 1<br>No. 2<br>No. 3 | 62.10<br>62.11<br>62.12<br>62.13 | Intended to fit engineering applications for lumber 5″ and wider, such as joists, rafters, headers, beams, trusses, and general framing |

**TABLE 9.9** Structural Decking Lumber Grades

| Structural decking grades | | | |
|---|---|---|---|
| **Product** | **Lumber grades** | **WWPA western lumber grading rules section reference** | **Typical uses** |
| Structural decking<br>2″–4″ thick<br>4″–12″ wide | Selected decking | 55.11 | Used where the appearance of the best face is of primary importance |
| | Commercial decking | 55.12 | Customarily used when appearance is not of primary importance |

**TABLE 9.10** Timber Grades

| Timber grades | | | |
|---|---|---|---|
| **Product** | **Lumber grades** | **WWPA western lumber grading rules section reference** | **Typical uses** |
| Beams and stringers<br>5″ and thicker, width more than 2″ greater than thickness | Dense select structural<br>Dense No. 1<br>Select structural<br>No. 1<br>No. 2 | 53.00 & 170.00<br>53.00 & 170.00<br>53.00 & 170.00<br>70.10<br>70.11<br>70.12 | Grades are designed for beam and stringer-type uses when sizes larger than 4″ nominal thickness are required |
| Post and timbers<br>5″ × 5″ and larger, width not more than 2″ greater than thickness | Dense select structural<br>Dense No. 1<br>Select structural<br>No. 1<br>No. 2 | 53.00 & 170.00<br>53.00 & 170.00<br>53.00 & 170.00<br>80.10<br>80.11<br>80.12 | Grades are designed for vertically loaded applications where sizes larger than 4″ nominal thickness are required |

### Lumber Dimensions

Attention to the dimensions of lumber is important in engineering and construction. Lumber dimensions are usually provided and sold in nominal measurements. Nominal measurements are the size of a piece of lumber when it has been rough cut prior to surfacing. The actual measurements of the piece of lumber are the final size after the drying and plaining processes are complete. Shrinkage is more pronounced laterally (across the grain) than it is longitudinally (with the grain), so a board sold as an 8-ft 2 × 4 is generally very close to the full 8 ft in length.

## *WOOD PRODUCTS*

Wood products, due to their precise design specifications, provide uniformity and predictability in their performance. These materials are made of nature wood strands, particles, veneers, and fibers that are bound together using adhesives. The various products that are created provide a range of applications, durability, and structural performance beyond traditional wood.

**TABLE 9.11** Appearance Lumber Grades

| Appearance lumber grades | | | | |
|---|---|---|---|---|
| | **Product** | **Lumber grades** | **WWPA western lumber grading rules section reference** | **Typical uses** |
| Highest quality appearance grades | Selects (all species) | B & BTR SELECT<br>C SELECT<br>D SELECT | 10.11<br>10.12<br>10.13 | SUPREME<br>CHOICE<br>QUALITY |
| | Finish (usually available only in Doug Fir and Hem-Fir) | SUPERIOR<br>PRIME<br>E | 10.51<br>10.52<br>10.53 | |
| | Special Western Red Cedar Pattern Grades | CLEAR HEART<br>A GRADE<br>B GRADE | 20.11<br>20.12<br>20.13 | |
| | Common Boards (WWPA Rules) (primarily in pines, spruces and cedars) | 1 COMMON<br>2 COMMON<br>3 COMMON<br>4 COMMON<br>5 COMMON | 30.11<br>30.12<br>30.13<br>30.14<br>30.15 | COLONIAL<br>STERLING<br>STANDARD<br>UTILITY<br>INDUSTRIAL |
| General-purpose grades | Alternate Boards (WCLIB Rules) (primarily in Doug Fir and Hem-Fir | SELECT<br>MERCHANTBALE<br>CONSTRUCTION<br>STANDARD<br>UTILITY<br>ECONOMY | WCLIB<br>118-a<br>118-b<br>118-c<br>118-d<br>118-e | |
| | Special Western Red Cedar Pattern Grades | SELECT KNOTTY<br>QUALITY KNOTTY | WCLIB<br>111-e<br>111-f | |

## GLUE-LAMS

Glulam is manufactured from wood laminations that are bonded together with adhesives. The grain of each lamination or "lam" runs parallel with the length of the member.

Glue-laminated timbers are manufactured to meet a range of design stresses. Beams are manufactured with the strongest lams on the bottom and top of the beam, where the maximum tension and compression stresses occur. Lower grade lumber is used in lower stressed zones of the beam.

Maximum allowable bending stress is the property of most concern with bending members. APA Glulam Design Specifications, Form EWS Y117 provide design stresses for all glulam produced by APA member.

Glulam beams are available in different levels of appearance. They are classified based upon application and aesthetic requirements. These beams may look different, but they all have the same structural characteristics for a given strength grade.

Standard glulam appearance classifications are shown in Table 9.12.

Glulam beams are versatile and possess different load-carrying capacities based on different combinations of lams (Fig. 9.24). The APA Data File: Glued Laminated Beam Design Tables, Form EWS S475 provides tables of load-carrying capacities for glulam. Glulam beams that have been manufactured in conformance with ANSI/AITC Standard A190.1 American National Standard for Structural Glue Laminated Timber are certified with the APA EWS trademark stamp.

**TABLE 9.12** Glulam Appearance Classifications

| Appearance classification | Application |
|---|---|
| Framing | Concealed application; provided in widths designed to fit flush with 2 × 4 and 2 × 6 wall framing |
| Framing-L | Concealed application, LVL laminations are used on the outermost surfaces |
| Industrial | Concealed application or where appearance is not important |
| Industrial-L | Concealed application or where appearance is not important; LVL laminations are used on the outermost surfaces |
| Architectural | Exposed to view |
| Premium | Custom order where finished appearance is of primary importance |

**FIGURE 9.24** Glulaminated beams.

## STRUCTURAL COMPOSITE LUMBER

**Structural composite lumber (SCL)** is a group of engineered wood products that are used for structural applications. These products are manufactured in blocks by layering dried wood veneers, strands, or flakes with moisture-resistant adhesives. The blocks are then cured in a heated press. These blocks, referred to as billets, are then cut into specified sizes.

SCL products remain stable and less prone to warping, splitting, or shrinking when compared to sawn timber as well as the ability to take greater loads. They can be used for many structural applications to include beams, joists, studs, columns, rafters, and I-Joist and I-Beam flanges. SCL products include Laminated Veneer Lumber (LVL), Laminated Strand Lumber (LSL), Parallel Strand Lumber (PSL), and Oriented Strand Lumber (OSL).

**Laminated veneer lumber (LVL)** is the most widely used SCL product used for structural applications. LVL is manufactured by bonding together rotary peeled or sliced thin wood veneers under heat and pressure. Prior to lamination, the veneers are dried and the grains of each veneer are oriented parallel to the long direction. This makes LVL stronger, straighter, and more uniform than solid timber. LVL is less prone to shrinking or warping. LVL can also support heavier loads and span longer distances than normal timber. LVL is used as headers and beams, rafters, scaffold planking, and the flange material for prefabricated wood I-joist. LVL can be manufactured to almost any length, restricted only by transportation to site.

**Laminated strand lumber (LSL)** is manufactured using flaked wood strands possessing a length-to-thickness ratio of approximately 150. These strands are combined with an adhesive, oriented and formed into a large mat or billet. The billet is then pressure-bonded together. LSL is commonly shaped into framing boards for floor joists and beams. LSL lumber is commonly shaped into framing boards for floor joists, studs, and support beam as well as millwork components.

**Parallel strand lumber (PSL)** is manufactured using veneers cut into long strands. These veneers are laid in a long parallel formation and bonded together with an adhesive to form the finished structural section. The length-to-thickness ratio of the strands in PSL is approximately 300. PSL is used for beam and header applications where high bending strength is required. PSL is also used as load-bearing columns and is manufactured in sizes comparable with sawn wood members. PSL beams can be stained or finished where they are used in exposed situations and aesthetics are important.

**Oriented strand lumber (OSL)** is manufactured in a similar manner to LSL using flaked wood strands. The wood strands used in OSL are shorter than those in LSL. The strand geometry for OSL results in length-to-thickness ratios of approximately 75. Combined with an adhesive, the strands are oriented and formed into a large mat or billet shape and then they are pressure-bonded together. OSL resembles oriented strand board (OSB) in appearance; however, the strands in OSL are arranged parallel to the longitudinal axis of the member. OSL is used primarily as structural framing for residential, commercial, and industrial construction. Common applications of OSL in construction include headers and beams, tall wall studs, rim board, sill plates, millwork, and window framing. OSL also offers good fastener-holding strength.

SCL is a highly predictable and uniform engineered wood product. The grain of each layer of veneer or flake runs in the same direction. This method of manufacturing produces a solid product that is almost free from warping and slitting. SCLs are typically used in applications where structural properties are important. These properties are evaluated using the methods outlined in ASTM D5456 "Standard Specification for Structural Composite Lumber." The grading, APA mill number, product evaluation, and ASTM Standard that the SCL product meets are indicated in the APA stamp (Fig. 9.25).

**Cross-laminated timber (CLT)** is manufactured using layers of kiln-dried lumber boards that are stacked in alternating directions. These layers are bonded together using structural adhesives that are compressed to form a solid rectangular panel. These panels consist of an odd number of layers typically between 3 and 7. These panels are capable of handling load transfer on all sides and possess excellent fire, seismic, and thermal performance.

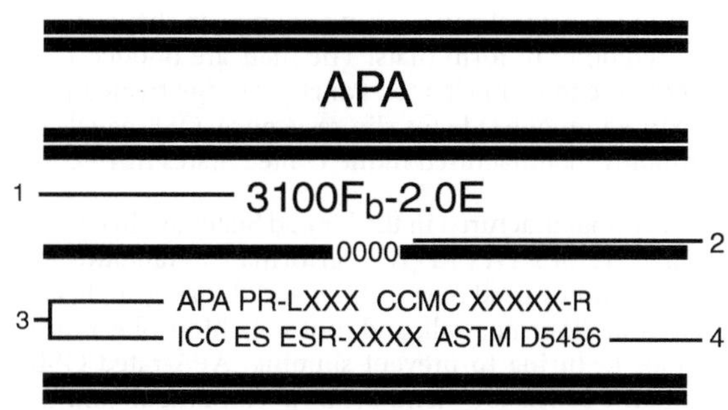


1. Qualified SCL grade (usually represented by design values).
2. APA mill number.
3. Product evaluation reports.
4. Standard specification for LVL.

**FIGURE 9.25** Sample APA product stamp for LVL lumber. (Downloaded from the APA website.)

## Structural Wood Panels

**Structural wood panels** are panel products composed primarily of wood that are used and approved for single-layer floors or for sheathing used to construct roofs, subfloors, and walls. Materials used for these panels include plywood, oriented strand board (OSB), particle board, and fiber board.

These panels are specified by bond classification, span rating, performance category, and grade.

**Bond classification** is a function of the moisture resistance of the glue used to bond the mats under the intended end-use conditions of the product. Structural wood panels have two bond ratings, exterior and exposure 1. Exterior panels have bonds that can withstand repeating wetting and drying cycles or long-term exposure to weather. Exposure 1 panels are not suitable for long-term exposure to weather. These panels are intended to resist weather exposure due to construction delay or low-severity weather conditions.

**Span rating** is an index number that identifies the recommended maximum center-to-center support spacing for the uses in roofs, walls, subfloor, and single floors. Spans are typically specified singly for wall applications and single floor (Floor 24 o.c.), while roof and subfloor are often combined in a fractional format. For example, a span rating of 32/16 designates a roof span of 32 in. and a subfloor span of 16 in.

Performance category of a panel is related to the range of panel thickness that is linked to the nominal panel thickness designations used in the International Building Code (IBC) and International Residential Code (IRC).

**Plywood** is manufactured using thin sheets of veneers (plies) that are arranged into layers to form a panel. Plywood always has an odd number of layers. The manufacturing process of plywood involves the cutting of the veneer from a log. Once cut to the appropriate width, the veneer is dried and graded. The veneers are then laid up in cross-laminated layers. It is possible to have two or more plies with the grain running in the same direction, but the layers will always be in odd numbers. Adhesive is applied to the veneers and they are placed in a hot press where they are bonded under pressure to form panels.

Wood shrinks and swells the most across the grain. Wood is strongest along its grain. By alternating the direction of the grain between adjacent layers, the shrinking and swelling are minimized in each direction. Alternating layers also maximizes the strength and stiffness of the wood.

**Oriented Strand Board (OSB)** is manufactured using a cross-oriented pattern of wood to create a stiff and strong structural panel. Thin rectangular-shaped wood strands are arranged in layers at right angles to each other to form mats. The mats are bonded together with waterproof adhesives under heat and pressure to form a panel. Panels are edge treated to guard against moisture penetration and may be textured on one side for slip resistance. OSB panels can be manufactured to any size but are most commonly manufactured in the United States in $4' \times 8'$ sheets.

OSB panels manufactured in the United States are in conformance with either the Voluntary Product Standard PS 2 or APA PRP-108 performance standards for use in floors, walls, and roofs.

OSB is commonly used in sheathing and flooring applications (Fig. 9.26). APA-rated OSB sheathing is used in subflooring, wall, and roof sheathing. Sheets are manufactured with one surface having an overlay or texturing to prevent slipping. APA-rated OSB Sturd-I-Floor is used for single-layer flooring and the panels are manufactured with tongue-and-groove edges.

Span ratings indicate the maximum permitted center-to-center spacing of supports, in inches, over which the panels should be installed in normal construction.

In APA-rated sheathing and Sturd-I-Floor, the span rating applies when the long panel dimension is across supports. The span rating for APA-rated siding panels is for vertical installation.

In APA-rated sheathing, the span rating looks like a fraction, such as 32/16 (Fig. 9.27). The numerator denotes the maximum spacing of support in inches when the panel is used for roof sheathing and

**FIGURE 9.26** OSB use as sheathing.

the denominator denotes the maximum on center spacing of supports when the panel is used for subflooring.

Sheathing panels with roof span ratings of 24 or greater may be used vertically or horizontally as wall sheathing over studs at 24 in. on center. APA rate sheathing may also be manufactured specifically for use as wall sheathing. These panels are identified with span ratings of Wall-16 or Wall-24.

APA-rated Sturd-I-Floor panels are designed specifically for single-floor applications under carpet and are manufactured with span ratings of 16, 20, 24, 32, and 48 on center.

APA-rated siding is produced with span ratings of 16 and 24 on center. Both panels and lap siding may be used direct to studs over nonstructural sheathing.

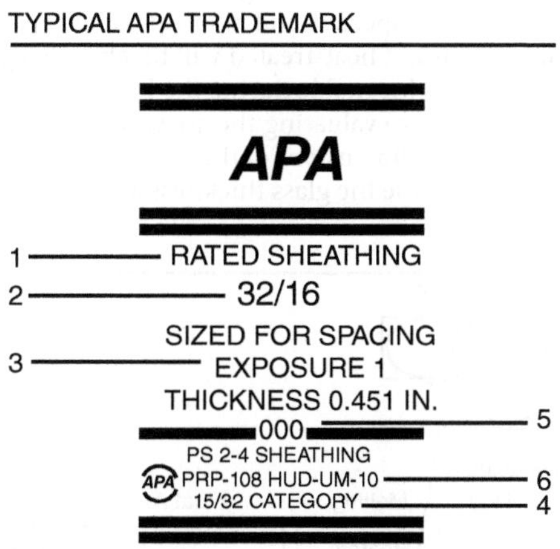


1 APA panel grade
2 Span rating
3 Exposure durability classification
4 Performance category
5 Mill number
6 APA's performance-rated panel standard

**FIGURE 9.27** APA trademark stamp. (Downloaded from the APA website.)

Properly handle and store OSB panels by protecting the ends and edges from physical damage. Keep the panels under cover and away from moisture. Avoid placing panels in contact with the ground and be sure when storing outside to place them on a raised and level surface to provide appropriate ventilation.

## GLASS

Glass, because of its transparency and heat resistance, as well as its pressure, breakage, and chemical resistance, is used as the primary fenestration material in commercial and residential buildings.

### Primary Ingredients

The major ingredient in glass is sand, more specifically, silica or silicon dioxide. The silica is mixed with soda ash, which contains sodium hydroxide or sodium carbonate, lime (calcium carbonate), small amounts of alumina, potassium oxide, and various other elements to control the color. The other ingredient is cullet or broken recycled glass from previous manufacturing processes. For construction most glass is clear, but for architectural features, colored glass has become popular.

The majority of the glass manufactured uses the float glass process (Fig. 9.28). Raw materials are weighed and mixed to the appropriate proportions to reflect the desired properties of the glass being manufactured. In addition to the raw materials, recycled glass or cullet is also added to the mix. These materials are mixed and then transferred to a furnace where they are melted. The molten glass is then transferred to the float area where it is floated on top of a bath of molten tin to form glass of desired thickness. The molten mass solidifies slowly while flowing over the bed of molten tin, after which it is annealed to remove stresses induced during the cooling process. The way in which the glass is cooled determines its strength. The annealing process slowly cools the hot glass to relieve internal stresses. This makes the glass more durable. Glass that has not been annealed can easily crack or shatter when subjected to small temperature changes or mechanical shock.

The glass is then inspected and upon passing it is cut to size and sent to storage (Fig. 9.29). Annealed glass that is not heat-treated will break into jagged irregular-shaped shards.

Breakage of glass is inevitable during the lifecycle of its use in a building. ASTM E1300 establishes standard procedures for evaluating the structural stability and probability of breakage in glass. The standards consider the dimensions of the window, the support condition, and the likely wind pressure experienced to determine the glass thickness appropriate to the application.

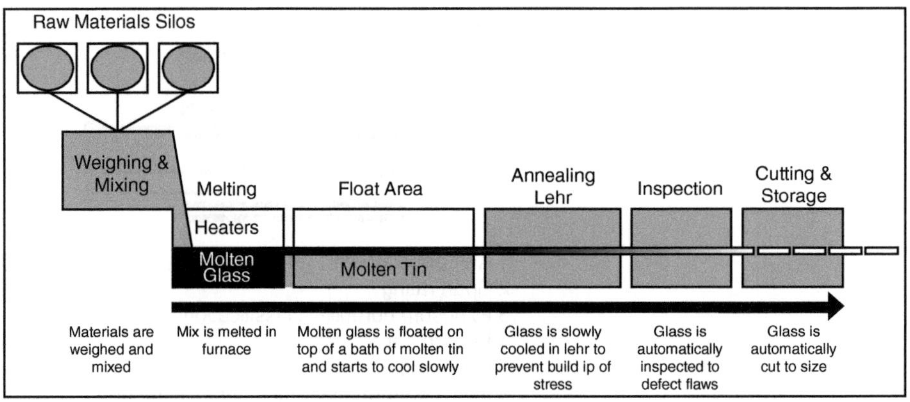


**FIGURE 9.28** Float glass process.

**FIGURE 9.29** Glass being cut to size. (Courtesy of Glass Enterprises, Inc.)

## Glass Properties

**Density** of glass is 158 pcf (2,500 kg/m$^3$). This is important to architects, engineers, and construction managers who need the weight of materials for engineering and construction site considerations.

| A 5/32″ (4 mm) thick pane of glass weighs 2.0 lb/ft$^2$ (10 kg/m$^2$). |
|---|

When glass is deflected, it has one face under compression and the other in tension. Glass possesses an extremely high compressive strength between 116 and 145 ksi (800–1,000 MPa). The resistance to tensile stress is significantly lower. The bending strength is a measure of the materials resistance during deflection. The bending strength for glass has different values depending on the heat treatment it has received: 40 MPa for annealed glass and between 120 and 200 MPa for toughened glass, depending on thickness, edgework, holes, and notches.

Glass is a perfectly elastic material. It does not exhibit permanent deformation, until breakage. The Young's modulus ($E$) expresses the tensile force that would theoretically have to be applied to a glass sample to stretch it by an amount equal to its original length. It is expressed as a force per unit area. For glass, $E = 7 \times 1{,}010$ Pa = 70 GPa.

When a sample is stretched under mechanical stress a decrease in its cross-section is observed. Poisson's ratio ($\mu$) is the relation between the unit decrease in the direction perpendicular to the axis of the effort and the unit strain in the direction of the effort. For glass in buildings, the value of coefficient $\mu$ is 0.22.

Linear thermal expansion is expressed by a coefficient measuring the stretch per unit length for a variation of 1°C. The coefficient of linear expansion for glass is $5 \times 10^{-6}$ in./in. °F ($9 \times 10^{-6}$ m/mK). Thermal conductivity determines the amount of heat required to flow through the cross-sectional area of the float glass sample in unit time at a temperature gradient. The thermal conductivity of glass is equal to 0.8 W/mK [0.0025 (cal/sec)/(cm$^2$ C/cm)].

Glass can be produced in large homogeneous panes that contain perfectly flat and parallel surfaces. The optical properties of glass are not affected by aging. The surface of glass is affected if it is exposed for a long time to highly alkali materials in conjunction with high temperatures. Float glass will also react to compounds that contain hydrofluoric acid under normal conditions. These are used for treating glass surfaces.

Heat-treating of glass is completed prior to being used in buildings for safety and improved durability. Tempered glass is produced by cutting annealed glass to the desired shape and size and then reheating it to 1200°F (620°C). The surfaces are rapidly cooled to create surface and edge compression in the glass and place the interior of the glass in compensating tension. Tempered glass is four times as strong as annealed glass of the same thickness. Once tempered, the glass cannot be cut or drilled, so all fabrication must be completed before the tempering process. To comply with safety glazing codes, a permanent label must be on the glass identifying the fabricator, the glass type, and the safety glazing standard that it meets. Fully tempered safety glass is often used for other applications where safety is desirable, because of its break pattern or when significant additional strength is needed to resist wind pressure, thermal stress, or both.

Heat-strengthened glass is cooled more slowly which results in less compression in the glass. Heat-strengthened glass is approximately twice as strong as annealed glass of the same thickness. However, the glass is designed to break into large pieces that are contained within the glazing system until it is removed. It does not qualify as a safety glazing material because of the break pattern that is similar to annealed glass. However, it does not have the tendency to break spontaneously like tempered glass. Once the glass is heat-strengthened it cannot be cut or drilled. The requirements for heat-strengthened and tempered glass are detailed under ASTM C1048 "Standard Specification for Heat-Strengthened and Fully Tempered Flat Glass."

Laminated glass is a common type of safety glass. Laminated glass is manufactured with two or more panes of annealed glass with a layer of plastic in between. The plastic is most commonly polyvinyl butyral (PVB). The glass is produced using heat and pressure to bond a thin layer of PVB between the layers of glass. Laminated glass is used in commercial and residential buildings due to advanced safety features. Laminated glass displays high-performance impact resistance and fragment retention. If the glass is broken, instead of shattering, it is held together by the plastic interlayer. The main safety feature of this glass is that if broken, the pieces of broken glass will adhere to the plastic instead of scattering. Advanced polymers enable the glass to withstand storms and hurricanes and even achieve decorative effects. Laminated glass also aids in the screening of UV radiation, resistance to fire, and possesses acoustic soundproofing. It is used in skylights, glass railings, facades, and floors. The ASTM C1172 "Standard Specification for Laminated Architectural Flat Glass" governs the requirements for laminated glass.

Standard window glass will fail at temperatures approaching 250°F. Fire-rated glass can remain competent to temperatures that exceed 1,600°F and is classified by the International Building Code as a life-safety building material. ANSI Z97.1 "Safety Glazing Materials Used in Buildings" establishes the specifications and methods of testing for the safety properties of safety glazing materials used for building purposes as does the Consumer Product Safety Commission 16 CFR 1201 "Safety Standard for Architectural Glazing Materials."

Fire-rated glass is categorized based upon the amount of time it remains in place and compartmentalizes fire and smoke. ASTM E119 "Standard Test Methods for Fire Tests of Building and Construction Materials" covers the testing procedures for glass fire ratings. Fire-rated glass is subjected to temperatures exceeding 1,600°F and based upon the amount of time it remains fixed in place it is assigned a rating of 20-, 45-, or 60-min fire rating. Fire-rated glass is also tested with the hose-stream test to simulate actual fire scenarios according to the specifications outlined in the NFPA 257 "Standard on Fire Test for Window and Glass Block Assemblies." The test simulates the rapid cooling of glass when exposed to water. This predicts the ability of the glass to remain intact during activation of sprinklers or the use of fire hoses. The results of these tests must be documented through labeling on the glass.

Fire-rated glass is divided into two categories: fire-*protective* and fire-*resistive* glass.

**Fire-protective glass** is not designed to prevent transmission of radiant heat and cannot exceed 25% of the total area of a wall.

Fire-protective glass also may be **subject** to additional testing standards, including
NFPA 252—Standard Methods of Fire Tests of Door Assemblies
NFPA 257—Standard on Fire Tests for Window and Glass Block Assemblies
Underwriters Laboratories (UL)—Standard for Safety Fire Tests of Window Assemblies
UL 10B—Fire Tests of Door Assemblies
UL 10C—Positive Pressure Fire Tests of Door Assemblies

Fire-resistive glass will not conduct heat and limits temperatures to 250°F or less on the protected side for the duration of its designated time rating. This glass is able to be used for enclosures and exit corridors.

**Fire-resistive glass** also may be subject to additional testing standards, including

ASTM E-119—Standard Methods for Fire Tests of Building Construction and Materials
NFPA 251—Standard Methods of Test of Fire Resistance of Building and Construction Materials
UL 263—Fire Tests of Building Construction Materials

**Fritted glass** is a decorative glass printed with ink that contains ultrasmall particles of ground-up glass. Once the frit is imprinted on the glass, the glass is dried and then fired in a tempering furnace. This transforms the frit into a hard, ceramic-type coating. The design is part of the glass which makes it a durable and functional design suitable for architectural and interior applications. Because the glass has lower transparency, it can be used to control the amount of solar light and heat that penetrates a space (Fig. 9.30).

Spandrel glass is a special form of opaque fritted glass that is used in glass curtain wall construction. It is used between floors of a structure to cover structural materials such as drop ceilings or concrete floor slabs. The frit coating is applied to the interior surface of the glass. The glass must be either tempered or heat strengthened to resist the thermal stresses developed due to the colored glass-absorbing heat (Fig. 9.30).

**Reflective glass** is also known as solar control glass. It is a clear or tinted glass that has a thin layer of metal oxide, approximately 0.000003″ thick, applied to the surface of the glass. The glass reflects light, thereby reducing the solar heat gain coefficient (SHGC) while allowing some of the visible light to pass through.

**FIGURE 9.30** Spandrel and fritted glass along with a shading feature on a curtain wall. (Photo courtesy of Architectural Glass Institute ©Joe Garvin.)

**Low-emissivity** (low-e) glass with improved thermal performance due to an ultrathin, almost colorless, and completely transparent low-e coating. The glass appears to be clear, but it reflects heat back to source. The glass reduces heat loss while providing maximum light transmittance. The coating also helps to resist UV rays.

Low-e coatings are often used on one of the lites of double-glazed window systems, thereby reducing the thermal transmittance of the glazing unit. Glass from different production batches or with different metal oxides used in the coating will appear to be different colors, making it difficult to match multiple lites of low-e glass.

The transparency of glass is a function of how much visible light passes through the glass relative to the amount of light striking the glass. This is called the visible light transmittance (VT). Clear glass has a value of VT between 0.80 and 0.90. This means that 80–90% of the visible light striking the glass surface passes through to the interior of the building. The remaining light is either reflected by the glass or absorbed and converted to heat. The higher the value of VT, the better the quality of the daylight that is passed through the glass. VT has a value between 0 and 1 but most glass values fall between 0.30 and 0.80.

**Tinting glass** is made by adding color pigments to the raw materials in the float process. This reduces the visible light transmittance of the glass. Standard colors include bronze, gray, dark gray, green, blue, and blue/green (Fig. 9.31). The density of the color is a function of the glass thickness. The thicker the glass, the denser the color and the less light is transmitted. Tinted glass also absorbs some of that solar energy, called heat absorption. The added heat causes thermal stress on the glass so it is important that tinted glass be tempered to avoid shattering.

**FIGURE 9.31** Tinted glass within a curtainwall system. (Photo Courtesy of R.A. Kennedy & Sons.)

To measure the reduction of heat gain from solar radiation by the tinted glass, a measure called the solar heat gain coefficient (SHGC) is used. It is the ratio of solar heat admitted through glass to the total heat energy striking the glass. It also accounts for heat that is conducted or radiated into the space due to heating of the glass itself. Clear glasses have SHGCs ranging from about 0.90 to 0.70, depending on the clarity and thickness of the glass. SHGCs for tinted glasses range from about 0.70 to 0.35, meaning that these glasses allow 70–35% of the solar heat energy striking the glass to pass through.

Insulated glass units refer to two or more lites of glass held apart by a spacer at the perimeter and sealed around the edges with an air space between to form a single unit (Fig. 9.32). An insulated glass unit reduces thermal transmittance, sound transmission, and the likelihood of condensation forming to the inside living area. Insulated glass has a useful life and deteriorates over time.

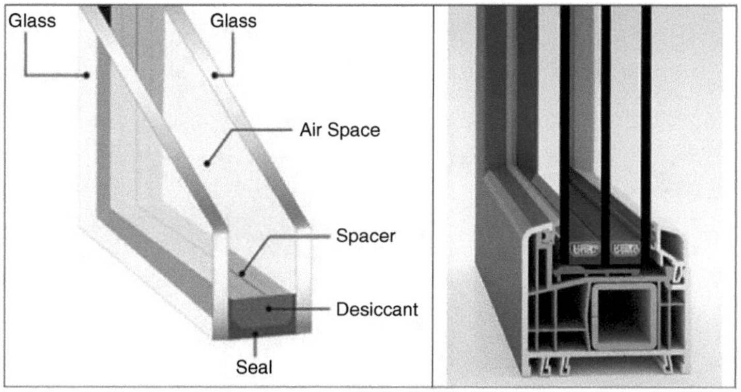


**FIGURE 9.32** Insulated glass window unit cross-section. (Courtesy of INTUS Windows.)

Most insulated glass units use a metal spacer that is filled with a desiccant to keep the enclosed air space free of visible moisture. The perimeter of the unit is then sealed with a sealant, like silicone. A standard unit has sealed air, i.e., gases such as argon. An argon gas filling is heavier than air and reduces the speed of convection currents in the gas, providing higher insulating values. The thermal performance of insulated glass reduces heat gain in hot weather and heat loss and condensation in cold weather.

Different types of glass can be incorporated into a unit. Tinted or other heat absorbing glass is the most commonly used in the outboard lite. Tempering or heat strengthening are required as the units receive high thermal stresses and are subject to breakage.

## Glazing for Windows, Doors, and Skylights

Buildings use glazing components and products throughout a structure. There are requirements associated with the testing and labeling of windows, doors, and skylights that are used in buildings. Exterior windows and doors are covered in Section 1709.5.1 of the IBC. This section requires windows and sliding doors to be tested and labeled in accordance with the North American Fenestration Standard/Specification NAFS-11 for windows, doors, and skylights.

Curtain wall and storefront use glass products in their assemblies. These are addressed in Section 1709.5.2 of the IBC. These assemblies are to be tested to 1.5 times the Allowable Stress Design load in accordance with ASTM E330, and the glass is to be designed in accordance with ASTM E1300-12AE01 (Fig. 9.33).

Exterior swinging doors are tested and labeled in accordance with NAFS-11 or tested to 1.5 times Allowable Stress Design load in accordance with ASTM E330. The IBC also permits garage doors to be tested to ANSI/DASMA 108-12 in lieu of ASTM E330.

Unit skylights are required to be tested and labeled in accordance with NAFS-11 by the IBC. Unit skylights are factory-manufactured fenestration assemblies intended to be installed in a single roof opening without intermediate framing members.

Special requirements and inspections for glazing systems located in high-wind areas are detailed in the IBC. The high-wind areas are determined by the exposure category of the building. Other special requirements relate to seismic issues. Glass that is used in glazed curtain walls, storefronts, or partitions must meet the seismic requirements of ASCE 7. The load resistance of glass under uniform load is determined in accordance with ASTM E1300.

**FIGURE 9.33** Curtainwall glass. (Photo courtesy of Architectural Glass Institute ©Joe Garvin.)

## *SAFETY GLAZING*

Impact is a critical challenge with glass. Glazing is tested in accordance with CPSC 16 CFR Part 1201 and shall comply with the test criteria for Category II. There are some exceptions for glass used in doors or enclosures for hot tubs, whirlpools, saunas, steam rooms, bathtubs, and showers. These are tested in accordance with ANSI Z97.1. This glazing must be identified by a manufacturer's designation specifying who applied the designation, the manufacturer or installer, and the safety glazing standard with which it complies. This identification once applied cannot be destroyed.

### Glass in Handrails and Guards

Glass used for handrails and guards must use laminated glass that is constructed of fully tempered or heat-strengthened glass. This glass shall comply with Category II or CPSC 16 CFR Part 1201 or Class A of ANSI Z97.1 with a minimum nominal thickness of ¼ in. The glass panels and their support systems must be designed to withstand the loads specified in the IBC Section 1607.8.

### Structural Glass Baluster Panels

Guards with structural glass baluster panels shall be installed with an attached top rail or handrail. The top rail or handrail shall be supported by not fewer than three glass baluster panels or shall be otherwise supported to remain in place should one glass baluster panel fail.

### Glazing in Athletic Facilities

Glazing in athletic facilities that are subject to impact loads as part of a wall or door section is required to meet human impact standards set by Category II of CPSC 16 CFR Part 1201 or Class A of ANSI Z97.1.

### Glass in Walkways

Glass that is installed as part of a floor or ceiling assembly as a walking surface and constructed with laminated glass shall comply with ASTM E2751 or with the load requirements specified in Chapter 16 of the IBC. These assemblies must also comply with the fire-resistance rating and marking requirements.

## *ASPHALT AND BITUMINOUS PRODUCTS*

Asphalt, because of its water-resistant qualities and good durability, is used for many building applications: to exclude water, provide a cushion against vibration and expansion, and for site paving.

### Asphalts for Dampproofing and Waterproofing

Dampproofing is generally only a mopped-on coating, whereas waterproofing is a built-up coating of one or more plies. Bituminous systems used for dampproofing and waterproofing may be hot applied or cold applied. ASTM D449 "Asphalt Used in Dampproofing and Waterproofing" specifies three types of asphalt. Type I, a soft, adhesive, easy-flowing, self-healing bitumen, is intended for use for underground construction, such as foundations, or where similar moderate temperature conditions exist. The softening point of Type I may range from 115 to 140°F. Type II may be used above ground; for example, on retaining walls or where temperatures will not exceed 122°F. The softening point of Type II may range from 145 to 170°F. Type III is an asphalt less susceptible to temperature than Type II, with good adhesive properties, and suitable for use above grade on vertical surfaces exposed to direct sunlight. D449 asphalts are suitable for use with an asphalt primer meeting the requirements of ASTM D41. In construction of membrane waterproofing systems with these asphalts, felts should conform to ASTM D226 or D250, fabrics to D173, D1327, or D1668, and asphalt-impregnated glass mats to D2178. For cold-applied systems, asphalt emulsions or asphalt mastic reinforced with glass fabric may be used. ASTM D1187 specifies asphalt-based emulsions for protective coatings for metal. D491 contains requirements for asphalt mastic for use in waterproofing building floors but not intended as pavement. The mastic is a mixture of asphalt cement, mineral filler, and mineral aggregate. D1668 covers glass fabric for roofing and waterproofing membranes. At the inside face of exterior walls, an asphalt felt protection strip may be installed between each furring channel and the wall to prevent migration of moisture.

### Bituminous Roofing

Hot asphalt or coal tar is used for conventional built-up roofing. The bitumens are heated to a temperature high enough to cause fusion with the saturate bitumen in roofing felts, thus welding the plies together (Fig. 9.34). The optimum temperature at the point of application for achieving complete fusion, optimum mopping properties, and the desirable interplay mopping weight is called the equiviscous temperature (EVT). There is usually a 25°F variance permitted above and below the recommended EVT by the manufacturer.

### Built-Up Roofing

For constructing a built-up roof, four grades of asphalt are recognized (ASTM D312):

1. Type I, for inclines up to 1⁄2 in./ft
2. Type II, for inclines up to 11⁄2 in./ft
3. Type III, for inclines up to 3 in./ft
4. Type IV, suited for inclines up to 6 in./ft, generally in areas with relatively high year-round temperatures

**FIGURE 9.34** Roof membrane with welded plies.

Types I through IV may be either smooth or surfaced with slag or gravel. Softening ranges are 135–151°F, 158–176°F, 185–205°F, and 210–225°F, respectively. Heating of the asphalts should not exceed the flash point, the finished blowing temperature, or 500°F. Coal-tar pitches for roofing, dampproofing, and waterproofing are of three types (ASTM D450): Type I, for built-up roofing systems; Type II, for dampproofing and membranes waterproofing systems; Type III, for built-up roofing, but containing less volatiles than Type I. Softening ranges are 126–140°F, 106–126°F, and 133–147°F, respectively.

## Roofing Felts

For built-up waterproofing and roofing, types of membranes employed include felt (ASTM D226, D227) and cotton fabrics (ASTM D173). Felts are felted sheets of inorganic or organic fibers saturated with asphalt or coal tar conforming to ASTM D312 and D450. Standard asphalt felts weigh 15, 20, or 30 lb per square (100 $ft^2$), and standard coal-tar felts weigh 13 lb per square. Cotton fabrics are open-weave materials weighing at least 3½ $oz/yd^2$ before saturation, with thread counts of 24–32 per inch. The saturants are either asphalts or coal tars. The saturated fabrics must weigh at least 10 $oz/yd^2$.

## Roll Roofing

Asphalt roll roofing, shingles, and siding consist basically of roofing felt, first uniformly impregnated with hot asphaltic saturate and then coated on each side with at least one layer of a hot asphaltic coating and compounded with a water-insoluble mineral filler (Fig. 9.35). The bottom or reverse side, in each instance, is covered with some suitable material, like powdered mica, to prevent sticking in the package or roll. Granule-surfaced roll roofing (ASTM D249) is covered uniformly on the weather side with crushed mineral granules, such as slate. Minimum weight of the finished roofing should be 81–83 lb per square (100 $ft^2$), and the granular coating should weigh at least 18.5 lb per square. Roll roofing (ASTM 224), surfaced with powdered talc or mica, is made in two grades, 39.8 and 54.6 lb per square, of which at least 18 lb must be the surfacing material.

There are three standard types of asphaltic shingles: Type I, uniform or nonuniform thickness; Type II, thick butt; and Type III, uniform or nonuniform thickness (ASTM D225). Average weights must be 95 lb per square (100 $ft^2$). For types I and III, the weather-side coating must weigh 23.0 lb per square, and for Type II, 30.0 lb per square. The material in these shingles is similar to that in granule-surfaced roll roofing (Fig. 9.36).

**FIGURE 9.35** Roll roofing with asphaltic coating.

**FIGURE 9.36** Installing asphalt shingles on a sloped roof.

## Asphalt Mastics and Grouts

Asphalt mastics are used for waterproofing floors and similar structures, but are not intended for pavement. These consist of mixtures of asphalt cement, mineral filler, and mineral aggregate, which can be heated at about 400°F to a sufficiently soft condition to be poured and troweled into place. The raw ingredients may be mixed on the job or may be premixed, formed into cakes, and merely heated on the job (ASTM D491). Bituminous grouts are suitable for waterproofing above or below ground level as protective coatings. They can also be used for membrane waterproofing or for bedding and filling the joints of brickwork. Either asphaltic or coal-tar pitch materials of dampproofing and waterproofing grade are used, together with mineral aggregates as coarse as sand.

Expansion and isolation joint filler strips are common on most building projects to fill linear void spaces. Asphalt-saturated cellulosic fiber, cork, and self-expanding cork are standard materials adhering to ASTM D 1751.

## Bituminous Pavements

Asphalts for pavement (ASTM D946, AASHTO M226, and AASHTO M323) contain petroleum asphalt cement, derived by the distillation of asphaltic petroleum. Previously, various grades were designated as 40–50, 60–70, 85–100, 120–150, and 200–300, depending upon the depth of penetration of a standard needle in a standard test (ASTM D5). Emulsions range from low to high viscosity and quick- to slow-setting (ASTM D977). Now, asphalt petroleum grading and asphalt mixture design predominantly follow Superpave protocols, an acronym for Superior Performing Asphalt Pavements. Superpave is an asphalt concrete mix-design method for determining the aggregate structure and proportions of aggregate and asphalt cement based upon desired performance.

## Structure of Asphalt Pavements

Asphalt pavements are constructed as layered systems (Fig. 9.37). This structure consists of a prepared subgrade, an aggregate subbase, an aggregate base, and an asphalt-bound surface layer. The asphalt structure generally consists of two or more courses: the surface wearing course and one or more underlying binder courses. The aggregate characteristics and amount of asphalt cement are designed specifically to match the needs of the binder and surfacing courses. Under the Superpave mix-design system (AASHTO M323), the allowable aggregate gradations are designated by the nominal maximum aggregate size (NMAS). NMAS is one size larger than the first sieve that retains more than 10% of the aggregate mass.

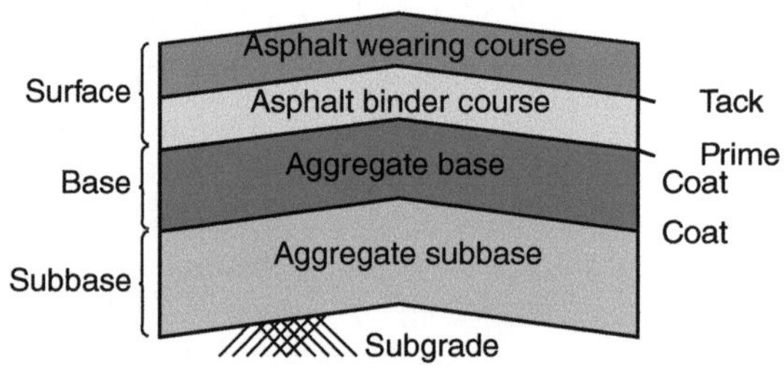


**FIGURE 9.37** Typical structure of a flexible pavement.

## Asphalt Mixing Plants

Asphalt cement and aggregates are heated and mixed to produce the hot mix asphalt concrete at asphalt mixing plants (Fig. 9.38). Some asphalt mixing plants are operated primarily as producers of multiple mixes for FOB (free-on-board) plant sales or to serve multiple paving spreads on small jobs. These plants must be able to easily and quickly change their production mix to meet the requirements of multiple customers. Other plants are high-volume producers serving a single paving spread; this is particularly true of portable plants that are moved from project to project. When liquid asphalt is mixed with the aggregate in traditional hot mix asphalt, the temperature of the asphalt should be in the range of 275–325°F. Warm mix asphalt is heated in the range of 225–250°F using additives to reduce energy and heating costs. Matching the temperature of liquid asphalt and aggregates will permit effective adhesion and minimize the effects of boiling or bubbling asphalt cement on the aggregate surface.

## Asphalt Paving

The mixing plant, trucks, asphalt paver, and compactors are synchronized for an efficient paving operation. Rear-dump trucks or live-bottom trucks transfer the mix directly into the paver hopper (Fig. 9.39). Mat thickness and cross slope are controlled by the paver screed. Roller compactors smooth and compact the asphalt mixture to a required density and profile. Three basic roller types are

**FIGURE 9.38** Asphalt mixing plant.

**FIGURE 9.39** Paving an asphalt mixer with a dump truck and paver.

**FIGURE 9.40** Compacting the asphalt mat with a steel.

used to compact asphalt-paving mixes: smooth drum steel wheel, pneumatic tire, and smooth drum steel wheel vibratory (Fig. 9.40). Paving adjacent to buildings and structures prevents a challenge to the operation and can reduce tonnage production.

# CHAPTER 10
# BUILDING CONSTRUCTION METHODS

*This chapter describes the construction processes of three different kinds of buildings—residential, commercial, and multistory (up to 12 stories). Most people are familiar with houses and may even have an understanding of how they are built. The activities required to build a larger commercial building, while having activities similar to those required to build a house, are more complicated. Most commercial buildings have similar designs—a one-story open shell with a concrete slab floor. Multistory buildings are a completely distinct class of buildings with specific building code requirements.*

## *MOBILIZATION*

A constructor's preparation to construct a building begins long before work begins on site. Some of the work begins before the constructor is even sure of being offered a contract. Specifically, bidding or negotiating, and soliciting prices from subcontractors, takes place before the contract is signed with the project owner. This is the "get work" portion of the constructor's model.

Once offered a contract to build a project, the contractor has a multitude of immediate tasks:

- Arrange financing for accomplishing the work. This involves an estimate of expenses and a cash flow analysis to identify the monthly cash requirements required to complete the work (Chapters 3 and 4).
- Prepare a detailed project schedule (Chapter 7) showing the work of each subcontractor so the project manager can eliminate workspace and task conflicts.

> It is better to plan strategically so problems do not occur, than to be in a problem-solving situation.

- Select subcontractors (buyout). Always match the project schedule with subcontractor work load and the subcontractor's reputation for performing quality work at the scheduled time.

> Low prices are not always the best measure for selecting subcontractors.

- Select material suppliers and finalize purchase agreements.
- Set up the project office. The size of the needed office space generally reflects the size of the project. Construction of a small residence might be managed out of the superintendent's pickup truck. A multimillion-dollar commercial building would require an on-site field office.

- Provide secure storage areas for tools, equipment, and materials.

> Plan material storage areas to minimize rehandling.

- Attend a preconstruction meeting with the owner, architect/engineer, and interested public agencies. The preconstruction meeting has several purposes, but the most important is to open the lines of communication between the contractor and owner. The topics discussed would normally include safety requirements, permits, payment schedule, actions required by the owner during construction, and relationships with neighbors. Each project is different and each preconstruction conference covers different topics.
- Begin the preparation of submittals and shop drawings, particularly shop drawings for long-lead-time items. Often the construction of a house does not require the submission of shop drawings because the builder has a standard list of materials and supplies used for house construction. For the construction of other kinds of facilities, shop drawings are required. The first shop drawings submitted are for the items to be installed first and the items with the longest lead times, such as structural steel and nonstandard mechanical equipment.

## BUILDING SYSTEMS

A building is composed of numerous subsystems. The main subsystems are

- **Foundation.** The foundation carries the live and dead loads of the building. Live loads consist of the people and movable equipment occupying the building as well as environmental loads such as positive and negative wind (hurricanes), snow, hydrostatic pressure from ground water, and/or seismic. Dead loads are the weight of the structure itself and the nonmoving equipment in the building.
- **Framing.** In the case of most residential construction, the framing material is 2 × 4 or 2 × 6 lumber. In a commercial building, the frame is often steel or concrete but wood is becoming common and reinforced masonry and reinforced concrete masonry units (CMUs) can be used. Steel is popular because with the rise of 3D modeling technology steel frames can be built precisely to specifications. The mass and thermal properties of concrete can reduce temperature changes and can significantly reduce energy costs.
- **Building envelope.** The roof, walls, and floor are distinct elements of the building envelope used to enclose the interior space. The envelope is the physical separator between the interior conditioned environment and the exterior climate: thermal, moisture, acoustic, and light. The envelope is designed to provide energy conservation, attenuate sound transmission, and offer fire safety.

Glass

Brick

Stone

- **Interior walls.** The interior walls may be part of the building's structural frame (load bearing). Load-bearing walls are usually positioned directly above one another for transmitting loads to the foundation. In most commercial buildings, interior walls are completely separate from the building frame. But for low-rise commercial construction (three stories or less), interior walls may be load bearing, similar to residential design practice. Nonstructural walls are often termed "partition," or "panel" walls. Structurally, there are three wall types:

1. Load-bearing—support the weight of the structure above.
2. Shear—resist in-plane lateral forces.
3. Partition—used to separate interior building space, supports its own weight.

> When performing renovation work, it is necessary to determine whether a wall is load-bearing or non-load-bearing. Any part of a removed load-bearing wall must first be replaced with suitable structural support.

Interior walls can be constructed using wood framing, light gauge steel framing, or CMUs. In commercial buildings, building codes require interior wall assemblies to offer passive fire protection. There are also fire code requirements (usually 2- or 3-hr rated) for area separation walls which are non-load-bearing and vertical walls between adjacent units in apartment buildings, condominiums, and townhouses.

- **Utilities.** Electricity, water, wastewater disposal, telephone, television cable, and data and communication systems are commonly required utility systems. Sophisticated buildings, such as medical facilities, have many more utilities. An example would be an oxygen system in a hospital.
- **Environmental systems.** These include plumbing, electrical, heating, lighting, air-conditioning, ventilation, and acoustics, along with the control systems to operate each. To minimize energy use, the goal is an "intelligent building."
- **Communications infrastructure.** Businesses rely on and homes need communication systems: voice, information processing, and signaling systems.
- **Transportation systems.** Most building transportation is vertical—elevators, escalators, and stairs. Hydraulic elevators are normally used in low-rise residential and commercial buildings. Taller buildings rely on traction motor elevators. Only buildings with extended horizontal dimensions such as airport terminals have horizontal transportation systems (people movers).
- **Life-safety systems.** Fire alarm and suppression are normally required in public building but some municipalities now require fire suppression systems in new homes. In addition, the National Institute of Standards and Technology (NIST) has since 9/11 placed increased focus on public safety communication systems.

## HOUSES

Consider the construction of a residence with a garage. The house has several bedrooms, a great room with a fireplace, a dining room, and a kitchen. The walls have 2 × 4 or 2 × 6 studs. Wood is the traditional exterior siding for houses but the siding could be brick, vinyl, or stucco and the roof could be asphalt shingles or clay tiles depending on the climate of the location. Water could be provided by a well or a connection to the local utility. The house would have a septic system if located in the county. A gas furnace provides heat and there is a central air-conditioning system.

## Site Preparation

The first on-site activity is typically site preparation or site clearing. This involves clearing only trees in the footprint of the structure (while saving as many other trees as possible) and removing unwanted debris. The builder of tract houses will, in most cases, clear the entire area to be developed.

Surveying to establish the exact location of the house will follow the clearing and site earthwork work, such as leveling of the house location. If a well must be drilled it can be done at any time; however, the contractor will delay construction of the septic system until last to ensure the field lines of the system are not damaged by heavy equipment or truck traffic. The ready-mix concrete trucks will need to have access to the foundation and early installation of the septic system could create an unnecessary obstacle to truck movements. Other service entrances such as electric power and telephone maybe constructed later for the same reason. The electric meter and interior wiring can be installed without having power brought to the building. The same is true for other communication services.

## Foundation

Houses and single-story buildings typically have shallow foundations but soil conditions dictate the foundation type. A 4- to 6-in. concrete slab-on-grade is the most common type of house foundation. It requires the least amount of lot preparation. A basement is simply a special case of a grade beam foundation.

The contractor must excavate to the correct elevation for the foundation footings and place the necessary water and drainage pipes in trenches below the slab. Such drainage systems normally drain into a sump or to the municipal system. Concrete footings would be required to support basement walls. There will be a shallow footing at the edges of the slab to support the structural walls of the house. A 6-in. slab constructed on a poor soil can support the live and dead loads of a residence.

There is a need for uniform support of the slab. If this is not provided, there is the potential for cracks and differential settlement. Therefore, there is typically a 4- to 6-in. layer of gravel or crushed stone uniformly spread and compacted under the slab to reduce the chance of cracking and damage to the plumbing pipes or electrical conduit placed in the slab.

A moisture (vapor) barrier will be placed between the concrete and the gravel. Moisture moving through a slab is the adversary of floor coverings, adhesives, and coatings.

Under slap piping completed.

Compacted stone under the slab.

Reinforcing steel bars (#4 bars) will be embedded in the concrete footings under the structural walls. The slab concrete will have 6-in. by 6-in. wire mesh. If the slab is on an expansive soil it can be strengthened to minimize movement by a post-tensioned system. Finally, concrete from a local "ready-mix" plant will be placed in the formwork. Figure 10.1 shows a completed slab-on-grade house foundation.

Slab post-tensioning system.

Following the grade beam construction, the basement walls are assembled. The forms and reinforcing steel are positioned, and the concrete is placed and vibrated in the forms. Concrete basement walls would be constructed by placing the rebar (probably #4, 9-in. on-center) and setting the forms (reusable). Tie-down bolts, for the first floor sill, will be placed in the formwork. These tie the frame of the structure to the concrete walls of the basement.

Basement wall with tie-down bolts

## Framing

Framing for residential construction is rapidly moving to prefabricated panelized systems. The panels are manufactured off-site and trucked to the project location. The scrap lumber remains at the prefabrication site. Prefabricated panel framing is a job of erecting and assembling panelized wall and roof system sections, rather than field measuring, cutting, and fabricating. A fabricator using a 3D CAD program can guarantee quality and accurately dimensioned assemblies. Studies of framing for houses have found a 75% reduction in man-hours when panelized systems are used. Prefabrication is the future of homebuilding because of its many advantages.

**FIGURE 10.1** Finished floor slab.

**FIGURE 10.2** Framing a residential slab-on-grade house.

Traditional construction of timber frame homes (Fig. 10.2) takes place on site where materials are custom cut and assembled (stick-built). The first step is bolting the sill to the top of the concrete. If it is not a slab-on-grade house but has foundation walls, the next step is installation of the floor joists for the first floor and placing the subfloor. Joists are either dimension lumber such as 2 × 8s, 2 × 10s, or 2 × 12s. The minimum size joists will depend on the type of wood (yellow pine, redwood, western red cedar), the grade of lumber, spacing of joists (16″ and 24″ on-center[1] are common), and the load. There are also manufactured I-joists made from plywood or oriented strand board (OSB). Common depths for I-joist are 9½″, 11⅞″, 14″, and 16″.

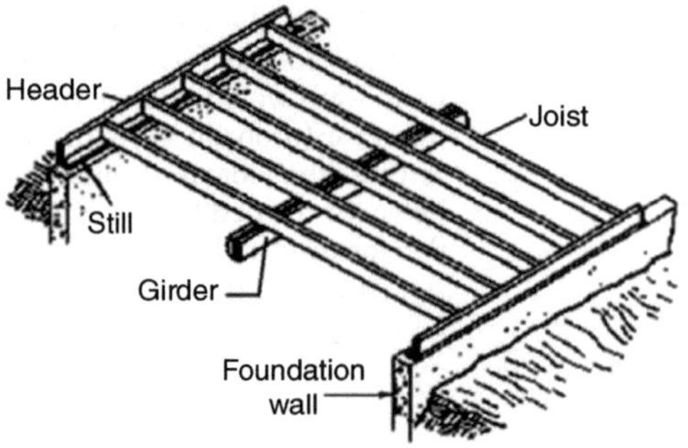


Still and joist system.

The subfloor is typically either plywood or OSB. The exterior and interior walls are raised on top of the subfloor. Homes without basements have frames bolted directly to the concrete floor slab (Fig. 10.3). Homes with basements have the floor joists and subfloor on the sill.

The first-floor frame is built on the subflooring. The floor for a second story rests on top of the first-floor frame. Constructing the frame for the second story follows the same process as constructing the frame of the first floor. The roof framing system, consisting of rafters and joists, is erected on top of the second-story frame. Figures 10.4 through 10.6 depict the house as framing progresses.

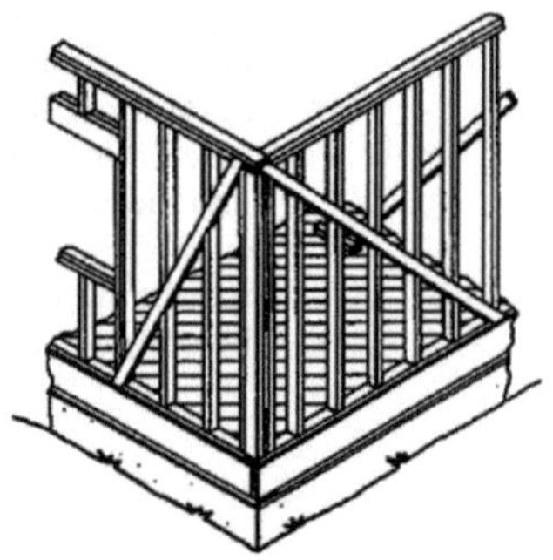

First floor frame on subflooring.

## BUILDING ENVELOPE

The building envelope contains the components used to create the outer shell of the structure. It has three primary functions:

1. **Structural support**—provides strength and rigidity for resistant to external and internal loads.
2. **Environmental control**—limits the flow of matter (dust and water) and energy into and out of the building. The envelope would include the insulation, sheathing, moisture barrier, exterior siding and roofing, windows, and doors.
3. **Finish** (aesthetics)—visual appearance creates a feeling of elegance or permanence.

[1] On-center (OC): from the middle of one element to the middle of the next.

**FIGURE 10.3** House framing bolted directly to the concrete floor slab.

**FIGURE 10.4** Rafter and roof joist framing.

**FIGURE 10.5** Most of the exterior sheathing completed.

**FIGURE 10.6** Ready for stucco and placement of the roof tiles.

## Walls

Taking advantage of advances in enclosure materials, modern buildings including homes employ hybrid and energy-efficient wall systems. Trombe, passive solar walls, are being used to heat buildings. The wall, with a glass external layer, is built on the winter sun side of a structure to absorb heat during sunlit hours and then slowly release the heat during the night. Trombe walls are most effective in areas receiving winter sun. Passive solar designs store, reflect, and distribute solar energy. These approaches do not involve the use of mechanical and electrical devices but necessitate the careful placement of windows and the use of thermal insulation.

The sheathing which is plywood or OSB is nailed to the house frame as soon as the work progresses enough to allow its installation (Fig. 10.5). Nailing the wall sheathing in place helps increase the stability of the structure. Together the frame and sheathing can create a shear wall. Shear walls act as rigid vertical diaphragms transferring the lateral forces from exterior walls to the foundation of the house. Following the wall sheathing, the windows and doors can be installed and the exterior siding of the building applied. Siding protects a home from weather and improves insulation.

Wood is the traditional exterior siding for homes. But vinyl siding because of it durability, color options, and low cost is very popular. Brick is another traditional exterior siding. With brick veneer there needs to be an impenetrable membrane installed between the veneer and the timber framing. In the south-western United States, stucco is common. It also requires an impenetrable membrane.

## Roof

Most roofs have a layer of roofing paper (felt or tar paper) nailed to the roof sheathing. Roofing paper has a protective coating of bitumen (asphalt), which serves as a water barrier to wind-driven rain or snow pushed under the shingles. The builder must know the required weight and required overlap of the roofing paper, in conformance with the local building code. The roof is shingled immediately following the placing of the roofing paper.

When a tile roof is installed, the roofing paper underneath the tiles is the primary waterproofing element. The tile serves as a sunscreen protecting the roofing paper and adds a cosmetic element to a house. Nevertheless, the material placed below the tile is the critical element affecting the service life of the roof.

Many homes have fireplaces, wood burning or gas, with a sheet metal duct used to carry flue gases. On the roof there is a wood frame built to look like a chimney. If there is a real masonry fireplace with a brick chimney, it must be constructed prior to roofing the building because the flashing needs to be placed properly around the chimney at the roof level to prevent leakage.

## Insulation

Insulating material will be placed under the first floor, in the exterior walls, and above ceilings adjoining unfinished attic space. This completes the insulation envelope around all sides of the climate-controlled spaces in the residence. The basement if used as family space would be insulated but the attic and garage are usually outside the insulation envelope. There is no reason to insulate between floors. Similarly, there is no reason to insulate the exterior walls of the garage unless the garage itself is heated. Insulation is rated in terms of thermal resistance, **R-value**, which indicates the insulation's resistance to heat flow. Higher R-values indicate greater resistance to heat flow. When the insulating materials are dense solids placed in direct mutual contact, their R-values are additive for the barrier. The U.S. Department of Energy publishes recommended total R-values for wood-framed houses. These values vary with the type of heating system installed but, in general, for cavity walls in southern states R-values from R13 to R15 are recommended. In northern states R-values from R13 to R21 are recommended.

**R-value**
*A numerical measure of the insulating characteristics of materials*

There are several types of insulation, not all of which are used in residential construction. The wall insulation would normally be blanket insulation in the form of batts or rolls, which are made from mineral fibers. They are available, with or without **vapor barrier** facing, in widths suited to standard spacing of wall studs and attic or floor joists. It is important to carefully cut and fit the material around plumbing pipes, wires, and electrical outlets as the loss in effectiveness can be significant. The vapor barrier must be placed on the warm side of the wall or ceiling. Cold climates on the interior side. In warm climates placement is on the exterior side.

**vapor barrier**
*Plastic sheets placed either inside or outside the insulation, depending on the climate, to protect the insulation and building from excessive moisture.*

Foam board or spray foam insulation is used on the inside of concrete basement walls to reduce heat loss into the concrete. When using foam board it is important to carefully seal all joints. Spray foam easily covers pipes, wires, and other utilities. Foam board sheets may also be installed outside the sheathing on homes with a stucco finish. The insulation below the first floor is placed between the floor joists. Normally, the joist braces are adequate to hold the insulation, but to properly hold floor insulation in place it may be necessary to add holders.

Vapor barriers are used in most parts of the country to keep the insulation dry because it does not insulate effectively if it becomes wet. Moreover, moisture can cause damage to the structure if it condenses inside the walls. A vapor barrier is placed under concrete basement floor slabs. Compacted crushed stone or coarse gravel should be placed below floor slabs cast-on-grade. The stone acts as a capillary break. Sheet polyethylene is placed on the stone and in direct contact with the floor slab.

The ceiling insulation is another matter. **Blown-in insulation** (pneumatic installed, loose-fill insulation) is usually used. Fiberglass blown-in loose fill insulation is placed above the ceiling of the structure's top story using special pneumatic equipment. It cannot be blown in until the wall and ceiling sheetrock is completed. The blown-in attic insulation is normally thicker and has a higher R-value than the insulation in the walls because heat rises. Loose-fill blown-in cellulose insulation is also available for enclosed walls.

**blown-in insulation**
*Type of insulation placed over ceilings. Placed by a blowing machine through hoses with compressed air.*

The purpose of the insulation is to limit the flow of heat. Homes in cold climates need insulation to retain inside warmth during the cold winters. Homes in warmer climates need insulation to keep the outdoor heat from flowing into cool air-conditioned space during the summer. The amount of insulation needed for a house depends on the climate, building materials, heating system, size and shape of the house, alignment of the house with the sun, and habits of the occupants.

The electric and plumbing rough-in should be done before the insulation is installed to prevent damage to the insulation during piping and wiring installation and to allow the insulation installer to place the material around and behind the wiring and electric boxes. The electric boxes and wiring must be placed accurately to meet code requirements. Insulation has flexibility and can be made to fit around obstructions such as electric wiring and plumbing pipes.

### Windows and Doors

Windows can be installed at any time after the structure's framing is in place. Window units are designed to be installed by attaching directly onto the framing of the building prior to the application of the building exterior material such as shingles, siding, or masonry. Most contractors do not want to install the exterior doors until they are ready to secure the building, but the doorframes should be installed immediately after framing. When the exterior doors are in place, entry becomes more of a nuisance, but there is less likelihood of pilferage.

### Interior Walls

Framing the interior walls, for a house, is an integral part of the framing operation. In the case of a timber framed house these are typically 2 × 4 studs covered with sheetrock panels (gypsum drywall). The panels are either nailed or screwed to the studs. In large commercial buildings, interior walls may not be part of the original construction process. Each building tenant has interior partitions constructed to suit how they want to divide their space.

Sheetrock consists of a plaster core with a paper covering on each side. The edges are slightly thinner than the rest of the board, to accommodate the tape and joint compound, which is used to hide the joints. As sheetrock is installed, cutouts are made for electric switches and wall outlets, light fixtures, windows, and any other items or wall features.

The nail or screw heads are covered by a layer of joint compound to create a smooth surface completely free of dents and blemishes. Covering the nails, screws, and joints is an operation called tape and paste, and must be done at least twice, preferably three times, to produce a quality surface (Fig. 10.7). The final product must dry for several days to produce a hard surface capable of accepting paint or other wall coverings.

## *UTILITIES*

All utilities within a building are constructed following the same general principles. Utilities are generally installed in two phases: (1) rough-in and (2) finish. The rough-in utility work can begin when the building framing is nearly complete.

The **utility entrances** to a house should be established during framing or if underground before casting of the floor slab. The electric company owns the overhead service drop, or underground service lateral supplying the

**utility entrances**
*The point where a utility company's service (electricity, gas, water, and sewer) attaches to the utility wires or pipes within a building.*

**FIGURE 10.7** Pasting the sheetrock joints.

power to house. There is no fuse, circuit breaker, or overcurrent protection, and the connection is energized at all times.

When the municipal water supply is connected to the house, a valve called a "curb-stop" is installed at the end of the water lateral. The curb-stop valve makes it possible to turn off the water supply to the house. To supply water to the plumbing of the house a copper or plastic water line connects to the curb-stop.

A sewer pipe will connect the house to the municipal sewer main on the street. This pipe must have a slight pitch so gravity will carry the waste to the municipal main. Natural gas is supplied through pressurized pipes. There is a valve at the connection point between the house gas lateral and utility's main.

Utility entrances to a house.

## Electrical

Rough-in of electrical system components is a matter of identifying the locations for all items within the building walls such as the main electrical panel, fixture mounts, switches, receptacles, and appliance outlets. This work occurs after the framing is in place but before placement of insulation and closure of the walls and ceilings with wallboard (Fig. 10.8). Note in Fig. 10.8 how the contractor has identified appliance locations on the slab. All necessary connecting cables will be routed through the framing members and inserted into wall and ceiling boxes. There may be other systems such as a central sound system, a security system, or computers requiring special wiring. All of these systems are wired at this stage. In many municipalities it will be necessary to have a building inspector check the work before it is closed-in to ensure the wiring meets electrical code requirements.

The electricians will return for the finish electric work after the sheetrock is in place and painted. During this second visit, they will install the lights, speakers, and building control systems. The electric power cable is installed and connected at this time.

## Water

With slab-on-grade construction, the water supply and wastewater drain lines must be installed prior to the construction of the slab. All above-slab pipes for the sinks and showers, washer, and outdoor faucets are installed during the framing process or as soon as the framing is complete (Fig. 10.8). Plumbing materials are rigid and require proper placement and, to flow properly, the drain parts must be at the proper elevations. Some items such as bathtubs must be installed during framing (Fig. 10.9). To prevent any accidental discharge of water into the building, pipes are not connected to the water supply at this point. Pipes will be capped and pressure tested. As with electrical rough-in, the building inspector will typically need to check the rough plumbing work.

> Many communities require the use of sprinkler systems in new residential homes.

The water service line, meter, and backflow prevention valve will be installed in the final stages of the project. The drain lines can be connected to the sewer system at any time. Finish plumbing work will be completed very late in the construction process.

## Mechanical

The ductwork rough-in (Fig. 10.10) for the forced air-heating and air-conditioning, and wiring for the mechanical system controls also follow the framing work. A house with a crawlspace may have the

**FIGURE 10.8** Electric and plumbing rough-in for a residence.

**FIGURE 10.9** Plumbing rough-in including the tub.

**FIGURE 10.10** Ductwork mechanical rough-in.

ductwork placed under the floor joists. The other option is to place the ductwork in the space above the ceiling. The ductwork joints must be well sealed with either a mastic or a butyl tape. System leakage is an energy waster. The local building code may dictate the need to use mastic. A builder working for an energy-conscious owner should seal the duct joints with mastic. At the same time, other minor items such as the exhaust ducts for the dryer are installed. Finish mechanical work will be accomplished at the same time as finish plumbing and electric.

Typically, electric, plumbing, mechanical, and communication system rough-in can be accomplished in a few days each, with crews of two to four workers, but the work duration depends on the expertise of the crews and the size of the house.

Water supply for a home will be from either the municipal water system or a well. Since most wells are not located close to the structure, a well can be drilled at any convenient time during construction. To avoid damage, the water service line to a municipal system might not be installed until the home is nearing completion.

Waste disposal can be by either the municipal waste system or a septic system. A septic system with drain field is usually not installed until there is no longer any need for heavy equipment to cross the area.

## ENVIRONMENTAL CONTROL SYSTEMS

Environmental control systems, such as lighting, heating and cooling, ventilation, and their control arrangements, support building occupant's well-being. The wiring and ductwork for these systems are installed during the utilities rough-in stage. As the house nears completion, the final work for each system is accomplished. For instance, the furnace and air-conditioning equipment do not need to be installed until the project is nearly complete unless positioning of HVAC units can be accomplished with great ease earlier. For attic placement of a unit, ease of placement is often dictated by the size of the access hatch. Air-conditioning compressor systems are installed outside the house. Lights and appliances are among the last items installed.

## INTERIOR FINISH

The interior surfaces are painted after the walls are finished and then the finish electric and finish plumbing work follows. Kitchen and bathroom cabinets with their countertops are installed after the walls are finished. The baseboards can be nailed in place before carpets are installed if ⅜″ spacers are used to hold the boards in position for nailing. With tile or linoleum, it is best to install the baseboards after the flooring. The baseboard should defiantly follow installation of laminate or hardwood flooring. Painters often like to prepaint the moulding before its installation.

The International Building Code (2018 ed.), section 803 specifies fire and smoke development limits for interior wall and ceiling finish materials. The section specifies similar limits for interior floor finish materials.

## COMMERCIAL BUILDINGS

Construction of commercial buildings is a growing market and well-designed energy-efficient buildings are cheaper on a first-cost basis as well as on a life-cycle cost basis. Many chain restaurants, department stores, grocery stores, office buildings, and warehouses are essentially the same standardized one-story building design. The predominant size for such structures is 10,000 $ft^2$ or less. The floor is a concrete slab-on-grade with exterior walls of concrete masonry or metal siding and a metal roof system.

### White Box Building

A white box is very popular with commercial real-estate owners because the tenant can customize the interior space to fit their specific business needs. These buildings are usually constructed for commercial purposes but they are also used for condominiums and townhouses. The box has a roof, exterior windows, doors, and basic heating, ventilation, and air-conditioning. There can be minimal lighting, plumbing, and other interior work. A white box is built ready to lease and the interior tenant improvements (build-out) will be completed once the lease agreement has been negotiated and executed.

## Finished Commercial Buildings

Large box-type buildings with simple floor plans, a "big box" can be constructed very rapidly using **tilt-up construction** techniques. If the building is occupied by a retail outlet, the front will be partially glass. The roof is approximately 30 ft above the floor and only heavy enough to support the local snow loads or wind forces in areas having high winds or hurricanes. Typically, the roof is flat and supported by rows of steel columns in a grid approximately 30-feet square. These buildings can be erected quickly. They are built as an empty shell, and the interior work is added according to the needs of the tenants. These are not low-quality buildings, but they are quite simple and easily built.

**tilt-up construction** *The building's concrete floor is constructed first then the walls are formed in place on the floor and raised into place by crane.*

Figure 10.11 shows such a building. While they are all different, each is constructed in the same manner. The horizontal dimensions of the buildings may change depending on the space requirements for the owner. If two stories are required in part of the building, the second story is inserted into the high exterior walls. The roof profile does not necessarily change.

The building envelope is the key to building energy use and thermal comfort. The important elements impacting energy use are as follows:

- Thermal performance
- Building orientation
- Permeability (air and moisture)
- Daylighting

A fast-food restaurant may be only 5,000–10,000 square feet, while a department store may be 40,000 square feet or more. The exterior finish may be anything from painted CMUs to exposed aggregate concrete panels. When these buildings are used as office space or restaurants, a drop ceiling may be installed to create a finished look in the commercial space. Used as a department store or warehouse, the building typically has no ceiling and the ducts for the air-conditioning are exposed, on the underside of the roof.

**FIGURE 10.11** Tilt-up commercial building.

**FIGURE 10.12** Structural steel framing and roof joists for a small commercial building.

Such structures contain all the systems of residential construction.

- **Foundation.** Grade beam to support the walls and roof and a slab-on-grade floor.
- **Framing.** The exterior walls are the framing, typically CMUs or metal (Fig. 10.12). The wall framing and/or interior columns support the roof joists.
- **Building envelope.** The insulation, exterior walls, and the roof make up the building envelope. Insulation is applied directly to the walls.
- **Interior partitions.** The interior of the building is built to suit the purpose of the tenant. For instance, a store would have a display area and a storage area, with smaller rooms for the administrative staff. A restaurant would have a kitchen, possibly a bar, and seating areas for patrons.
- **Building utilities.** Electricity, water, wastewater disposal, telephone, television cable, and communication systems such as sound systems or closed-circuit television are installed in two stages, rough-in and finish, just as done in a home construction.
- **Environmental control systems.** These include heating, lighting, air-conditioning and ventilation, and acoustics, along with their respective controls.
- **Transportation systems.** Many of these buildings are one story and do not require transportation systems.
- **Fire suppression.** Fire sprinklers are required in commercial buildings.

## TILT-UP CONSTRUCTION

Tilt-up construction is a fast and economical method for delivering large box-type buildings. The technique minimizes the cost of formwork because the building's concrete floor slab serves as the casting yard for the tilt-up wall panels (Fig. 10.13). The slab should be smoothly troweled and level.

> Smoothness is important because the slab surface will be reproduced, in reverse, on the face of the wall panel during casting.

The casting process reduces the cost of materials and labor for wall construction. To improve the quality of casting slabs, contractors are using laser screeds. Ride-on trowels are used to ensure the floor slab has a smooth surface finish. The wall panels can be designed to be load-bearing or non-load-bearing. Exterior walls are often load-bearing. Noninsulated panels are standard. This building process works well for shell-type buildings. Compared to other building methods, the panels are formed, constructed, erected, and supported very quickly.

**FIGURE 10.13** Slab-on-grade building floor serves as the casting surface for the tilt-up wall panels.

**FIGURE 10.14** A 2- to 3-foot gap between the outside dimension of the building's floor slab and the final interior face of the exterior tilt-up panels.

All panels can be cast on the building's floor slab if there is sufficient room. At this stage, the slab-on-grade has no joints so the wall panels will be smooth. The floor slab joints will be saw cut after the tilt-up panels are lifted into place. If site conditions restrict the available slab area, the panels can be constructed on top of one another. Panels can be successfully stacked up to eight panels on top of each other.

A commercial concrete curing bond-breaking compound is used as the curing compound for the floor slab. A second coat is applied later to serve as the bond breaker between the floor slab and the tilt-up panels. The wall panel concrete is normally specified as 3,000 psi compressive strength with a 4-in. slump, and panels range from 6- to 8-in. thick.

The floor slab is constructed so there is a 2- to 3-foot gap between its outside dimension and the final interior face of the exterior tilt-up panels (Fig. 10.14). This gap provides space for construction of a strip footing around the perimeter of the building. After the wall panels are erected, the gap between the outside dimensions of the floor slab and the final interior face of the exterior tilt-up panels is filled with concrete to complete and tie the slab to the wall. Reinforcing bars extend outward from the previously placed floor slab and the tilt-up panels have exposed reinforcing bars (Fig. 10.14), so when the concrete is placed in the space between the slab and the interior of the wall panels, the structural systems are tied together.

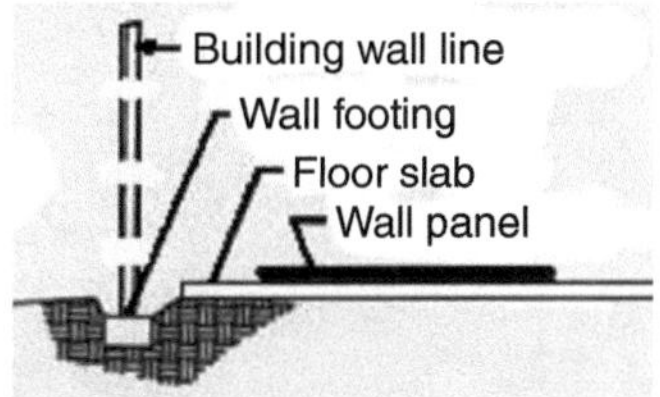


## Casting the panels

Following the cure period for the floor slab, side forms for the panels are positioned. Once the forms are in place, the floor is cleaned and the bond breaker applied. After the bond breaker application, the crews must be careful not to contaminate the clean surface. Now the reinforcement, lifting inserts (hardware), bracing hardware, embedments are placed, and window and door blockouts set. The inserts provide a means to lift the panels and to support the panels temporarily during construction. The number of inserts depends on the size of the panel.

Panels are cast face down because the lifting hardware must be in the exposed upper face. Knockouts are created for possible later placement of a door or window in a panel. The knockout is a grooved outline about 1 in. deep in the panel section. No reinforcing steel or electrical conduit is placed within the knockout area. If in future a tenant should request a door, it is possible to easily remove the concrete in the knockout section by saw cutting. The panels are designed to be structurally sound with the knockouts removed.

Each panel has a particular function and position in the overall structure. Contractors try to cast the tilt-up wall panels as close to their final position as possible. They do not have to be directly adjacent to their erected location, but they are laid out on the slab in a pattern to minimize crane time during the erection. For efficiency of the lifting operation, panels are erected in a clockwise rotation, starting from a chosen building corner. To ensure a proper placement sequence, the erection of certain panels at the correct place and time may require some panels to be rigged and hoisted over or around others still in their horizontal cast position on the slab.

> Careful planning improves panel lifting efficiency.

A carefully planned panel placement map is used to lay out the panels on the building slab. Steel concrete edge forms having an angle shape are commonly used for casting the panels, but dimension lumber is an option. Edge forms are spaced so there is a minimum of 2 ft between individual panels. The use of wood cleats, spaced between 36 and 48 in. apart along the edge form and located at each corner, is a common method for securing the forms to the floor slab. The cleats are attached to the slab by drilling 3/16-in. hole and driving in three 16d double-headed nails. These just fit and can easily be pulled out when the forms are stripped. Chamfer strips are positioned along each panel edge so the edges are not damaged when the panels are lifted and placed. Placement of concrete is fast and easy because it is done on the ground like a slab placement. The mix should be consolidated to ensure flow around the reinforcing steel. For most projects, a trowel finish is satisfactory.

The final phase is the erection of the concrete panels. Hoisting the panels requires the use of a crane (Fig. 10.15). It is imperative to follow the proper lifting sequence so the crane can erect as many panels as possible in the minimum amount of time and, thereby, reduce construction cost. Before the temporary braces are attached to maintain the panel in an upright position, each panel must be leveled and plumbed. The braces should be designed for a 60-mph wind; note the braces in Fig. 10.15.

After the panels are erected, the roof system can be placed on the structure. At the same time, the space between the floor slab and the wall panels is filled with concrete, securing the wall panels to

FIGURE 10.15 Hoisting a wall panel from the casting floor.

FIGURE 10.16 Tilt-up building with an internal steel frame for a second floor.

the floor. Once these operations have been completed, the temporary braces can be removed and joints saw cut into the floor slab to prevent random cracking. With two-story buildings there can be a steel frame to support the second floor within the tilt-up walls (Fig. 10.16). The transfer of loads between the internal steel frame and the concrete panels is attained by the use of embedded steel plates cast into the panels. Three- and four-story tilt-up construction has become common.

## MULTISTORY BUILDINGS

The nature of multistory building construction is very different from homes and one-, two-, or three-story commercial building construction. While multistory buildings contain all of the same systems and follow the same general sequence of construction operations, the accomplishment of those operations is technically different from the techniques and methods used in residential and small commercial buildings. The mechanical, electrical, transportation, and life safety systems are much more complex in multistory buildings.

During the first half of the 20th century, there was basically one way to design frames for multistory buildings—a heavy steel skeleton. During the second half of the century, structural steel and reinforced concrete design changed. Now multistory buildings are constructed using different designs such as rigid frame, shear wall, wall-frame, framed-tube, and bundled-tube.

The multistory building design seeks to meet the requirements of the building occupants. These buildings have large areas of open space on each floor so the area can be divided easily into individual workspaces. Furthermore, the building must provide a structurally sound and architecturally pleasing environment. Structural engineering and the work of an architect is as much art form as set principles.

## Building Foundation

Foundation footing steel

Small buildings such as houses have shallow foundations. Some smaller buildings sited on good ground and not very heavy can be supported by shallow foundations. These could be spread footings or open footings. Each individual footing carries the concentrated load of a single column and spreads the concentrated weight over a large contact area. Construction of such footings is simple. Taller and heaver buildings have deep foundations. The design of the foundation varies according to the kind of soils and bedrock at the location, and whether the structure is in a high-risk seismic zone. To support a building of moderate height, a raft (mat) foundation may be feasible. But for the heavy, concentrated loads, imposed by taller buildings, caissons or piles are usually required (Fig. 10.17). The use of concrete precast or steel H-piles is common practice. Steel H-piles can take very heavy loads. Groundwater levels are also an important foundation consideration, especially if the foundation work requires excavation for a basement or a lower level parking garage.

**FIGURE 10.17** Driving concrete piling for the foundation of a multistory building.

***Plumbing Connections.*** An excavation company may have to be engaged to bring the water and sewage services into the building. However, in some locations the utility companies handle the connections to their systems. It is not unusual for permits to be required.

The feeder water line will be brought from the service to the meter location. Much of the connection work or at least the piping for the connections will penetrate the basement walls and the foundation walls or slab so the piping must be completed before the concrete is placed.

## Building Frames

The purpose of the building frame is to economically support the dead and live loads of the finished structure, and to support the external loads caused by wind, weather, temperature change, and earthquakes or hurricanes. The building structure must meet all these loading requirements and keep vibration to an acceptable level. At the same time, floor space must be used to provide convenient access to each floor. Figure 10.18 shows the erection of the steel skeleton for a new building.

***Shear Wall Buildings.*** Shear wall buildings achieve strength by replacing the braced frame diagonal bracing with rigid walls to resist distortion. A majority of residential and light commercial buildings use horizontal diaphragms (roof and floor systems), in conjunction with vertical shear walls, to support gravity loads, resist lateral loads, and provide structural stability. The rigid vertical diaphragm walls are capable of transferring lateral forces. Masonry walls do not work for this purpose because the mortar lacks the required tensile strength. Many smaller buildings have shear walls made of plywood or OSB. Structural panel diaphragms built of dimension lumber take advantage of the capacity of wood to absorb impact loads (Fig. 10.19).

**FIGURE 10.18** The steel skeleton for a multistory building.

**FIGURE 10.19** Dimensioned lumber multistory shear wall building.

***Braced-Frame Buildings.*** The braced-frame is one of the more economical methods of designing a steel frame building. The vertical members are the columns, the horizontal members are the girders, and the diagonals are the braces. The diagonal braces work in both tension and compression. Any force distorting the square shape of the frame pulls one of the diagonals and pushes on the other in each square opening of the frame. Without the diagonal braces, the structure could distort easily. Wind, for instance, would push the building out of its square shape.

A variation of the braced frame is achieved by changing the length and placement of the diagonal braces so they span more than one floor. This design usually has the disadvantage of increasing the difficulty of hiding the diagonal braces. They interfere with window placement and change the appearance of the building if they are external. But the diagonals can be used as an architectural feature (Fig. 10.20). A braced-frame building is suitable for buildings up to about 50 stories.

***Rigid-Frame Structures.*** A rigid-frame structure has no diagonal braces. There are only "rigid" connections between vertical or tapered columns and horizontal girders. The rigid-frame design approach is ideal for reinforced concrete buildings (Fig. 10.21) since there is inherent rigidity in rein-

**FIGURE 10.20** Braced-frame multistory building with exposed diagonals.

**FIGURE 10.21** A concrete rigid-frame building under construction.

forced concrete joints. In the case of a concrete rigid-frame structure, the ready-mix concrete for the elements may be placed monolithically.

With multistory steel frame building (Fig. 10.22), the columns and girders are large, heavy members needed to support a concrete floor slab. The column and girder connections must be strong so they do not distort and do not require any diagonal braces to keep the shape of the building square.

The rigid-framing system is not suitable for buildings of more than 25 stories. To lift and position steel structural components of both single and multistory buildings, it is necessary to have a mobile crane to lift the heavy steel members during frame assembly. To reach the connection points safely, the workers will need aerial lifts, both scissor and boom types are usually used (Figs. 10.12 and 10.23).

Prolonged exposure of steel structural components to elevated temperatures (fire) degrades their structural capacity. Therefore, the members of a steel frame building (columns, beams, and girders), by fire code, must meet mandated fire endurance levels. Fire codes seek to prevent loss of structural stability for a prescribed time period.

Spray-applied fire-resistive material (SFRM) products are the most common type of fire protection for structural steel. After the steel members are erected, the low-density fiber or cementitious compounds are spray applied to the steel members of the building. SFRM is being applied to the lower steel in Fig. 10.22. Note the lower steel is now white.

FIGURE 10.22 A steel rigid-frame building under construction.

FIGURE 10.23 A rigid-frame building under construction.

## Construction Challenges

The economical height of a shear wall structure is affected by the type of wall material and the magnitude of expected seismic forces. A building with a structural steel skeleton built around the central core with steel columns and steel girders for each floor can have concrete shear walls instead of steel diagonal bracing.

At the time the upper floors are being framed, exterior shell and utility rough-in of the lower floors can proceed. The core of the building carries the elevators and the utilities. Most multistory commercial buildings are constructed without any interior finish work because the tenants will later specify an interior design to suit their requirements.

Multistory buildings present many engineering challenges. The challenge of designing water system, fire suppression, and wastewater disposal for the upper floors often leads to construction difficulties. Electrical and communication systems are major arrangements passing horizontally and vertically. Vertical transportation for both people and materials is a greater challenge for a 12-story building than for a 2-story building, and often requires banks of elevators.

A common construction challenge of multistory buildings is the inherent lack of construction site workspace. Although there are some exceptions, multistory buildings are often built in crowded urban areas. Mobilization of equipment must be done during times of minimum traffic. The movement of large equipment such as cranes or trucks carrying structural steel and other materials requires careful attention to coordination planning. There is little or no storage space at the site, so all materials are delivered to the site when needed. The contractor's coordination efforts seek to ensure the steady arrival of materials to keep the work progressing on schedule. Management of such a project should consider using pull scheduling.

As with all buildings, construction begins with the foundation, probably a deep foundation. The building will be supported on piles or caissons. As soon as the foundation is complete, the erection of the steel frame of the building and the placing of the concrete floor slabs begins, and continues in a repetitious manner until all floors are completed. Since all floors are often designed with the same shape, the erection of a 12-story building is a matter of repeating the same process 12 times. The placement of the exterior cladding follows the construction of the structural system and progresses upward from the ground level. When the roof and exterior cladding of the building are in place, the interior finish work can begin. It is not uncommon for companies to move into a building and open for business before all of the available office space is completed.

## Completing the Frame

Designers specify floor systems for structures based on the proposed use of the building. There is no single appropriate floor solution. Often open web steel joists are placed between the girders of a steel frame building. These joists are used to support reinforced concrete floor slabs, typically 2½–3½ in. thick. Before the joists are unloaded at the project site, they should be inspected for physical damage. A forklift or crane will be needed to unload the bundles and lift individual joists into position. The size of the needed lifting equipment will depend on the size and weight of the joists.

The joists will be positioned roughly 4 ft apart on the steel girders of the building frame. Their final positioning on the girder is adjusted to a joist spacing of 4 ft 1¼ in. center to center. This spacing permits a full 4×8-foot sheet of plywood form to be placed between two joists. With the joists properly located, "roll bars" are connected between them to support the plywood form. The flat steel roll bars have two handles for rotating the bars into slots on the top chord of the joists. The bars will support the plywood forms, the dead weight of the concrete, and a construction load of 40 pounds per square foot. They should be placed 21 in. apart when used for casting 2½–3½ in. thick slabs. Plywood sheets, ⅜, ½, or ⅝ in., are positioned on the roll bars (Fig. 10.24). The sheet thickness will depend on the slab thickness.

***Composite slabs for steel frame buildings.*** A shallow floor of lightly reinforced concrete is cast on the steel joists. The slabs are normally reinforced using an upper layer of wire mesh and, occasionally, additional reinforcing bars for heavy loads. The roll bar formwork supports can be done by personnel riding aerial lifts on the floor below. Shoring is usually not required and the plywood sheets can easily be stripped for reuse. Precast concrete units may be used in conjunction with steel beams. The units may be solid or hollow-core, and with tapered or bluff ends. They are normally prestressed.

**FIGURE 10.24** Roll bar formwork supports for bar joist supported concrete floor slabs.

The top chord of the joist will become embedded in the concrete floor slab. This creates composite action between slab and joist. With the forms and necessary blockouts in place, the iron workers can layout the steel mesh and any necessary reinforcing steel.

The concrete for the slab is pumped to the upper floor locations and lightly but thoroughly vibrated to assure full encasement of the top joist chord. Stay-in-place prestressed, precast concrete panels can also be used to form building slabs.

Preparing to cast a concrete floor slab.

> Concrete placement for a slab should start at a perimeter and proceed continuously. If placed by bucket, do not dump in separate piles and then level.

***Concrete slabs for buildings.*** In multistory flat-plate building construction (a one- or two-way system) with loads supported directly on columns or load-bearing walls, no beams or column capitals (Fig. 10.25), the freshly cast slab for a particular floor is supported by formwork with a system of shores and reshores extending one or more additional stories below. Flat-plate concrete floor building is an economical construction approach because of the simple repeating formwork system.

The sequence of operations for construction of concrete floor slabs in a concrete framed building of the flat-plate type is as follows:

Finishing the floor slab.

- Install formwork and slab reinforcement.
- Place the liquid ready-mix concrete.
- Strip the formwork and shores to allow the slab to deflect and carry its own weight.
- Remove reshores from the lowest level in the sequence.
- Place reshores in the story where the formwork was stripped.
- Cycle the formwork for the floor placement above.

> Concrete should not be deposited more rapidly than it can be spread, struck off, consolidated, and bull floated.

**FIGURE 10.25** Casting concrete floor slabs for a concrete framed building.

**FIGURE 10.26** Water impermeable membrane on insulating panels.

The reshoring distributes the applied construction loads to multiple stories. The liquid concrete is heavy so sufficient levels of support, in many cases more than one floor, are needed to carry the weight. The anticipated construction loads, the curing time of the concrete mix (strength gain speed), and the rate of floor construction are factors used to determine the number of stories below slab placements requiring reshores.

***Roof.*** The roof slab can be normal weight or structural lightweight concrete. In place of a monolithic concrete slab, load-bearing precast concrete panels are sometimes used. The precast panels are reinforced or they can be prestressed or post-tensioned. When precast panels are employed, a covering or leveling layer of lightweight or structural lightweight concrete is sometimes applied.

The thermal performance of a low-rise (large footprint) building is determined by the roof system. Insulated panels are, therefore, placed on top of the concrete roof slab and then a liquid foam is sprayed on top of the insulating panels to create a water impermeable membrane (Fig. 10.26).

The liquid coatings are highly elastic, self-flashing, and can be easily applied to contoured surfaces. However, care must be exercised during application because development of a uniform thickness is essential to performance. Roof quality is dependent on this application. The waterproofing membrane is the most important component of the roofing system. It keeps water out of the building.

While it is termed a flat roof, there must be sloping flow paths directing water to discharge points. Establishment of the necessary fall on a flat roof is critical to effective drainage. Membrane roof systems are another popular choice for commercial buildings. Adhesives are used to hold the sheets to the roof deck. Metal roofing systems can also be used. Corrugated galvanized steel is often used for low-slope roofs. However, metal roofing has a much higher installation cost.

Owners seeking Leadership in Energy and Environmental Design (LEED) certification for their projects may specify a "green roof." With a green roof the membrane will have to be strong enough so water in the upper growing medium layer will not seep through. The design of a green roof will place a root barrier under the growing medium layer and below the root barrier a drainage layer to remove excess water. To protect the membrane, there may also be a slab of lightweight concrete.

Roof sloped to drain.

***Walls.*** The walls of a building have three primary functions: (1) support the load-carrying capacity of the structural frame, (2) separate the interior and exterior environments (air, moisture, heat, sound), and (3) present an appearance (finishes). Finish may not be necessary in all areas (above suspended ceilings). A possible fourth function is to provide conduits or space for routing HVAC, utilities, and communication services. Wall and glazing insulation are critical in determining HVAC loads for multistory buildings. When the exterior wall is nonstructural, it is termed a curtain wall.

Exterior walls for commercial building are often constructed with steel studs to which a gypsum board cover is attached on the interior side. Because steel is a thermal conductor, an insulation layer (board) is placed on the outer face of an exterior wall. The insulation is placed to prevent exterior heat and cold from reaching the framing (Fig. 10.27).

The rigid insulation could be molded expanded polystyrene (MEPS). MEPS is commonly referred to as "*beadboard*" and is available in large sheets. Extruded expanded polystyrene (XEPS) is another board-type insulation material similar to MEPS but more expensive. XEPS is more consistent in

**FIGURE 10.27** Exterior insulation following erection of the structural frame.

density than MEPS, making it better suited for use on roofs. It also has excellent resistance to moisture absorption. Either of these insulation types can be fastened to the structural sheathing or framing with broad-head nails, wide plastic washers, and/or adhesive caulk.

When exterior rigid insulation is specified for wall assemblies, it does not significantly change standard construction practices. It may, however, be easier for the team preforming the task to work from both sides of the wall. Joints in the insulation should be offset and staggered. All insulation joints should be tight and gaps larger than ⅛ in. need to be filled with insulation or closed-cell spray polyurethane foam. Finally, a protective or finish layer is affixed to the exterior side of the wall system. The water control layer of the wall can be on either side of the insulation.

Installing exterior insulation.

Glazing often makes up a major part of the exterior wall surface of a commercial building (Figs. 10.28 and 10.29). Glass offers an infinite combination of color, thickness, and opacity. The two

**FIGURE 10.28** Glass faced commercial building.

**FIGURE 10.29** Installing glazing in a commercial building.

**FIGURE 10.30** Assembled window unit being lifted for installation on an upper floor.

most common thicknesses used in commercial construction are ¼-in. monolithic (single-glazed) and 1-in. insulating glass (double-glazed, two panes of glazing). There are more expensive triple- and quadruple-glazed windows sometimes specified. These are used in very cold regions.

Double-glazed windows have two ¼-in. glass layers with a ½-in. airspace between. The air in the space can be atmospheric air, but panels are available with inert internal gasses. Filling the space with argon (most common) or krypton gas (most expensive) is done to reduce thermal transmittance (U-factor). A lower U-factor means the glass has a greater resistance to heat flow, that is, better insulating properties. U-factor ratings are used for windows, doors, and skylights. The National Fenestration Rating Council (NFRC)–recognized rating is for the whole window (glazing, frame, and spacers). However, there are also center-of-glass U-factors. These are a measure of the performance of only the glazing.

The work of installing building glass (glazing) has traditionally been undertaken by a specialist glazier. It is also possible to purchase complete window units (Fig. 10.30) which the general contractor can install.

## Building Interior

The interior of a building must conform to codes designed to provide minimum standards for safety, health, and general welfare as they relate to occupancy of a building. These will include local fire codes and possibly energy efficiency requirements or benchmarks. Besides the local codes there are the Americans with Disabilities Act (ADA) requirements with power-operated door requirements for pedestrian use and detailed water closet dimensions.[2]

Interior tasks present a special workflow *scheduling* situation. There is a need to think and schedule in terms of *effective workspace* and *managed workflow*. Many of the specialized work tasks require time-consuming preparation and set-up time. Once time has been consumed in preparation, the specialty trade must have the necessary unimpeded time to perform the work.

| To improve productivity, schedule work tasks based on full use of preparation time. |
|---|

[2] Americans with Disabilities Act and Architectural Barriers Act Accessibility Guidelines, Chapter 6.

Similarly, cleanup is important. First, it creates a safe work environment. Second, it minimizes damage to completed work. A dusty workspace can ruin the best paint job.

The building contractor usually does not have the opportunity to influence design, unless it is a design–build project. But sharing of ideas can bring future work. Solid wood doors are expensive and beautiful but do not hold up well to commercial use. Similar to wood doors, rolled carpet is a less expensive option but carpet tiles can be easily replaced, particularly when replacing the high-wear carpet areas in a commercial building.

| Use construction knowledge to gain future negotiated work. |
|---|

The space between the floor of the level above and the finished ceiling—the sandwich—is used for housing various building systems: structural support, plumbing, electrical, HVAC, and fire protection. The contractor must carefully plan and coordinate the work of the specialty subcontractors installing the plumbing, electrical, and HVAC systems in this space. With the load-carrying components of the structure in place, construction of the interior walls separating individual spaces can proceed. The interior wall will form separate spaces and provide the supporting structure for plumbing and electrical systems.

***Partition walls.*** The frame of a partition wall in a commercial building is commonly made of light galvanized steel sections. Wood framing is usually used in houses. Metal stud framing significantly impacts other trades. Because of the impact of space usage on all trades, planning is important. The frame is installed first and then it is covered with gypsum board. The gap between the two sides of the walls can be filled with insulating material if needed. Prepunched holes (Fig. 10.31) in the metal framing sections permit electrical and plumbing to pass through the inner part of the wall.

Steel studs are available in a variety of widths, lengths, and gauges (lower gauge numbers indicate thicker steel). Walls containing the piping for plumbing fixtures often require greater widths. Also consider the locations of access panels behind faucets, drains, and other parts of the plumbing system. Access panels are needed to easily reach shut-off valves and drain cleanouts. Wall-mounted plumbing fixtures will require mounting brackets to be installed within the wall.

The steel tracks (plates) for the metal studs are attached to the concrete floor slab using 1¼-in. hex-head concrete screws. First a hole is drilled through the metal tracks and into the concrete the full length of the screw. If installing the studs one at a time (stick building) it is best to first secure the floor and ceiling tracks. Stud spacing for interior walls can be 12, 16, or 24 in.

| Make sure the upper and lower tracks create a plumb wall. |
|---|

**FIGURE 10.31** Metal stud framing.

Panelized steel wall framing components assembled in a manufacturing environment are available. Steel framing panelization meets exacting tolerances, is light weight, and reduces job site labor. Panels are typically shipped unsheathed.

Gypsum boards are screwed into wall framing using 1¼-in. self-tapping drywall screws when there is no specification. Screws should be spaced every 8 in. along edges where two sheets meet on a stud and 12 in. on-center elsewhere; always check local codes.

After the gypsum panels are installed, joint tape or joint compound is used to cover the seams and screw holes. Lastly, the taped and compounded area is sanded to provide a smooth finish.

The project specifications should state the required level of Gypsum Board Finish,[3] which are as follows:

Panelized steel framing.

Level 0—No taping or finishing is required.
Level 1—All joints and interior angles shall have tape set in joint compound.
Level 2—All joints and interior angles shall have tape embedded in joint compound and wiped with a joint knife, leaving a thin coating of joint compound over all.
Level 3—All joints and interior angles shall have tape embedded in joint compound and shall be immediately wiped with a joint knife, leaving a thin coating of joint compound over all. One additional coat of joint compound shall be applied over all joints and interior angles. All joint compounds shall be smooth and free of tool marks and ridges.
Level 4—All joints and interior angles shall have tape embedded in joint compound and shall be immediately wiped with a joint knife, leaving a thin coating of joint compound over all. Two separate coats of joint compound shall be applied over all flat joints and one separate coat of joint compound shall be applied over interior angles. Fastener heads and accessories shall be covered with three separate coats of joint compound. All joint compound shall be smooth and free of tool marks and ridges.
Level 5—All the operations in Level 4 are required. Additionally, a thin skim coat of joint compound, or material manufactured especially for this purpose, is applied to the *entire surface*.

## Plumbing

The plumbing work includes all pipes, fittings, and appliances used to bring water into the building, and needed to drain and provide waste disposal. This last part is termed the drain-waste-vent (DWV) system. The pipes for the water system can be copper, galvanized iron, or chlorinated polyvinyl chloride (CPVC). The use of polyvinyl chloride (PVC) pipe and fittings is often restricted by code. CPVC can withstand higher temperatures than standard PVC. PVC pipes also lack sound attenuation and the noise of water cascading through sanitary and storm PVC piping can be unacceptable.

PVC is an attractive piping choice because of its low cost and ease of installation. Nevertheless, before considering its use for a multistory residential project or hotel, carefully assess the following:

- Cost of the needed firestopping when plastic pipe penetrates walls and floors
- Thermal expansion relief
- Hangers and supports
- Sound transmission

[3] *Recommended Levels of Gypsum Board Finish* (GA-214-10ε), a consensus document by the Association of the Wall and Ceiling Industry (AWCI), Ceiling & Interior Systems Construction Association (CISCA), Drywall Finishing Council (DWFC), Gypsum Association (GA), and the Painting and Decorating Contractors of America (PDCA).

The water supply system of a multistory building of more than two stories cannot usually be supplied by the operating pressure of public water systems. Multistory building over two stories require pressure booster systems; booster pumps, roof tank, or hydropneumatic storage tanks.

The building drainage system will have separate stacks for waste and soil (sewage) drainage. The building's soil drainage system should discharge into the public sewer by gravity unless there are below-ground basements. By code, multistory buildings will have sprinkler systems on each floor and possibly standpipe systems. Standpipe systems are designed to provide a prepiped water system within the building for fire department use. Fire hose threads used in the standpipe systems must be compatible with local fire department hose threads.

***Plumbing Rough-In.*** Once the walls have been framed, the drains and vent pipes, and the water lines, can be installed. Sinks, faucets, or other fixtures will, however, not be installed at this time. Vent pipes may run level, but some codes require a slight slope toward the main drain. Before the drywall can be attached to the framing, the gas lines will also need to be routed in the wall space. The gas work may be handled by the plumber but in some locations it is the work of different specialty contractor.

Off-the-floor water closet systems can be installed more economically than conventional floor-mounted systems in multistory buildings. The savings is more pronounced when there are batteries of water closets on each floor. The necessary clear distance between walls for horizontal adjustable carriers is 8½ in. for units on one side only and 11 in. for back-to-back units on the two walls. Vertical adjustable carriers require 8 in. for single units and 11 in. for double units.[4] This is a difference compared to floor-mounted units installed in either a single or a back-to-back unit as only 4 in. is required between walls.

Off floor water closets.

Manufactured bathroom pods are available for projects having multiple identical requirements, hotels, schools, hospitals, even multiunit residential buildings. These are ready-to-install pods with steel framing finished interior and pretested plumbing and electric systems. The use of Building Information Modeling (BIM) allows for precision in overall fit within the pod and to the systems in the building structure (Figs. 10.32 and 10.33).

Flammability is of particular concern when systems are installed in building plenum spaces (the space above a suspended ceiling). This space permits air to circulate throughout a building, and supports heating and cooling system operations. Similarly, a plenum can be under a raised floor. This is common in computer installations. However, fire and smoke from one part of the building can spread quickly through a plenum. Building codes and local rules, consequently, prohibit the placement of combustible materials in the plenum space.

Plumbing is almost always placed in the plenum of multistory buildings. But when plastic piping is used, any pipe passing through floor levels must be in an approved fire stop assembly. In addition, there are floor-to-floor fire separation (fire resistance) ratings to maintain.

Building codes and standards specify strict limits on the flammability and smoke potential materials passing through the plenum. ASTM E84 and UL723 are used to evaluate the burning characteristics of construction materials and are extensively referenced in building codes. Materials placed in the plenum space must have a flame spread index = 25. Smoke can spread quickly to other interconnected parts of a

[4] Minimum Space Requirements for Enclosed Plumbing Fixture Supports, Plumbing and Drainage Institute (PDI).

**FIGURE 10.32** Back-to-back installation of bathroom pods.

**FIGURE 10.33** Bathroom pod set in the building frame.

building through plenums. Consequently, for commercial building a smoke-developed index (SDI) of 50 is required for any material placed in a return air plenum. In the specifications, these requirements are combined and written as "piping running through plenum spaces, will have a 25/50 flame/smoke rating."

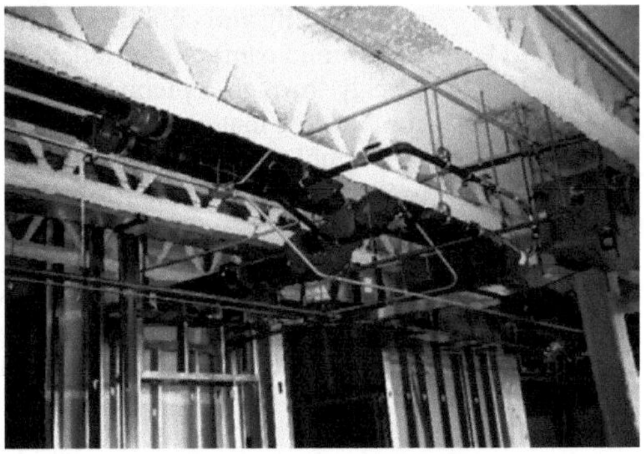

Utilities in the ceiling plenum.

Many other specialized plumbing systems for special gases and liquids are required for hospitals and manufacturing facilities. These can require special materials and construction procedures. Some buildings have rainwater harvesting systems. These systems channel rainwater from the building's roof to storage tanks. The water is then used in evaporative cooling systems, toilets, landscape irrigation, and for other nonpotable water loads. Such systems add a further piping requirement to the building's plumbing network. The International Association of Plumbing and Mechanical Officials (IAPMO) has a *Green Plumbing and Mechanical Code Supplement* (2015).

Before all of the "rough plumbing" is concealed by insulation, wall boards, or ceiling tiles the work must be tested by an approved method and inspected. Similarly, systems under the foundation should be tested by an approved method before the concrete is placed.

Ready for finish plumbing.

***Finish Plumbing.*** After the interior finishing is completed including the laying of floors, but may be not carpets, finish plumbing work will begin. Some contractors want the plumbing finish work completed before carpet is installed. While others install the carpet and then cover it for protection. The plumbing finish work includes removal of the caps on drain and water supply pipes needed for rough-in testing. With the caps removed, the installation of fixtures, drain assemblies, water supply valves and lines, sump pumps, and back flow preventers can proceed.

Fixtures are installed and hooked to supply and drain lines. The sinks, water closets, and tubs will be caulked and sealed before testing of connections. Hot water heaters will be installed and adjusted. Supply and drain lines, when needed, will be connected to dishwashers, garbage disposals, and ice makers. While some plumbers hook up gas appliances, others do not. All work must be performed per the project plans and all applicable building codes and/or municipal ordinances.

## Electrical

All building wiring must meet or exceed code requirements for the project location. In the United States, the National Electrical Code (NEC) is the regionally adoptable standard for the installation of electrical circuitry and equipment. Nonetheless, some local authorities have their own regional regulations.

The utility company distribution system moves power at high voltages. There are four major distribution-level voltage classes: 5, 15, 25, and 35 kV. The 15-kV voltage class is the most prevalent. For small commercial buildings, the utility uses a transformer to lower the voltage to 120/240, 120/208, or 277/480 V before it is fed through a meter into a building. The meter, owned by the utility, provides a record of the building occupant's electrical energy consumption. Electrical energy is commonly measured in kilowatt hours (kWh). Within the building, all wiring, panels, and devices are the property of the building owner. To control the distribution of power to the building's tenants in a small commercial building, a distribution board or panel board will have a main service breaker and a series of breakers for each of the individual circuits in the building.

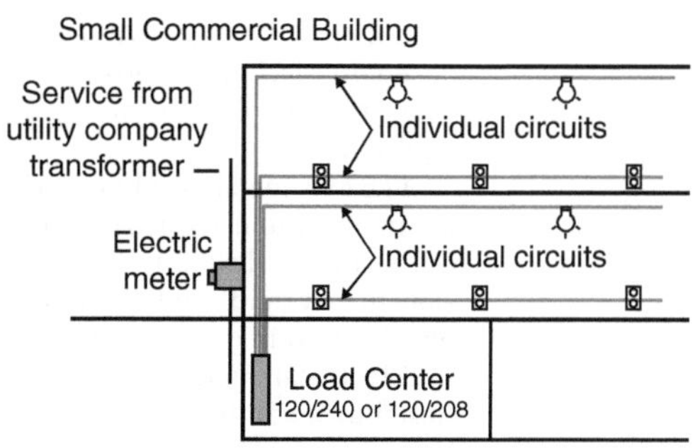


Electrical circuitry.

Depending on the tenant, a large multistory building may have a large electrical load. In such cases, the building owner may purchase power at high voltages to minimize the cost of distribution and interior infrastructure. The owner would then provide a step-down transformer or transformers to lower the voltage as needed. The transformer will feed a branch panel. Each branch panel is used to control circuits to parts of the building. Circuits from the panel feed specific electrical requirements: lighting, outlets, pieces of equipment.

Small buildings typically have emergency battery packs in safety light fixtures for egress signage. The fire alarm system and other life safety equipment can have an integral battery pack in case of a loss of power. However, larger buildings usually rely on standby generators to provide power to life safety equipment. This is specified by code in many areas. Furthermore, hospitals and other critical care facilities need to continue to operate when there is a power outage. Therefore, standby generators, automatic transfer switches, and uninterrupted power supplies would be part of the building's electrical system.

Generator outside a medical building

Because equipment is vulnerable to damage from lightning and power line anomalies, grounding is essential for personnel safety and system reliability. Solid grounding (SG) is the standard for transformers and generators in commercial buildings operating at 600 V or less. SG means the generator or transformer is directly connected to the ground through a conductor of negligible resistance and reactance.

***Electrical Rough-In.*** New construction wiring is done before the walls and ceilings are closed-in and floor coverings placed. In exposed locations, electrical wires are run through the rigid or flexible electrical conduit. The electrical wires are pulled or "fished" through the conduit during installation. This is different from cable, which is a group of bundled wires encased inside a protective sheathing. Each of the separate wires in a cable has its specific purpose in connection with a circuit. There are several types of conduit with their use based on mechanical protection, corrosion resistance, and costs. The conduit will terminate at an electrical box or a piece of fixed equipment.

Flexible metal conduit (FMC) is used in the construction of commercial and industrial properties to protect electrical wiring. Because of its flexibility, FMC is placed in areas where it is not possible to fit a rigid conduit. However, FMC is not waterproof and can only be used in dry environments.

Electrical metallic tubing (EMT) or "thin wall" is a rigid tube but it can be bent with a conduit bender. Because it is not threaded, EMT is installed with couplings and fittings. It is used for exposed indoor wiring runs in light commercial construction. The NEC requirements for EMT require supports every 10 ft or less and within 3 ft of fixtures. As the name implies, it is thin wall and a missed directed nail driven into a wall will penetrate EMT.

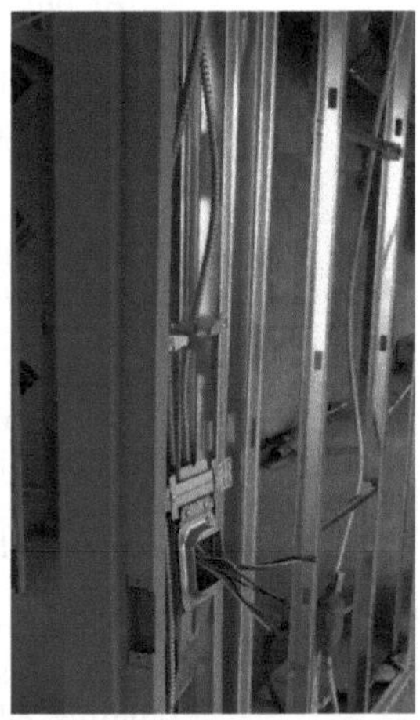
Electrical wires at a switch box.

Rigid metal conduit (RMC) is suitable for indoor and, if it has been galvanized, it is suitable for outdoor, wet, or dry conditions, and embedment in concrete. It provides a channel termed an "Electrical Raceway System" for electrical wires or cables. An RMC raceway is recognized as an equipment grounding conductor providing shielding from electromagnetic interference (EMI). It is also the standard for providing protection from physical impact.

Electrical conduit.

Rigid polyvinyl chloride (rigid PVC), similar to plastic plumbing pipe, is lighter than steel conduit and less costly. The conduit tubing and fitting can be glued together. A glued assembly will be watertight. It can moreover, after being heated, be bent. However, rigid PVC is nonconductive so an extra bonding (grounding) conductor must be pulled into each conduit.

Electrical junction boxes serve as containers for electrical connections. They are not a special type of box but are standard electrical boxes used to enclose wire splices. The box protects the connections, provides a safety barrier, and deters tampering. Both metal and plastic boxes are available. When using metal conduit a metal box is required to complete the grounding. All wiring splices should be contained in a junction box. The standard boxes for ceiling- and wall-mounted lighting fixtures of less than 50 pounds are octagon, and standard-size round boxes range from $1\frac{1}{2}$ to $2\frac{1}{8}$ in. deep. By code, the front of the box must be closed by a cover.

When electrical circuitry passes through fire-resistant-rated walls, partitions, floors, or ceilings, the NEC requires the penetrations to be fire stopped by approved methods so the fire-resistance rating is maintained.

The number of electrical rooms, space dedicated to electrical equipment, will depend on the size of the building and number of floors (Fig. 10.34). As floor square footage and power needs increase, multiple electric rooms may be necessary. With a multistory building it may be best to have electric rooms on each floor to serve lighting, receptacles, and equipment, both normal and emergency. By the NEC code (Article 700) there are emergency system wiring and equipment standards for electrical safety circuits and equipment. The standards seek to enable the evacuation of buildings having hotels, theaters, and healthcare facilities where a large number of people assemble. By standardizing electrical components and integrating the electrical devices into fewer structures, it is possible to reduce the size of electrical rooms.

NEC dictates the minimum amount of space needed around the equipment for access, operations, safety reasons, and conduit installation. In multistory buildings, branch equipment rooms should be stacked.

**FIGURE 10.34** Installing equipment in a multistory building electrical room.

***Finish Electrical.*** Lighting fixtures, switches, and receptacles are installed after the paint, drywall, and plaster work is completed. Any utility or appliance connections to heating and air-conditioning are also made. Once everything is wired, the power is turned on and checked for **short circuits** or other problems. If there are no problems the building inspector will be contacted for a final inspection.

**short circuits**
*An "unintended" low-resistance current path in the wiring circuit causing the current to follow unintended wiring. If there is no fuse or breaker, the current will damage the circuit.*

## ABOVE CEILING COORDINATION

Mechanical, electrical, and plumbing (MEP) systems all occupy the above ceiling space of a commercial building (Fig. 10.35). The extensiveness of pipes, wires, and ductwork includes specialty systems such as fire detection/protection and communication/datacom. Many of the systems are similar in nature but different specialty contractors often install the individual services. While the coordination of the MEP contractors needing to work in the same limited space is critical to efficient delivery of a project, the coordination process can be time consuming and expensive. The process involves both

- Scheduling access to the area
- Delineation of system locations in three dimensions

Without careful planning of these activities, the project will experience schedule delays and there will be increased cost.

The design consultants providing the drawings and technical specifications for these systems typically work independently or design each system independently. Their designs seldom incorporate thinking about constructability issues or even consider the workspace required to install the system. The general contractor and the project specialty subcontractors often have to take the diagrammatic drawings indicating location and routing paths for a system and by overlaying each optimize space utilization. To be successful in developing a plan and work schedule, communication between all participants is critically important. Once a plan is developed, all the effected specialty contractors should sign off on the allocation of space so that all can proceed with the necessary shop drawing and fabrication of materials.

Because it is easier for piping and electrical circuitry to be worked around the ductwork, one manual approach has been for the HVAC contractor to provide the ductwork layout first. The proposed HVAC ductwork layout in the initial iterative review is a comparison with the building's structural frame and shear walls. Then the initial duckwork layout is checked against the water, drain,

**FIGURE 10.35** Systems in the ceiling space.

waste disposal, and the DWV system. The plumbing is chosen second because of the requirement to slope graded lines and waste lines to allow for gravity flow of the sanitary system. Finally, the proposed layout is checked against electrical system space requirements. The checking process will require several iterations before a final layout for all systems is developed. The review and checking is not just about three-dimensional space for the physical components of the individual systems but must consider the space necessary for all needed pipe and duct *supports* and *insulation*. Finally, there is the human element and the need for space to perform the installation work.

This historical approach offers a path forward using 3D modeling to eliminate conflicts. The use of BIM technology software has become standard practice for commercial projects. It provides graphic location and clearances for installation of the MEP systems. When project documents are provided in 2D format a *proposed* 3D model can be developed for each of the independent systems. Then the independent models can be merged with a software integration tool into a common 3D model. With the use of clash detection software physical interferences are identified and the system models modified until all conflicts are resolved. Those responsible for performing the work are provided a visualization of space utilization. The visual picture is much easier to understand than a 2D drawing. After the resolution of interferences, fabrication and installation plans can be produced for the systems.

> The future of confined space utilization is with BIM modeling.

# CHAPTER 11
# FOUNDATIONS

*Building systems of all kinds rely on strong foundation support to ensure the longevity of the structure. Foundation systems are as varied as the buildings they support. The foundations transfer building loads to the soil below in many ways. They are a function of the type of soil, the bearing capacity, and the location of rock and the groundwater table. Some systems are simple, shallow footings while others include deep piles. These systems and preparing to build foundations are addressed.*

## STRUCTURAL EXCAVATION

Foundations are systems that support structures and transmit the load of the structure to the ground surface. A properly prepared foundation is required for a sound, stable, and safe structure. Excavation is required to reach the specified depth of the foundation surface. The depth of the foundation depends on many factors. The engineering properties of the soil are critical. These include the bearing capacity, the soil type, the depth of frost penetration, and the location of the groundwater table. Other considerations include the possibility of excavations in the near vicinity and the size of the structure.

Foundation preparation begins with clearing, grubbing, and stripping of the areas upon which a structure will be built. These steps remove materials having poor engineering properties that could impact the construction of a sound foundation.

Clearing and grubbing involves the removal of trees, brush, and large roots to expose the soil prior to excavation operations. The clearing operations require the removal of trees, vegetation, felled timber, brush, and debris. The grubbing activity removes below ground interferences, such as stumps, roots, logs, and debris (Fig. 11.1). The plans and specifications for any building project should delineate the clearing and grubbing limits, the depth of removal, and the method of disposal.

**FIGURE 11.1** Clearing and grubbing for a residential foundation.

In the process of foundation preparation, foundation stripping generally follows clearing and grubbing. Stripping consists of the removal of sod, topsoil, boulders, vegetation, roots, and any unsuitable materials that remain on or just beneath the surface after the completion of clearing, grubbing.

The foundation is considered a subsurface structure. Excavation for the foundation will consist of open excavations (Figs. 11.2 and 11.3) and trench-style excavations. The depth of the excavation depends on the structure that is being built. Prior to any excavation, it must be determined if there are any underground utilities in the area.

If the excavation is to a deep depth, different types of soil and rock may be encountered while digging the foundation. It is important to carefully review the geotechnical report prepared for the structure to identify the material in the subsurface. The types of soil and rock encountered will determine the type of equipment needed to complete the excavation and the support required to keep the excavation open and safe for the duration of the foundation construction. Rock on the site may range from soft rock that is easily ripped to very hard rock that will require blasting. Each of these conditions will require different equipment, preparation, and permitting.

The adequate assessment of the depth and quantity of groundwater conditions is important in structural excavations for foundations. Understanding groundwater conditions is required for the selection of groundwater control measures for the excavation operation and for maintaining a stable foundation. Control of groundwater influences the manner of excavation and the slopes to which the excavation can be cut without support and how excavated material must be handled.

Structural excavations face a few construction challenges. These are related to groundwater, slope stability, and the weather conditions during construction. Groundwater poses challenges and must be controlled if present in any area of an excavation. All excavation and foundation work must be accomplished in the dry and therefore a dewatering system may be required if seepage is expected in the excavation. Care should be taken to accurately assess the groundwater conditions in the location of the foundation excavation.

If groundwater seepage occurs into the excavation, the source of the water must be determined. Is it a brief incident such as a heavy rain or large snow melt event? Is it actual groundwater seepage that is exceeding the capability of the dewatering system installed? Excavation operations must be halted

**FIGURE 11.2** Open excavation for residential foundation. (Photo courtesy of Brian Bralynski.)

**FIGURE 11.3** Shallow trench excavation.

if excess water is entering the work area as these conditions present a danger to the crew. Workers should not re-enter the excavation until the seepage is controlled.

Even if groundwater is not seeping into the excavation, care must be taken to minimize the potential buildup of excess hydrostatic pressure in the subsurface. This pressure can cause issues with slope stability as well as cause problems for excavation support structures and the final foundation structure. Careful visual observations should be made during the excavation process for signs of trouble in the soil. These include uncontrolled seepage, piping of soil material from the foundation area or the soil slope, as well as any movement of the slope.

During the excavation of soil slopes, erosion must be kept to a minimum and the contractor must have a plan to manage all soil erosion on site. Erosion protection of exposed slopes may be required as well as the installation of silt fences to minimize soil migrating from the site and into adjacent areas or waterways.

Structural excavation quantities will vary depending upon the depth and area of the foundation. Soil from the excavation may need to be stockpiled during construction for use in other areas or as a backfill material. These stockpiles should be located in convenient locations and be subjected to the same erosion control measures as the rest of the site. Some municipalities require that if a stockpile will be onsite for a long duration that they may need to be protected by vegetation. The specifications will detail the requirements for stockpiling of excavated material.

Once the excavation is completed, the subsurface area where the foundation will be placed should be reviewed. Checks should be made to identify that all the utilities and conduits that must be placed under the foundation are in the correct location.

The area excavated for the foundation requires final preparation prior to the placement of the foundation. The area is to be cleaned and leveled and the bottom of the excavation should be compacted to the required density. If the foundation is on rock, the area should be as level and clean as possible to receive the foundation material.

## DEWATERING

The dewatering system is required to operate for the duration of time that there are people working in the excavation. Having a saturated soil at the bottom of an excavation will not allow the soil to properly compact and the foundation work will be delayed. This soil must be drained completely or removed and replaced before the placement of any concrete footings or backfill are placed.

Controlling groundwater during excavation operations is extremely important. The dewatering process provides temporary reductions in groundwater levels for structural excavations that must extend below groundwater level. There are many systems that can be used to keep water out of the excavation during construction and then away from the foundation once it is built. Some are basic and simple while others require more detailed design and analysis.

The **sump pump** is the most basic form of groundwater control (Fig. 11.4). Groundwater is collected by a sump and then it is pumped away from the excavation. This method is an effective and economical way to drawdown small quantities of water. The soil type plays a large role in the efficiency of this method. In well-graded soils and fissured rocks, the sump works well. In fine-grained soils, the pump might draw soil particles from the excavation causing settlement issues. Consideration must be given to the discharge and disposal of water from the sump as it may have a high sediment load, limiting how and where the water may be discharged.

**Well points** are used to lower groundwater levels in excavations of shallow depth (Fig. 11.5). This system provides a safe working environment during the excavation operation. The system consists of small diameter wells placed around the area to be dewatered. These wells are attached to a header pipe and the header pipe is connected to a well point pump. The pump creates the vacuum that extracts the water from the ground. The well points are installed around the excavations and are pumped using

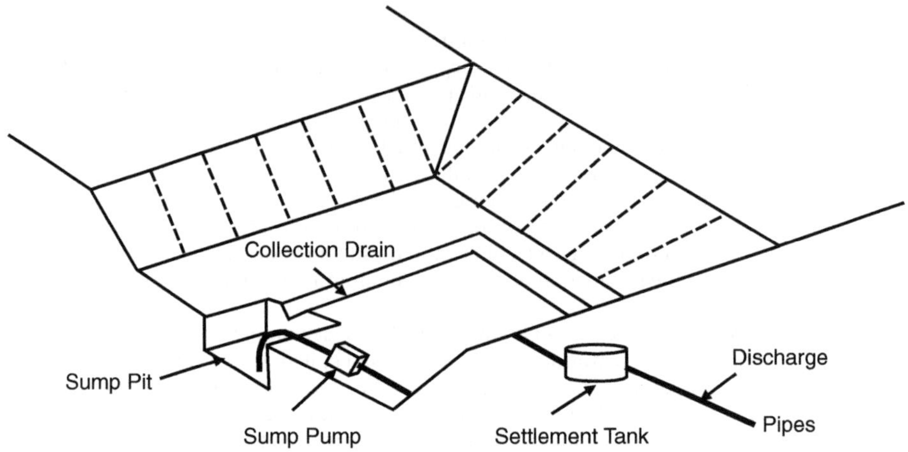


**FIGURE 11.4** Sump pump system for dewatering.

**FIGURE 11.5** Well point system.

powered pumping systems attached to headers, water discharge pipes, and power supply generators. These systems are also monitored using controls.

A **deep well system** is comprised of bored wells that are each pumped using a submersible borehole pump (Fig. 11.6). The pumping operation from the wells lowers the groundwater level and creates a drawdown around the well itself. Large drawdowns can be achieved with this method by having several wells acting together over the entire excavation. The amount of drawdown is only limited by the depth of the wells and the subsurface conditions. These wells are typically installed around the excavation and the water pumped is collected by surface pipes. The wells get their power from power supply generators and are monitored from the surface.

**Eductor** systems use wells to lower the groundwater level to provide safe working conditions by controlling the pore water pressure in low permeability soil (Fig. 11.7). Eductors use air in the wells to create a vacuum that draws water out of the soil.

They are typically used to improve the stability of the side slopes and base of excavations in soils that would be difficult to dewater with well points or deep wells.

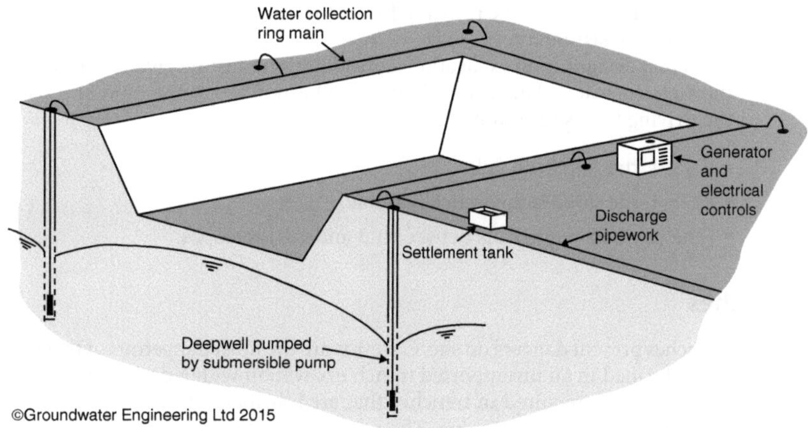


**FIGURE 11.6** Deep well system. (Downloaded from Groundwater Engineering Ltd.)

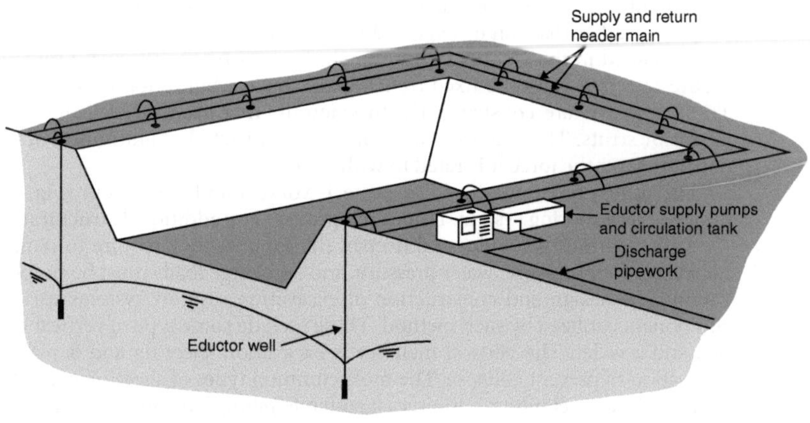


**FIGURE 11.7** Eductor system. (Downloaded from Groundwater Engineering Ltd.)

## SUPPORT OF EXCAVATION

The main goals of an excavation support system are to minimize the excavation area, keep the sides of deep excavations stable, and ensure that movements will not cause damage to any neighboring structures or utilities in the surrounding ground.

To keep the surrounding ground stable, the failure mechanisms of slopes must be considered and that depends on the type of soil and the cut. Slopes can fail in the form of a rotational slump in homogeneous clay depositing soil into the excavation from the toe to a section of the slope. Cohesionless soils such as sands and gravel fail as a translational slide and all types of soils can fail by slipping along a plane of weakness.

A review of the forces at work on the slopes show that the driving forces that cause instability are the components of soil weight that is acting downslope. The forces of the soil strength that act in the opposite direction are the resisting forces the slope provides. Slope failure occurs when driving forces exceed the resisting forces.

When designing an excavation, consideration must be given to these two forces. This is done with the factor of safety. The factor of safety (FS) is defined as the ratio of resisting forces (or moments) to the driving forces (or moments).

- FS ≤ 1, the slope will fail.
- FS > 1, the slope is theoretically stable.
- The usual FS required is between 1.3 and 1.5 for slopes.

## Shallow Trenches

Trenches present dangers on site. Cave-ins are the most dangerous of these hazards and no work should be performed in an unsupported trench excavation with a depth 5 feet or greater. A trench protection system may be required in trenches that are less than 5 feet deep if the type of soil encountered may cave in when the trench is dug. The primary function of excavation support in a shallow trench is to protect people working in the trench from the ground caving in around them. The secondary function of the trench is to provide support to nearby structures and allow equipment access to the work.

When designing and selecting a trench support method, there is no one-size-fits-all solution. The first steps are to read the plans, specifications, and geotechnical reports to understand the constraints and conditions that will be encountered. Specific attention must be paid to the water content of the soil, the weather that can be expected in the area, the depth of the cut, and other operations that will be conducted in the vicinity of the trench. Various types of trench protection systems are available. Figure 11.8 illustrates the fixed trench box. This configuration is also referred to as a trench shield. These supports are constructed with sidewalls that are braced using steel or aluminum spreader beams or struts. The sidewalls are typically constructed of aluminum or steel of different thickness depending on the force it is rated to withstand.

The trench box is designed to protect workers and equipment (Fig. 11.9). A trench box is not designed as a shoring system. It does not provide any additional structural support to the excavation.

As excavations get wider and deeper, the support they require to remain structurally sound will increase. Earth pressure, water pressure, and surcharge loads must be supported to ensure a safe excavation. The design and construction of excavation support systems varies, but there are a few key components utilized by each method. These include some type of vertical member, a horizontal member, and a waler. The vertical member is excavation sheeting and is placed against the sides of the excavation to prevent collapse. The most common types of sheeting used in building construction are timber piling, steel piling, and precast concrete piling. The horizontal member is typically a strut system that is designed to support the load in the horizontal direction. The struts are structural elements that transfer load from the wall to an adjacent parallel structure. The other element is the wale or waler. This element is a horizontal member that presses against the vertical member and distributes the load

### General Trenching and Excavation Rules

- Keep heavy equipment away from trench edges.
- Keep surcharge loads at least 2 ft (0.6 m) from trench edges.
- Know where underground utilities are located.
- Test for low oxygen, hazardous fumes, and toxic gases.
- Inspect trenches at the start of each shift.
- Inspect trenches following a rainstorm.
- Do not work under raised loads.

**FIGURE 11.8** Trench shield. (Photo courtesy of Kiewit.)

**FIGURE 11.9** Trench boxes protecting utilities. (Photos courtesy of Allan Myers and Kiewit.)

of a strut to vertical member. This allows a compressive span between the struts. Each support of excavation system has specific applications depending on the construction taking place.

If the excavation is narrow, the bracing of the excavation can be accomplished using internal struts. These double-sided bracing support systems provide bracing across the excavation using a system that transfers the applied load to the support structure. The system in Fig 11.10 consists of a vertical member that supports the lateral earth pressure and the surcharge load. This vertical method is some form of sheeting. The load is transferred via the waler to the struts which are a horizontal member. These members have equal and opposite forces at each end and maintain the equilibrium of the system (Fig. 11.11). All forces, even the surcharge loads placed at the surface, are treated as horizontal forces.

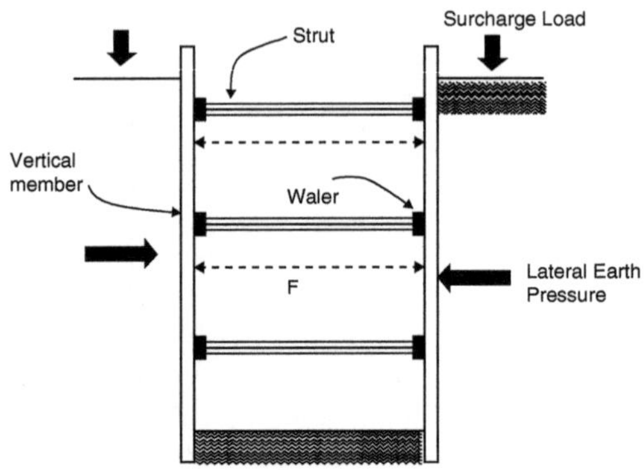


**FIGURE 11.10** Internal strut braced wall system.

**FIGURE 11.11** Internal strut braced wall system.

## Single-Sided Bracing Supports

As excavations get wider, a single-sided bracing support system may be more appropriate to protect the excavation. These systems are constructed in many different forms and consist of a vertical member that supports the lateral earth pressure and surcharge load and transfer it to the walers. The vertical member is driven into the ground surfaced and secured using the appropriate toeing method. The walers transfer loads to the inclined struts or raker beams. The forces in the rakers are supported by a kicker block or a foundation slab which transfers the loads to the ground. The force in the raker beam (F) is greater than the horizontal force applied by the lateral force (P) of the soil pressure and surcharge load. The raker system introduces an upward vertical force onto the vertical members as well as the kickers (Fig. 11.12). It is critical to have both adequate toeing and grounding of the kickers to ensure the whole stability of the system.

The raker system works very well to support excavations; however, the raker itself is located within the excavation. This limits movement of people, equipment, and material in the footprint of the excavation. To eliminate any obstructions in the excavation, tieback systems can be used. These types of excavations rely on the sheeting method to support the lateral earth pressure and surcharge load and transfer it to the walers. The walers transfer the load to the tiebacks. The forces in the tiebacks are supported by a grouting in the soil or rock. The force in the tiebacks *(F)* is > the horizontal force of the lateral force *(P)* due to the soil pressure and surcharge load. The tieback system introduces an upward vertical force onto the vertical members as

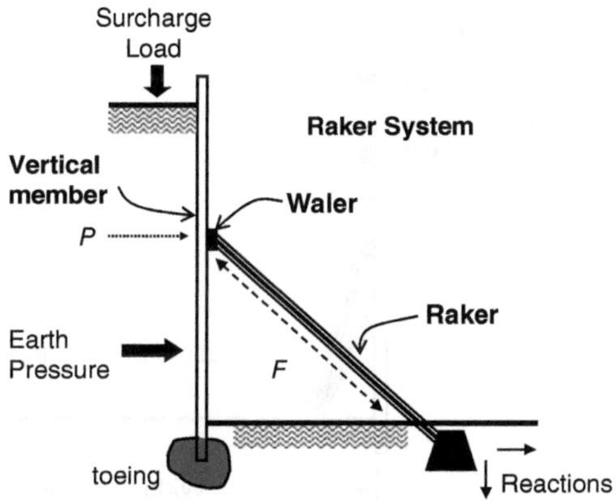


**FIGURE 11.12** Single-sided braced excavation.

well as the grouted anchor of the tieback. Adequate anchoring of the tiebacks coupled with adequate ground conditions at the bottom of the vertical members contributes to the whole stability of the system.

The tieback anchors can replace internal braces to provide an excavation free of obstructions. The tiebacks are constructed by drilling a small diameter shaft into the ground behind the sheeting system. Steel tendons are inserted into the shaft and grout is pumped into the shaft to anchor the tendons as the drill casing is retracted. Hydraulic jacks are used to lock off the tieback to the sheeting system. A major benefit of the tieback system is that multiple levels of tiebacks allow for deeper excavations that are unobstructed (Fig. 11.13).

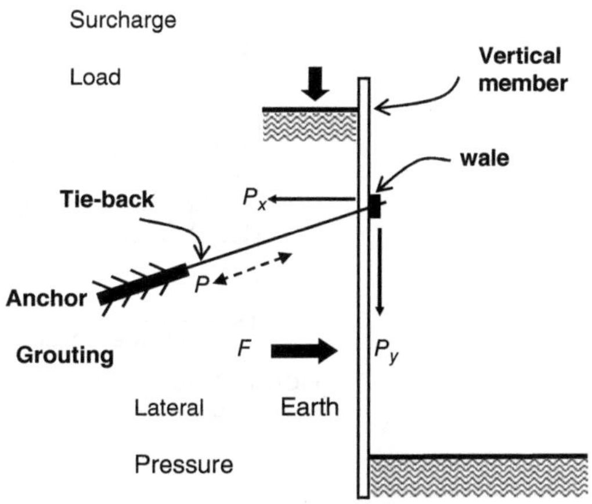


**FIGURE 11.13** Anchor or tieback system.

## Limitation on the Use of Diagonal Bracing

In very deep excavations, it may be uneconomical to use rakers. This is because complete excavation of the interior of the site is required to allow construction of kickers needed to support the first line of rakers, as well as the need to use tall berms to provide temporary support to the vertical wall. There are other alternatives that can be used that are based upon a combination of systems. These combinations include the use of tiebacks, struts, and rakers at different levels (Figs. 11.14 and 11.15).

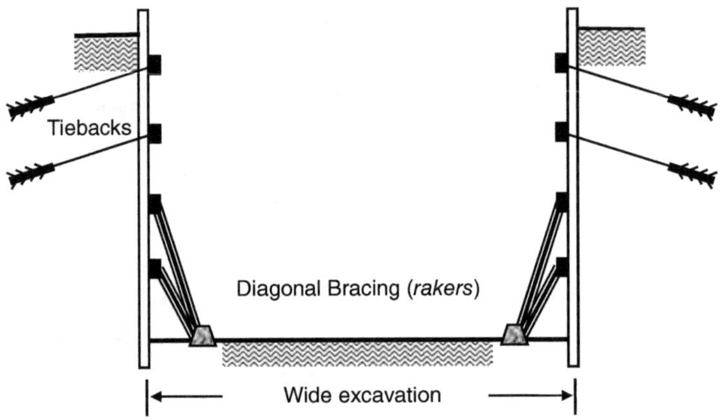


**FIGURE 11.14** Combination bracing for a wide excavation.

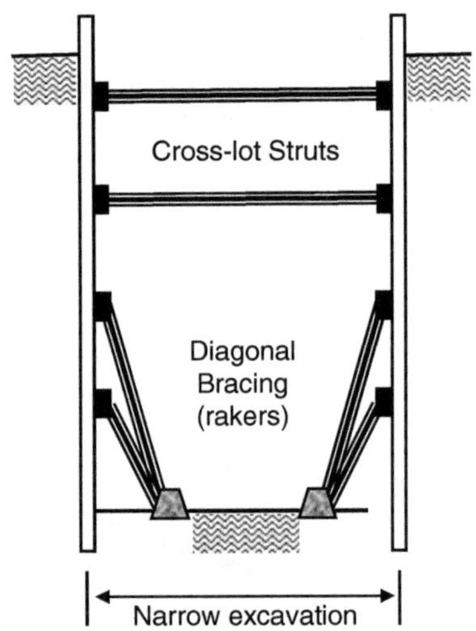


**FIGURE 11.15** Combination bracing for a narrow excavation.

## Cantilever Supports

Excavations using the cantilever system of support are constructed with no bracing, only the vertical member (Fig. 11.16). This type of excavation support is typically used for shallow depth excavations. This type of support system is limited due to the deflection under earth pressure. The cantilever system can only be used where there is no risk of settlement behind the vertical member endangering the stability of adjacent structures. All types of cantilever support walls will deflect unless they are completely uneconomical for widespread use. The most commonly used vertical member in this type of system is a steel sheet pile.

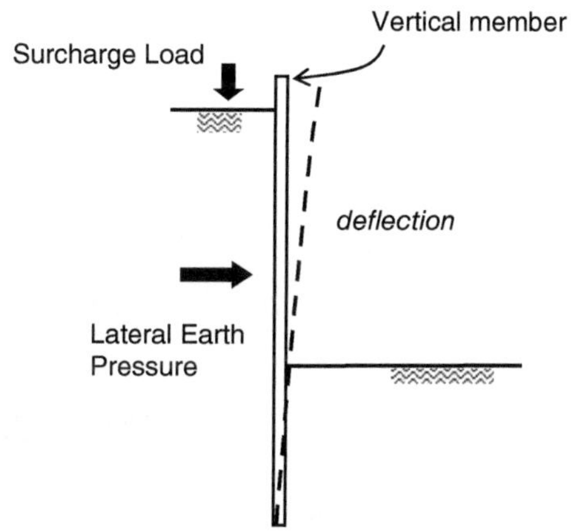


**FIGURE 11.16** Cantilever system.

## Soldier Piles and Lagging

When building deep foundations a common support of excavation system used is soldier pile and lagging walls. These types of support structures can be used to stabilize slopes and to retain earth while building construction is underway. Soldier pile and lagging systems are efficient and can be used to support most types of soils, making them a cost-effective choice for builders.

Typical soldier pile walls are constructed by vibrating, driving, or drilling a steel beam or pile vertically into the ground surface, typically an H-pile. The pile is then held in place with a grout column at the bottom of the excavation. Other types of vertical members can be used depending on the needs of the project, including precast concrete piles, standard pipe sections, and even micropiles. These walls form a continuous support structure around the excavation. The spacing of the vertical elements will depend upon the forces the wall must resist, but typically are between 6 and 10 ft apart. After the piles are placed, the excavation operation can begin. Excavation occurs in lifts that are typically about 5 ft in depth, depending on the soil type. The use of a soldier pile system is not suitable in noncohesive soils, as they may move through the lagging. Lagging is installed between the soldier piles and then the excavation is completed (Fig. 11.17). Depending on the type of soil encountered, voids may be experienced behind the lagging material. If this is the case, backfilling and compacting of material may be required to ensure limited movement behind the wall.

The horizontal lagging in this system is used to keep the soils from collapsing into the excavation and is installed in between the steel piles. When the excavation exceeds the depth that can be cantilevered, soldier pile and lagging systems can be created using bracing or tiebacks for deflection control and strength.

The excavation support can be a temporary measure or part of a permanent structure. Temporary walls most often use wood lagging while the permanent structures may use precast concrete or even shotcrete as the lagging. The lagging material plays an important role in the transfer of horizontal pressure from the soil to the soldier pile and to prevent soil from flowing between the soldier piles.

The soldier pile system can be constructed to support a variety of depths of excavations. These systems can be designed as cantilever-type support structures up to a height of about 12 ft. Beyond that depth or if wall stability needs to be increased, tiebacks can be installed into the retained soil. In addition to tiebacks, steel struts or walers can be placed between the soldier piles for increased lateral support (Fig. 11.18).

FIGURE 11.17 Simple soldier pile wall. (Photo courtesy of Allan Myers.)

FIGURE 11.18 Laterally supported soldier pile walls. (Photos courtesy of Kiewit.)

The use of this type of excavation support system depends upon the depth of the excavation and the soil type. This system is not well suited for soft clay materials or situations where the groundwater table is within the depth of the excavation. The presence of water or weak soils impacts the installation of the lagging material causing major challenges to building the structure.

When the soil conditions are right, there are many advantages to using this type of excavation support system. These include the ability to install the system quickly and at a lower cost than other support systems. In addition, the materials that are used to construct these systems are varied and flexible and this versatility allows for adjustments in construction methods and materials to be made in the field if necessary. In addition, improper backfilling could result in ground losses that cause surface settlements.

## Steel Sheet Pile

Another method to support deeper excavations is the use of a steel sheet pile wall. Sheet piles can also be constructed from timber or reinforced concrete, but steel is the most common material. This method is used in soils that are inappropriate for soldier pile wall excavation support. These include soils that would make the placement of wood lagging difficult or that are potentially unstable when they are exposed during excavation.

Sheet piles are effective when seepage into the excavation must be controlled or if structural elements must be constructed below the water table. Sheet piles are highly effective in cutting off concentrated flow through pervious layers within or below the excavation. They also protect against the possibility of a soil blow-out condition or other source of ground loss in the excavation. Because of this characteristic, they are used often in waterfront structures and to complete excavation operations in the water (Fig. 11.19).

Sheet piles have several advantages to their use. They provide a high resistance to driving stresses that are encountered during installation. The sheet pile sections are lightweight, can be reused on several projects, and have a long service life. The length of the piles is easily adapted by either welding or bolting sections together.

The construction of a sheet pile wall involves laying out a sequence of sheet pile sections and ensuring that sheet piles will interlock. Inspection of the sections to ensure they are not bent or damaged prior to installation will save time during construction. The individual sheet piles are then lifted and driven or vibrated to the desired depth. The adjacent sheet pile with the interlocks is then driven and locked in place. These steps are repeated until the wall perimeter is completed (Fig. 11.20).

Steel sheet piles are rolled steel sections that consist of a web with integral interlocks on each edge. The interlocks contain a groove and a flattened tongue section. The tongue fits into the groove of the adjacent sheet.

Commonly used steel sheet piles types are shown in Fig. 11.21.

**FIGURE 11.19** Sheet pile cofferdam for bridge pier construction. (Photo courtesy of Kiewit.)

**FIGURE 11.20** Sheet pile supported excavation. (Photo courtesy of Kiewit.)

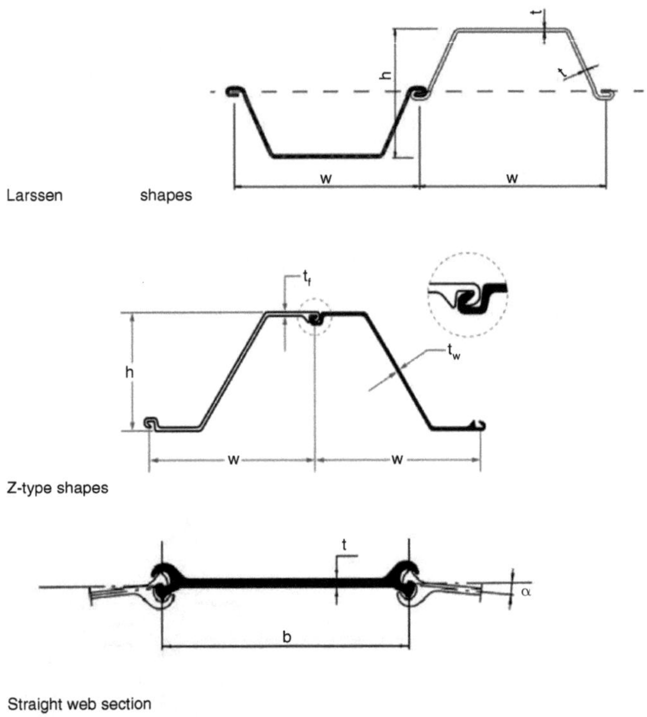


**FIGURE 11.21** Typical sheet pile sections.

## FOOTINGS

A footing is defined as a structure constructed under the base of a wall or column for the purpose of distributing the load over a larger area. The footing may be built using brick, masonry, or concrete. The footings can be found in various shapes and sizes depending on their use and the soil type.

### Individual Footing

Individual footings are the most common type of foundation used for building construction. These footings are typically used for shallow depths and when the structural loads are carried by individual columns.

The shape of an individual footing is either square, rectangular, or circular, though the latter is less common. The size of the footing is determined based upon the load on the column and the safe bearing capacity of soil. The design of the footing must ensure that the bearing capacity of the soil is not exceeded, provide adequate resistance to overturning and slide, and ensure that the ground settlement is minimal.

Individual footings are economical and easy to construct as they require less excavation (Fig. 11.22). A challenge with these footings is that the soil needs to be stable around the area of the footing. Due to the fact that these footings may be spread far apart, they also present a challenge with differential settlement.

There are a few types of individual footings. They are either a simple footing, stepped footing, or sloped footing.

**FIGURE 11.22** Individual footings. (Photo courtesy of Brian Bralynski.)

A **simple footing** is built using concrete that can be reinforced. The thickness of the footing is constant and the shape can be square, rectangular, or circular. This type of footing is economical, but it can reach large sizes due to the amount of load it must bear.

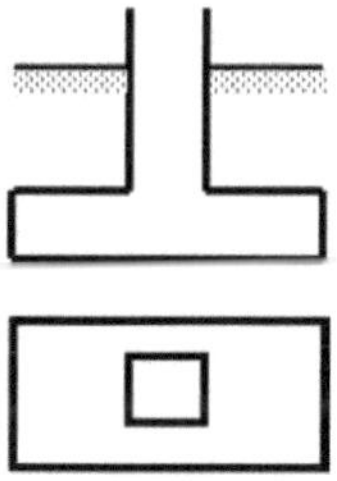

Simple Footing

A **stepped footing** is built in layers until it achieves the width required by the design. This technique is common in the construction of residential properties.

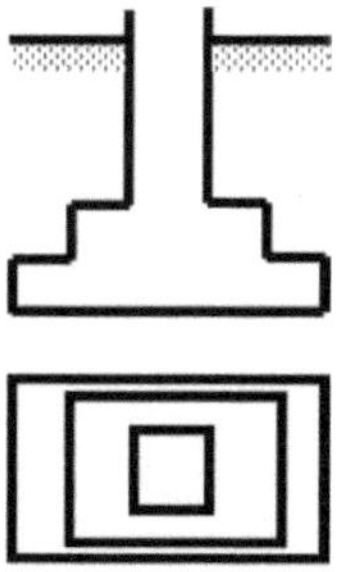

Stepped Footing

A **sloped individual footing** is more economical than the simple footing as it requires less concrete and reinforcing bars due to its shape. The challenge with this type of footing is the time and effort that is expended to construct it. During construction, a stiff concrete mix must be used to maintain the 45° slope on all sides.

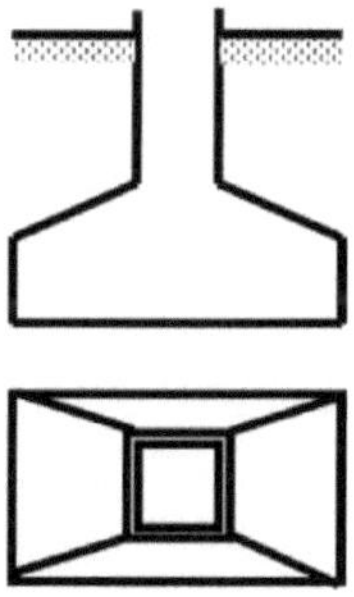

Sloped Footing

Individual footings work well when the load of the structure is low and the columns are not spaced close together. The soil bearing capacity must be high at the shallow depth of the footing in order to be economical to construct.

> For a column with a vertical load of 6,250 lb and a safe bearing capacity of 2,500 lb/ft$^2$, the area of the footing required will be 6,250/2,500 = 2.5 ft$^2$.
>
> For a square footing, the length and width of the footing will be 1.58 ft x 1.58 ft.

A **combined footing** is a special case of an individual footing. Combined footings are used in foundation structures when column loads are not uniform or the bearing capacity of the soil is low causing the overlap of adjacent individual footings. While similar to individual footings, the structural design of combined footings is different.

There are three basic types of combined footings that depend on the shape. These are rectangular, trapezoidal, and T-shaped combined footing (Fig. 11.23). The design intent behind the combined footing is the uniform distribution of loads under the area of the footing. This requires that the centroid of the footing area is the same as the centroid of the column or structure that is supported by the combined footing.

The use of combined footings is efficient and economical when columns are placed in close proximity to one another. The combined footing can help solve difficult design challenges such as when the column is close to a property line and an individual footing would cross that line or if the dimensions of one side of the footing are restricted due to another obstacle, like a water line.

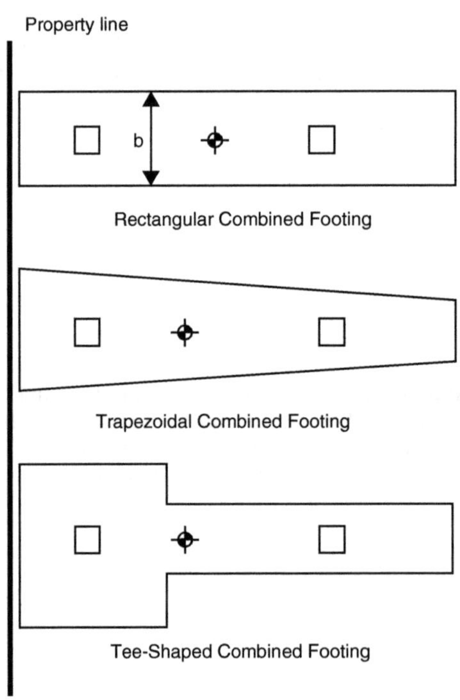


**FIGURE 11.23** Strap footing.

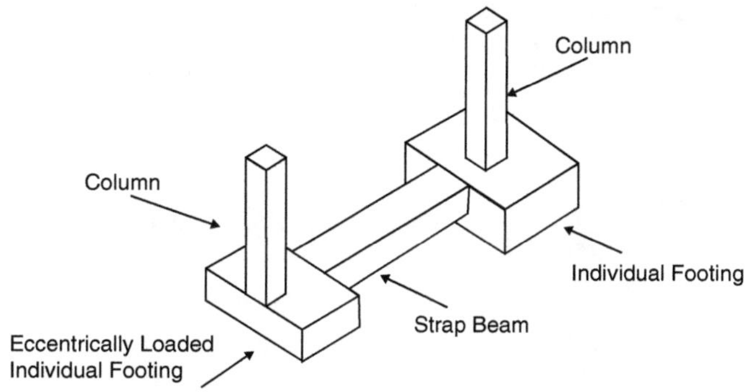


**FIGURE 11.24** Combined footings.

The trapezoidal combined footing is primarily used when column loads are unequal and the resultant forces of these loads would not coincide with the center of a rectangular combined footing. The trapezoidal shape allows for the centroid of the footing to correspond to the centroid of the resultant loads, thereby ensuring the distribution of the loads will be uniform.

Another possibility if a combined footing is not feasible is the use of a strap footing (Fig. 11.24). This type of footing can be successfully used when a column is close to the property line or other obstruction and the soil bearing pressure is unable to support the combined footing load.

The strap footing consists of two or more individual column footings that are connected by a concrete beam. The concrete bean is defined as a strap beam and cantilevers over the footings. The beam assists in distributing the load of eccentrically loaded individual column footings and improves the overturning resistance of the footing.

## Spread Footings or Strip Footings and Wall Footings

A spread footing is a type of shallow foundation used to support the column and walls of a structure. The spread footing also transmits and distributes the structural load to the soil beneath it. This type of footing is used when the soil is dry and has a bearing capacity sufficient to support the weight of the structure over the base area of the footing.

These should not be used on soils where there is any possibility of flow of water above the bearing layer of soil which may result in scour or liquefaction.

Spread footings have a base wider than a typical load-bearing wall foundation. The wider base of this footing type spreads the weight from the building structure over more area and provides better stability.

Spread footings used on the perimeter of a building are typically constructed of reinforced concrete and between 16–24″ wide and between 6–16″ thick and will vary according to the load. The concrete used in spread footings typically has a compressive strength between 2,000 and 5,000 psi depending on the load it supports. Continuous spread footings can support wall structures constructed from brick, concrete masonry units and stone, as well as columns (Fig. 11.25). These footings can be installed below ground level or below the frost line to prevent heaving due to freezing and thawing actions. Due to their versatility and their design, they are less likely to fail than individual footers.

Spread footings reduces cracking caused by settlement of a structure wall. The footing has continuous contact with foundation, thereby reducing the foundation failure risk. These types of foundations are easy to construct and provide cost savings in design, construction, and quality control.

A form of spread footings is strip footings. These are a continuous strip, typically of concrete, that is centrally formed under a column or load-bearing walls to provide support. The strip serves as a flat base upon which the wall or column can be constructed. The width of the wall is designed to ensure

**FIGURE 11.25** Spread footing supporting a concrete masonry unit wall. (Photo courtesy of Brian Bralynski.)

**FIGURE 11.26** Perimeter foundation strip footing for residential application. (Photo courtesy of Brian Bralynski.)

the spread of the load on the foundation area of soil without exceeding the soil bearing capacity. Continuous shallow foundations can use a strip footing to create a perimeter foundation where the load of exterior walls is positioned on the footings (Figs. 11.26 and 11.27).

The location and the size of the footing are proportional to the overall width of the wall. For the design of a strip footing, the size of the footing depends upon the manner in which the load will be transferred to the soil below. A rule of thumb regarding load transmission indicates that the load is transmitted to the ground at a 45° angle from the base of the wall to the soil. The width of the foundation must be three times the width of the wall or column the footing is supporting. The depth of the footing must be equal to or greater than the width of the wall or column supported.

Spread footings are used frequently in residential applications and may not need reinforcement to meet code, although nominal reinforcement should be used for flexure. The lower the load-bearing capacity of the soil, the wider the spread footing needs to be to avoid dramatic settlement. Spread footings are also poured below the frost line. The footers make for a level and smooth line upon which the concrete foundations are then poured to form a continuous wall supporting the perimeter of the home.

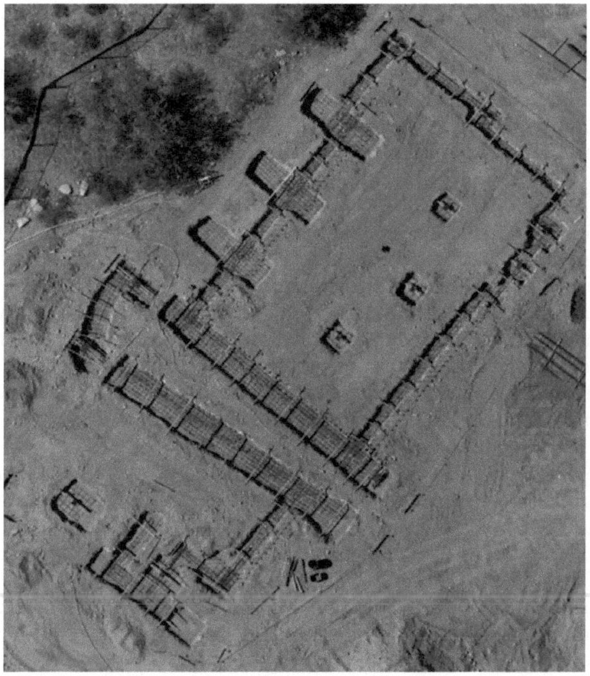

**FIGURE 11.27** Aerial photo of perimeter foundation strip footing for commercial application. (Photo courtesy of Beth Lemon.)

**TABLE 11.1** Soil Characteristics

| Soil type | Typical allowable soil bearing capacity (lb/ft$^2$) | Drainage capability |
|---|---|---|
| Bedrock | 4,000–12,000 | Good |
| Clean gravels | 3,000 | Good |
| Dirty gravels | 3,000 | Good |
| Sand | 2,000 | Good |
| Silty sand | 2,000 | Medium |
| Silt | 1,500 | Medium |
| Clay | 1,500 | Poor |
| Organic material | 0–400 | Poor |

## SHALLOW FOUNDATIONS

A **mat** foundation, also referred to as a **raft** foundation, is a combined footing type of foundation that rests directly on the soil and extends over the entire footprint of the structure. The foundation supports the building structural loads from columns and walls and transfers all the loads to the surface below. Mat foundations might be more economical in situations where individual footings would have to cover more than half the building area.

These types of foundations are used when the bearing capacity of the soil is low to reduce the stress on the soil. Unlike a spread or individual footing, the mat foundation distributes the load over the entire building footprint and not in concentrated locations. The mat foundation is also good to use in areas where there are concerns of movement in the subsurface. These foundations are used in the construction of commercial and residential structures.

Take, for example, a building has a footprint that is 50 ft × 50 ft, whose wall, column, and floor loads total 315 tons, and that is supported by a raft foundation, then the stress transmitted to the soil is 0.5 tons per square foot. If the same building was supported by nine individual footings of 3 ft × 3 ft, the footing area available to spread the load over would be 81 square feet. This would yield almost 4 tons per square foot of stress being transferred to the soil. This illustrates that using a mat foundation is an excellent way to lower the stress transmitted to the soil.

Mat foundations are straightforward to construct. The soil is excavated to the appropriate depth and compacted to the required density as indicated in the design documents. A waterproofing membrane is then laid over the ground surface if required. Reinforcement is placed in the excavation and the concrete is then placed to the desired thickness, finished and cured. If the foundation needs additional support, the foundation may be stiffened by ribs or beams built in during construction, which will add extra strength and rigidity.

The mat foundation provides numerous advantages over other footing types. In addition to handling larger loads, the spreading action of the foundation enables structures to be built on soil that possesses a lower bearing capacity. The mat foundation system is also a more economical option when considering the space in a floor plan. When more than half of the foundation footprint would be covered by individual footings or if the walls of a structure are close enough together that a spread or individual footing would overlap, a mat foundation is more economical to use.

Mat foundations tend to be cheaper and quicker to use than traditional footings. There are a number of reasons why this is the case:

- The foundation and floor slab is combined, which saves time and materials.
- Less excavation is required.

The main disadvantage is that they can be prone to edge erosion if they are not treated properly.

The **flat plate raft** (Fig. 11.28) is used primarily for lightweight structures built upon soil that is not compressible. The reinforced concrete slab used in this foundation is a minimum of 6 in. thick. The reinforcement is provided in these foundations in a mesh form and is in both directions and the top and bottom.

The thickened plate under columns system is used when the column loads are very high. The reinforced concrete mat slab thickness is increased along these column lines to provide additional support.

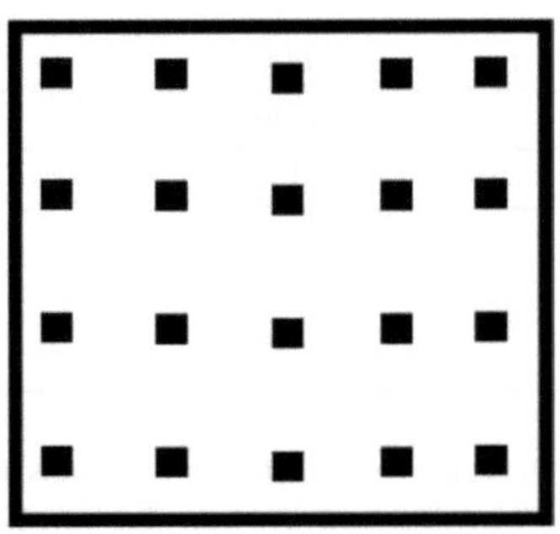

**FIGURE 11.28** Flat plate raft plan view.

The **two-way slab and beam** (Fig. 11.29) foundation is used when the column loads are unequal and the spacing between them is large. Beams are placed perpendicularly to each other and are connected by the reinforced concrete slab. The columns are constructed at the intersection of the beams.

In situations where the soil is highly compressible and the groundwater table is high, the **piled raft** (Fig. 11.30) foundation is suitable. The mat foundation is supported on piles that are inserted in the soil below. Given the strength and expense of these foundations, they are most commonly used for high-rise structures or where the soil settlement is predicted to be high.

The **cellular mat** (Fig. 11.31) type of foundation is also referred to as a box mat foundation due to the manner in which the mat is constructed. They are used in loose soils. The structure involves the

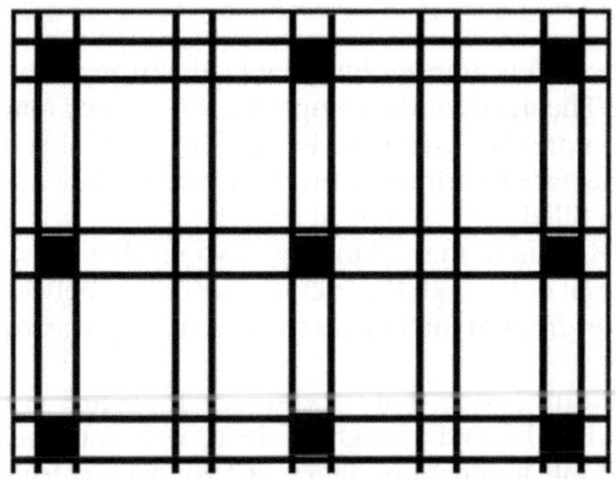

**FIGURE 11.29** Two-way slab and beam plan view.

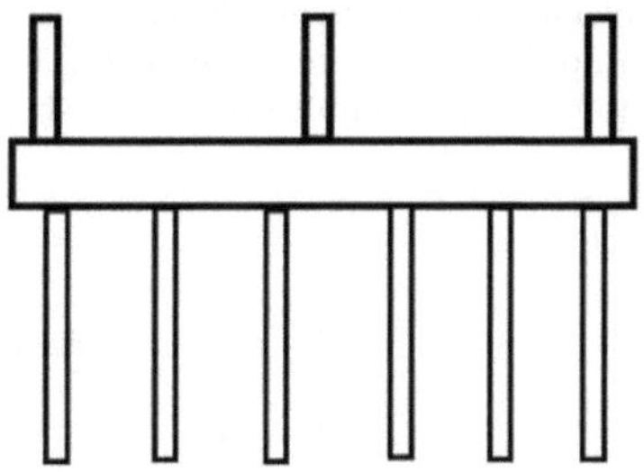

**FIGURE 11.30** Piled raft elevation view.

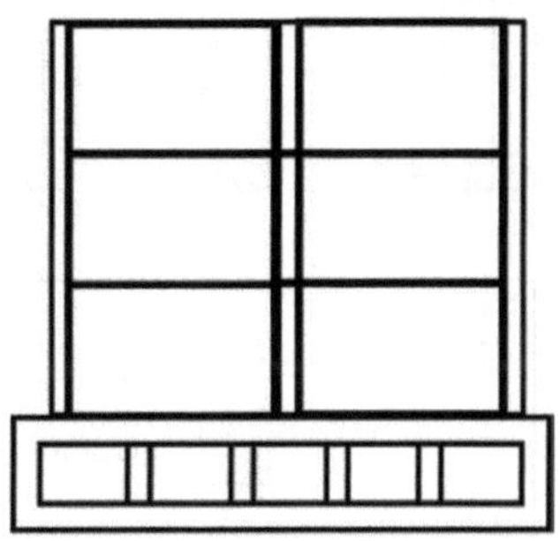

**FIGURE 11.31** Cellular mat elevation view.

forming of boxes where the walls of each box act as a beam. The walls of the boxes are connected on the top and bottom by concrete slabs.

The difference between a shallow foundation and a deep foundation is where the building loads are transmitted to the soil. When the soils below a structure are not capable of supporting heavy loads, a deep foundation can be used to build upon. A deep foundation carries loads from a structure through weak compressible soils on to stronger and less compressible soils or rocks at depth. The deep foundation is connected to the building that it supports and is most often used in the support of taller, heavier structures. Deep foundations are versatile and can be used to support new or even reinforce existing structures. Major types of deep foundations are piles, piers, and caissons, with subcategories in each.

## DEEP FOUNDATIONS

Pile foundations are used to transfer building loads through columns to competent soil deep below the ground surface. The depth of the competent soil or rock may be 20–150 ft from the ground surface. Pile foundations are also used to resist uplift of a building due to lateral loads applied. Piles are installed in the subsurface in primarily two basic ways. They can be driven into the ground or holes can be dug and then filled with reinforced concrete.

Piles transfer their load to the soil through two mechanisms, either end-bearing at the bottom of the competent material or through the friction that occurs between the soil and the surface of the pile. End-bearing and skin friction forces help to prevent the occurrence of any differential settlement of foundations.

Piles may be built in clusters that are connected together at the ground surface using a pile cap. The pile cap evenly distributes the weight of the column into the pile (Fig. 11.32).

When classifying pile foundations, there are two major categories of foundations. These are either displacement piles or replacement piles. When a pile is driven or vibrated into the ground it displaces the soil around it and hence is called a displacement pile. If a pile is placed or inserted into a previously drilled hole it is replacing the soil; hence, it is called a replacement pile.

Piles that are **driven** into the subsurface to support a structure are used when the soil conditions under the building are not strong enough to support a shallow foundation system. Timber, precast

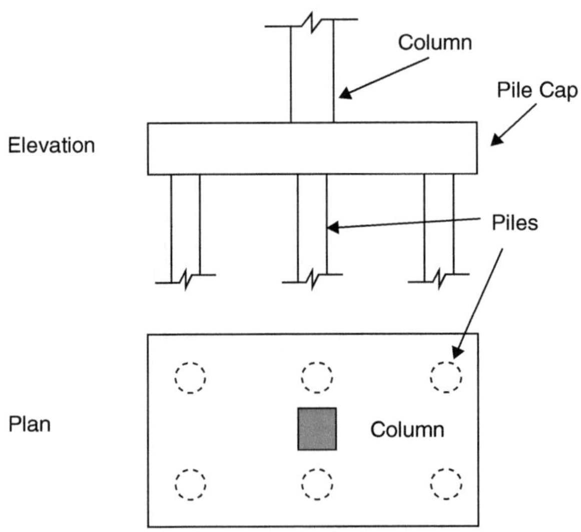


**FIGURE 11.32** Pile cap.

**FIGURE 11.33** Driving precast concrete piles. (Photo courtesy of Rob Webb.)

concrete, structural steel sections (mostly H-piles), and steel pipes are all materials that can be used for pile foundations (Fig. 11.33). Driven piles are typically 12–36 in. in diameter or width and thus smaller in size than drilled shafts. When selecting the material for the pile, it is important to review the geotechnical report to determine the subsurface soil profile, the location of the groundwater table, and whether the soil is corrosive. In addition, it is important to identify the location of existing utilities, overhead restrictions or clearances, site accessibility, and the required length to which the pile will be drilled. Driving a pile is a noisy operation and so the local noise ordinances for the project site should also be investigated.

Each of the material used for piles has unique characteristics. Concrete piles are the most versatile when it comes to the variety of sizes and shapes in which they can be constructed. Concrete piles are also less expensive than steel piles, but they require larger equipment to handle and drive them into the ground. Steel piles, typically H-piles, are suitable for piles that require a lower load capacity. These piles are more expensive than other pile materials, but they possess major advantages. They are driven without displacing a large amount of soil and can be socketed into bedrock. They are lighter weight relative to other pile types and possess high corrosion resistance. Timber piles were the first type of pile used in construction and have a long history of use. The challenge with timber piles is that their length is somewhat limited, they are hard to anchor, and can be easily damaged during the driving operation. They are suitable in situations that require low load bearing and are mostly used in revetments, docks, and temporary construction environments.

Regardless of the type of materials used for the pile, the most common method to drive the pile into the ground is through the use of an impact-type hammer. There are various types of hammers that can be chosen and most often the economics is what drives the hammer selection as well as the type of soil. Site restrictions, soil types, and the type of pile that is being driven are other considerations when determining the method to drive the pile. If the soil type is appropriate, piles can be vibrated into the ground.

## Piers

A pier foundation is constructed of cylindrical columns to support and transfer large superimposed loads to firm strata. This type of deep foundation system only transfers load through end-bearing. Based upon the way the pile transfers load, piers are generally more shallow depth than other deep foundations and can be used for small loading situations. Piers are of larger diameters and most often constructed of concrete.

Drilled shaft or caisson foundations are versatile foundation systems. In their simplest form they are constructed by making a cylindrical excavation, placing a reinforcement cage if necessary, and then placing concrete into the excavation. With the appropriate equipment the diameter of the shaft may extend to up to 20 ft and up to 300 ft in depth. For most building applications, shaft diameters between 3 and 12 ft are more common. The length of the pile depends upon the location of competent soil strata. Drilled shafts transfer load through a combination of skin friction and end-bearing resistance. Drilled shafts provide resistance to overturning and lateral forces due to their large diameter of reinforced concrete and transfer larger loads than pile foundations.

Because of the size versatility of a drilled shaft, it may be used in place of a driven pile group, thereby eliminating the need for a pile cap. The construction of drilled piers uses augers and other types of excavation equipment to construct the shaft. This eliminates the need for a pile hammer and the noise and strong ground vibrations that are associated with driving piles.

Drilled shafts may also be constructed by under-reaming the shaft bottom to create a bell (Fig. 11.34). The most common shape for a bell is at 60°, but 45° bells are also constructed. This type of foundation is used less frequently due to the challenges associated with the soil cleanout when excavating the bell. It is often more cost effective to install a deeper straight shaft.

This type of foundation system is not appropriate for certain soil conditions. Deep deposits of soft clay and water-bearing granular material cause difficulties for the excavation process. They are also not suitable for situations where the soil is prone to caving or an artesian aquifer is present.

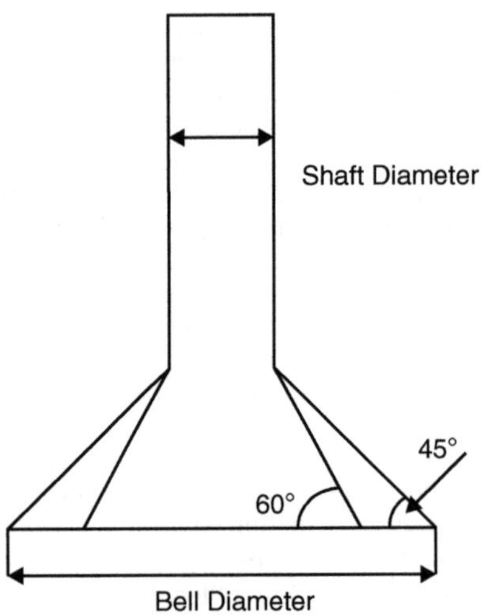


**FIGURE 11.34** Drilled shaft.

# CHAPTER 12
# ELECTRICAL CONSTRUCTION

*This chapter contains information about electricity, electrical systems, and the manner in which electric power is incorporated into construction projects for use by building occupants. The focus is on the core elements of electrical theory, means and methods used in all types of construction, electrical distribution, electrical devices, and low-voltage specialty systems.*

## *BASIC ELECTRICAL THEORY*

Electrical systems contain some of the most technologically complex components in construction, and understanding electrical systems requires a basic understanding of electrical theory and physics.

### Key Terms

To understand electrical systems, it is necessary to understand the basics of electricity and a few key terms.

**Alternating Current.** The flow of electrons which reverses in flow direction at regular intervals.

**Ampere.** A unit of measure for the flow of electrical current. One ampere is the amount of electrical current provided when one volt of electrical pressure is applied against one ohm of electrical resistance.

**Circuit.** The flow path of electrons from a current source. The point at which electrons enter a circuit is referred to as the "source" and the point at which they leave is referred to as the "return."

**Current.** The flow of electrical charge through a medium is referred to as "current." There are two types of electrical current: Direct Current (DC) and Alternating Current (AC). Direct current flows in one direction and alternating current periodically changes direction.

**Cycle.** The change in an alternating electrical sine wave from zero, to a positive peak, to zero, to a negative peak, and back to zero.

**Direct Current.** A steady flow of electrons moving continually in the same direction along a conductor from a point of high potential to a point of lower potential.

**Ohm.** A unit of measure for resistance. Every electrical conductor provides some resistance to the current flow. One ohm is the amount of resistance limiting the current flow to one ampere in a circuit with one volt of electrical pressure.

**Pole.** One or two points on a magnet at which its magnetic attraction is concentrated.

**Polarity.** In electrical systems, polarity refers to the flow of electrons from negative to positive "poles" or terminals. Referring to direct current, current flows in one direction only and one pole is negative and one pole positive. In alternating current, the direction of electron flow is periodically reversed and the poles alternate between positive and negative.

**Resistance.** The opposing force offered by a circuit or component of a circuit to the passage of electrical current.

**Volt.** A unit of electrical pressure or electromotive force which causes current to flow in a circuit. One volt is the amount of pressure required to cause one ampere of current to flow against one ohm of resistance.

**Voltage.** The force which is generated to cause current flow in an electrical circuit (also called electromotive force or electrical potential).

**Watt.** The unit of measure for indicating the electrical power applied in a circuit. It is calculated by multiplying the current (in amperes) by the electrical pressure (in volts) which causes it to flow.

**Watt-Hour.** A unit of electrical energy indicating the amount of work done in an hour by a circuit at a steady rate of 1 W.

Electrical systems can be compared to water flowing in pipes. To understand a rudimentary electrical system, one should picture a mechanical system consisting of a pump pushing water through a closed system of pipes. An electrical current is similar to the water flowing through the pipe. In this comparison, the systems are analogous in the following ways:

- The pump is like a battery or electrical source moving the electricity into the path.
- The pipe is like the wire of an electrical system, it provides a path for the flow of electricity.
- The pressure generated by the pump drives the water through the pipes, like the voltage generated by the battery, driving electrons through the circuit.
- The sediment and dirt in the pipe system slows the flow of water creating a pressure difference from pump to end of a pipe. In a similar way, resistance in an electrical system resists the flow of electricity and creates a voltage drop from end to end. This electrical energy loss across the system results in the release of heat.

Ohm's law describes the relationship between these core concepts in an electrical system: when electrical current is flowing through a conductor, such as a wire, the intensity of the current (in ampere) equals the electromotive force (in volt) driving it, divided by the resistance of the conductor (in ohm). Thus, the flow is in proportion to the electromotive force, or voltage, as long as the resistance remains constant. Simply put, it is a measurement of the work required to move a unit charge between two points.

When expressed mathematically, Ohm's law is: $V = IR$
where $V$ = Voltage
$I$ = Current
$R$ = Resistance

Ohm's law can be applied to electrical circuits once the circuit type is determined to be "Series" or "Parallel." Series and parallel circuits are the two types of electrical circuits. They are differentiated by the nature of the flow path of the electrons from source to return.

In a series circuit, the current is constant but voltage may vary because the path is in-line with the power source. This means the components in a series circuit are connected along the same conductive path so the current must flow through each component, but because the components all have some resistance, voltage is lost ("dropped") across each of the components. This also means every device in a series circuit must be functioning so the current can travel from the source to the return without disruption. If a single device in a series circuit is not functioning, the flow is disrupted and the circuit is broken. An example of a common type of series circuit is a string of holiday lights where if one bulb burns out or is broken, the following lights in the circuit will not work, the circuit is broken.

Alternatively, in a parallel circuit, the current's path of travel branches off from the source but voltage is common throughout the circuit. Therefore, the current travels along multiple paths so it does not travel across all devices to reach the next, thus reducing the total resistance. In this case if one device is not functioning in the circuit, the circuit is not broken and current continues to travel to subsequent devices in the circuit.

## BUILDING INFORMATION MODELING (BIM)

Contractors, for work coordination purposes, utilize software to develop both two- and three-dimensional models of project work. This is done by drawing representations of the various building components, to precise scale, within a virtual model of the structure. This phase of BIM is called "gross coordination" where the intent is to populate each respective trades' installed systems within the building without an effort to review constructability or interferences between proposed components.

"Finite Coordination" introduces the concept of **clash detection**, which is the process of determining if the proposed systems will all fit within the defined space and not conflict with each other during or after installation. The clash detection process is generally managed by the general contractor or construction manager; however, the majority of the clash detection and remediation is done by the trade subcontractors.

**clash detection**
*Using a 3D project model to identify physical interferences between building systems.*

The tolerance of the clash detection software is a parameter defined by the contract documents or the general contractor. The tolerance can have a significant impact on the ability of the BIM team to resolve conflicts.

> It is critical to define a clash tolerance acceptable to the "**entire team**" (the GC and all specialty contractors) at the start of the project but the tolerance can be relaxed as a potential solution to resolve issues in the field.

Tolerances are typically set at 0–1 in. but should be based on the level of congestion and the number of systems.

The information contained within the model can be observed with a simple, viewing software. This software can be installed on any laptop and some tablets for field review of the 3D model. "**Fly-throughs**" can easily be demonstrated and system conflicts resolved with precise dimension references for use in the field. In addition, isometric details can be produced for complex areas such as electrical rooms, mechanical spaces, and information technology rooms.

**fly-throughs**
*A video created using 3D software to show the physical relationships of project systems installed according to plans.*

The data coordinated in the model is directly transferred into many types of computerized layout tools to support estimating and work in the field. If used properly, the computerized layout process is considerably faster and more accurate than traditional layout methodologies and can be used for all aspects of electrical work layout including deck inserts, hanger supports, underground, switchgear layout, in-wall, and overhead.

### BIM Participants

In many cases, only the "major subcontractors" are required to participate in the BIM process; however, the definition of "major" varies by general contractor or construction manager and project. It is necessary to identify the BIM participants early and ensure the general contractor engages the appropriate parties to ensure the process is effective and efficient. At a minimum, the following entities should participate in BIM:

- Electrical
- **Low-voltage** systems
- Mechanical/HVAC
- Sheet metal
- Plumbing
- Fire protection
- Structural steel

**low voltage**
*Technically speaking, low voltage is anything below 600 V. But in the context of building construction as used here common nominal low voltages include 12, 24, and 48 V.*

- Structural concrete/decking
- Concrete
- Framing
- Miscellaneous metals
- Supplemental steel
- Pneumatic tube
- Ceiling grid

In most cases, the structural steel is very accurately placed in the model but the following items are often omitted:

- Spray fireproofing for structural steel members (usually 1½ in. beyond the steel dimension)
- Connection points (e.g., mechanical or welded connections affecting the overall dimension of the steel)
- Miscellaneous metals (sometimes supplied/installed by the structural steel contractor but often not a part of their modeling/coordination scope)
- Supplemental steel members

It is critical to ensure inclusion of these often neglected items. They must be appropriately addressed and adequate accommodations for their "real" dimensions made during modeling. Particular attention should be given to thickness of fireproofing, accessibility to the steel members, size of joints and joint materials, supplemental members, and nonstructural members.

## Concrete and Composite Decks

Because most electrical support systems attach to the building structure, it is important to consider the type of concrete deck and attachment system in the modeling process. The **flutes** on composite metal/concrete decks are rarely placed in the 3D model for clash detection.

**flute**
*A fabricated fold or bend in a steel deck panel which projects downward from a horizontal plane to form a repetitive groove.*

> Flutes play a critical role in the placement of inserts for electrical supports.

Load capacity of a deck insert is reduced when it is not placed in the center/lower portion of the flute. In some cases, placement in the upper portion or angled portion of the flute may reduce the load-bearing capacity of the insert to a level below the requirement for the utility support. To resolve this issue, it is imperative to include the deck material in the model with precise dimensions.

This will allow accurate placement of the inserts in the field so the minimum distance between inserts can be provided while still retaining the proposed load capacity of each insert. Failure to correctly model the flutes of a composite metal/concrete deck will result in significant rework of insert locations by the field installers and detailers.

## Framing

Framing is of critical importance to the electrical BIM process as the framing members support virtually all of the electrical devices and associated infrastructure. Electrical equipment is often mounted to framing, and conduit/cable systems traverse framing members to feed devices installed within the framing. Therefore, it is imperative to include the following components during the BIM clash detection process:

- **King Studs**/Kong Studs
- Headers
- Duct openings
- Conduit openings
- Top of walls and deck attachments
- Exterior connections (e.g., framed wall to exterior wall)
- Sheer wall connections/transitions
- Expansion joints
- Seismic joints
- Shafts and shaft walls (shaft walls must also be identified for the purposes of stud size and attachment as they are often too small to allow for any utility attachment)
- In-fill framing

**king stud** or kong stud
*A full height vertical stub of a window or door opening, running continuously from the bottom sole plate to the top plate.*

Fireproofing of any framing members must likewise be considered in the model and an appropriate addition to the dimensions of each framing member must be made based on the size and type of firestopping material.

**"No Fly Zones," "Halos," or Access Locations:** The terminology may vary, but access, for installation, code, or maintenance, must be considered for every required utility during modeling. Typical areas of concern are as follows:

- Electrical junction boxes
- Cable tray (clearance on sides, top and length of run)
- Combination fire/smoke dampers
- Variable air valves (VAV)
- Mechanical and plumbing valves
- Smoke or heat detectors
- Duct smoke detectors
- CCTV cameras
- Speakers
- Low-voltage junction boxes
- Fire protection piping (sway requirements by code)
- Top of wall terminations at beams
- Deck insert distances (minimum edge distances)

Both the clearance required for installation and the clearance required for future operation of the devices should be considered and dimensioned in the model. It is also critical to ensure the access panels are appropriately placed in framed (hard-lid) ceilings and correspond to the access location in the model.

## Shop Drawings and Product Data

Prior to the start of a modeling effort, precise information about specific equipment is required as certain components (both electrical and for other trades) can and will drive the coordination effort. To ensure the correct information is used, approved product data submittals and shop drawings for the following items should be obtained and distributed to the BIM team prior to starting any BIM activity:

- Electrical switchgear
- Electrical **transformers**
- High/Medium voltage electrical distribution equipment
- Electrical panels, modular panels, integrated switchgear
- Generators
- Uninterruptible power supplies
- Paralleling and control gear
- Generator radiators and exhausts
- Fuel tanks
- Air handling units
- Pumps
- Water softeners
- Storage/Chemical tanks
- Humidifiers
- Variable air valves

**transformer**
*A device used to convert power at one voltage to another voltage by carrying the electricity through coils.*

### 2-Dimensional Details

In addition to the 3D Computer Model, the following 2D details/efforts should be considered on every project:

- Seismic and gravity insert drawings (composite of all trades)
- Sleeve drawings (composite of all trades)
- Access panel location drawings (composite of all trades)
- Point load drawings (composite of all trades)
- Cable tray fill ratio and access drawings (identify fill for all and accessibility in inches for the entire length of tray)

## *RACEWAYS*

Electrical raceways are channels designed to house and protect electrical conductors in both temporary and permanent installations. This section identifies common raceways and fittings used in electrical construction, their respective applications, and installation methods.

### Conduit

Electrical conduit is the most common type of raceway used in commercial, industrial, and heavy/highway/civil construction. It can come in several forms, based on the application and installation location. The following list identifies the types of electrical conduit allowed by the National Electrical Code (NEC):

- Polyvinyl chloride conduit (PVC)
- Electrical nonmetallic tubing (ENT)
- Electrical metallic tubing (EMT)

- Intermediate metal conduit (IMC)
- Galvanized rigid conduit (GRC)
- Flexible metallic conduit (Flex)
- Liquid tight flexible conduit (LT Flex)

***Polyvinyl chloride conduit (PVC).*** Commonly called PVC conduit, polyvinyl chloride conduit is comprised of rigid polyvinyl chloride molded into tubing, often with plasticizer additives to allow for flexibility in the conduit installation. PVC conduit is manufactured with two nominal dimensions: (1) the outside diameter of the conduit tube, ranging from ½ to 6 in. in most electrical applications and (2) the length of the conduit segment, typically in 10 or 20 ft segments. In addition to diameter and length of the conduit segment, PVC is often classified by schedule, which refers to the thickness of the conduit. In electrical construction, Schedule 40 and Schedule 80 PVC conduits are used. They have the same outside diameter but different inside diameters at each nominal size due to the increased wall thickness of the Schedule 80 vs. Schedule 40. As a result the Schedule 40 conduit is lighter per linear foot than Schedule 80 conduit.

PVC conduit is very light, flexible, and can be further manipulated using specialized heating tools which allow a considerable amount of conduit bend to occur without the use of premanufactured conduit fittings.

PVC conduit is manufactured in segments with one end having the same outside diameter as the nominal dimension of the conduit, and the other end having an oversized end having the same inside diameter as the nominal dimension. This design allows for conduit segments to be inserted into each other, similar to a male/female connection. This method is also used in the various types of PVC conduit fittings to allow conduit segments to be easily cut to size and fit together. The conduit segments and fittings are secured in place by applying PVC adhesive to the male portion of the conduit prior to insertion into the female segment or fitting, thus eliminating the need for a conduit coupling in many portions of the conduit system. When dry, the PVC adhesive provides a weather-tight seal in the closed conduit system.

PVC conduit systems can utilize several manufactured fitting types for various purposes. The following list identifies some of the common fittings and their application:

- **Elbow**. Typically manufactured in 30, 45, and 90 degree angles with a standard or long radius, elbows are used to transition or bend the conduit system laterally or vertically.
- **Coupling**. A coupling is used to connect two segments of conduit with the same outside diameter.
- **Conduit Body**. Conduit bodies are manufactured in several types with varying entry/exit points and combinations and are used to provide a junction point for the purposes of pulling or accessing wire through the conduit system.
- **Male Adapter**. Similar to a coupling, but manufactured with one smooth end and one threaded end. Male adapters are used to connect a segment of a PVC conduit system to a fitting with female threads of the same diameter.
- **Female Adapter**. Similar to a coupling, but manufactured with one smooth end and one threaded end. Female adapters are used to connect a segment of a PVC conduit system to a fitting with male threads of the same diameter.
- **End Cap**. PVC End Caps are used to close the end of a segment of PVC.
- **PVC Spacer**. Used in underground applications, PVC Spacers secure multiple conduit segments in a trench and maintain a desired distance between conduit segments during earth or concrete backfill.

Because of its composition, PVC conduit is extremely effective to resist corrosion and is, therefore, often used for underground installations (Fig. 12.1) or those exposed to weather. This includes the following applications:

- Underground feeders and distribution
- Underground branch or lighting circuits

**FIGURE 12.1** PVC conduit installation underground.

- Underground low-voltage systems (Backbone and Branch)
- Exposed in corrosive environments

Nevertheless, PVC conduit does not withstand compressive forces as well as other types of electrical conduit and can be easily damaged by impact forces, so it is often encased in concrete when buried in areas subject to heavy loads such as roadways or driveways.

***Electrical Nonmetallic Tubing (ENT).*** Similar in composition to PVC, ENT is manufactured from composite plastic materials with additives to allow for maximum flexibility and low weight. ENT conduit is manufactured with: (1) the outside diameters of the conduit tube, ranging from ½ to 2 in. and (2) since ENT is manufactured in flexible coils or reels it comes in coil or reel lengths. ENT is manufactured in the colors blue, yellow, or red, which typically denote its use for Branch Circuits, Communication Circuits, or Fire Alarm Circuits, respectively; however, this color coding may differ in each design.

ENT conduit is very light, more flexible than PVC. The utilized fittings are secured via adhesive, similar to PVC.

Because of its inherent flexibility, ENT does not require the use of elbows or angled fittings used in other conduit systems, but it does utilize connective fittings. The following list identifies some of the most common fittings and their application:

- **Coupling.** A coupling is used to connect two segments of conduit with the same outside diameter.
- **Adapters.** ENT Adapters are used to mate ENT with another conduit system, such as PVC or EMT conduit or boxes. Adapters can be mechanical (pressure/friction) or threaded.
- **Bridge.** ENT bridges are used to secure ENT in place when installed in concrete.
- **Stub Up or Down Fitting.** ENT stub up or down fittings secure the ENT segment in place at a terminal location (floor or ceiling).

ENT is approved for use in indoor applications and in concrete encasement (Fig. 12.2). Due to its size limitation of 2 in. outside diameter, it is most commonly used in the following applications:

- Concrete Deck Branch Circuits
- In-Wall Branch Circuits, Metal, or Wood Stud

**FIGURE 12.2** ENT in a concrete deck (prior to concrete placement).

- Overhead Concealed Branch Circuits
- In-Wall and Overhead Low-Voltage Systems

***Electrical Metallic Tubing (EMT).*** EMT is manufactured from galvanized steel or aluminum. It is sized by the nominal dimension of the outside diameter and is generally manufactured in 10-foot segments.

EMT conduit is more durable and heavier than PVC but not as heavy as IMC or GRC and can be bent by hand, in sizes up to 1¼ in. Because of its relative strength, durability, and low weight per foot, EMT is the most commonly used conduit system for commercial, interior electrical systems.

EMT conduit systems can utilize several manufactured fitting types for various purposes. The following list identifies the most common fittings and their application:

- **Elbow.** Typically manufactured in 30, 45, and 90 degree angles with a standard or long radius, elbows are used to transition or bend the conduit system laterally or vertically.
- **Set-Screw Coupling.** A set-screw coupling is used to connect two segments of conduit with the same outside diameter and is secured by tightening a screw to apply pressure on the exterior of the inserted conduit.
- **Compression Coupling.** A compression coupling is used to connect two segments of conduit with the same outside diameter and is secured by twisting the coupling to expand the coupling interior and apply pressure to the inserted conduit.
- **Set-Screw Connector.** Similar to a set-screw coupling, a set-screw connector uses a screw to secure the inserted conduit and connects that assembly to a terminal point (box, panel, etc.) via a threaded end with a locking hub.
- **Compression Connector.** Similar to a compression coupling, a compression connector secures conduit by applying pressure to the conduit wall and connects the assembly to a termination point (box, panel, etc.) via a threaded end with a locking hub.
- **Conduit Body.** Conduit Bodies are manufactured in several types with varying entry/exit points and combinations and are used to provide a junction point for the purposes of pulling or accessing wire through the conduit system.

**FIGURE 12.3** EMT with set-screw fittings.

Because of its composition, EMT conduit (Fig. 12.3) is extremely durable and effective resisting the effects of fire and smoke, but is susceptible to the effects of corrosion. It is, therefore, often used in interior commercial and industrial applications. This includes the following applications:

- Overhead Feeders and Distribution
- Overhead Branch or Lighting Circuits
- In-Wall Branch or Lighting Circuits
- Overhead Low-Voltage Systems (Backbone and Branch)
- In-Wall Low-Voltage Systems

***Intermediate Metal Conduit (IMC) and Galvanized Rigid Conduit (GRC).*** Both IMC and GRC are manufactured from galvanized steel. These conduit types are sized by the nominal dimension of the outside diameter and are generally manufactured in 10-foot segments.

IMC conduit is thicker and heavier than EMT, but roughly 30% lighter than GRC. Due to its relative strength and thickness, smaller sizes of IMC can be bent by hand, but it is much more common to use mechanical/hydraulic bending tools for bending of the conduit segments. GRC is exclusively bent utilizing a tool.

IMC and GRC conduit systems can utilize several manufactured fitting types for various purposes, but unlike EMT, IMC and GRC are threaded utilizing a mechanical or hydraulic threading machine, so fittings utilize threaded components. The following list identifies some of the most common fittings and their application:

- **Threaded Elbow.** Typically manufactured in 30, 45, and 90 degree angles with a standard or long radius, elbows are used to transition or bend the conduit system laterally or vertically.
- **Threaded Coupling.** A threaded coupling is used to connect one segment of conduit with a fitting and is secured by screwing the coupling on via the threads.
- **Threaded Union or Three-Piece-Coupling.** A threaded union or three-piece-coupling consists of two threaded components and an expansion fitting between them. This allows for the connection of two conduit segments on either side of the coupling since the fitting itself can spin to tighten on the threads.

**FIGURE 12.4** IMC conduit with threaded fittings.

- **Threaded Connector.** Similar to threaded coupling, a threaded connector uses threads to secure the inserted conduit and connects that assembly to a terminal point (box, panel, etc.) via a threaded end with a locking hub.
- **Threaded Conduit Body.** Conduit bodies are manufactured in several types with varying entry/exit points and combinations and are used to provide a junction point for the purposes of pulling or accessing wire through the conduit system.

Both IMC and GRC are used in heavier duty applications subject to physical damage, corrosion. They are extremely durable and effective to resist the effects of fire and smoke; thus, they are also used when a higher fire rating is necessary. IMC (Fig. 12.4) is primarily used for indoor commercial and industrial applications, while GRC is used in both indoor and outdoor commercial, industrial, and civil applications. This includes the following applications:

- Overhead and Underground Feeders and Distribution
- Overhead and Underground Branch or Lighting Circuits
- Mechanical and Industrial Equipment Feeders and Connections

***Flexible Metal Conduit (FMC) and Liquid Tight Flexible Metal Conduit (LFMC).*** These are both manufactured by helically coiling a self-interlocked, ribbed strip of aluminum or steel to form a hollow tube through which electrical or low-voltage wire or cable can be pulled. LFMC adds a plastic, water-proof, corrosion-resistant coating to the same manufacturing process to allow for outdoor use and exposure to corrosive elements.

FMC and LFMC are both manufactured in coils or reels and are relatively light, extremely flexible, and can be manipulated easily by hand. While conduit is in the name of these systems, they are technically classified as raceways, not conduit, by the NEC.

Similar to ENT, FMC and LFMC do not require bending or elbows so the fittings used are primarily for the stabilization, connection, and penetration of the system. The following list identifies the most common fittings:

- **Coupling.** A coupling is used to connect two segments of conduit with the same outside diameter. FMC and LFMC couplings can be set screw (similar to EMT couplings), screw-in (where the coupling itself has threads to attach to the interior of the FMC/LFMC), or snap-in (a friction fitting).
- **Connector.** A connector is used to connect a segment of conduit with a terminal point (box, panel, etc.) with the same outside diameter opening. FMC and LFMC connectors can be set screw (similar

**FIGURE 12.5** LFMC feeding a rooftop air handling unit.

to EMT couplings), screw-in (where the coupling itself has threads to attach to the interior of the FMC/LFMC), or snap-in (a friction fitting), all with a male threaded end with locking bushing for the terminal side.

- **Stub Up or Down Fitting.** FMC and LFMC Stub Up or Down fittings secure the FMC/LFMC segment in place at a terminal location (floor penetration, ceiling penetration, junction box, or panel) when a degree of bend is necessary in lieu of a standard connector.

FMC and LFMC are used in heavier duty applications subject to vibration, physical damage, or corrosion. Both are extremely durable and effective to resist the effects of fire and smoke; thus, they are used when a higher fire rating is necessary. However, LFMC can produce a toxic smoke when exposed to fire conditions. This factor is often considered in the selection process. FMC is primarily used in indoor commercial and industrial applications, while LFMC is used in both indoor and outdoor commercial (Fig. 12.5), industrial, and civil applications. This includes the following applications:

- Mechanical equipment connections, particularly those where the equipment may vibrate consistently while in use.
- Electrical equipment connections where seismic or gravity forces could cause movement.
- Branch and feeder applications where unique bending or flexibility are required.

## Cable Tray

A Cable Tray is a cable support channel, often with a "U" shape open top. Trays can be manufactured from steel, aluminum, or plastic. They are designed specifically with the intent of organizing and managing insulated conductors in an accessible environment. Cable trays can be used for electrical and low-voltage cabling, either independently or together, and are manufactured in several types with different uses.

Cable trays can have multiple configurations including a solid bottom, open bottom, channels with dividers, a wire mesh or "basket" style (Fig. 12.6) tray, ladder tray (for vertical support of cables or transitions from horizontal cable tray systems), and covered tray. Configurations vary based on the application and location of the installation, but the primary benefit of a cable tray system is that the associated electrical or low-voltage system cables can be regularly accessed with relative ease, unlike a closed conduit system.

Cable tray systems of all material types utilize similar fittings but vary on the means of connection due to the type of material. In most configurations, the following are commonly used:

- **Horizontal Radius Fittings.** Used to turn the cable tray (similar to the bending of a conduit) on a horizontal axis, radius fittings are manufactured in standard degree radii such as 30, 45, and 90 degree or can be custom manufactured to meet field conditions.

**FIGURE 12.6** Basket-type cable tray overhead installation.

- **Vertical Radius Fittings.** Used to change elevation of the cable tray on a vertical axis, these fittings are manufactured in small degree radii or custom manufactured for field conditions.
- **Horizontal Tee Fittings.** Used when three segments of tray meet in an intersection and are shaped like a "T" for that reason.
- **Clips and Clamps.** Clips and clamps are mechanical connectors used to connect cable tray segments via friction, pressure, or mechanical elements such as screws. These can also be used to connect horizontal or vertical gravity and seismic support elements, such as threaded rod, to the cable tray.

While cable trays are extremely versatile and can be used for power and low-voltage applications, both indoor and outdoor, the limiting application factor is often the contents of the tray such as wire type and wire insulation type, which may not always have similar versatility. Therefore, cable trays are most commonly used for low-voltage system applications in commercial construction and in power distribution applications in industrial, mission critical, and manufacturing construction.

## Wireway

Wireways are troughs of steel or aluminum. They have hinged or screw-type removable covers designed to house insulated conductors in an accessible environment. Like cable trays, wireways can be used for electrical and low-voltage cabling, either independently or together, and are manufactured in several types with different uses. Conductors are laid into wireways after the wireway has been installed as a complete system.

Wireway systems (Fig. 12.7) utilize fittings similar to those used in cable tray systems, with the primary difference being that a wireway system is a "closed" system, meaning it is not accessible unless the cover is opened or removed. In most configurations, the following are commonly used:

- **Horizontal Radius Fittings.** Used to turn the wireway (similar to the bending of a conduit) on a horizontal axis, radius fittings are manufactured in standard degree radii such as 30, 45, and 90 degree, or can be custom manufactured to meet field conditions.
- **Vertical Radius Fittings.** Used to change elevation of the wireway on a vertical axis, these fittings are manufactured in small degree radii or custom manufactured for field conditions.
- **Horizontal Tee Fittings.** Horizontal tee fittings are used when three segments of wireway meet in an intersection and are shaped like a "T."
- **End Caps.** Because wireways are closed systems, the end cap is used where the wireway terminates without connection to another segment or electrical component.

**FIGURE 12.7** Wireway installation above distribution boards.

The use of wireway systems is limited to indoor use in dry locations not exposed to physical damage or corrosive elements. While there are variations for exterior use, such as a NEMA 3R or NEMA 4X type wireway, these are uncommon. Therefore, wireway systems are most often used in electrical or mechanical rooms where several conductors converge and require access or routing in a manner where conduit systems are not practical or efficient such as above multiple panels (entering or exiting panel boards with many crossing conductors).

## Surface Raceway

A Surface Raceway consists of a surface-mounted assembly manufactured of metal, plastic, or composite material which is secured to a wall or ceiling and protects insulated conductors. Surface raceways can be used for electrical and low-voltage cabling, either independently or together, and are manufactured in several types with different uses. Like a wireway, a surface raceway is a closed system, but unlike a wireway, surface raceways are generally not easily accessible as they do not contain hinged or removable covers. Similar to conduit systems, surface raceway systems require the pulling of wire through them to install the conductors whereas cable tray and wireway systems allow for the placement of the conductors before the covers are installed.

Surface raceways can have multiple configurations including channels with dividers to separate electrical and low-voltage systems and may also include knockouts for devices or preinstalled devices for either electrical or low-voltage systems such as receptacles or data ports. Configurations vary based on the application and location of the installation, but the primary benefit of a surface raceway system is that the associated electrical or low-voltage system cables and devices can be installed in renovation settings or when walls or ceilings are not-accessible for conduit or in-wall or ceiling installations.

Surface raceway systems utilize fittings similar to those used in wireway systems; however, vertical radius or offset fittings are not typically used because surface raceway systems are limited to use on flat surfaces. In most configurations, the following are commonly used:

- **Horizontal Radius Fittings.** Used to turn the surface raceway (similar to the bending of a conduit) on a horizontal axis, radius fittings are manufactured in standard degree radii such as 30, 45, and 90 degree, or can be custom manufactured to meet field conditions.
- **Horizontal Tee Fittings.** Used when three segments of surface raceway meet in an intersection and are shaped like a "T."
- **End Caps.** Because surface raceways are closed systems, the end cap is used where the surface raceway terminates without connection to another segment or electrical component.

**FIGURE 12.8** Surface raceway in a laboratory.

The use of surface raceway systems is limited to indoor use in dry locations not exposed to physical damage or corrosive elements, and these systems are most often used where conduit or cable systems cannot be concealed in walls or ceilings to feed convenience devices. Therefore, these systems are most often in renovation environments or installed on or adjacent to furniture requiring power or communication devices where the furniture does not include internal rated raceways for such applications. Examples of this are laboratory benches, office furniture, shop furniture or tables, and manufacturing or testing tables and benches. In these cases, the most common surface raceway systems (Fig. 12.8) are Plugmold (including devices) or Wiremold (including devices). These are available in configurations allowing for installation in wet or corrosive environments such as wash-down areas, sterile areas, and areas exposed to hazardous chemicals or materials.

## Electrical Duct

Electrical Duct is a rectangular channel manufactured from steel material which is secured to a wall, ceiling, floor or embedded in a concrete structure and protects insulated conductors that generally require circuit separation. Like wireway, electrical duct is a closed system and contain hinged or removable covers. Electrical duct allows for the placement of the conductors before the covers are installed.

Electrical ducts can have multiple configurations including channels with dividers to separate electrical and low-voltage systems and may also include knockouts for devices or preinstalled devices for either electrical or low-voltage systems such as receptacles or data ports. Configurations vary based on the application and location of the installation, but the primary benefit of an electrical duct system is that the associated electrical or low-voltage system cables and devices can be installed in renovation settings or when walls or ceilings are not accessible for conduit or in-wall or ceiling installations.

Electrical duct systems utilize fittings similar to those used in wireway systems; however, Vertical Radius or Offset fittings are not typically used because electrical duct systems are limited to use on flat surfaces. In most configurations, the following are commonly used:

- **Horizontal Radius Fittings.** Used to turn the surface raceway (similar to the bending of a conduit) on a horizontal axis, radius fittings are manufactured in standard degree radii such as 30, 45, and 90 degree, or can be custom manufactured to meet field conditions.
- **Horizontal Tee Fittings.** Used when three segments of surface raceway meet in an intersection and are shaped like a "T."
- **End Caps.** Because surface raceways are closed systems, the end cap is used where the surface raceway terminates without connection to another segment or electrical component.

**FIGURE 12.9** Electrical floor and wall duct in a medical setting.

The use of electrical duct (Fig. 12.9) systems is limited to indoor use in dry locations not exposed to physical damage or corrosive elements. These systems are most often used where conduit or cable systems cannot be concealed in walls or ceilings to feed convenience devices. Therefore, electrical duct systems are most often used in electrical rooms (when combining multiple circuits above a piece of switchgear) or in medical environments to provide electrical connections to equipment such as MRI machines, CT machines, Integrated Operating Room Equipment, and Booms.

## RACEWAY AND CONDUIT SUPPORTS

When conduit is installed overhead, which is typical of most building applications, the conduit must be secured in place with a conduit support system. Conduit support systems are used to support the gravity load of the conduit. Generally, lateral or seismic movement is not considered. Conduit support systems, therefore, require an attachment point, typically the structure above the conduit or adjacent wall structure, a connection to the attachment point, and a connection to the conduit itself. Conduit support systems can be designed to support virtually any of the aforementioned raceway systems and can do so in a number of configurations, from a single conduit to a multiple conduit rack.

### Attachments

The method of attachment for the support system is driven by the structure of the building. The most common attachment types are:

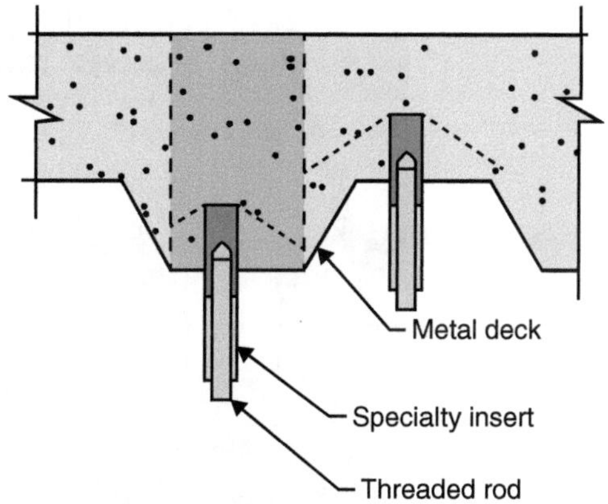


**FIGURE 12.10** Concrete insert and threaded rod in composite concrete/metal deck.

- Concrete Deck Inserts
- Concrete Anchors
- Shot-Pin Anchors
- Mechanical Steel Attachments

***Concrete Deck Inserts*** are manufactured from a combination of steel and composite materials and are designed to receive various sizes of galvanized threaded steel rod. The concrete deck insert (Fig. 12.10) is installed prior to the concrete installation in either a structural concrete deck or composite concrete/metal deck and the concrete is then placed around the insert. Once the concrete cures, the threaded rod is installed in the insert and the conduit is then attached to the threaded rod. Concrete deck inserts are used in support systems for many other trades, including mechanical, plumbing, and fire protection, and consequently, the use of concrete deck inserts requires a significant coordination effort prior to installation to ensure the quantity of inserts does not reduce the load-bearing capacity of the concrete structure to which it attaches, and to ensure the associated utilities to be hung from the inserts are not in conflict with each other during installation.

***Concrete Expansion Anchors*** are very similar in concept to a concrete deck insert; however, they are designed to be installed after concrete is placed and has hardened, which allows for greater flexibility in installation and a lesser coordination effort prior to the installation. Concrete anchors (Fig. 12.11) are installed by drilling a hole in the concrete and inserting the receiving portion of the anchor into the drilled hole. The anchor then expands when the threaded rod is installed and the expansion creates a friction force between the anchor and the concrete capable of supporting the associated gravity load.

***Shot-Pin Anchors*** are another method of attaching anchor systems to concrete and metal deck structures after the concrete is placed. Like an expansion anchor, the shot-pin anchor does not require a significant coordination effort prior to concrete placements and there is greater flexibility as to where it can be installed. However, the shot-pin anchor systems generally have the lowest load support rating of the three concrete attachment systems and, therefore, are generally used only for single-hung conduit/cable or a small quantity of conduit/cable or box attachments.

Shot-pin anchors (Fig. 12.12) are literally "shot" into the concrete deck utilizing a powder-actuated tool or "gun." They use friction to carry the weight of the associated supports. A shot-pin anchor attaches to a threaded rod using an eyelet or mechanical connection such as a rod coupling or support bracket.

**FIGURE 12.11** Concrete drill-in anchor and threaded rod in composite concrete/metal deck.

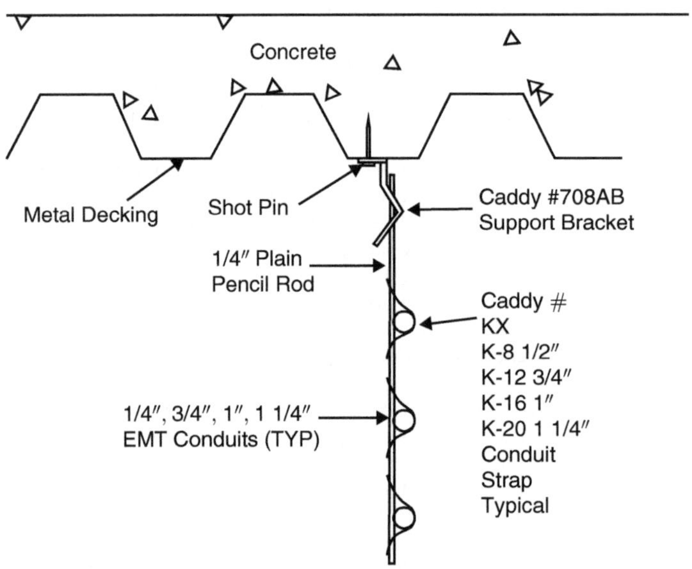


**FIGURE 12.12** Multiple conduit hanger configuration with shot-pin attachment.

In the absence of a concrete or composite deck, it is necessary to attach gravity support components to other structural members such as steel or wood. In these cases, a mechanical attachment is used to attach the hanger system to the structure. The most common form of attachment is a steel beam clamp which uses a set-screw and friction to attach to the flange of a steel beam and mechanical components to attach to a threaded rod to which the conduit or raceway attaches.

## Rack Systems

The most efficient installation of conduit and raceway is when they are routed and grouped together for as long a distance as possible from source to destination. This can include both feeders, branch, and low-voltage systems, and to do so, a rack system is utilized. A rack system can be attached overhead or to an adjacent wall structure.

When attached overhead, the rack system utilizes the aforementioned anchor attachments and threaded rod, but groups two or more anchor and rod assemblies together and connects those assembles with steel C-Channel or "Strut." The raceway, conduit, or cable rests on the strut and is secured in place with a strut-clamp or other type of mechanical attachment. Often called a "Trapeze" rack (Fig. 12.13) because of its similarity to an acrobatic trapeze swing, this type of rack system is most common in overhead electrical installations.

If an overhead attachment is not possible, a rack system can still be used to group and support multiple raceways, conduit, or cables together by supporting the rack system from a wall structure. In these cases, the rack system can attach to two walls and utilize C-Channel or "Strut," similar to a "Trapeze" rack or they can be cantilevered (Fig. 12.14) from a singular wall and utilize C-Channel or Angle Iron to support the electrical components. This is less common for large electrical racks because wall structures are not generally designed to bear large loads; however, it is a common practice for small conduit or raceway installations.

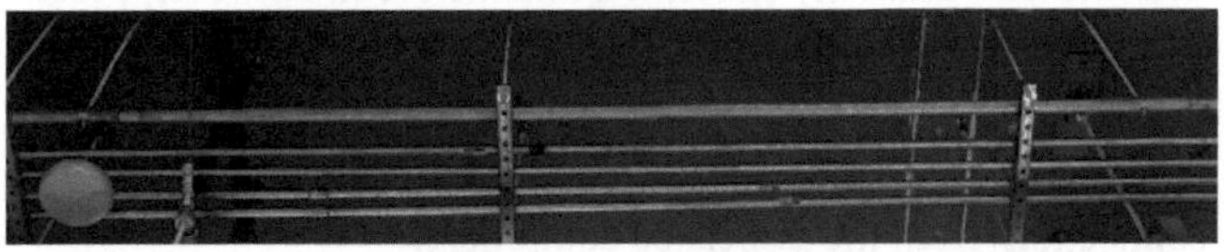

**FIGURE 12.13** "Trapeze" rack system with overhead attachments.

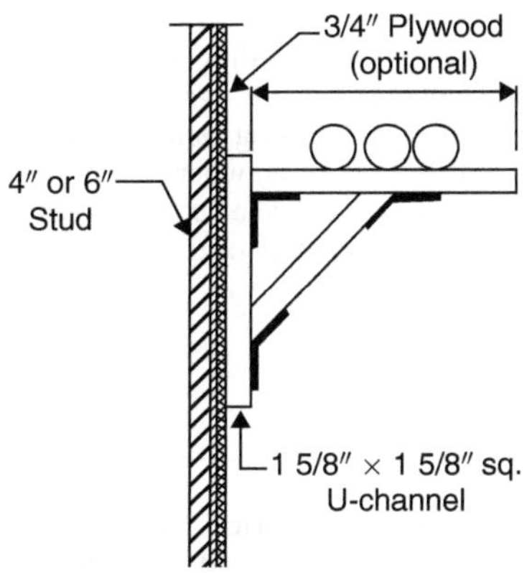


**FIGURE 12.14** Cantilever wall-mounted rack system.

## SEISMIC BRACING

To mitigate the effects of seismic events (earthquakes), seismic bracing can be added to electrical systems to minimize system damage while also ensuring the critical electrical systems remain operational during seismic motion events. Seismic bracing is a regional requirement but some of the same bracing elements can be used to prevent damage during other natural disasters such as severe wind events, hurricanes, and tornados. Local codes define seismic bracing requirements.

***Seismic Engineering.*** All seismic bracing systems must be engineered to meet the local codes and applicable seismic zone requirements. Structural engineers determine the center of gravity (COG) of a building and most associated seismic bracing calculations are based on the location of the item being braced with respect to the COG.

### Raceway Bracing

Raceways are braced using similar materials as conduit supports such as strut or threaded rod but the brace locations are considerably less frequent than the gravity supports. While gravity supports are generally installed at 10 ft intervals or less, seismic bracing of raceway systems is generally placed every 30 ft or at greater distances.

Bracing is done on both the transverse (perpendicular) and longitudinal (parallel) planes to allow the electrical system to move with the building's structural members and not bear the force created by the structure's movement.

### Equipment Bracing

Electrical equipment is generally anchored to the structure mechanically which means the equipment itself remains relatively stable in a seismic event. However, because the electrical components (conduit, cabling, and fittings) attached to the equipment are also attached to other structural members of the building, such as a ceiling or adjacent wall, flexible connections are used to allow for movement of the attached components. In addition, cable installations are installed with a sufficient "slack" or excess cabling, to allow for movement of the cable components without tearing the terminations or connections from the associated equipment.

### Light Fixture Bracing

Light fixtures hung from the building structure using threaded rod, aircraft cable, or pendants will sway considerably during a seismic event. To prevent the damage this sway may cause to the fixtures themselves or the adjacent building components the fixtures may contact when swaying, "sway bracing" is installed in the direction(s) where the movement can occur. Sway bracing usually uses aircraft cable or threaded rod to stabilize a fixture or series of connected fixtures and prevent the sway by attaching them to other members of the building not subject to movement in the same direction as the fixtures.

## WIRE

Inside of a raceway system, electrical conductors are installed to carry the electricity and electrical signals from point to point. The conductors are most commonly in a wire form, made of malleable, conductive metals such as copper or aluminum so they can be bent to the form of the raceway system and have minimal resistance to the electrical current.

Wire generally has five distinguishing characteristics. These must be considered prior to use:

- Material
- Insulation
- Wire size
- Wire color
- Ampacity

## Material

The two most common materials used in the manufacturing of electrical wire are aluminum and copper. Copper has a very high conductivity and, by cross-sectional area, is more conductive than aluminum. However, aluminum is considerably less expensive by weight and is, consequently, much more widely used when large conductors are required such as in transmission and distribution systems, whereas copper is much more common in branch wiring systems, low-voltage systems such as data and fire alarm.

Copper electrical wire can be solid, meaning the wire is a solid piece of annealed copper, or stranded, meaning the wire is comprised of multiple strands of smaller copper wire wrapped helically around each other to create a larger wire. Aluminum wire is generally solid.

## Insulation

In most applications, electrical wire is wrapped with a coating of insulation material. The insulation is designed to provide protection to the wire for various environmental conditions and to protect the surrounding infrastructure (e.g., the raceway housing the wire) from the properties of the wire and electricity. The insulation on the wire is stamped with nomenclature describing the performance characteristics of the material. The most common characteristics of wire insulation are listed below and use the following nomenclature:

- T: Thermoplastic Insulation, a fire-resistant material
- H: Heat-resistant up to 167°F
- HH: Highly heat-resistant up to 194°F
- W: Wet-location-rated or damp-location-rated but also suitable for dry locations
- X: Flame-retardant synthetic polymer insulation
- N: Nylon-coated for resistance to oil or gasoline

The letters, in combination, mean that the wire insulation maintains all the noted properties. For example, THHN is a thermoplastic, high heat-resistant, nylon-coated insulation.

The most common types of wire used in residential, commercial, and industrial applications are as follows:

- THHN—Thermoplastic, High Heat-Resistant, Nylon coated
- THWN—Thermoplastic, Heat and Water-resistant, Nylon-coated
- THW—Thermoplastic, Heat and Water-resistant
- XHHW—Cross-Linked High Heat Water-Resistant Insulated Wire (Fig. 12.15)

## Wire Size

Wires are measured by cross-sectional area. Small wire is measured by the America Wire Gauge (AWG) and large wire by circular mills. The AWG wire size designation ranges from 0000 (also referred to as 4/0, pronounced "four aught") for the largest wire of 0.46 in. or 11.684 mm to 40 for the

**FIGURE 12.15** 500 MCM XHHW wire.

**TABLE 12.1** Wire Insulation Color Is an Indicator of Voltage and Application

| Function | Color code (120/280/240 V) | Color code (270/480 V) |
|---|---|---|
| Three Phase Line (L1) | Black coating | Brown coating |
| Three Phase Line (L2) | Red coating | Orange coating |
| Three Phase Line (L3) | Blue coating | Yellow coating |
| Neutral (N) | White coating | Gray coating |
| Protective Earth or Ground (PG) | Green, no coating, green and yellow | |
| Single Phase Line | Black coating | Red (for 2nd hot) |

smallest wire with a diameter of 0.00314 in. or 0.0799 mm. The AWG gauges are used to describe both solid and stranded so the gauge size can represent either the singular diameter of the solid wire or the sum of the cross-sectional areas of the individual strands in stranded wire.

Wire sized above the AWG gauges is measured in circular mills, which is a unit of area, equal to the area of a circle with a diameter of one mil (one thousandth of an inch). Large wire, used for feeders, distribution, and transmission, is commonly sized in MCM, which stands for "Thousand Circular Mils." The most common wire sizes of this type range from 250 to 1, 250 MCM, with 50–150 MCM as a typical increment between sizes.

## Wire Color

The color of the wire insulation is a distinguishing factor for all wire around the world, but the color function varies by country. In the United States, wire insulation color is dictated by voltage and application, where phase conductors, neutral conductors, and ground conductors are distinguished. Table 12.1 shows the standard wire color coding used in the United States.

## Ampacity

Ampacity is the overall current-carrying capacity of the wire, measured in amperes (A). The ampacity of the wire is a function of the cross-sectional area and associated resistance for that size wire; thus, larger diameter wires have a much higher current-carrying capacity than smaller sized wire.

## CABLE

Cable is an extremely efficient electrical material concept because it combines two or more insulated wires in a single "jacket." The jacket acts like a raceway to contain the multiple wires, but is flexible and allows for a much quicker installation of smaller conductors.

Cable jacketing is made from either metallic or nonmetallic materials (such as polyvinyl chloride or nylon) depending on the location and application of the installation. The two most common types of cable used for electrical distribution and branch circuits in the United States are Metal Clad or MC cable and Nonmetallic cable or NM cable.

MC cable (Fig. 12.16) is manufactured with an aluminum jacket and multiple conductors preinstalled within the jacket. It can be produced using virtually all gauges of the AWG and multiple combinations of wire quantity and size. The nomenclature used to designate the wire size and quantity is "Size in AWG / Quantity." For example, an MC cable with two (2) No. 12 AWG wires would be designated as "12/2 MC." The application of insulation types and colors remains the same within the MC cable as it would in common raceways, and the wires themselves still carry the associated markings.

NM cable is similar in internal composition to MC, but is wrapped in a nonconductive and heat-resistant thermoplastic sheathing. NM cable (Fig. 12.17) is also produced using virtually all gauges of the AWG and multiple combinations of wire quantity and size. The nomenclature used to designate the wire size and quantity is "Size in AWG / Quantity." However, unlike MC cable, the nomenclature for NM cable can vary by application and sheathing composition. NM cable is most commonly used in interior and residential branch wiring applications when using standard NM cable. Other configurations include:

- Underground Feeder Cable (type UF-B), which can be buried directly in dirt and is moisture resistant, corrosion resistant, and fungus resistant.
- Service Entrance Cable (Type SE/SER/SEU), which can be used for above-ground service wiring or interior feeders. SER has reinforcement tape and SEU has a concentrically applied neutral. Both are UV resistant.

Both MC and NM cables are produced in hand coils, coil packs, or reels, and are bundled or secured using staples, ty-wraps, or brackets. Cable of all types is generally cut by hand, utilizing a rated hand tool or mechanical cutting means and requires little or no preparation following cutting.

## ELECTRICAL BOXES

Electrical boxes are utilized to house devices (device box) or for the transition and splicing of wiring (junction box). In either application, the same box type can be used, but is determined by the quantity of devices or quantity of wires or cables entering/exiting the box. The box material varies by

FIGURE 12.16 MC cable of different configurations.

**FIGURE 12.17** Romex in panel.

application and construction type as well. Plastic boxes are generally used in single family, or low-rise multiunit residential or hospitality, and metal boxes are used for most other applications. Some variations of box types do exist for specialty applications such as corrosion-resistant gasketed boxes (outdoor applications), PVC-coated metal boxes (toxic or explosion proof applications), and boxes with threaded hubs (vapor tight conduit applications) but these are less common.

## Device Boxes

Electrical device boxes are used to house electrical devices such as receptacles, switches, and smoke detectors. Device boxes are generally rectangular and can range in size from a single device, also called single-gang, to four device, also called four-gang (Fig. 12.18). However, some variations in shape are made to accommodate a specific device. Octagonal, or "octagon," boxes are used for smoke detectors and ceiling fans. In addition, large, multi-gang rectangular boxes are manufactured for oversized devices such as wall clocks or televisions, or larger ganged devices such as receptacles and switches in high-density applications.

Boxes are mounted to the wall or ceiling members with an integral box bracket or external bracket and a mechanical means of connection such as a screw or nail (Fig. 12.19).

Wire or cable enters the box using manufactured "knockouts," which are small, perforated circles or rectangles punched in the box during manufacturing to allow for easy, consistent removal of portion of the box where the wire or cable will enter or exit. Knockouts can also be made in a box using a knockout tool or hole saw, provided the quantity of surface area removed from the box does not exceed the code requirements. Devices are generally designed to be flush with the surface of the finished wall, floor, ceiling, or furniture in which the box was installed, but because there are so many variations in surface finish depth, an extension ring or "trim ring" may be required to extend the

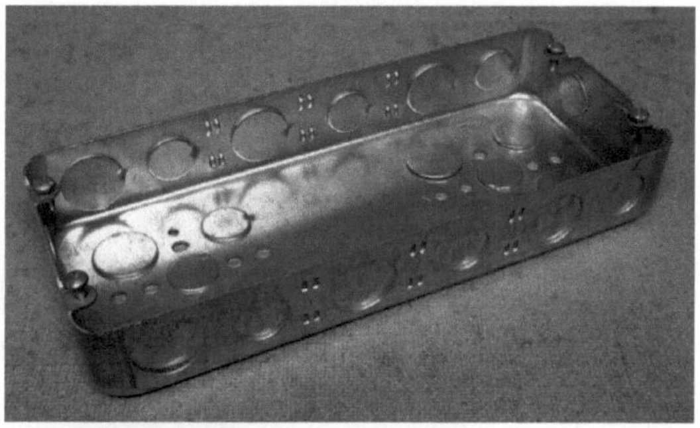

**FIGURE 12.18** Metal box.

**FIGURE 12.19** Single-gang plastic box with nails.

surface of the device outward to the finished surface. The device plate (Fig. 12.20) or device cover is then installed to cover the remaining gap between the device and box and ensure all exposed electrical components are safely covered.

## Junction Boxes

Unlike Device Boxes, Junction Boxes (Fig. 12.21) do not house a device, but instead are used as a junction point during the pulling of wire or to house a splice of wires. Junction boxes can range in size from a small, 4 × 4 in. box to a large, enclosure housing multiple feeders or utility cables.

Junction boxes are sized according to the quantity and type of conduit, cables or wires entering and exiting the box, and can be manufactured of multiple materials including metal, plastic, concrete, or a combination of these materials. Because a junction box is used as a pulling transition point or to house a splice, the junction box interior must be accessible. Therefore, junction boxes have removable covers which are screwed or snapped into place. A junction box cover can be gasketed for moisture or vapor protection and is made of the same material as the box itself.

**FIGURE 12.20** Finished single-gang receptacle with plate.

**FIGURE 12.21** Various junction boxes.

## Box Supports

Not all boxes have integral supports or can be supported directly to an adjacent structural member (Fig. 12.22). In some cases, supports must span between structural members such as studs or be attached to a structural member such as a beam or concrete deck (Fig. 12.23). The support of the box can vary by application, but must consider all associated loads such as conduit, cable, or wiring

**FIGURE 12.22** Metal stud mounted box.

**FIGURE 12.23** Junction box hung from concrete composite metal deck.

entering and exiting the box. Critical systems may also require additional bracing for seismic or gravity loads to ensure continuity of an emergency or critical power system.

## CONDUIT BODIES

In a conduit system, a junction box is not always required when the sole purpose of the junction point is to act as a transition point for a wire pull. Transitions are often necessary when wire runs are too long to easily pull the wire or cable and would place too large a tension on the wire during the act of pulling. Transition points can also be required when the degrees of bend in a conduit system exceed the code maximum.

**FIGURE 12.24** Conduit bodies with covers.

In these cases, a conduit body can be used in place of a junction box. A conduit body is manufactured of the same material as the conduit itself and is designed to connect directly to the conduit segments. The conduit body has a larger interior area than the conduit entering and exiting it and has a removable cover to allow wire to be accessed during the pulling operation.

Conduit bodies (Fig. 12.24) are manufactured with different configurations of conduit entering/exiting and are named based on the conduit direction and configuration. The following are the most common types of conduit bodies used in electrical installations:

- Type LB (conduit entering and exiting perpendicular at 90 degrees but on the same plane)
- Type LL (conduit entering and exiting perpendicular but on planes 90 degrees to the right)
- Type LR (conduit entering and exiting perpendicular but on planes 90 degrees to the left)
- Type C (conduit entering and existing on the same plane)
- Type T (two conduits entering on one plane and one conduit at 90 degrees from that plane)
- Type X (four conduits entering at 90-degree intervals from each other)

## ELECTRICAL DEVICES

Electrical devices, in the most simplistic terms, are the interface between the end user and the electrical system. The electrical devices provide a common interface between electrical fixtures, appliances, tools, and equipment, and the electrical distribution system. These devices can be separated into two major categories: receptacles and control devices.

### Receptacles

Receptacles (Fig. 12.20) or "outlets," as they are commonly called, allow for the insertion of a male "plug" into the female receiver.

There are six common types of electrical receptacles in the United States:

1. **Standard (15 A in residential, 20 A in commercial).** A standard receptacle for residential use has an ampacity of 15 A and 20 A in commercial construction. These are the most common types of

receptacles found in electrical construction. Receptacles in the United States have three prong openings: hot, neutral, and ground. This is commonly called a "grounded" or "grounding" receptacle because it allows the grounding wire to connect to the third prong in the outlet. A 20 A receptacle can be differentiated from a 15 A receptacle visually because it has a horizontal notch on the left prong opening.

2. **GFCI (Ground Fault Circuit Interrupter).** A GFCI receptacle is a short-circuit protection device. It is designed to terminate power when it detects a short circuit or ground fault on the circuit. Electricity is designed to flow from the hot prong, through the device plugged into the receptacle and back through the neutral prong. If the GFCI receptacle senses power flowing anywhere else in the system (e.g., to ground), it shuts the power off. GFCI receptacles are a code requirement of any area in a wet location, such as a bathroom or exterior, and provide added protection for the user.
3. **AFCI (Arc Fault Circuit Interrupter).** AFCI receptacles are designed to prevent electricity from "jumping" from one wire to another, or "arcing." An electrical arc can start a fire in adjacent materials without tripping a standard breaker, so AFCI receptacles are designed as a front-line fire protection device to stop electrical fires before they start. The United States requires both AFCI receptacles and AFCI breakers in all bedrooms of homes built after 1999 or homes renovated after 1999.
4. **USB (Universal Serial Bus).** USB receptacles provide a power interface to electronic devices that have Universal Serial Bus connectors. The devices can be integral to a standard receptacle or can stand alone.
5. **Smart.** Smart receptacles are either Bluetooth, RF, or Wi-Fi equipped and are designed to interface with other "smart" components in the electrical system or home. "Smart" generally refers to the connectivity of the device and its ability to communicate with, and be controlled by, other devices such as a smartphone, tablet, or computer.
6. **Switched.** Switched receptacles are controlled by a switch integral to the device (e.g., a switch and receptacle in the same housing) or a receptacle controlled by a switch remotely, via wired switch leg. These receptacles can be useful when a device requires a receptacle but it is not desirable to keep the device powered all the time.

The aforementioned receptacles are the most common in the United States but several other variations exist. Receptacles are manufactured in accordance with standards set by the National Electrical Manufacturers Association (NEMA) and are defined by their current ratings, voltage, and blade configuration and width. NEMA connectors are named with a numeral, hyphen, and numeral with letters preceding and following the numeric code (e.g., NEMA 5-15 or NEMA 1-15R). The nomenclature has the following meanings:

- Preceding Letter: Denotes Straight-Blade or Locking. This means the "prongs" inserted into the receptacle can either be easily slid in and out of the receptacle or will lock into place or unlock with a twisting motion. An "L" prefix, denotes curved-blade prongs, with twist-locking connectors. Receptacles without an "L" prefix are straight-blade, nonlocking.
- First Numeral: Denotes the particular configuration of the receptacle. This includes quantity of current-carrying terminals or "poles" and connected wires, voltage, and if it is single phase or three phase.
- Numeral Following Hyphen: Denotes the current rating for the device, in amperes.

Table 12.2 identifies the different types of receptacles manufactured under the NEMA standards.

## NEMA Receptacle Configurations

Receptacles are installed within the device boxes noted in the prior section and covered so no electrical components are exposed. Receptacles can be "ganged," in a multi-gang box, which means the multiple receptacles can be installed in a single box and grouped together with a single faceplate.

**TABLE 12.2** Receptacles Manufactured Under the NEMA Standards

| | | 15 AMPERE | | 20 AMPERE | | 30 AMPERE | | 50 AMPERE | | 60 AMPERE | |
|---|---|---|---|---|---|---|---|---|---|---|---|
| VOLTAGE | NEMA | Receptacle | Plug | Receptacle | Plug | Receptacle | Plug | Receptacle | Plug | Receptacle | Plug |
| 125V | 1 | 1-15R | 1-15P | | | | | | | | |
| 250V | 2 | | 2-15P | 2-20R | 2-20P | 2-30R | 2-30P | | | | |
| 125V | 5 | 5-15R | 5-15P | 5-20R | 5-20P | 5-30R | 5-30P | 5-50R | 5-50P | | |
| 250V | 6 | 6-15R | 6-15P | 6-20R | 6-20P | 6-30R | 6-30P | 6-50R | 6-50P | | |
| 277V, A.C. | 7 | 7-15R | 7-15P | 7-20R | 7-20P | 7-30R | 7-30P | 7-50R | 7-50P | | |
| 125/250V | 10 | | | 10-20R | 10-20P | 10-30R | 10-30P | 10-50R | 10-50P | | |
| 3∅250V | 11 | 11-15R | 11-15P | 11-20R | 11-20P | 11-30R | 11-30P | 11-50R | 11-50P | | |
| 125/250V | 14 | 14-15R | 14-15P | 14-20R | 14-20P | 14-30R | 14-30P | 14-50R | 14-50P | 14-60R | 14-60P |
| 3∅250V | 15 | 15-15R | 15-15P | 15-20R | 15-20P | 15-30R | 15-30P | 15-50R | 15-50P | 15-60R | 15-60P |
| 3∅Y 120/208V | 18 | 18-15R | 18-15P | 18-20R | 18-20P | 18-30R | 18-30P | 18-50R | 18-50P | 18-60R | 18-60P |

Receptacles are manufactured in two basic styles (standard or Decora) and are available in multiple colors as are the receptacle plates.

## Control Devices

Control devices interrupt electricity between points so users can identify and control where electricity flows within the system. The most common type of control device is the simple "toggle switch." A toggle switch (Fig. 12.25) is manually operated and uses a rocking mechanism to open and close the contacts within the switch, interrupting the flow of electricity to the devices when open.

Toggle switches are installed between the power source and the controlled device. However, there may be a need for multiple switches in a configuration such as a room with multiple entrances or exits

**FIGURE 12.25** Two single-pole standard ivory toggle switches installed in a four-gang box.

or a stairwell with upper level and lower level switches. In these cases, multiple switches can be installed in a "three-way" (two switch locations) or "four-way" (three or more switch locations) configuration. The switches used in these configurations have more terminals on which wires can terminate to accommodate for the additional "legs" of wiring that are required to each switch.

Switches can also have added functionality beyond simply opening and closing the contacts via toggle. The below list identifies several types of available switching technology. Each of these items is, at its core, a switch but the functionality noted is how the switch is actuated.

- **Dimming.** Modern dimmers turn power on and off rapidly to reduce the power flowing to the device, thereby reducing the output of the lights.
- **Radio Frequency Receivers and Transmitters for Wireless Controls and Switching.** Sends and receives radio frequency signals which are processed to provide actuation direction to the switch.
- **Motion Sensing Technology.** Motion sensing switches (sometimes called occupancy sensors) use infrared, ultrasonic, or microwave technology to sense motion and signal the switch to open or close.
- **Fan Controls.** Contains multiple configurations of switches within the single unit to control the fan speed and lighting.
- **Timers.** Uses a time clock or mechanical mechanism to actuate the switch.

Like receptacles, switches are installed within the device boxes noted in the prior section and covered so no electrical components are exposed. Switches can also be "ganged," in a multi-gang box, which means that multiple switches can be installed in a single box and grouped together with a single faceplate. Switches are manufactured in two basic styles (standard or Decora) and are available in multiple colors as are the switch plates.

## *FIRESTOPPING*

Firestopping is a generic term used to maintain the fire protection rating for a component of construction through which electrical utilities must traverse such as conduit through walls, cable tray through walls, and electrical boxes installed in fire-rated walls.

Firestopping is an engineered system and has three primary criteria:

1. "F" Rating—Expressed in hours, the amount of time a barrier can withstand the effects of flame.
2. "T" Rating—Expressed in hours, the amount of time a barrier can withstand the effects of temperature transfer.
3. "L" Rating—Expressed in cubic feet, the amount of air or cold smoke that can leak through a penetration.

Several types of systems exist to ensure electrical components penetrating building components have one or more of the aforementioned rating criteria. The following are the most commonly used systems.

## Moldable "Putty" Pads

Moldable "putty" pads are the system of choice when applying firestopping to electrical and low-voltage devices because they are inexpensive and easy to install. A moldable "putty" pad (Fig. 12.26) is a single-piece intumescent flexible pad installed by hand and, like clay, can be molded around an object such as an electrical device box. The following considerations must be made during the installation process:

The moldable putty material must remain as a single piece when applied to a box or device. Simply put, voids cannot be filled by placing smaller pieces in crevices or gaps.

- The moldable putty material must be placed flush with the face of the device so there is no annular space between the face of the material and the rated wall.
- The moldable putty material must maintain the same rating as the wall assembly (all rating requirements).

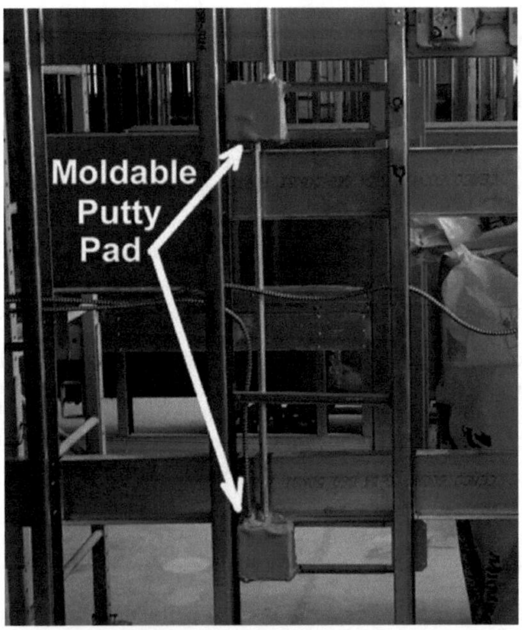


**FIGURE 12.26** Moldable putty pad on electrical device box attached to a metal stud.

- The moldable material must completely cover any entry points into the box (e.g., conduit, cable, etc.) with no visible annular space.
- It is a recommended practice to wait as long as possible to install the putty pads as they have a tendency to lose their adhesive qualities the longer they are exposed to the elements.

## Thru-Wall Penetrations

Thru-Wall penetrations for conduit and cable are typically made by utilizing sleeves (Fig. 12.27), two-part penetration systems, fire caulking, "pillow" systems, or a combination of these methods. Sleeves are small segments of conduit cut to slightly longer than the width of the wall to be penetrated. Electrical conduit or cable can then be installed inside the sleeve, as it is oversized from the material to be installed, and then an intumescent firestopping material is installed around the contents of the sleeve to eliminate any annular space. Sleeves are commonly used because they are the most simplistic system and very cost effective; however, once the contents are installed and fire caulking material is placed, there can be no modifications to the sleeve system.

To allow for future modifications in installed systems, a two-part firestop system can be used to achieve the same basic function as a sleeve. In these systems, one part is installed on either side of the penetration and the parts attach, via clamps or threads, to seal the connection. These systems are considered re-enterable (Fig. 12.28), meaning the contents can be changed over time, making this a

**FIGURE 12.27** Sleeve system with low-voltage cabling and fire caulk.

**FIGURE 12.28** Two-part re-enterable fire-rated system with low-voltage cabling on cable tray.

**FIGURE 12.29** Conduit penetrations through a masonry block wall with a pillow system.

very useful system during construction and after since they can accommodate changes in contents without significant rework.

"Pillow" systems are named because the devices used to fill the annular space surrounding the penetration resembles small pillows (Fig. 12.29). These devices are composed of similar material to the moldable putty pads and fire caulking, but also contain composite materials that, when heated, expand the pillow to create a firmer seal within the penetration. Pillow systems can be used for virtually any type of raceway penetration, but unlike sleeves and fire caulk, pillows can be removed, one by one, to accommodate for future installations.

Installation considerations for horizontal thru-wall penetrations include:

- Offsets are required when routing wall-mounted conduit through a rated wall to ensure the entire circumference of the conduit penetration can be caulked.
- Annular space (e.g., the space surrounding the penetration) must be within the size tolerance of the approved UL firestopping system for the application.
- Attachment hardware penetrations may require firestopping measures depending on the size of the hardware used and how many penetrations exist in the rated wall.
- Penetrations through smoke zones may also require adherence to an "L-Rating" requirement which may require the augmentation of the existing system or a separate firestopping system entirely.
- "Ganged Thru-Wall" penetrations (e.g., multiconduit) must have a total penetration area of less than that which is shown in the UL system.

## Thru-Floor Firestopping

In the majority of cases, the thru-floor penetrations occur in electrical rooms, tele-com rooms, and mechanical spaces such as shafts. These areas are often designed with a rated envelope for both fire and smoke to ensure their ongoing functionality of building systems during a fire event.

Consequently, the methodologies used to firestop penetrations in these envelopes must be predicated on a T-rating equal to an F-rating. The F-rating is a "Flame" rating which is expressed in hours and the number indicates the specific length of time a barrier can withstand fire before being consumed or before permitting the passage of flame through the opening. The T-rating is the "Thermal" rating which is expressed in hours and the number indicates the length of time the temperature on the nonfire side of the penetration does not exceed 325°F (163°C) above the ambient temperature. This ensures that the temperature on the side of the rated barrier away from the flame does not reach the flash point of any materials on that side of the rated barrier.

In the case of a thru-floor penetration (e.g., structural concrete, composite concrete/metal deck, or others), these ratings are required to be equal. This is accomplished using a two-part system (Fig. 12.30).

**FIGURE 12.30** Two-part vertical fire-rated system with EMT conduit risers.

In this system, conduit sleeves are placed in the concrete placement where the concrete surrounds the conduit sleeve at the full circumference to the depth of the concrete slab, and fire caulking material (intumescent material) is placed around the void surrounding the conduit sleeve.

## NORMAL POWER DISTRIBUTION SYSTEMS

The Normal Power Distribution System refers to the portion of the electrical system that transmits electricity, produced by a generation source such as a utility provider, to the end user for consumption. Power distribution systems are considered "Normal" when they have no redundancy [such as a generator or uninterruptible power supply (UPS)] and are subject to the operation of the generation source.

### Primary and Secondary Power Distribution

Distribution, in the context of the electrical system, is analogous to a road in a highway system. Like a road, the distribution portion of an electrical system is designed to move a high volume of electricity with relatively few interruptions from one location to another. While some users do produce their own power via private generation sources, most of the electricity is produced and transmitted by utility companies. Utilities transmit their electricity from the generation source to consumers over their "Primary Distribution Network." The point at which the electricity is received from the electric utility provider is referred to as the Primary Point of Connection (PPoC). Following the primary point of connection, the power system is referred to as the "Secondary Distribution Network." Extending the analogy of highways, the PPoC would be the off-ramp of the freeway, while the Secondary Distribution System is the local streets of the destination.

To minimize the size of infrastructure required to transmit electricity over large distances, utility providers utilize high-voltage (greater than 600 V) means of transmission in their primary distribution networks. However, because most electrical systems operate at voltages lower than 600 V (120 V/208 V, 120 V/240 V, or 277 V/480 V), a transformer is required to step-down the voltage from the high-voltage utility distribution to the common voltage normal distribution system. This transformer is referred to as the "Secondary" transformer since it is the point at which the secondary distribution network begins. The secondary transformer is often owned and maintained by the utility company and is generally the point at which the ownership and liability of the power distribution system transfers from the utility to the end user/property owner. Some exceptions to this do occur when the demand for power is so large for a facility that it is necessary to have high-voltage distribution in its normal power system. In these cases, the utility essentially extends their primary network to a user's substation, and the high-voltage substation becomes the beginning of the secondary network. The user then provides their own step-down transformers to provide voltages under 600 V to their system.

The secondary network consists of a main electrical service, distribution boards, transformers, motor control centers, panel boards, and/or load centers. The main electrical service is usually the point at which the electricity consumption is measured via meter or current transformers, both measuring the electrical current transferring from utility to user. The main electrical service also includes the first point of overcurrent protection and means of disconnect between the primary and secondary networks, often in the form of a main circuit breaker. The main circuit breaker protects both the primary and secondary networks from the failure of the other, and also provides a means by which all electricity can be shut off at the beginning of the secondary network.

Commercial secondary distribution networks usually operate at a voltage of 480 V from the main electrical service to the panelboard or load center. This means the switchboards, distribution boards, and motor control centers operate at 480 V and step-down transformers are then used between the equipment and a branch panelboard or load center so power at lower voltages, such as 120 V/208 V, is used for branch devices.

An electrical distribution system is depicted in a "single line" or "one line" drawing, which uses a single line to depict a path for power flow between entities in the system. Figure 12.31 is a sample of a normal power single-line drawing for a commercial electrical service and normal power distribution system.

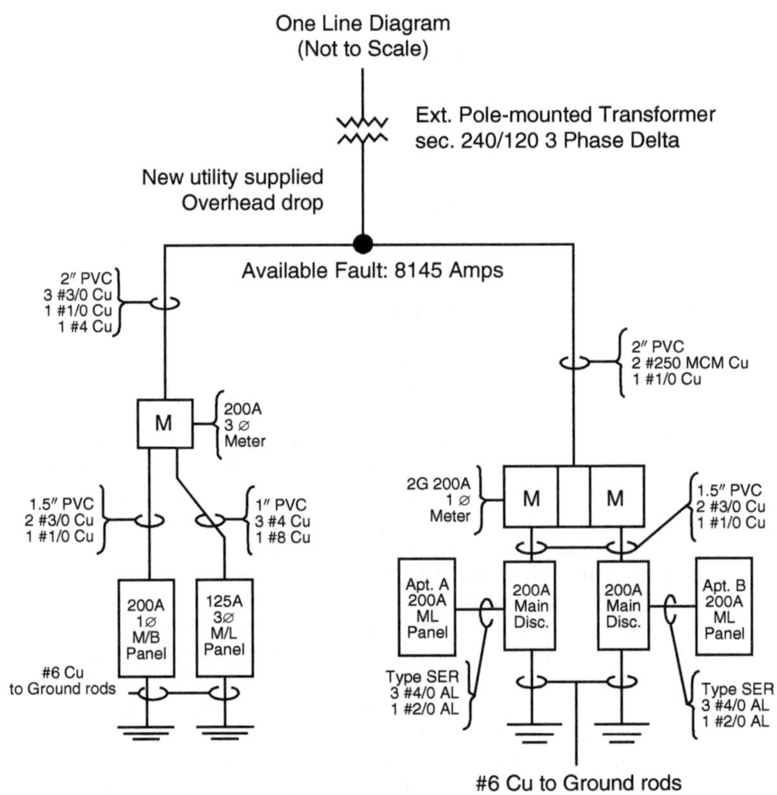


**FIGURE 12.31** Sample of a normal power single-line drawing.

## Feeders

All circuit conductors between the load side of the service equipment and the line side of the final branch circuit overcurrent device are feeder conductors. The definition of a feeder also includes the conductors from the source of a separately derived system or other nonutility power supply source and the final branch circuit overcurrent device.

A typical electrical system may have many types of feeders supplying many different types of loads. Feeders supply a combination of continuous and noncontinuous loads, motor feeders, outside feeders, or feeders to separate buildings, and, in many cases, there may be feeders from more than one voltage system on the same premises. DC system feeders may also be present.

Feeders are typically the largest conductors in an electrical system because they carry the largest volume of electrical current across the normal power distribution systems.

For clarity, the conductors between the utility service point and the service disconnecting means are service conductors not feeder conductors, although they carry the entire volume of the electricity needed for the electrical system. Special service conductor rules apply because these conductors do not have short-circuit or ground-fault protection other than what is provided on the primary side of the utility transformer.

| Service conductors are not feeders. |
|---|

In addition, "Branch" circuits are not feeders. A branch circuit is defined as *the circuit conductors between the final overcurrent device protecting the circuit and the outlet(s)*. The conductors on the load side of the final branch circuit overcurrent device are branch circuit conductors, not feeder conductors, no matter how large the circuit rating.

Feeder conductors are sized based on the voltage and load they are designed to carry and distance over which such a load is to be carried. Feeder conductors are sized according to the American Wire Gauge (AWG) and measured in Thousand Circular Mils (kcmil). Because feeder conductors can be manufactured in both copper and aluminum, the material type affects the ampacity of the conductors. Table 12.3 identifies the ampacity rating for feeder conductors in both copper and aluminum.

Applicable codes define the allowable insulation types for each feeder conductor, based on application, and this also has a minor effect on the load rating of the feeder conductor. However, the primary drivers behind the selection of feeder size are ambient temperature and length of the feeder.

## Bus Duct

Bus Duct (also referred to as Busway) is another type of feeder system that eliminates the need for conduit and conductors when carrying larger amperage loads. A bus duct (Fig. 12.32) is a sheet metal enclosure (duct) which houses busbars manufactured of aluminum or copper which carry the electrical load in the same way wire conductors do.

Bus ducts have many advantages over traditional conduit and cable feeder systems including:

- **Compact Design.** Busbars and the outer duct enclosure are both pressed flat and consume considerably less space than a conduit and wire system to carry the same electrical load.
- **Reliability.** Bus Duct systems are manufactured in segments and mechanically connected together; thus, there are fewer points of potential failure. In conduit and wire systems, there are multiple junction points through which the wire conductors must be pulled and that can often cause minor damage to insulation or conductors which can degrade performance over time.
- **Heat Absorption.** Because the design is so compact and the busbars are enclosed by a metal casing, the busbars can absorb heat more efficiently and the heat of transmission is distributed more quickly than in a conduit and wire system.

**TABLE 12.3** Ampacity Rating for Feeder Conductors

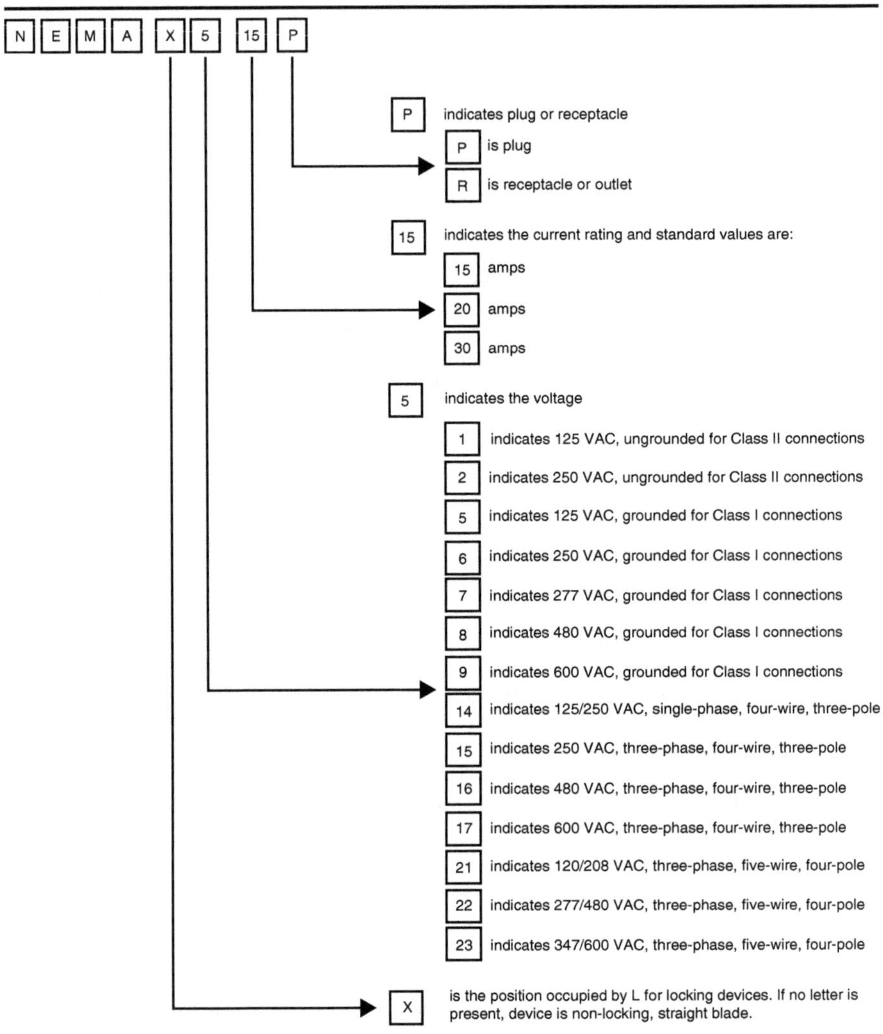


- **Expansion and Flexibility.** Bus duct systems are effectively modular and can, therefore, be modified easily after installation to accommodate changes in building construction or future needs, while conduit systems must be replaced. In addition, some bus duct systems also offer integrated overcurrent protection devices or "bus plugs." These allow for direction connection and disconnection of equipment at any time such as in a factory setting where manufacturing equipment may be replaced periodically.
- **Cost.** Bus duct systems are very cost effective because they are easier to install and support than a traditional conduit system and often weigh considerably less.

Because of these advantages, bus duct systems are the preferred option for large feeder loads when the feeders traverse a structure vertically or horizontally above-ground (e.g., not buried or encased in concrete as conduit systems often are).

**FIGURE 12.32** Bus duct supported from a concrete deck.

## *BRANCH POWER*

Continuing the analogy of electrical elements to roadway systems, Branch Power is analogous to a local street in a roadway system. When the distribution portion of the electrical system takes the power to the "exit" of the highway, the panelboard or load center can be considered the traffic signal at the exit ramp. These control the power at that junction before distributing it to the local streets or "Branch Power" portion of the system.

Branch power is the segment of the electrical system used to distribute power to the consuming device or appliance and does so at the required voltage. This requires the power to be transformed down from the distribution voltage to the line voltage (usually 120 V). This is accomplished with a transformer (Fig. 12.33). Transformers take power at one voltage and convert it to another voltage by carrying the electricity through coils. The transformer coils are conductive materials, such as copper or aluminum. The coils use the electrical current to produce a varying magnetic flux which induces an electromotive force across a second coil wound around a common core. Electrical energy is transferred between the two coils, without a conductive connection between them, which creates an induced voltage effect and alters the voltage of the output current to the desired voltage. Transformers are sized in Kilo-Volt-Amps, which is the equivalent of a kilo Watt and are specific to voltage and phase. Transformers can be "Step-Down," converting the voltage from a higher to a lower voltage, or "Step-Up," converting a lower voltage to a higher voltage (this may also be referred to as "buck/boost").

Transformers follow a distribution panel or breaker in the flow of the system and will precede the branch panelboard or load center.

### Panelboards and Load Centers

Panelboards and Load Centers are functionally the same; they both provide overcurrent protection and control of branch circuits. According to Underwriters Laboratories (UL) and the National Electric Code (NEC) there is no real distinction between panelboards and load centers. However,

**FIGURE 12.33** 15 KVA transformer.

panelboards are generally used in commercial and industrial applications, whereas load centers are used in residential applications. Some manufacturers will distinguish the two by the types of breakers each can receive where panelboards are manufactured with bolt-on breakers and load centers with snap-in breakers.

Panelboards and load centers consist of a back box (Fig. 12.34) or "can," an interior, and a cover. The back box is installed with the framing members during the electrical rough-in phase and all branch infrastructure is terminated in the back box. Once wire is installed (in conduit systems) or cables are installed (in cable systems), the branch wire is terminated on the assigned circuit breaker.

Most panelboards contain a main circuit breaker which provides the overcurrent protection for the panel and all branch circuits. This breaker provides protection against electrical fire or damage caused by a fault in the system. In essence, a circuit breaker operates on electromagnetism energized by the power flowing through the circuit it is protecting. When the current in the circuit exceeds the level tolerated by the breaker, the breaker "trips," breaks the circuit, and terminates the flow of power in the circuit.

The main circuit breaker is sized according to the total demand load on the panel. To determine the size of the main circuit breaker, the following information is necessary:

- Total demand load, in kilowatts or watts
- Maximum voltage of the system
- Number of phases in the system

Then the voltage is multiplied by the square root of the number of phases and finally the NEC safety factor of 25% is applied but multiplying the previous value by 1.25. The result of these calculations is divided by the load to the panelboard to determine the main breaker size in amperes.

A "Safe Load," as defined by the NEC, operates at 80% of the circuit breaker capacity used to protecting it. This applies to "continuous" loads, those placed constantly on a breaker. Noncontinuous loads or loads drawn for short-intervals can consume 100% of the breaker rating protecting the circuit. For example, a breaker sized at 50 A will protect a continuous load of 40 A or a noncontinuous load of 50 A.

Some panelboards and load centers do not contain a main circuit breaker and are termed "Main Lug Only." This means they contain conductive lugs, on which the feeder terminates, but the panel is

**FIGURE 12.34** Panelboard back box and interior with main and branch breakers.

not controlled by a main circuit breaker. This is common for panels used as distribution panels where the overcurrent protection is provided upstream via a circuit breaker.

## Branch Circuit Breakers

Branch circuit breakers operate in the same fashion as the main circuit breaker of a panelboard or load center to protect the branch circuits. But branch circuit breakers are sized at smaller intervals to protect smaller, individual circuits. Branch circuit breakers range in size from 15 to 100 A and are sized according to the connected load (loads must be 80% or less of the breaker size) but must also match the ampacity of the connected conductors.

Branch circuit breakers can also provide additional protection, such as Arc-Fault or Ground-Fault, where desirable or mandated by code (such as residential bedrooms or wet locations, respectively). In these cases, the breaker will be marked with the associated protection nomenclature and visually distinct from a standard breaker.

## Branch Circuits

Branch circuits can be installed in a conduit or cable system and will consist of wiring rated for the ampacity of the devices intended to be used on the circuit. Branch circuits generally serve three types of loads:

1. **Lighting Branch Circuit.** A grouping of light fixtures designed to be controlled and protected together.
2. **Convenience Power Circuit.** A grouping of electrical receptacles designed to serve the same function or area such as an office, room in a home, or area in a warehouse.

**FIGURE 12.35** MC cable branch circuits terminated in a panelboard.

3. **Dedicated Power Circuit.** A circuit designed to serve a specific branch device load and be isolated or protected individually such as a piece of office equipment, appliance, or machinery.

Branch circuits terminate in a panelboard (Fig. 12.35) or load center located relatively close to the circuit so the conductors can remain small and maintenance operations are conveniently close to the areas served by the circuit.

## *LIGHTING*

Lighting systems provide a controlled, artificial source of illumination and have a significant impact on the functionality and feel of an indoor or outdoor space. Lighting systems generally reside on the branch power portion of the electrical system and consist of two main components—the luminaire and the control.

The luminaire, or light fixture, is essentially the housing inside of which a light source resides. Light fixtures can produce light from multiple types of light sources, and are distinguished by their light source. The most common light sources in modern light fixtures are:

- **Light-Emitting Diode (LED).** Sometimes called Solid-State-Lighting, emits light when a current passes through a semiconductor, releasing energy and causing the emission of photons to produce light.
- **Fluorescent.** Fluorescent lamps use glass tubes containing mercury vapor which, when ionized, emit photons producing light at ultraviolet frequencies. The UV light is then converted into visible light when it passes through a phosphor coating on the inside of the glass tube.

Lighting adds warmth to an entrance.

- **High-Intensity Discharge (HID).** HID lamps produce light when an electric arc is created between two electrodes (typically made of tungsten) housed inside a transparent or translucent tube. The tube is filled with gas which enables the initial strike of the arc and, when the gas is heated and mixed with a metallic admixture, evaporates and produces light.
- **Incandescent.** Incandescent lamps use a wire filament that, when heated to a high temperature via the flow of electrical current through it, glows to produce visible light. The filament exists inside a glass or quartz housing. The housing can be filled with an inert gas or have a vacuum. Either will protect the filament from oxidation.

Each of the aforementioned light sources varies in cost of production, energy efficiency, environmental impact, and cost of installation, making the desired application of the light source the key driver in its selection.

Light sources provide different levels of intensity, measured in candlepower or "Candela." Candela is the base measurement of luminous intensity and measures how bright the light source is and at what distance the light source can be seen from any point in a single direction. In the most simplistic terms, a single Candela is the equivalent to the light of a single candle and a candle can be seen, with the same intensity at the same distance in any direction (without obstruction).

In addition to producing differing levels of intensity, light sources produce differing levels of light output, measured in "Lumens." Lumens (lm) are now the most common measurement for lighting systems and are a unit of measure for luminous flux. Luminous flux is the total amount of visible light the naked eye can see and differs from candlepower, in that it can be measured when obscured.

When applying the intensity and output to building construction, it is important to consider the effects of light on the surfaces of the building. Light sources produce differing levels of "Illuminance," measured in "Lux" (lx), which is the amount of light on a surface per unit area. For example, a light fixture producing 100 Lumens and positioned to shine on 1 square meter of surface will produce 100 Lux on the surface. However, if the light source were raised and now shines on 10 square meters of surface, it would produce 10 Lux on the total surface.

Most lighting systems are designed and commissioned in terms of Lumens, but sometimes the other units of measure are considered or presented when considering types of lighting sources, architectural features, and desired look and feel in a space.

Lastly, light sources also produce different color temperatures which are measured in degrees of Kelvin (K). Color temperature is a method of describing the appearance of the light provided by a light source and typically ranges from 1,000 to 10,000 K where 1,000 is the "softest" or "warmest" light and 10,000 is the "harshest" or "coldest" light. The color temperature of a light source is assigned using the basis of correlated color temperature (CCT) which is derived from the process of heating objects. For example, if a piece of metal is heated to a high temperature, the metal appears to glow and, depending on the temperature in Kelvin to which it is heated, the object will appear to exhibit different colors such as yellow, orange, blue, or white. The CCT system in lighting is meant to mimic the Kelvin temperature of the metal object.

## Functions

Lighting is used to provide three key functions:

1. **Ambient Lighting.** Ambient lighting is the general lighting produced in an area that permeates through the entire space. This can also be referred to as "general lighting" and is usually nonspecific in its use such as overhead lighting in an office space, corridor, or other general use space.
2. **Task Lighting.** Unlike ambient lighting, task lighting is specific to a particular function or activity. Task lighting is localized and can usually be provided from multiple sources allowing users to adjust the brightness or angle of the lighting. Examples are reading lamps adjacent to a bed or chair or a boom light above a medical procedure chair or table.

3. **Accent Lighting.** Accent lighting is used to draw attention to a specific object or area, making it the focal point of a space. Accent lighting supplements the ambient lighting and is usually used for a decorative or architectural purpose such as illuminating a picture in a museum or a sign or tree in a landscape.

A complete lighting system will incorporate elements of each of these functions and utilize different light sources, output levels, and colors to produce the desired functionality for the space. However, achieving this desired outcome in a building requires different methods of light fixture placement.

Unless a light fixture is a temporary or movable fixture, such as a table lamp or floor lamp, the installation is generally permanent so the style of light fixture is an equally important consideration in a lighting system. Therefore, light fixtures are produced in many forms to accommodate varying light sources in virtually any type of building condition. The most common fixture types are:

- **Recessed Troffer.** These are rectangular in shape, installed flush to the finished surface of a ceiling, and usually installed in a ceiling-grid system to provide ambient lighting.
- **Recessed Can Downlight.** These are circular in shape and installed flush to the finished surface of a drywall or ceiling-grid system to provide ambient or accent lighting.
- **Recessed Cove Light.** These are usually linear light fixtures. They are installed behind a built-up-cove to conceal the light fixture itself, but illuminate the cove and adjacent wall or ceiling to provide accent lighting.
- **Surface Sconce.** These are installed on the surface of an interior or exterior wall and are usually decorative in design. Sconces can be up-lights or down-lights (direction of the light) and can be used to provide ambient or accent lighting.
- **Linear Surface Lighting.** These are long, linear fixtures used to provide ambient or accent lighting in areas with long, steady surfaces such as corridors and auditorium walls.
- **Surface Pendant.** These are fixtures with the light source installed on the end of a large pendant used to extend the light source from its mounting location. These are typically used in areas with high ceilings and can be decorative or functional in design. Pendants can provide ambient, accent, or task lighting.
- **Chandelier.** These are decorative in nature and, like a pendant fixture, are hung from a pendant, cable or other extension apparatus to extend the light source down from its attachment location. Chandeliers are used for ambient or accent lighting.
- **Track Lighting.** These are channels of plastic or metal ("tracks) into which surface light fixtures are installed. The channel acts as both the raceway and mounting surface for the light source. It allows for the surface mounting of the channel and the easy directional adjustment of the light source. Track lighting is used for ambient, accent, and task lighting.
- **Under-Cabinet Light.** These are generally concealed fixtures installed under or within millwork to provide accent or task lighting to the millwork or counter surface.
- **High-Bay or Low-Bay Fixtures.** These are high-intensity fixtures used to illuminate large spaces such as warehouses or industrial/manufacturing floors. They are typically attached to high ceilings and provide ambient and task lighting.
- **Strip Lights or Industrial Strip Lights.** These are surface-mounted linear fixtures in incremental footage (usually 4-foot increments) strips and used in manufacturing or industrial areas. Strip lights can be hung from pendants or mounted directed to a surface (wall or ceiling) and provide ambient and task lighting. Strip lights can also be used for accent lighting in certain applications such as signage or wayfinding.
- **Exterior Pole-Mounted Fixtures.** These are fixtures mounted to a vertical pole of varying heights and the light source is contained in a "head." Pole-mounted fixtures are used to provide ambient lighting in areas such as parking lots and roadways.
- **Exterior Bollards.** Exterior bollards are cylindrical tubes inside of which light sources are placed to provide architectural accent lighting or ambient lighting for exterior paths of travel.

- **Landscape Lighting.** Landscape lights can be manufactured as up-lights or down-lights and are used to illuminate focal points of landscaping or pathways (ambient) for access and egress.
- **Step-Lighting.** These are recessed light fixtures embedded into the rise or run of a step or staircase to provide ambient or accent lighting for the steps.
- **Exit Sign and Combination Exit/Emergency Light Fixtures.** The most ubiquitous light fixture in nonresidential construction. Exit signs provide directional wayfinding to the exit of a building and may also provide functional illumination in the form of emergency flood lights or "bugeyes." Exit signs and combination emergency exit fixtures can have their housing flush or surface mounted, and are "backlit" or "edge-lit," but almost always protrude from the surface of a ceiling or wall in some way.

All of the aforementioned light fixture types can come with virtually any of the most common light sources and in many combinations of intensity, output, and color. Lighting designers and/or electrical engineers consider each of these elements when determining what light fixtures should be used in any building.

Several other types of light fixtures exist for specialty applications such as theaters, stadiums, traffic signals, and healthcare but all operate on the same basic elements of light source, housing type, and mounting location, depending on the application.

## Controls

Lighting systems are all controlled in some fashion and the sophistication of the control needs will determine the method of control. The most common light fixture control is a simple toggle switch which turns power to the light fixture on or off through the switch. But, providing switches to every light fixture may not be practical in large facilities or in locations where people are not always available to control the lights. Several types of automatic lighting control systems exist to provide these functions in more efficient ways.

Time-controlled devices (sometimes called timeclocks) operate a light fixture or lighting circuits on a set schedule and turn lights on or off based on their programming. Sometime controlled devices have a "countdown" function and provide power to the lighting until the timer completes its countdown. Other time control devices operate within a range of days and hours and provide power to the lighting based on the scheduled program of days of the week and hours of the day.

Photoelectric devices act like a switch to control a light fixture or lighting circuits by measuring ambient light provided by the environment and actuating the switch depending on how much environmental light is present. The technology in these devices varies, but each variation operates on the same principle that artificial light fixtures are not required when the environment produces the programmed light levels, and artificial light fixtures are required when the environment does not produce the programmed light levels.

Motion sensing devices also act like a switch to control a light fixture or lighting circuit by detecting motion, via internal sensor, to determine the presence of an occupant of the space the lighting serves. This technology varies by device but senses the motion or occupancy and generally works by turning on the power when motion is detected and turning off the power after a specified period of no further motion detection.

Networked lighting control systems use a combination of each of the above components, networked to a computerized control panel, to ensure maximum efficiency of the system and to allow additional programming. These systems may integrate the lighting with other systems such as building management systems (BMS), access control and security systems, surveillance systems, fire alarm systems, or nurse call systems. They are usually networked using low-voltage wiring and may even provide power via ethernet (also called PoE) to newer LED-style light fixtures. The networking of the light fixtures and controls allows for constant alteration of the programming to adapt and evolve with the needs of the building occupants.

For example, a lighting control system networked with a security system and surveillance camera system may automatically turn on light fixtures in the area of an intrusion alarm to ensure the surveillance cameras have optimal lighting to record the event and authorities have optimal lighting to pursue the intruder. Similarly, a lighting control system networked with a building management system and fire alarm system may be programmed to illuminate the desired paths of travel away from the fire and notify the evacuees of the location.

Modern networked lighting control systems are virtually limitless in their capabilities, provided they are wired for connectivity or capable of wireless communication.

## EMERGENCY POWER SYSTEMS

The core difference between a normal power distribution system and an emergency power distribution system is that the emergency power distribution system has a secondary or redundant power source and a means of switching from the normal source to the emergency source. In most circumstances, the redundant source of power is a generator, but other apparatus may be used such as an uninterruptible power supply (UPS) or battery.

In most applications, the transfer of power source is automatic, utilizing an automatic transfer switch (ATS). The ATS utilizes sensors to constantly monitor the normal power source and determine if the designed normal power is flowing to the electrical service. In the event of a power loss, the sensors trigger the actuation of the ATS and activation of the emergency power source, and reroutes the power from the normal source to the emergency source. The design of the ATS makes it impossible for power to flow from both the normal and emergency source simultaneously, unless designed to do so in some special applications. During operation of the emergency system, the sensors in the ATS continue to monitor the normal power source to determine if flow of electricity has been restored and, when it senses restored power, the ATS will transfer the system back to the normal source and disengage the emergency source.

Alternatively, in a less complex system or one not requiring immediate emergency power, a manual transfer switch (MTS) can be used. An MTS requires the power to be rerouted manually, by actuation of the switching mechanism, and restored the same way. An MTS can be considerably less expensive for a situation where there is no compelling need to automatically transfer power.

The transfer switch can be "Open Transition" or "Closed Transition," which refers to the manner in which the power is transferred. In an open transition, or break-before-make, switch, the transfer switch breaks contact with one power source, before making contact with another. This is the most common type of transfer switch and is designed to prevent the back feed of power from the emergency source (generator) to the normal source (utility). While this is the safest method of power transfer, it inherently results in a small interruption of power. In a closed transition, or make-before-break, switch, the transfer switch makes parallel contact with both the normal and emergency power sources and does so to ensure there is no interruption of power, even for a fraction of a second. The closed transition switch type is used most often in systems where power cannot be interrupted, for even a fraction of a second, and no UPS is used.

Figure 12.36 depicts a single-line drawing in a basic emergency power system, where the emergency power source is a generator controlled by an ATS.

### Generators

Generators are the most common source of emergency power in an electrical system. Generators use mechanical energy, created by an engine, to force the movement of charges present in the wire of its windings through an electrical circuit. Like an automobile engine, a generator engine produces mechanical energy which is then converted to electrical output by an alternator.

Generator engines can be supplied by multiple types of fuel, the most common of which are gasoline, diesel, propane, and natural gas. Smaller engines typically operate on gasoline while larger

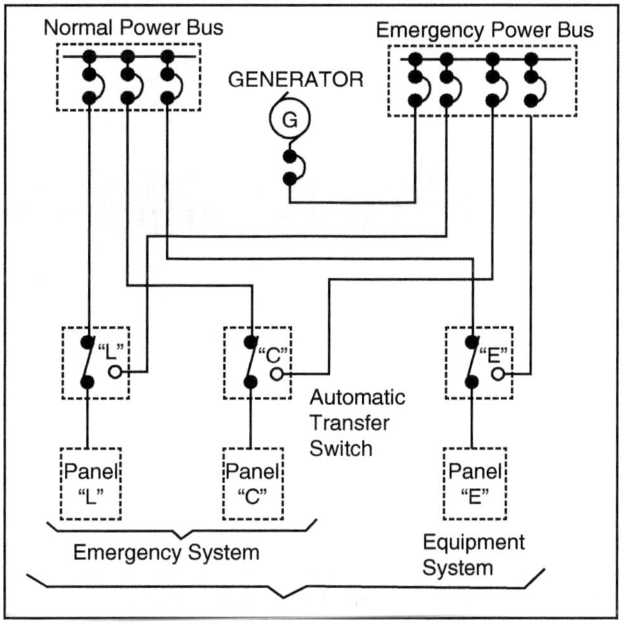


**FIGURE 12.36** A single-line drawing of a basic emergency power system.

engines operate on diesel, propane, or natural gas. The consumption of generator fuel varies by fuel type and connected load (how much power is actually drawn from the generator while running) and thus the exhaust systems can also vary by fuel type and load capacity. Exhaust systems are designed to reduce air pollutants created by the generator engine and can also vary by jurisdiction and location. The generator enclosure houses all of the components of the generator and also provides sound attenuation for the unit.

Generator output is measured in Kilowatts or Megawatts, and is manufactured in common increments. The following are the most common generator size ranges and applications:

- 5–15 kW (portable generator units for temporary power, construction work, residential standby power)
- 20–50 kW (fixed residential and small commercial units, large portable/towable applications)
- 75–300 kW (fixed commercial and industrial)
- 350–3,000 kW (fixed commercial, heavy industrial, healthcare, data, financial)

## Control and Paralleling Switchgear

In complex emergency systems, such as data centers, financial processing facilities, and healthcare facilities, it is sometimes necessary to ensure there are no power fluctuations of any kind and power remains static, regardless of source. In these cases, complex control switchgear can be used to sense and mitigate power fluctuations of any type and ensure clean power is provided consistently.

Control switchgear is a broad term for multiple components that can be added to the emergency power system to manage power quality. Control switchgear uses a series of sensors to measure fluctuations and variations that may precede a surge or outage. Relays in the system then send signals to various components designed to mitigate the predicted issue which may include routing power

**FIGURE 12.37** Three paralleled generators installed and serving a single load.

through a surge suppressor, engaging a UPS, engaging a generator, or other apparatus. In most critical power systems, the sensors operate in terms of milliseconds (a thousandth of a second) or deciseconds (one-tenth of a second) to engage the equipment and ensure minimal damage or power loss.

Control switchgear can also be used to manage the power use when operating a system from an emergency power source, or multiple emergency power sources. The most common application of this is the concept of parallel generation and load management.

Parallel generator is the utilization of two or more generators (Fig. 12.37) simultaneously to create a large electrical capacity and is done because generators are limited in capacity to their size engine. Consequently, multiple generators can be used in a system to achieve larger load capacity. When this is done, the generators are placed in "parallel," meaning they serve the same loads and the capacity of each is additive (e.g., (2) 2-mW generators result in 4 mW of capacity).

Load management, is the concept of adding or removing loads and throttling the generator output to accommodate the needs. In a load managed system, the control switchgear uses sensors, placed throughout the distribution system, to identify the demands on the system and send signals back to the controls to either increase or decrease capacity of the generator or increase or decrease the demand from the system. In essence, if the system is drawing more power than the generators are producing, the generators throttle up until their output meets the demand and vice versa. If the output of the generators is at a maximum, the concept of "load shedding" can be applied. Load shedding is an integral concept to the design of emergency power systems whereby all loads on the system are prioritized and the highest priority loads are given power the longest, while the lowest priority loads are only provided with emergency power if the system has the capacity to do so. This is all managed through electronic programming at the control switchgear and can be modified as the system changes.

## Uninterruptible Power Supply

Critical power systems frequently use UPS to bridge the gap in time between the loss of the normal power source and the engaging of backup power source. In the most complex systems, this can be measured in seconds, but even a second of disruption to a power source could be critical to the operation of a healthcare life support system, operation of a data center, or operation of a financial transaction hub.

A UPS (Fig. 12.38) is essentially a large, self-charging battery system where the battery is continuously recharged through the normal power source. The UPS system produces DC power, which is continuously converted back to AC power, via an inverter. In the event of a power failure, the battery provides power back to the inverter, which then outputs AC power to the system.

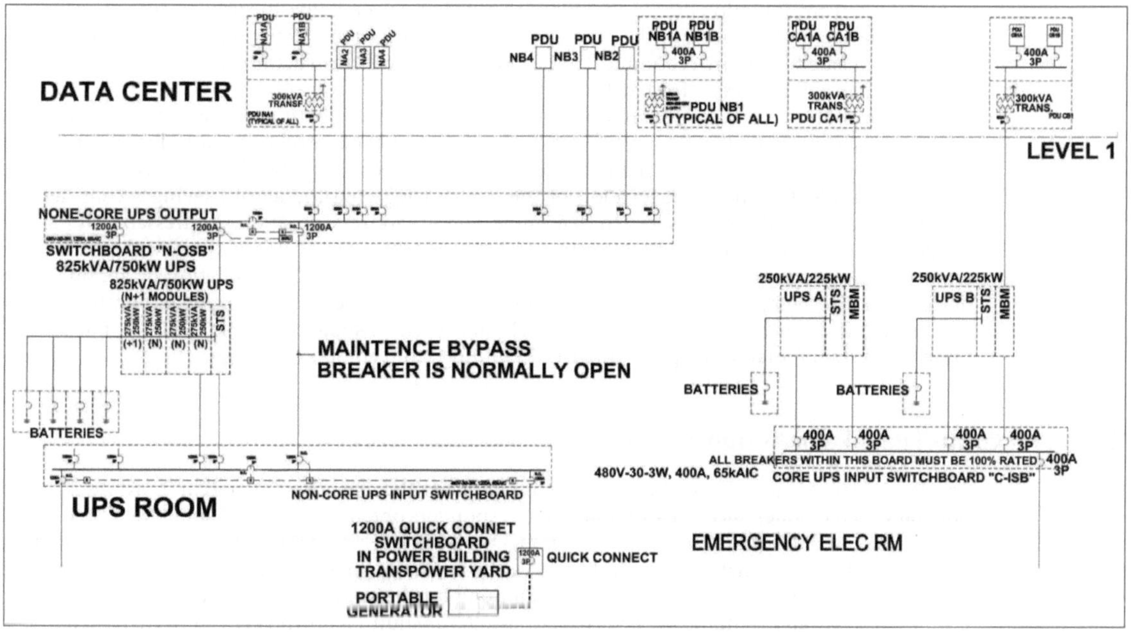


FIGURE 12.38 Single-line drawing of an emergency power system with UPS.

A complex system of sensors and controls, as described in the prior section, anticipates power fluctuations or loss and engages the UPS immediately, while simultaneously engaging the emergency power source. The UPS supplies power to the system prior to the loss or fluctuation of the normal power source so there is no interruption in supply. When the emergency power source is online and power becomes available, the control switchgear then reroutes power from the normal to the emergency source, while the UPS is still supplying power to the system, again to ensure there is no interruption in the power supply. Once normal power is restored, the same process is then reversed.

## FIRE ALARM

Fire alarm systems are designed to provide notification of a fire to both building occupants and first responders and to help guide evacuation and response efforts during a fire. Fire alarm systems can range in complexity from a simple initiation and audible notification, such as a residential smoke detector, to a fully addressable, automated voice annunciation system with audio/visual notification and intelligent user interface. Fire alarm systems often include some element of mass notification or public address system, integrated into the system when the building size or complexity requires more than a simple notification of an event.

Fire alarm systems are categorized as either "conventional" or "addressable" and include both "initiation" and "annunciation" devices. Initiation devices are those that trigger the alarm state, such as a smoke detector or pull station. Annunciation devices are those that provide the notification to the occupants and first responders of the fire event, such as a horn, speaker, strobe, or combination of the audio and visual components.

In a conventional fire alarm system, all devices and components in the system are wired to the same cable that connects to the main fire alarm control panel. The control panel will display a message or signal when any initiation device or devices activate but does not give the user any specificity to the activated devices or the nature of the event. Because of its lack of specificity, a conventional fire alarm system is considerably more economical to install as both the infrastructure and components require less work to install and less infrastructure to function.

An addressable fire alarm system provides unique, individual identifiers to each device or component in the system, called an "Address," and the main fire alarm control panel will display a message clearly identifying the device(s) that initiated the alarm. This specificity allows an occupant or first responder to identify the exact location of the device triggering the alarm, saving a considerable amount of time for the responders to determine how to address situation. Addressable fire alarm systems require considerably more infrastructure than a conventional system because there are limitations to the quantity and types of devices that can reside on a circuit (initiation or annunciation devices) and the components themselves are more expensive. Because of the cost implications of an addressable fire alarm system, they are generally not used in small or simple facilities, but are the preferred system for large campuses, high-rise buildings, and complex facilities.

## Fire Alarm Control Panel and Annunciators

The Fire Alarm Control Panel (FACP) is the brain of any fire alarm system and, in most cases, is the point at which maintenance personnel and first responders obtain information about the system or an alarm. The FACP (Fig. 12.39) usually has an LCD or LED interface which allows the user to identify the system status (normal, alarm, trouble, etc.), the nature of an alarm, and device(s) that triggered the alarm. The FACP also contains the programming for the system which may interface with other utilities throughout a building to perform actions such as shutting down HVAC units to limit air flow, closing smoke barrier doors, actuating rollup doors, starting a generator, and many more. Because a modern, addressable FACP contains most elements of a personal computer, its ability to interface with other building systems is virtually limitless.

## Fire Alarm Control Panel

An FACP can also include several functional options to further aid occupants and first responders. Examples of the most common elements are:

- **Battery Backup.** This is a code requirement in most jurisdictions and allows the system to operate in the event of a power loss.

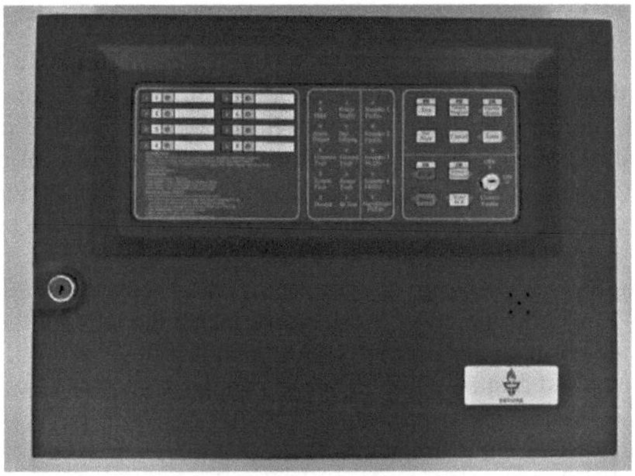

**FIGURE 12.39** Fire alarm control panel.

- **Remote Dialer.** This is an interface with a telephone system. It automatically dials a monitoring company or first-responder network the moment an alarm is triggered without requiring a human being to make the call.
- **Communication.** This can include modules for fireman's phones (simple analog phones used by first responders to communicate to the field), area of refuge (two-way communication with people gathered or trapped in a designated area of refuge), or public address (the ability to give direction to occupants and responders via the system's speakers).

The FACP is generally placed in a fire-rated room. The enclosure may also house other critical building systems such as elevator controls, emergency power annunciators, security system controls, and telephone/data equipment. In large facilities and campuses, the FACP may reside in a room called the "Fire Command Center" which is a designated, rated room for first responders to gather and have as much command and control of the building as possible.

In large or sprawling facilities, it is not always practical to have a singular location for command and control, so additional annunciators may be installed. Remote annunciators and graphic annunciators can provide critical information to multiple points throughout the building or campus. A remote annunciator is a small panel providing a similar display as the main FACP but does not have the same control/user interface. It is primarily for the communication of a situation, not a response effort, so this can decrease the size of panel and enclosure required when compared to the FACP.

Conversely, a graphic annunciator is often quite large because it visually displays the alarm system and condition by overlaying the system information on a floor plan or map of the building or campus, which can be a significant benefit to a first responder unfamiliar with the facility.

## Termination Cabinets and Remote Panels

Fire alarm systems of moderate size and complexity are designed with a singular head-end (the FACP), and several "nodes" on the network which take the form of a fire alarm termination cabinet (or terminal cabinet) or remote fire alarm panel. The devices on the floor are grouped and wired by function (initiation or notification) and some notification devices may require more complex wiring than others (speaker systems, voice command systems, etc.). The wiring required to connect each of the field device circuits to the FACP would be impractically large so a remote termination location is required. This is the point to which the field device wiring is run and then cross-connected with the backbone wiring running from the FACP device to the termination cabinet or panel.

Backbone wiring for a fire alarm system is similar to a structured cabling system, as it is a higher capacity wiring solution than the horizontal branch wiring and can group the signals from multiple horizontal circuit cables into a single run of cable.

## Fire Alarm Communication Paths (Conduit, Cable, and Wire)

The National Fire Protection Agency (NFPA) identifies seven "Classes" of communication and power infrastructure for fire alarm systems (A, B, C, D, E, N, and X). It further differentiates each class by performance in three key areas:

1. **Survivability.** The pathway's ability to continue to function during an event.
2. **Maintenance.** The ability to identify, isolate, and repair elements of the pathway.
3. **Reliability.** The lifespan and long-term functionality of the pathway.

The NFPA does not define the classes in terms of components, but rather in terms of performance in an event. In the simplest terms, the class of infrastructure describes how the infrastructure should behave in an event and considers behavior in three areas:

1. **Protection from Damage.** How the communication pathway is protected such as cable jacketing, conduit, conduit, and a rated shaft and what protection is provided in terms of hours in a fire condition.

2. **Redundancy.** If and how the system continues to operate, in whole or in part, when a problem arises.
3. **Supervision.** If the system can self-verify and self-report faults or issues through its design or elements such as resistors or data "hand-shakes."

The classification infrastructure is governed by the type of construction, local code, and desired operation by the building owner.

## Initiating Devices

Fire alarm initiating devices trigger the alarm condition and fall into two categories: manual or automatic. Manual devices require a person to trigger the device and are usually placed along egress pathways so they can be triggered while people are simultaneously evacuating. These devices include:

- Manual Pull Stations (Fig. 12.40)
- "Break Glass" Stations
- Emergency Stop/Actuation Buttons (usually in manufacturing or industrial environments)

Automatic devices are triggered by environmental conditions and do not require a person to initiate the alarm. These devices include:

- **Smoke Detectors.** Detects the presence of smoke above a concentration threshold.
- **Duct Smoke Detectors.** Detects the presence of smoke in an HVAC duct system via air sampling.
- **Heat Detectors.** Detects heat above a threshold.
- **Flow Valve Detector.** Detects water flow from a sprinkler system line.
- **Flame Detector.** Detects the presence of flames.
- **Beam Detector.** Detects smoke particulates when a light beam it produces is disrupted by the smoke.
- **Gas Detector.** Detects concentration of toxic gases or outgassing caused by fire.

**FIGURE 12.40** Manual pull station.

**FIGURE 12.41** Combination horn/strobe device.

### Notification Devices

Notification devices inform the occupants of a facility of the alarm condition and the need to evacuate. In more sophisticated systems, notification devices can also provide specific direction by voice command, via speaker. Notification devices can be audible, visual, or a combination of audible and visual. Examples of notification devices are:

- **Horn**, produces a high volume sound mimicking a horn.
- **Strobe**, produces a high visibility light.
- **Horn/Strobe**, combines the horn and strobe in a single device (Fig. 12.41).
- **Speaker**, can play analog or digital audio produced by the system or play voice commands.
- **Speaker/Strobe**, combines the speaker and strobe in a single device.

Both initiation and notification devices must meet the same survivability, maintenance, and reliability criteria as the system infrastructure and other components for the system to be classified uniformly, according to the NFPA. In a system with disparate classifications of components and infrastructure, the element with the lowest classification for survivability, fixability, and reliability will govern the classification of the system as a whole.

## STRUCTURED CABLING

A structured cabling system refers to the infrastructure on which the data and telephone signals reside and have smaller, standardized elements. The standardized elements are governed by the Telecommunications Industry Association (TIA), the American National Standards Institute (ANSI) with a published standard called the ANSI/TIA-568.

The six standard elements of a structured cabling system are:

1. Entrance Facilities
2. Telecommunications Room
3. Equipment Room
4. Backbone Cabling
5. Horizontal Cabling
6. Work Area

## Entrance Facilities

Entrance Facilities encompass the telecommunications service entrance to a building or system and any campus interconnect or cabling between buildings. Like electrical, telecommunication utility providers bring their utility service to the user and transition the service to user-owned and user-maintained infrastructure. The point at which these services transition between utility provider and user is called the Point of Demarcation. Both the utility provider and the private network install equipment at the point of demarcation. The installed equipment allows the utility to communicate and provide services over the private network.

## Equipment Room

The Equipment Room houses the equipment which works inside the facility, such as the main telephone switch(s) and network switches, as well as the main-cross-connect, intermediate-cross-connects, or horizontal-cross-connect.

## Backbone Cabling

The Backbone Cabling provides interconnection between the point of demarcation, the equipment room, and the telecommunications room(s)/closet(s). The backbone cabling (Fig. 12.42) can consist of large, multipair copper cable, fiber-optic cable, or both. Cables are terminated on terminal blocks and are wall mounted or rack mounted inside the respective rooms.

**FIGURE 12.42** Fiber-optic backbone cabling terminated in a rack-mounted fiber-optic patch panel.

## Telecommunications Room

The Telecommunications Room or "Closet" houses both the horizontal and backbone cables and is the location at which both terminate. The backbone cabling distributes services throughout the system but not directly to the user station. The station cabling, or horizontal cabling, is run from the telecommunication room/closet to the device. The connectivity between the horizontal cabling and backbone cabling exists because of cross-connects at the telecommunications closets. This is the location at which the backbone cabling and horizontal cabling are both converted to a common interface via patch cables and terminal equipment.

## Horizontal Cabling

Horizontal Cabling runs from the telecommunication room/closet to the telecommunication device outlet. Horizontal cabling is typically a "Category" cable, which is an unshielded twisted pair cable system of different signal bandwidths. Category cables are numeric (e.g., Category 5 and Category 6, Fig. 12.43, are examples) and increasing category numbers denote higher bandwidth capacity.

## Work Area

Work area components connect to the outlet and utilize the system to communicate across the network. This would include items such as computers, telephones, printers, and other networked peripheral devices. The connection between the device and the outlet is standardized (e.g., patch cord, modular cords, and adapter cables), as are the protocols used in communication.

## Wi-Fi or WLAN (Wireless Networking)

Wi-Fi and WLAN (Wireless Local Area Network) are interchangeable terms for a data network utilizing radio frequency (RF) technology to provide network connectivity in place of a hard-wired network system. Wi-Fi systems still generally receive their internet service from a wired connection to a router. It is then also hard-wired to a managed network switch and subsequently hard-wired to the wireless access point. The wireless component typically refers only to the elimination of the horizontal cabling component, not the entire network.

Wi-Fi networks utilize antennae to create a radio wave, at a specific frequency (typically 2.4 or 5 GHz), which when supplied to the antennae create an electromagnetic field with the ability to propagate through a space. The antennae are contained within an access point (AP), which serves as the main infrastructure component to a wireless network. The access point is used to broadcast a wireless signal which can be detected by wireless devices (laptops, tablets, phones, etc.) via a wireless network

**FIGURE 12.43** Category 6 horizontal cabling terminated in a rack-mounted patch panel.

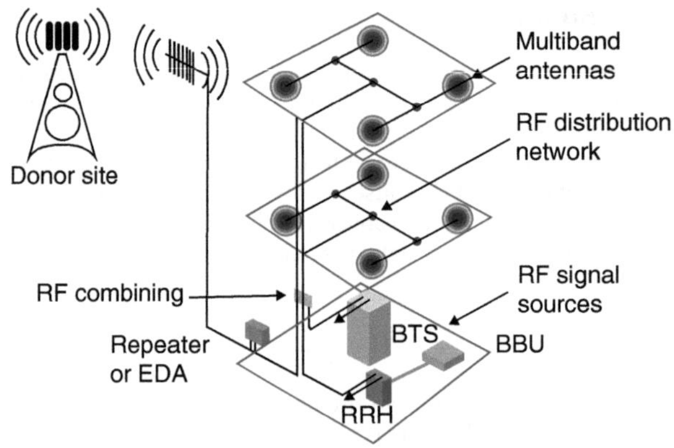


**FIGURE 12.44** Distributed antenna systems (DAS).

adapter. The receiving devices then "tune" into the signal in a fashion similar to how car radios tune into broadcast radio stations.

Each access point has a limited range, measured in square feet, and various elements' interior and exterior environment can interfere with the broadcast signal of the system. Therefore, the spacing of access points varies in every system and is designed using predictive software which accounts for all environmental elements possibly affecting the signal.

Once designed, access points are placed throughout the space requiring coverage and can be powered via 120 V connection or Power-Over-Ethernet (POE), which supplies power via the same network cabling used to supply the access point with its data signal.

Figure 12.44 is a sample network topology diagram for a simple Wi-Fi or WLAN system.

Distributed antenna systems (DAS) provide in-building wireless coverage for RF devices such as cellular telephones, public safety personnel radios, and other wireless systems. A DAS is a dedicated system for receiving external RF signals and re-amplifying them inside buildings using stand-alone hardware and infrastructure (typically low-voltage cabling and radio equipment at the "Head-End"). DASs are capable of supporting multiple RF bands in one single infrastructure including simultaneous cellular carriers.

Two types of DAS installations are most commonly used: passive or active systems. Passive DAS were the first generation and used as "leaky" or radiating coaxial cable as their antenna. The cable itself was the antenna inside the building and amplification was done at the head-end only and created major troubleshooting issues when power levels needed adjustment(s). Active DAS (Fig. 12.45) uses nonradiating coax along with discrete multiband antennas (MBAs) to provide predictive RF coverage and uniform power. The head-end distributes signals via fiber optic to remote units which in turn amplify the signals much closer to the end user and reduce signal power issues throughout a building or campus. These types of systems have therefore become the more popular of the two in recent years.

A DAS is typically installed by building owners for the combined use of various signal sources over one system (cellular carriers, public safety, and enterprise wireless). This type of system (one where multiple cellular carriers and other signals are carried on a singular infrastructure) is referred to as a "Neutral Host" system.

Conversely, many cellular carriers may provide or supplement the cost of a proprietary DAS system within a building or facility to obtain exclusivity of service. A DAS is typically installed by a private cellular carrier (AT&T, Verizon, or others) for their sole network use is referred to as a "Single Carrier" DAS. Other carrier or building owner systems are forbidden to use this type of DAS.

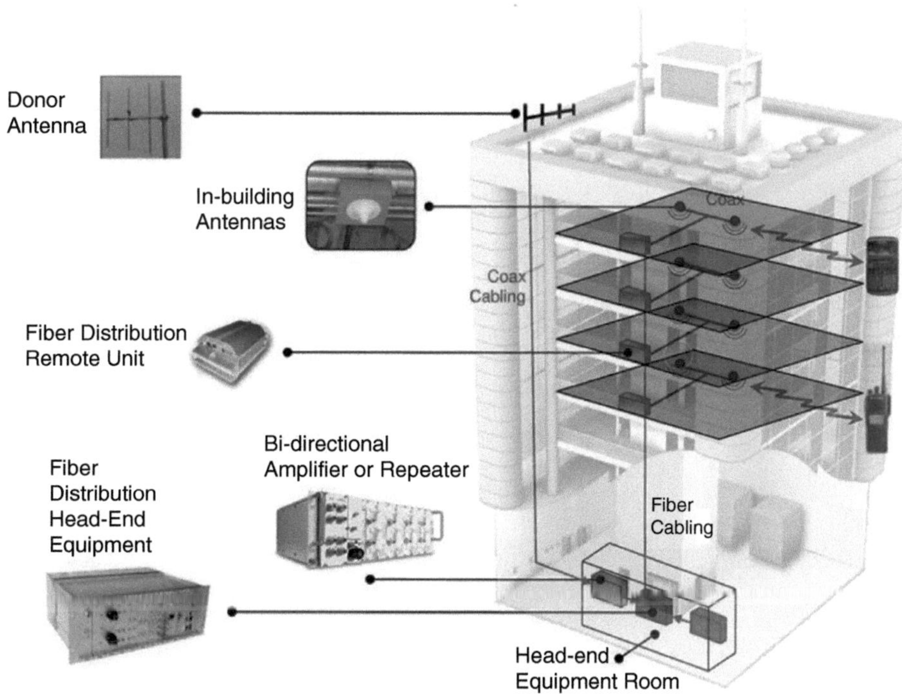


**FIGURE 12.45** Active DAS infrastructure sample.

Prominence of cellular telephone and data use has driver DAS systems to become considerably more popular for consumer use. In addition, regulatory requirements for in-building public safety radio coverage by municipalities following the 9/11 Terrorist Attacks and Hurricane Katrina disasters have driven a sharp increase in DAS networks as well.

# CHAPTER 13
# MECHANICAL PIPING SYSTEMS

*Mechanical and Electrical systems within buildings account for a large percentage of the contractual obligations for the contractor. Most often these installations are handled by specialty contractors that must be coordinated effectively to ensure a successful project, one that meets the schedule and budgetary constraints while maintaining the quality mandated in the specifications. Largely dependent on the program of the building, no two systems are alike other than in function alone. Care and consideration must be given to every building system aspect. This chapter covers CSI divisions 21, 22, and 23 as it pertains to piping systems within the built environment.*

## PIPING SYSTEMS USED IN THE BUILT ENVIRONMENT

Pipe can be used both inside and outside of buildings, above and underground. The location and function of the piping largely determines the type of material and the connection methods used in mechanical construction. Piping can be fused, welded, braised, glued, and screwed, again depending on the function of the finished product. Piping is classified by its thickness as well, using two common indicators, inside dimension and outside dimension. The Construction Specification Institute lists piping in multiple divisions including but not limited to division 21, 22, and 23. Table 13.1 details a list of commonly used pipe materials, their division, and their use.

The International Code Council (ICC) has developed model construction codes that all 50 states utilize, including codes dedicated to trades. The International Mechanical Code (IMC), the

**TABLE 13.1** Typical Piping Systems in Buildings

| Material | | Typical use | CSI division(s) |
|---|---|---|---|
| Plastic | PVC | DWV | 22 |
| | CPVC | Supply distribution | 21, 22 |
| | PEX | Supply distribution | 21, 22 |
| | HDPE | Municipal distribution/wastewater | 21, 31 |
| Copper | | Supply distribution | 21 |
| | | HVAC | 23 |
| Black iron | | Fire suppression | 21 |
| Cast iron | | DWV | 21 |
| | | Sanitary and storm wastewater | 31 |

International Plumbing Code (IPC), and the International Fire Code (IFC) all include standards that apply to piping used in the construction industry.

## PLUMBING SYSTEMS—CSI DIVISION 22

The control of water for irrigation, wastewater, and drinking dates to the Minoan civilization nearly 5,000 years ago. The Romans and Greeks also utilized systems to transport water to locations where needed, all through rudimentary physics. Systems in antiquity all used gravity flow systems which are still in use today. Wastewater systems in most residences and small commercial operation utilize gravitational flow and basic vacuum physics to transport waste "downhill" to either a septic tank or sanitary sewer network. Once received by the sanitary network, pumps are introduced to force the wastewater to treatment plants. Plumbing systems can be broadly defined as water supply and drainage both in and out of buildings. The list below is not exhaustive, but offers a jumping off point:

1. Water supply distribution for consumption
2. Water treatment
3. Waste collection and treatment
4. Storm water collection and disbursement
5. Building equipment supply, including HVAC, food service, pools, processing, etc.

Potable water is defined as water meeting minimum standards for human consumption defined by local departments of health according to the EPA. There is a National Primary Drinking Water Regulation (NPDWR) established by the EPA. These are legally enforceable primary standards and treatment techniques that apply to public water systems. The standards protect public health by limiting the number of contaminants allowed in water meant for public consumption. While not free of contaminants, the water is deemed safe for public consumption and the burden is on the contractor and subcontractor to maintain water quality before, during, and after construction. In most cases, potable water is provided from public water systems owned by the local municipality or water districts.

Providing water is among the chief concerns for any building contractor and should be addressed before construction begins. The local health and utility departments should provide both water quality analysis and volume capabilities to a project site. The mechanical engineer designs the plumbing systems based on data provided by these utilities. A fire protection engineer will also need the data to properly design suppression systems. Water obtained from a municipal water supply starts with water mains on a network complete with pumps, piping, and a large water reservoir, whether it is a lake, river, or deep well.

### Potable Water Distribution

Water mains are primarily made of plastic pipe. High-density polyethylene (HDPE) comes in a variety of thicknesses and colors and is quickly becoming the pipe of choice for both new installations and replacements. HDPE pipe can carry potable water, wastewater, slurries, chemicals, hazardous wastes, and compressed gases. HDPE pipe ranges in size from ¾″ all the way up to 63″. Smaller diameter pipe usually represents higher pressures and shorter distances. Pipe is usually delivered to the jobsite in long lengths, up to 20′ and heat fused together in the field. This method allows for no compromise in flexibility and can be forced underground using wet jet methods and other trenchless technology.

## HDPE

Piping made from polyethylene is a cost-effective solution for a broad range of piping problems in municipal, industrial, marine, mining, landfill, duct, and agricultural applications. Polyethylene is strong, extremely tough, and very durable. HDPE pipe can result in lower life cycle costs due to its ability to withstand corrosion and rust. Its connections are fuse welded to create a stronger bond than the original pipe and continually maintains optimum flow rates. HDPE is the material of choice for trenchless technology. It is used in directional boring, plowing, river crossings, and slip lining. Because HDPE is highly flexible even when welded over long distance, fewer fittings are needed. It has an allowable bending radius of 20–25 times outside diameter of pipe. Savings are also realized because the weight of the pipe is less than other types and smaller equipment can handle the pipe. The nomenclature that HDPE uses is the DR, or dimension ratio. **DR** stands for dimension ratio which is the average outside diameter of a **PE pipe** divided by its minimum wall thickness; therefore, the smaller the number, the higher the pressures it can withstand. Table 13.2 lists typical HDPE types, sizes, and uses.

C900 is the American Water Works Association (AWWA) standard for cast-iron-pipe-equivalent outside diameter polyvinyl chloride (PVC) pressure pipe and fabricated fittings for potable water transmission and distribution, covering nominal pipe sizes from 4 through 12 in. C900 pipe and fittings must comply with the requirements of the Safe Drinking Water Act. PVC compounds and products must be in compliance with NSF/ANSI 61 certification for potable water service. C900 pipe is delivered in standard laying lengths of 20 ft. C900 pipe is suitable for use in waste water force main services. C905 is a comparable standard for potable water PVC pressure pipe, designed for pipe and fittings with nominal sizes over 12 in.

## PVC

Polyvinyl chloride (PVC) pipe is the method of choice for drainage, waste, and venting (DWV) installations in the built environment. PVC comes in three standard thicknesses: schedule 10, 40, and 80. The most common use for plumbing applications is schedule 40, while schedule 80 is used in electrical installations. PVC pipe is extremely versatile and requires the least amount of training for installation. Most connections are fused together using solvent welds with PVC cement. Most often in buildings, PVC is used for DWV applications. The drainage stacks connect to each fixture and collect soil and waste to deliver to the sewers. The IPC dictates that as the fixtures connected to the system grow in number, the size of the pipe must increase to accept the larger volume of waste. The

**TABLE 13.2** HDPE Piping Systems

| Designation | Pressure | Typical use | |
|---|---|---|---|
| DR 11 | 300+ PSI | Water mains | 22 |
| DR 14 | 250–300 PSI | | |
| DR 18 | 200–250 PSI | | |
| DR 25 | 150–200 PSI | | |
| C900 | | Water mains | 4″–12″ |
| | | Sewer force mains | |
| C905 | | Water mains | Over 12″ |
| | | Sewer force mains | |

International Plumbing Code defines waste pipes as those that carry used water away from buildings for treatment without fecal matter. Soil pipes are defined as those that contain fecal matter. Most residential homes do not differentiate between waste and soil pipes; this designation is usually reserved for commercial applications. According to the U.S. Environmental Protection Agency (EPA) the average household uses 150 gallons per person per day, or roughly 12,000 gallons per month. Most residential neighborhoods collect wastewater from a single point at each residence, where it officially becomes a part of the municipal system. It is prudent to install a cleanout at the point of service transition. Homes built prior to 1980 will have cast iron as the underground waste piping, usually without a cleanout. As these pipes corrode and become clogged, plumbers will replace the cast iron with PVC and include a cleanout. A typical four-person household will have a 4″ PVC pipe as its largest wastewater main prior to connecting to a municipal system.

Traditionally, a centralized wastewater treatment facility is used to manage sanitary systems in urban area. According to the EPA, approximately 75% of U.S. households rely on municipal wastewater treatment systems. Sanitary sewer systems are inherently complex, and include gravity fed pipe systems, lift stations, pumps, etc. to transport influent to a central plant. In regions with older systems, sanitary sewer overflows (SSO) and combined sewer overflows (CSO) are examples of the deterioration and inadequacy of wastewater treatment infrastructure. The U.S. EPA has estimated that an average of 50,000 overflows occur each year. Maintenance costs and replacement costs are often not factored into decisions to support infrastructure. Municipal utilities have the difficult task of managing plants according to maintenance needs, demand-related upgrades, available funding, and future revenue projections. The dynamic nature of municipal utility planning and maintenance gives rise to inadequate performance measurements if relying on monetary decision-making alone.

The balance of homes in the United States will use a stand-alone septic system, in which there are new regulations to adhere. Septic systems must have adequate drainage area in the drain field area to allow the effluent to percolate properly. The majority of the treatment is handled in the baffled tank where solids will settle to the bottom and utilize biochemistry to remove contaminants from the water before it is returned to the water cycle. The septic systems are categorized as either aerobic or anaerobic.

### Interior DWV PVC Systems

Sizing of the wastewater systems in residential buildings is a very important endeavor because most homes are gravity systems that rely on water volume to operate properly. Special consideration must be given if extra efficient fixtures are installed. Low flow faucets and showers will result in a reduction of overall flow. Because gravity systems rely on a minimum flow in order to adequately collect all effluent in a system, it is important to size DWV systems correctly and ensure proper slope is installed on buried horizontal pipes.

## *WATER SUPPLY*

Currently, the average person uses 150 gallons of water per day. This number is derived from toilet usage, showers, drinking, and cooking. It has been projected that water scarcity will be a priority in the near future, causing changes to be made to fixtures and installation standards. This creates a conflict between conservation of water and current practice in plumbing standards. Recycled water, either from rainwater runoff or gray water capture from buildings, will need new systems installed to make it viable in the future. Many municipalities have made the decision to do this as a complete system, installing reclaimed water systems on the community level. These systems are often referred to as purple pipe systems, as the color has associated itself with reclaimed water and not to be consumed as drinking water.

Domestic water supply categories include plumbing facilities, food service, laundry, and heating and cooling systems. Fire protection may be included on smaller facilities where the water is shared with the public water service. Exterior water supply includes irrigation, pools, fountains, etc. Water demand for plumbing facilities and fixtures depends on the occupancy of the building and actual needs for the facility. Service Fixture Units (SFU) and Drainage Fixture Units (DFU) are used to correctly size the systems. Copper, plastic, galvanized steel, and stainless steel are common materials used for potable water systems.

Stainless steel can be used in lieu of copper pipe if elements within the water necessitate a stronger resistance to corrosion. Stainless steel will be joined via welds or threaded connections. Either way, stainless steel is seldom used due to its high costs. Hot-dipped galvanized steel is used outside to resist rusting and in nonconditioned spaces to resist freezing. Plastic pipes rated for potable water service include polyethylene (PE), acrylonitrile butadiene styrene (ABS), polyvinyl chloride (PVC), chlorinated PVC (CPVC), and cross-linked polyethylene (PEX). PEX and CPVC piping are most commonly used for hot water transmission. PEX can be used for all water supply applications and eliminates the need for fittings.

## FIRE SUPPRESSION PIPING

Water supply for fighting fires has been in use for over 100 years. Originally, piping was made of wood, basically hollowed-out trees to carry municipal supplied water from lakes and rivers to streets where it was accessed to fight fires. Each time a fireman tapped the line, a marker was placed on a plan and a plug was inserted to mark the spot. These plugs were the precursors of the modern-day fire hydrant we have in place today. Today, fire suppression is required on any space that will house public for gatherings or be placed into service as public spaces. Fire suppression systems are designed to provide enough volume and pressure to adequately fight fires, given the occupancy and use of the space. These designs are conducted based on the National Fire Protection Association (NFPA). There are several standards that make up the fire codes, but the most notable are listed below.

- NFPA 1 National Fire Code
- NFPA 101 Life Safety Code
- NFPA 13 Standard for the Installation of Sprinkler Systems
- NFPA 25 Standard for the Inspection, Testing and Maintenance of Water-Based Fire Protection Systems
- NFPA 70 National Electric Code
- NFPA 72 National Fire Alarm and Signaling Code

Fire protection piping systems must have an adequate water supply, either provided by a municipal water supply, well, or tank reserved for fire suppression purposes. Most commercial fire protection applications utilize black iron or galvanized steel piping compression fittings. Some installations allowed by code will be constructed of CPVC. CPVC piping for fire protection service will be orange in color according to applicable standards. CPVC installations are easier to install because standard connections are glued and cut with a handheld crimp cutter. Black iron pipe installations must be tapped and threaded on site, which is a lengthy and arduous process. Commercial applications require black iron pipe. It is best for a fire protection subcontractor to establish a single place on a construction site to prefabricate material on the ground and transport these into service one at a time. Fire protection systems will be accompanied by a fire pump designed to boost pressure and distribute high volumes of water as necessary to areas under duress. The network of fire suppression pipes will be outfitted with sprinkler heads on branch lines. These heads are designed to react to heat and release water on the fire. Fire suppression systems will have multiple risers within a system to separate the system into manageable sections to serve the building. Each riser is equipped with a valve and detection device that is tied to the fire alarm system. These risers are monitored by a central service that engages the local fire department to assist with fire event.

## FIRE SUPPRESSION SYSTEMS

There are many different types of fire suppression systems but in general the types can be broken into three main categories: portable systems, gaseous systems, and automatic sprinkler systems. The type of suppression system used in a building will depend on the building type and occupancy type of the structure, as well as the contents within the structure and/or the use of the structure. For example, a structure that contains a restaurant may include a foam suppression system within the exhaust hood over the stove, while a structure with electronic equipment or computers may utilize a dry chemical system. Portable systems do not require any piping systems; however, pipe systems are required for both the gaseous and automatic sprinkler systems.

Gaseous fire suppression systems employ inert gases and chemical agents to extinguish fires by displacing oxygen in an area. The use of gaseous systems is limited because a risk of suffocation exists when gas is introduced into enclosed spaces.

The Automatic Sprinkler System is the most effective and efficient means for controlling fire within a structure. There are six primary types of fire protection systems, each used for different applications, and each effective at suppressing fire.

**Wet Pipe System:** A full time water supply provides water, under pressure to the system piping. This is a simple and reliable system that is the most common type of fire suppression system.

**Deluge System:** All of the sprinkler heads within the suppression system are open but water is not present in the piping until the system operates. This type of system is used for special hazards where rapid fire spread is a concern because it provides a simultaneous application of water over an entire area.

**Dry Pipe System:** When a sprinkler head opens, the water supply that is under pressure forces gas out of lines, and allows water to flow. Installed in spaces where the ambient temperature are cold enough to freeze the water in a wet pipe system.

**Preaction System:** A hybrid of the wet, dry, and deluge systems. A heat or smoke detector preactivates the system prior to the introduction of water into the fire suppression system piping. Used in locations where accidental water discharge is undesired, such as in museums, libraries, and/or data centers.

**Foam/Water System:** The fire suppression system discharges a mixture of water and foam concentrate and suppresses fire by the removal of oxygen. This type of system is used within special hazard areas, such as with flammable liquids storage and airport hangars.

Figure 13.1 shows a basic ½″ copper pipe dedicated for medical gases. Copper pipe can be used for potable water supply, medical gases, fire protection, HVAC process water, and although seldom,

**FIGURE 13.1** ½″ copper pipe for medical gases.

drainage, waste, and venting for plumbing purposes. Copper pipe comes in four standard thicknesses depending on the required purpose, K, L, M, and DWV. K is the thickest and therefore the strongest. The most common are L and M and typically used for domestic water supply.

Figure 13.2 shows various sizes of copper tubes for a variety of uses. The copper tubes are generally marked according to their purpose and arranged per the drawings. Adequate support is necessary to ensure quality standards and water and/or gas pressure should be taken into consideration. In the figure, connections have been brazed or sweat-soldered to copper fittings to route the tubing into its proper destination. Also worthy of note is the red packing material as it goes through the CMU on the right. This is fire caulking, designed to protect the opening from fire spread from one side to the other through the wall.

Figure 13.3 shows several piping combinations, typical of a commercial building. It should be noticed that all the piping systems shown are self-supported; building codes do not allow for the work of one

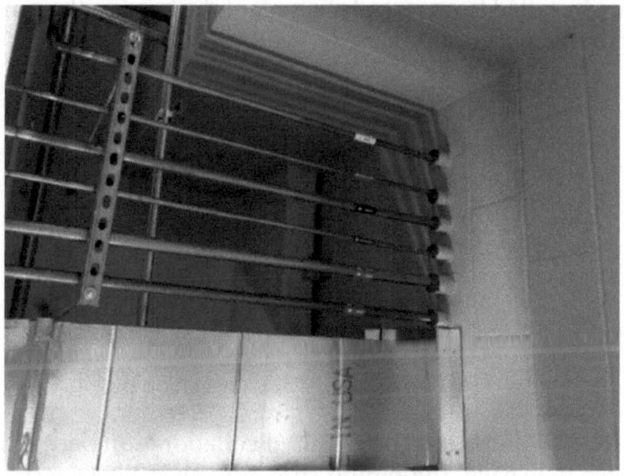

**FIGURE 13.2** Copper piping of various sizes.

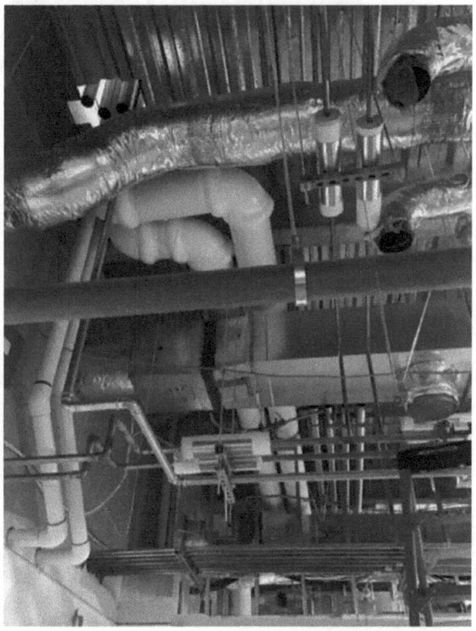

**FIGURE 13.3** Typical piping types in commercial buildings.

trade to be supported by another trade's work. Therefore, all supports, brackets, braces, and Unistrut should be anchored to the structure, in this case steel decking from above. Anchors can also be attached to steel frames, joists, and beams, as well as concrete or CMU walls. Another item of note within Fig. 13.3 is the insulated buffers between the copper tubing and the galvanized Unistrut. This is necessary to prevent a chemical reaction in dissimilar metals that would advance corrosion. Coordination efforts by all trades must be admired due to the intricate nature of the layout of systems. There are no less than six trades pictured in the illustration and it begins to show the cooperation needed on a standard project. Insulation is installed on rain leaders extended from the above floors to prevent condensation.

Figure 13.4 shows black steel piping for fire protection connected with compression fittings to withstand the pressures required for fire suppression piping. Fire suppression piping is generally placed into service at high pressures for lengthy periods of time without fluctuation to be ready at any given moment without notice. Also pictured in Fig. 13.4 is a flexible extension to a sprinkler head. These are installed to ensure placement of the head in the center of a ceiling tile and to offer protection against minor seismic activity.

**FIGURE 13.4** Fire protection piping.

Figure 13.5 illustrates the required collaboration between trades. While the sheet metal ductwork dominates the space, the piping shown is most likely serving each terminal unit within the HVAC

**FIGURE 13.5** Trade collaboration within systems.

network. Drains are shown on each insulated pipe and must be accessible to the facility manager in case of emergency. The picture captures the intricacy of space planning with regard to systems within a building. Thus, the necessity of information modeling within complex structures to properly layout piping systems is revealed.

Figure 13.6 shows several pipes insulated to prevent condensation from forming unwanted water intrusion on the building occupants below the ceiling. Often, water is mistaken for a leak in the exterior structure when in fact it is borne out of neglected pipes and their condensation. As mentioned previously, note the support provided for the piping is for the HVAC process water only and not attached to any other trade.

**FIGURE 13.6** Pipe insulation.

Figure 13.7 highlights the importance of insulation and the planning required to properly install the pipes. The outside diameter of the pipe is of particular importance along with the wrapped insulation to know the final dimensions of the installation.

**FIGURE 13.7** Piping coordination.

Figure 13.8 shows high-pressure bolted flange connections on piping connected to large pumps. Pumps are used to increase pressure of fluids and transport large volumes through a network of pipes. Most connections within pumped networks are bolted flange connections. The connections may also have shock absorbers to withstand pressure fluctuations as demand for production water changes.

FIGURE 13.8 High-pressure bolted flange connections.

## CHILLERS AND BOILERS

Large campuses with heavy demands for air conditioning and heating will rely on central plants to provide chilled water and steam for heating and cooling purposes. These are often referred to as central plants. This temperature-sensitive water must be distributed evenly and quickly across long distances. The network is a large combination of pumps, heat exchangers, chillers, tanks, and other equipment (Fig. 13.9). Any endeavor to create a central energy plant involving chilled water and

FIGURE 13.9 Chiller.

steam boilers will undoubtedly have a large portion of the work dedicated to MEP piping. A recently completed project in Oklahoma involved a large nationwide contractor to replace the chilled water and steam boilers on a college campus. The contractor utilized BIM to coordinate all the piping, but due to the complex nature of the project, hired a specialized MEP Project Manager to coordinate only those trades. Piping systems within central plants are categorized by the service provided and medium conveyed. They will have low, medium, and high temperature and pressure classes. The arrangement of pipes is also of chief concern, whether the system is in series or parallel with pumps and other routes or networks. Systems can be gravity fed but are most often closed, forced flow systems. Steel, copper, and plastic are the most common piping systems utilized in central plants. A forced system refers to those networks that use pumps and compressors to increase or decrease the pressures on the network and move vast amounts of fluid in a short amount of time.

## PUMPS

Pumps are categorized by their operating principles and casing design. Positive displacement, centrifugal, and turbine pumps are common in building construction. Centrifugal pumps are the most common type used in central plant operations. Pump performance is measured on standard curves published by the manufacturer and based on pressure and necessary volume. Figure 13.10 shows a turbine pump coupled with an electric driver mounted above.

**FIGURE 13.10** Turbine pump coupled with an electric driver (black) mounted above.

Booster pumps are auxiliary pumps which are used to increase or maintain pressure in a potable water system. These pumps are used most often in systems that have an inconsistent demand for water. The booster pump is in standby mode when the demand is low. The pump will be activated when the demand is high. The pump maintains the necessary water pressure to meet the increased demand. Figure 13.11 shows a series of booster pumps used to move water through the black iron pipe system.

The goal of many pumps is to simply move liquids. Circulation pumps are used within plumbing systems to accomplish this. These pumps are used to circulate special liquids, liquefied gases, and potable water. Circulation pumps are normally centrifugal but also include turbine pumps.

### Fire Pumps

These pumps are used to provide the required water pressure needed to meet the hydraulic design requirements of the fire sprinkler system. These pumps are used in multistory buildings or in fire

**FIGURE 13.11** Booster pumps, black iron piping, and compression fittings.

sprinkler systems that require a large volume of water flow (Fig. 13.12). These pumps are required in fire protection systems whose water supply is sourced from a ground level water storage tank.

Fire pumps are powered using an electric motor or a diesel engine. If the local building code requires power independent of the local electric power grid, these pump systems may be powered by an emergency generator. A fire pump is triggered to operate when the pressure in the sprinkler system drops below a certain pressure.

**FIGURE 13.12** Electric-driven horizontal split case fire pumps.

Types of fire pumps include:

- Horizontal split case
- Vertical split case
- Vertical inline
- Vertical turbine
- End suction

A jockey pump is another component in the fire sprinkler system. It is a small pump that maintains pressure above the pressure settings of the larger fire pump. The function of this pump is to maintain pressure in the fire protection piping system so the larger fire pump does not need to run. A jockey pump package for installation typically includes a pump, motor, and a controller.

The primary code that governs fire pump installations in North America is the National Fire Protection Association's (NFPA) NFPA 20 Standard for the Installation of Stationary Fire Pumps for Fire Protection.

## PREFABRICATED ASSEMBLIES

The utilization of prefabricated assemblies has become popular in the building industry. It not only saves time and money, but also increases quality and improves the safety. By prefabricating plumbing components, on site installation costs and time are reduced. By welding, soldering, and assembling a unit in a controlled environment, the quality control on the installation is increased. These assemblies can be pressure tested prior to being brought to the site for installation. This will reduce rework on the site. Prefabrication also reduces the need for storing material and handling the material on site. This improves workflow and increases productivity. The use of prefabrication reduces the risk of injury to the workers installing the systems on site as the time spent working at heights or in cramped spaces is minimized.

The challenge with prefabricating piping systems or piping connections is involved with the correct sizing, location, and coordination of all system components as well as the trades installing these systems. The ideal prefabrication projects involve any piping or ductwork assembly that is able to be repeated throughout a facility. This might be a hospital, hotel, dormitory, or office building. Prefabrication can also include large building systems that the piping will be fit into upon arrival on site. The strategy of prefabrication can be scaled to any size project or system.

Figures 13.13 and 13.14 highlight two different types of prefabricated systems. Figure 13.13 illustrates coordination between multiple trades to complete the complete assembly of a ceiling HVAC and plumbing rack. Figure 13.14 shows a prefabricated "box" system that was prepared off site. As the penetrations are prefabricated, care must be taken to ensure proper fit and alignment with prefabricated items. The connections shown without insulation are connections made in the field and will be fully wrapped prior to backfilling.

## PIPE INSULATION

Piping insulation for thermal protection includes fiberglass, mineral wool, and foam plastic to maintain the temperature of water for chilled, cold, hot water, and steam condensate. Insulation will vary between ½″ and 2″ thick depending on the heat transfer resistance required. In spaces with relative high humidity, or in warmer climates where local air is inherently humid, cold water pipes and those pipes that are exposed to nonconditioned spaces should also have insulation to avoid unwanted condensation. Often insulation is used as an acoustic treatment on piping to limit the noise pollution in quiet spaces, such as a conference room or library.

**FIGURE 13.13** Multitrade prefabricated assembly.

**FIGURE 13.14** Piping installed to a prefabricated system.

Figure 13.15 illustrates pipe insulation for chilled and heating hot water and directional flow labels for maintenance purposes. Figure 13.16 shows a large diameter pipe wrapped in fiberglass insulation.

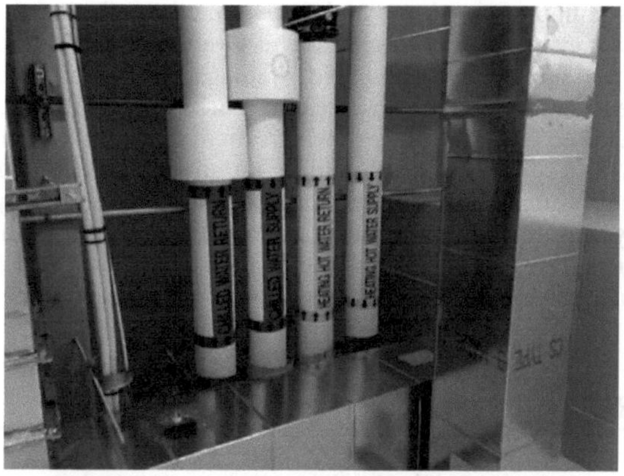


**FIGURE 13.15** Pipe insulation and directional flow labels.

**FIGURE 13.16** Fiberglass wrapped insulation.

## EXPANSION AND CONTRACTION IN PIPING

Pipes will carry varying temperatures of liquid from one source to another and often pass through various ambient temperature zones during the path of travel. Flexibility should be built into the piping system to allow for expansion and contraction of the piping, separate from the building movement. Air bellows and water hammer arrestors, or shock absorbers, are installed to protect against pipe bursts due to fluctuating pressures and movement.

## BACKFLOW PREVENTION

Usually, piping systems are designed for flow in a single direction. Pumps are designed to increase pressure and allow flow in a single direction. Specialized valve known as pressure-reducing valve limits the amount of pressure and flow going through to prevent high pressures on piping systems. Backflow prevention devices such as check valves allow water to only flow in one direction, regardless of flow pressure. Many household drainage systems are equipped with check valves to prevent unwanted sewage backflow in case of a blockage downstream. Pressure will build up and cleanouts will provide access to the pipes downstream of check valves to allow for proper maintenance of piping systems. A vacuum breaker should be installed at branch connections to allow for atmospheric or ambient pressure to enter the system when piping systems drop below a certain level. This will allow piping systems to utilize physical principles of plumbing systems on gravity-fed systems. Most plumbing fixtures have internal air gaps to provide this atmospheric pressure valve. If not installed, an air admittance valve is necessary to introduce ambient air into the system for proper flow.

## PLUMBING FIXTURES

Water closets, or the standard toilet, contain an integral trap to prevent gases from escaping during a flushing event. The most basic water closet has a tank full of water located above the empty pipes to use gravitational flow to express the waste. In commercial applications, a power-assisted diaphragm valve will be installed to introduce a set volume of water at a pressure consistent with the plumbing system to evacuate the toilet.

Bidets are basically small wash basins that provide potable water for personal hygiene. While popular in Europe, they are still found in upper quality restrooms around the country.

Urinals are similar to water closets in way of material and function, but only receive fluid waste from humans. There are low flow urinals and waterless urinals that utilize the fluid generated by the user to properly drain the fixture. Waterless urinals utilize an oil cartridge as a trap to seal unwanted odors from entering the restrooms. However, this adds another layer of building maintenance that a standard full flush urinal doesn't have. Installers must be aware of the ultimate goals of the facility to properly install urinals.

In most commercial applications, a bank of water closets and urinals will be provided on a single wall to provide a single collection waste pipe to connect to the building system (Fig. 13.17). This is known as a wet wall, and it is becoming more common to prefabricate this section of plumbing for delivery to the jobsite in a packaged system complete with wall hung carriers for the fixtures. This

**FIGURE 13.17** Bank of water closet piping.

adds a layer of quality control that site installation methods often lack. The carriers must be able to carry at least 300 pounds and are often created from steel brackets that house not only the fixtures but also provide support within the wall for the waste and water supply piping simultaneously.

What is the difference between a lavatory and a sink? A lavatory is meant for hand washing and a sink is not. However, there are a few more distinctions that separate the two. A sink will not have an overflow drain designed to introduce air to the system for faster drainage. Other than that, the terms are often interchanged and provide a source for potable water meant for human consumption and washing.

Bathtubs are meant for filling and washing and must be designed to withstand not only occupants, but the volume of water needed to fill it. They can be made from porcelain china, cast iron, or fiberglass. The standard bathtub is 5′ long and 2′6″ deep and will accommodate most humans. A trend toward stand-alone tubs and larger jet-assisted tubs provides extra comforts to the user.

Showers can be an integral part of a larger plumbing fixture and can be constructed of a single-molded fiberglass tub unit. Often, they are only nozzles designed to increase the pressure and reduce flow at the head unit and surrounded by waterproof materials such as ceramic tile. Special care should be given to the substrate, and protection of the drains is important to ensure a quality installation.

## PIPING COORDINATION

The coordination process involves the collaboration of different stakeholders on a project. The MEP design as developed by the architect must be installed according to the plans in coordination with all the other building systems. The level of sophistication associated with the coordination of the MEP trades has improved with the adoption of software technology.

The coordination process involves the reviewing of the detailed design that has been prepared by the appropriate subcontractor. These subcontractors each prepare detailed shop drawings that are 2D in nature. While 2D drawings can show the overall layout of a system, they are not able to illustrate possible clashes or conflicts between systems. 3D simulations and models assist with this process by comparing all systems within a project simultaneously and detecting any conflicts. The 3D modeling approach allows for the conflicts between systems to be detected prior to the subcontractor venturing in the field to complete the installation. This prevents rework as well as costly changes to the systems. Figure 13.18 shows the piping and systems coordination in a large biotech pilot plant facility. The smooth installation of all the systems was coordinated using extensive 3D modeling and coordination.

**FIGURE 13.18** Complex MEP coordination facilitated by the 3D modeling process.

The key to any coordination process is to ensure that each discipline collaborates. 3D modeling facilitates this process. The model enables the designer, the mechanical, electrical, and plumbing contractors, as well as the structural engineer the opportunity to compare data, drawings, and details related to the interaction between systems and how they will be constructed. The 3D process also allows for the collaboration and coordination between all trades and disciplines to happen much earlier in the project. The earlier the collaboration the better the final product.

The MEP system interacts with all building systems in some way and therefore coordination between the systems and the trades installing those systems is a critical activity. The mechanical, electrical, plumbing, and fire protection services must be installed within the constraints of the architectural features as well as the structural system. These systems provide valuable services to the occupants of the building and must function properly, safely, and reliably.

The multidisciplinary collaboration that occurs during the MEP coordination involves the architectural features, the building envelope, the structural requirements such as column and beam locations, as well as the MEP design.

The MEP components in major buildings have been estimated to cost up to 60% of the total cost of the entire project. There is a lot at stake when coordinating these systems. Accomplishing this coordination prior to entering the field ensures minimal chances for system conflicts in the field. When completing the coordination for MEP, three areas of knowledge are required: design, construction, and operations and maintenance.

When conducting the coordination process, the goal is to remove conflicts between the physical components associated with the construction of each system. Specific focus is placed on the most congested areas within a structure to prevent the interface of any building service systems with any building systems.

When looking at coordinating the service systems in a building, identifying their locations in the building is the most important step. To accomplish this, each system needs to be assigned both horizontal and vertical coordinates that are within the constraints of the structural components and the architecture. This requires the involvement of various trades and equipment.

Mechanical systems will involve the HVAC equipment and associated distribution systems. This will include ductwork, roof top air-handling units, cooling tower, and associated piping. In the coordination process, the mechanical specialty contractor will complete the shop drawings and their labor force will be responsible for the installation of the equipment according to the plans.

Electrical systems will involve all the wiring, panels, switch boards, cable trays, and transformers, as well as any supporting equipment required to keep these items in place. The electrical specialty contractor will complete the shop drawing and qualified electricians will install the equipment according to the design and all applicable building codes.

Plumbing systems involve both delivery and sanitary piping systems. These systems will include piping, pumps, sumps, traps, and tanks. The plumbing specialty contractors will complete the shop drawings and the plumbers will install the systems according to the design and all applicable building codes. There may be two separate contractors for the plumbing and the sanitary system.

Fire safety systems are designed according to the local fire safety code based upon the type of building and the occupancy for which it is designed. These systems include sprinklers, piping, fire sprinkler pumps, and fire detection systems.

External systems that require coordination include the building envelope and the roof system, especially where there are penetrations through these surfaces to accommodate any other system.

The coordination process involves detecting and resolving and clashes between systems prior to the start of construction. To begin the coordination process the most complete and updated design drawings for each separate discipline must be reviewed together. These drawings must accurately reflect the locations for each of the components associated with the system.

The use of 3D modeling or Building Information Modeling (BIM) in these types of coordination efforts is an excellent tool. The model enables the visualization prior to construction of the architecture and structural features which will constrain the location of the building service systems. There are different models, most notable the Architectural and Structural BIM models, that must be complete before the MEP BIM model can be used to determine possible challenges with the installation of systems. Coordination using these models can begin in the preconstruction stage, thereby assisting with early collaboration between the various specialty trade contractors, fabricators, consultants, and engineers.

In the BIM approach a software clash detection model is run that seeks out clashes and conflicts between systems such as piping with ductwork or piping with a floor beam. Detecting these clashes prior to entering the field or having these elements fabricated will save both time and money on the project.

Clashes can occur between any building systems. Examples of some common piping clashes that are detected through the coordination of BIM models include:

- Plumbing to Structural—common clashes include piping to foundation footing; piping to columns; and piping to beams.
- Plumbing to Mechanical—common clashes include piping to ductwork of all sizes.
- Plumbing to Electrical—common clashes include piping to light sources or cable trays.

MEP coordination with a BIM-based design begins in the schematic design phase and must be finished prior to the release of the construction documents. Using the BIM models and associated clash reports the general contractor and all specialty contractors can review the design collectively and work to resolve the clashes detected using the MEP BIM coordination model.

When using BIM to complete the coordination process, be certain that the appropriate MEP BIM template is being used for the systems that will be coordinated. Ensure that all trades on the project are incorporated, to include mechanical, electrical, plumbing, and sanitary, and fire protection services.

Ensure that the architectural model that will be used for the coordination is the validated model for the project. Ensure that the coordinates of the MEP and architectural models are the same prior to linking the two models together.

In the MEP BIM model, care must be taken to ensure all disciplines and trades are included in the model. Coordination and collaboration that must occur for the MEP disciplines with the building structure include level by level, room by room, and both in the horizontal (plenum) and vertical (core) spaces. Collaboration between elements outside the building includes the mechanical, electrical, plumbing, and sanitary, and fire protection systems.

To determine if there are clashes or coordination issues between systems, the clash detection application within the software is completed. Standard software models will run clash detection between any two disciplinary models that are coordinated together. The software will generate a report that indicates the number and location of clashes that may have occurred between the coordinated models. The clash detection operation generates a report that will identify the element that clashes with another in the system. Using the BIM software this clash can be fixed and then another clash detection process can be completed. If new clashes emerge based upon the fix, the process is repeated until all clashes are solved or identified and reported. Some fixes are simple and can be directly handled within the model. Others may require altering the design or moving the location of system components. This may require more substantial review by the designers, engineers, and contractors. However, this is not happening during the construction process, so the time spent resolving the challenges is well spent.

The use of the BIM software and clash detection programs enables the model of each system to be coordinated during the design phase. This enables clashes to be fixed in the model prior to releasing the design, resulting in a clash-free design.

Large amounts of time and money are wasted if clashes between systems are found in the construction stage. These conditions will have to be rectified on site at a greater cost and most likely causing a delay. Completing MEP BIM coordination prior to the construction phase will reduce the amount of project rework and limit waste on site.

The collaboration of all the trades, contractors, designers, and engineers during the design phase ensures that each stakeholder is able to explain their needs and specifications to the other team members. Due to the early collaboration the design can be reviewed more effectively. There will also be fewer mistakes that will need to be rectified during the construction phase.

When modeling the facility in the software program, any clashes or conflicts that are found in the model are resolved within the model. These modifications are less expensive to complete in the model than they are in the field. The changes are easier and more affordable to complete in the model. The level of cost savings depends on the conflict. For example, the average cost of fixing a field conflict in a typical commercial building project was found to range from \$500 to \$3,500 for minor rerouting and \$2,000 to \$25,000 for major conflicts and design changes.

Improper coordination of the major building systems prior to the construction phase leads to rework and the waste of materials caused by these errors. In addition to the material wastes, time and money are also wasted when work must be completed, or errors need to be remedied in the field.

## PIPING COMMISSIONING

The commissioning of the plumbing system will begin upon the completion of the preliminary punch list items by the subcontractors. The goal of commissioning plumbing systems is to ensure that these systems are functioning in accordance with the requirements outlined in specification Division 22. The commissioning process assists the operating staff in training and orientation with the building plumbing system. The process also serves to create test protocols and record data that will advance the plumbing system from substantial completion to complete operation.

The general steps associated with plumbing commissioning are:

- Installation Verification
- System Start-Up
- Functional Performance Testing

Through each of the steps, the staff responsible for the operation of the system must be trained. This will be accomplished during each of these phases. The team responsible for the commissioning includes owner's representatives, construction and installing subcontractors, test and balance subcontractor, external commissioning agents, and anyone responsible for maintaining the system. The equipment manufacturer's representatives may be required to be present during the start-up process according to the specification. They may also be required to provide equipment training.

Common types of plumbing systems that undergo commissioning include:

- Chilled Water System
- Service Water Heating and Distribution
- Compressed Air System
- Sump pumps
- Tap primers
- Fire Protection Systems

Submittals associated with the commissioning process include the actual commissioning plan as prepared by either the commissioning agent or the subcontractor installing the system. The subcontractors or commissioning agent must supply the functional performance tests procedures for all the systems. As part of this submittal, a narrative description of the system must be included. The commissioning plan should outline the organization, scheduling, team members, and documentation pertaining to the overall commissioning process.

Installation verification of the system is the first step in the commissioning process. The subcontractor who completed the plumbing systems work is responsible for ensuring all the work is in place according to the specifications and is ready for start-up. This includes any change orders or punch list items. The verification process includes all piping specialties including balance, control, and isolation valves. Other requirements associated with installation verification depend upon the building and systems being installed. Some typical piping concerns include:

- Control sensor types and location
- Identification of piping valves, equipment, and controls
- Major equipment such as pumps, valves, gauges, and thermometers

Documentation related to prestart-up tests being performed shall be submitted to the commissioning team. This would include any tests performed by the manufacturer prior to installation. If any of the work is found to be incomplete, incorrect, or not functioning properly, the subcontractor responsible shall correct the issue prior to the system start-up work begins.

Before the start-up process begins, a plan must be submitted by the subcontractor who installed the system. The plan will be reviewed by the entire commissioning team and approved before the subcontractor begins the start-up process. Common start-up plans include items such as:

- Flushing and cleaning of pipe
- Filters, strainers, and screens

- Valve/damper positions
- Electrical tests
- Pressure tests
- Safety tests
- Chemical treatment
- Manufacturer's tests

Certain members of the commissioning team will be present for the start-up process to observe the proper functioning of the system. If any deficiencies are encountered during the start-up process, these are noted by the commissioning team. It is the responsibility of the subcontractor to correct any deficiencies and then demonstrate to the commissioning team that the system is operating according to the specifications.

Certain systems may require test and balance work. This process will occur after systems have successfully completed start-up. System and equipment deficiencies observed during this activity are to be noted and corrected prior to any further performance testing.

The final part of the commissioning process involves the functional performance test. This begins once the operational testing, adjusting, and balancing of the piping systems have been completed. The training required for these systems should have also been completed prior to the performance testing.

The training plan should at a minimum include the following elements pertaining to the system:

- Purpose of equipment
- Principle of how the equipment works
- Important parts and assemblies
- How the equipment achieves its purpose and necessary operating conditions
- Most likely failure modes, causes, and corrections
- On-site demonstration

The Functional Performance Testing process is focused upon ensuring the system operates in accordance with the specified design requirements and intent. A testing protocol is developed for each individual system. The subcontractor will implement the protocol for the commissioning team to observe. The commissioning team will evaluate the process, identify any deficiencies, and recommend modifications based upon the system performance. The system and the associated equipment performance will be evaluated over a range of loads and functional levels and the performance documented.

# CHAPTER 14
# HVAC

*The Heating, Ventilation, and Air-Conditioning (HVAC) system in a residential or commercial building involves a large capital investment to build, operate, and maintain. This system modifies the thermal comfort and indoor air quality of the building. HVAC systems heat and cool air as well as control the humidity. These systems also distribute and circulate the air through the building and play a major role in indoor air quality. Key aspects associated with new construction of HVAC systems involve understanding the owner project requirements and meeting appropriate code and standards requirements. The selection of equipment and distribution approach must meet the load demand, intent, and context of the building. These products and materials are detailed in CSI section 23 and addressed throughout the chapter. In addition, these systems must comply with local codes as well as the relevant American Society of Heating, Refrigerating and Air-Conditioning Engineers (ASHRAE) and International Energy Conservation Code (IECC) standards.*

## HVAC SYSTEMS

HVAC systems are a large component of the construction budget as well as the cost of operation and maintenance of a building. Determining the appropriate system and equipment to use in a structure depends on the actual heating and cooling loads for the planned project, the design criteria, and energy efficiency goals. The selection of equipment and the system depends upon the owner's performance requirements and the goals they are striving to meet with the building. Is the building seeking LEED or ENERGY STAR ratings? Or is it striving to hit net zero energy goals? These goals greatly affect the selection of the system. In addition to the performance goals, the HVAC system must meet local codes and standards associated with energy efficiency and performance. These standards are ANSI/ASHRAE 62.1 (Ventilation for Acceptable Indoor Air Quality) and 62.2 (Ventilation for Acceptable Indoor Air Quality in Residential Buildings) and 90.1 Energy Standards for Buildings Except Low-Rise Residential Buildings as well as the Chapter 4 Commercial Energy Efficiency IECC code. In addition to energy efficiency and building codes, these systems must meet fire codes. The National Fire Protection Association (NFPA 90A) Standard for the Installation of Air-Conditioning and Ventilating Systems covers construction, installation, operation, and maintenance of air-conditioning and ventilating systems.

HVAC performance metrics are a starting place for design. Efficiency and the level of performance are measured using different criteria. Some common HVAC metrics include:

- Annual Fuel Utilization Efficiency (AFUE)—annual fuel output energy to input energy including pilot loss
- Coefficient of Performance (COP):
  - Cooling—the ratio of the rate of heat removal to the rate of energy input
  - Heating—ratio of the rate of heat delivered to the rate of energy input

- Seasonal Coefficient of Performance (SCOP)—total output of a device during a normal operating season
- Energy Efficiency Ratio (EER)—ratio of net equipment cooling capacity in Btu/hr to the total rate of electric input in watts
- Seasonal Energy Efficiency Ratio (SEER)—total cooling output of an air conditioner during normal annual usage period for cooling (Btu) divided by the total electric energy input (watts)

HVAC systems are comprised of four main components:

- Source—the unit that produces the heating or cooling effect
- Distribution—circulate the effects from the source to the zone in the building
- Delivery—introduce the effect into the various spaces
- Control—provide for beneficial operation of a system

## HEATING FOR BUILDINGS

Space heating or adjusting the ambient air for comfort is of utmost importance within buildings and can be accomplished in a variety of ways. At the most basic level, a fire in a wood burning stove produces heat to warm a room and surrounding spaces if the building materials are good heat absorbers (Fig. 14.1). Off gassing must be considered, and ultimately efficiency plays a role in heating. Coal fired or wood burning stoves are not a very efficient use of fuel to create heat. To increase efficiency, materials are introduced to provide cleaner burning products with limited waste. Furnaces are self-contained units that utilize fuel at a high level and deliver heated air directly where needed. Water is also utilized to transport heat and does so quite efficiently. The heating systems are categorized based on British Thermal Units delivered per hour (Btu/hr) and the fuel source.

**FIGURE 14.1** Wood burning stove.

## BOILERS

Boilers are a standard source of heating used in residential and commercial applications. They heat water creating either steam or hot water for use in the heating system by burning combustible fuel or using electricity. The generated steam is distributed through the building using pipes that feed steam radiators. The distribution of the hot water is through baseboard radiators and radiant floor systems. The hot water can also run through a coil that is used to heat air. Commercial boilers are pressurized systems that disperse the heat to maintain a comfortable indoor space. There are different types of boilers that burn different fuels with varying levels of efficiency.

### Electric Boilers

An electric boiler contains a water tank that contains electrical heating elements (Fig. 14.2). The boiler uses electricity from the main power supply of the building to heat the water through the heating elements. When the water reaches the appropriate temperature, an electric pump moves the water through the heating system in the building. The water is recirculated to the boiler and heated once again.

Electric boilers are energy efficient because no heat is lost by the generation of waste gases that exhaust through flues or chimneys. Because waste gases do not exist, nearly all of the heat created is used to heat the building. Electric boilers are up to 99% efficient. While efficient, the cost of electricity in an area may be greater than other fuels, so the cost to run the system may not be less.

**FIGURE 14.2** Residential boiler.

### Condensing Boilers

Condensing boilers are extremely energy-efficient system fueled by the combustion of gas or oil. Steam is produced as an exhaust gas during the combustion process. These units condense the water vapor in the exhaust gas and recover the latent heat of vaporization. The condensed water is discharged through a condensate drainage system. The recovered heat would have otherwise been wasted. While the effectiveness of the condensing process varies depending on the temperature of the water returning to the boiler, the efficiency is greater than a noncondensing boiler.

## Pulse Combustion Boilers

The pulse combustion process is a highly efficient way to burn fuel. In this context a pulse is defined as one cycle of ignition and combustion of a gas/air mixture within a specially designed chamber. The process is naturally aspirated not requiring a blower motor for operation. This reduces electrical consumption. The system uses an assist fan during the pre/post purge cycles only. This fan shuts down once combustion has been achieved. The only electricity required is for purge cycles, powering fuel valves, controls, and safety features which is what makes it so efficient to operate.

## Cast Iron Sectional Boiler

Cast iron sectional boilers consist of multiple sections that are connected. They typically range in size from 35,000 to 14,000,000 Btu/hr. The steam pressure does not exceed 15 psi and the hot water pressure does not exceed 160 psi. Temperatures within the boiler are at a maximum of 250°F. Cast iron retains heat well; however, it is a brittle material that is not conducive for high-pressure steam production activities. They are highly resistant to corrosion.

Residential boilers are installed as a prepackaged unit. Larger commercial size sectional boilers are either prepackaged or built on-site. Sectional boilers can be constructed from cast aluminum or cast iron. Additional sections can also be added for greater capacity. These boilers are typically easier to install and fit, making them a good choice for tight locations.

## Water Tube Boilers

Water tube boilers are pressurized systems capable of generating saturated or superheated steam. Water passes through boiler tubes while the combustion gases pass over the tube surfaces creating the steam. These boilers are utilized where high steam pressures are required.

## Fin Tube Boilers

Fin tube boilers are pressurized equipment used to heat water or convert water into steam under controlled conditions. Water passes through boiler tubes while combustion gases pass over the tube surfaces. The difference with a fin tube boiler is that the tubes are fitted with fins that increase the area available to transfer heat. Fin tube boilers provide primarily residential and commercial hot water for domestic hot water as well as space-heating needs.

## Fire Tube Boilers

Fire tube boilers are pressurized equipment used to heat water or convert water into steam under controlled conditions. Combustion gases are circulated inside boiler tubes transferring heat to the water outside. These boilers are typically utilized in applications where required steam pressures are not greater than 300 psig.

Fire-tube boilers have different equipment options based upon the number of times the combustion gases travel the length of the pressure vessel. This is termed a "pass." Each time the combustion gases travel through the tubes they do so in the opposite direction. This requires an area for the gases to reverse course, termed a "turnaround zone." These zones can be dryback, which means the area has a refractory lining or water-backed where it is water-cooled.

**FIGURE 14.3** Commercial boiler system being installed.

## Feedwater Pumps

A feedwater pump for a boiler supplies water to the boiler. The water used in the boiler is either freshly supplied or returning condensate. The condensate is produced from the condensing of the steam the boiler produces. Feedwater pumps supply the freshwater while the condensate is typically recycled and fed back to the boiler using a condensate return pump system.

Boiler feedwater pumps operate at high pressures due to the high temperatures associated with the feedwater. The chemical composition of the boiler feedwater is important when selecting the type and material for the pump to avoid corrosion. Figure 14.4 shows feedwater piping and valves for a chiller.

The pumps used most often for boiler feedwater applications are the stamped stainless steel vertical multistage inline pump and the multistage horizontal ring section pumps. The pressure capacity demands and other requirements of the boiler system will determine the type of pump selected. For pressure requirements under 1,000 feet of head, the vertical inline pump system typically performs well. As the pressure increases, the ring section pump is more appropriate.

**FIGURE 14.4** Feedwater piping and valves for a chiller system.

### Deaerators

Prior to entering the feedwater pump the water may be required to undergo treatment to remove dissolved substances, including oxygen, from boiler feedwater. The deaerator removes oxygen and carbon dioxide from the feedwater. This ensures pure water enters the boiler and reduces the likelihood of pitting from occurring. Pitting occurs when dissolved oxygen reacts with the boiler tube walls. Pitting can force a shutdown of the boiler. Ensuring pure water enters the boiler ensures a long life.

## FURNACES

A furnace operates on the principle of forced air heating. The furnace transfers heat to the air which is sent through ductwork of a building and out vents. Furnaces are generally either fueled by electricity or natural gas.

### Electric Furnaces

Electric furnaces rely upon an electrical ignition to begin the heating process. The ignition activates heating elements that contain electric resistance coils. The heating elements activate in stages to avoid overloading the electrical system in the building. As current passes through the coils, they heat the air that blowers move over the elements, each of which is typically rated at 5 kW.

All-electric furnaces have an AFUE of 95–100%, meaning higher running costs can potentially be recouped through efficiency. Electric furnaces are considered safer than gas and oil options due to no danger of gas leaks and no actual fire present in the furnace.

### Gas-fired Furnaces

Natural gas furnaces use a pilot light to ignite a series of burners inside a combustion chamber. These burners create heat that enters a heat exchanger. The heat is then transferred to the air, raising it to the temperature set by the thermostat. Once the air achieves the desired temperature, the blowers distribute the air through the ducts. Propane furnaces operate on a similar principle.

New gas furnaces have an AFUE rating of 89–98%. Gas is the cleanest burning of nonrenewable energy options; however, they carry the danger of carbon monoxide leaks. This requires the installation of carbon monoxide detectors in most buildings. The expected lifespan of a gas furnace is approximately 25 years if well maintained.

## FUEL-FIRED HEATERS

Fuel-fired heaters are self-contained heat-generating units that are automatically controlled. They are designed to operate and transfer heat by using a burner, fan, and heat exchanger. These units consist of gas- or oil-fired duct furnaces, unit heaters, direct and indirect fired heaters, and gas-fired makeup air units. These units are typically overhead mounted. Heat is directed downward using louvers installed in the ducts. When used in conjunction with a supply fan, heat is radiated downward. These types of heaters are found in warehouse, manufacturing facilities, and loading docks. In general, they are useful to heat any space that is not serviced by the building's main HVAC system.

## *HEAT EXCHANGERS FOR HVAC*

A heat exchanger is a system used to transfer heat between two or more fluids. Heat exchangers are used in both cooling and heating processes. The fluids may be separated by a solid wall to prevent mixing or they may be in direct contact.

Heat exchangers are either given a hot fluid to provide heating or a cold fluid to provide cooling.

- A fluid can be either a liquid or a gas
- Heat always flows from hot to cold
- There must be a temperature difference for heat to flow
- Most heat exchangers for HVAC purposes use convection and conduction

Coil heat exchangers use one or more tubes that run back and forth a given number of times. These tubes separate the two fluids used in the heat exchanger. The tubes contain one fluid while the other fluid flows outside. Using a heating example, heat is transferred from the hot fluid flowing inside the tube to the tube wall through the convection process. The heat conducts through the wall of the pipe and transfers heat to the outer fluid through convection. The coils found in HVAC systems are commonly finned tubes. These have a larger surface area to transfer heat.

Plate heat exchangers use thin plates of metal to separate the two fluids. To improve the heat transfer, the fluids flow in opposite directions. The heat from the highest temperature fluid is plate transferred onto the plate wall via convection. The heat is then conducted through the plate to the other side. The other fluid which is at a lower temperature carries the heat away by convection. Plate heat exchangers are used in the ducts of air-handling units to exchange thermal energy between intake and exhaust air streams. These exchangers are typically made from thin sheet of aluminum.

## *CENTRAL COOLING EQUIPMENT*

Space conditioning also includes air conditioning, or removing unwanted heat from spaces to create cool, dry air to meet comfort standards. Cooling equipment does not create cold air; it simply removes the unwanted heat from spaces by circulation, ventilation, and filtration. Air is circulated through buildings and across heat exchangers to remove heat and humidity to result in air that feels drier and cooler than outside ambient air. There are several methods utilized within buildings to remove heat, from the basic window hung packaged air conditioner to large-scale chillers and cooling towers. All equipment works on the same principles of heat transfer, that all heat moves from a warmer medium to a cooler medium, and that heat transfer can be captured and utilized to deliver the air comfort levels desired by engineers. The basic metric of air conditioning is the "ton" of cooling, which is equal to the amount of energy required to melt a 2,000 pound ice block over a 24-hr period. As if that was not arbitrary enough, engineers use a variety of mediums to achieve this, namely air-to-air systems, and water-to-water systems. Air conditioning is simply a large heat exchange network designed to deliver heat units to a sink and move forced air across a cooling evaporator to ductwork designed to deliver it to building occupants for their comfort. This section reviews standard equipment to achieve these goals.

### Refrigerant Compressors

Refrigeration compressors provide air conditioning, heat pumping, and refrigeration for large-scale facilities and equipment. These systems work by taking in low-pressure gas on the inlet and compressing it mechanically. How the gas is compressed is what differentiates compressors. The compression creates a high-temperature, high-pressure gas that is key to the refrigeration cycle that is the foundation of air-conditioning systems.

## Refrigeration Cycle

The refrigeration cycle or heat pump cycle is the model describing the transfer of heat from regions of lower temperature to regions of higher temperature (Fig. 14.5). This is the operating principle behind air conditioners, heaters, and it defines the operating principles behind refrigerators, air conditioners, heaters, and heat pump devices.

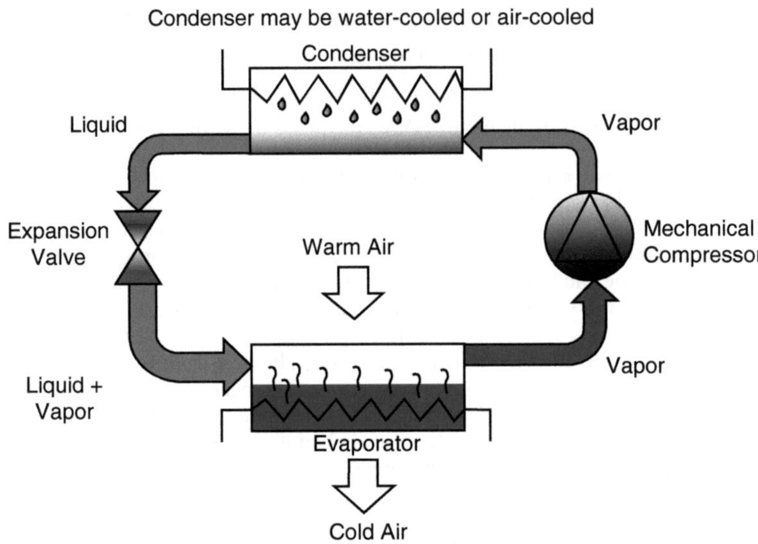


**FIGURE 14.5** Refrigeration cycle.

The cycle involves specific equipment that transforms the fluid into various states. First, the refrigerant passes through an evaporator. The evaporator uses warm air to heat the fluid and converts it completely to vapor. The fluid then moves through a compressor. The compressor increases the pressure and temperature. At this point the fluid is at a superheated level. The fluid then passes through another evaporator. This transfers heat to the ambient air and condenses the fluid to a liquid. After the fluid passes through an expansion valve, the pressure of the fluid is decreased. This cools the fluid and turns the liquid into a liquid/vapor mixture.

## Centrifugal Compressors

Centrifugal compressors exert a centrifugal force on the refrigerant through the rotating action of an impeller wheel. This occurs in a round chamber of a fixed volume. These compressors are best utilized to compress large volumes of refrigerant to relatively low pressures. To generate high compressive forces, these compressors use two or more stages in series. These compressors are simple in design, possess few moving parts, and are energy efficient.

## Reciprocating Compressors

Reciprocating compressors use a piston-and-cylinder arrangement to provide compressive force similar to a combustion engine. The reciprocating motion of the piston compresses the refrigerant inside the cylinder. These compressors have a low initial cost and possess a large power output range. They can also reach extremely high pressures. However, they have high maintenance costs, potential vibrational issues, and are not typically designed to run continuously at full capacity.

### Rotary Compressors

Rotary compressors have two gear-like rotating elements between which the refrigerant is compressed. These compressors are very efficient because the actions of taking in refrigerant and compressing refrigerant occur simultaneously. They have very few moving parts, low rotational speeds, and low initial and maintenance costs. The limitation with these compressors is that they can only compress smaller volumes of the gas and produce less pressure than other types of compressors.

## *CONDENSERS*

Condensers operate at lower pressures than evaporators and take advantage of the compression cycle results by utilizing the heat generated by compressing refrigerants. The heat is then removed to a sink and the resulting liquid is moved to an evaporator where the compression cycle is reversed. The expanding lower pressure results in a cooling effect that is captured in closed systems to interact with warmer medium to then transfer the heat. A standard condensing unit is placed where the heat is expelled into the atmosphere or heat sink medium using a network of closed loop refrigerants. A variety of condensing equipment types are available, which are reviewed below.

### Air-Cooled Condensers

An air-cooled condenser (ACC) is a direct dry cooling system. These systems use ambient air to cool and condense steam. The ACC condenses steam inside air-cooled finned tubes. The cool ambient air flowing over the finned tubes removes the heat from the steam causing condensation. The condensate then returns to a boiler. ACCs are used for **thermal power plants**.

### Water-Cooled Condensers

The water-cooled condenser removes heat from vapor and transfers the heat to water that runs through it. The vapor condenses on the outside of the tube and transfers the heat to the water inside of the tube.

An evaporative condenser converts refrigerant contained within a closed loop from a heated vapor to a cooled liquid form. This occurs by the evaporation of water that is sprayed on the exterior surface of the refrigerant tubing. The evaporative system is a hybrid between an air-cooled condenser and a water-cooled condenser. These condensers can efficiently run on fans with lower horsepower, thereby consuming less energy. They are also easier to install and maintain than an ACC or water-cooled unit because they operate with fewer components. They are suitable for locations with smaller floor space.

## *CHILLERS*

Chillers have become an essential HVAC component of a wide variety of commercial facilities and represent the single largest consumer of electrical usage in most facilities (Fig. 14.6). Chillers facilitate the transfer of heat from an internal environment to an external environment. This heat-transfer device relies on the physical state of a refrigerant as it circulates through the chiller system.

Chillers function on the principle of vapor compression or vapor absorption. These systems provide a continuous flow of coolant to the cold side of a process water system at a desired temperature. The coolant is then pumped through the process, removing heat and then flowing back to the return side of the process water system. The chiller system circulates the refrigerant through an evaporator, compressor, condenser, and expansion device of a chiller moving through the entire refrigeration cycle.

**FIGURE 14.6** Commercial chiller.

## Absorption Chillers

Absorption chillers use the collection of waste heat from other processes or equipment to drive a thermodynamic process that allows water to be chilled and distributed for HVAC needs. These systems do not use conventional refrigerants. Water in the system is usually mixed with either ammonia or lithium bromide.

An absorption chiller system is comprised of a condenser, a generator, an evaporator, an absorber, and a heat exchanger. To begin the process, the water mixed with lithium bromide is stored in the absorber. The mixture is then pumped through the heat exchanger. It is then sent to the generator tank that is located at the top of the chiller. The generator collects all of the waste heat from outside or from other systems in the building. This heat separates the lithium bromide and water. The water gradually releases heat and becomes a vapor rising to the top of the chiller where the condenser is located. The lithium bromide sinks to the bottom.

The lithium bromide is piped back to the absorber, where it originally began. The vapor in the condenser travels through a cooling tower which is at a lower air pressure. This causes the vapor to become water and travel into the evaporator and waits to be mixed with the lithium bromide in the absorber again.

The absorption chiller produces chilled water while consuming just a small amount of electricity to run the pumps. It will continue removing heat from the building as it goes through the heating and cooling circulation.

## Centrifugal Chillers

A centrifugal chiller uses the vapor compression cycle to chill water. This chiller removes heat collected from the chilled water and heat from the compressor and transfers it to a water loop. The water loop is cooled by a cooling tower. These chillers have relatively few moving parts, are easy to maintain, and provide high-cooling capacity in a small footprint.

The centrifugal chiller utilizes the four basic components of the refrigeration cycle. A heat exchanger removes heat from the entering chilled water. A shell and tube heat exchanger is the most common configuration. This heat transforms the refrigerant from a liquid to a gas. Chilled water circulates in the tubes of the heat exchanger while the refrigerant completely submerges the tubes in the shell.

A centrifugal compressor converts rotational kinetic energy to the dynamic energy of fluid flow. Refrigerant enters the pump impeller and is accelerated, flowing radially outward, then exits. The process raises the pressure of the refrigerant by converting kinetic energy into pressure and heat.

Another shell and tube heat exchanger is used as a condenser to remove heat from the refrigerant gas. The pressurized refrigerant condenses to a liquid. The removed heat increases the temperature of the condenser water. This water transports the heat to the cooling tower where the heat is released to the atmosphere. Lastly, once the refrigerant condenses to a liquid, it passes through a device to reduce pressure.

### Air-Cooled Chillers

Air-cooled chillers are used for commercial and industrial facilities. The heat is absorbed from process water and then discharged. These chillers have a compressor, condenser, evaporator, and expansion valve that operates to chill water and absorb heat. These systems use air to fuel condenser cooling. These systems are used in outdoor applications primarily as the excess heat is released to the atmosphere.

### Water-Cooled Chillers

Water-cooled chillers are used with a cooling tower or adiabatic system. Water-cooled chillers are generally smaller units and are typically placed indoors without as much need for ventilation. Water-cooled chillers are also often used more effectively in high-temperature areas as they are cooled using an external source.

## *ENERGY RECOVERY SYSTEMS*

Energy Recovery Systems use the energy in building air that would normally be exhausted to treat the incoming ventilation air (Fig. 14.7). In the summer, the system will dehumidify and precool incoming ventilation air. This happens by transferring removed heat into the exhaust airstream to cool the condenser coil at lower temperature. During the winter, the system uses the waste heat to preheat and humidify incoming air.

Rotary wheels systems consist of a rotating cylinder that is filled with an air-permeable material. The system is located between the ventilation and exhaust airstreams. As the wheel rotates, it draws in heat energy and releases it into the colder airstream.

Flat plate exchangers consist of alternating layers of plates that are separated and sealed with a heat or energy recovery core-treating incoming air.

Heat pipe technologies: Heat pipes are air-to-air heat exchangers consisting of metal tubes charged with a refrigerant (R-134a for most commercial applications).

**FIGURE 14.7** Energy recovery system controls.

## COOLING TOWERS

A cooling tower in simple terms is a **heat exchanger**. Air and water are brought into direct contact within the tower. This contact causes a small amount of water to evaporate and the temperature of the water is reduced.

The cooling tower is an important component within many HVAC systems (Fig. 14.8). Water that has been heated from air-conditioning condensers or other processes throughout the building needs to be cooled. The heated water is pumped to the cooling tower through a piping system. The water is sprayed through nozzles onto a cooling fill material. This spraying process slows the rate of flow of the water moving through the cooling tower. It also exposes the water to a larger surface area, increasing the air-water contact. The air in the tower is pulled using an electric motor-driven fan, causing a small amount of water to evaporate. This creates the cooling action. The cooled water is pumped back into the building and is ready to absorb heat and start the process again.

Cooling towers are designed and manufactured in different forms and sizes to perform different functions. The various types of towers are uniquely suited to meet the requirements of a facility. In forced draft cooling towers, the airflow is supplied by an exhaust fan. This fan is located at the base of the cooling tower and causes overpressure. Axial and centrifugal fans are used in the forced draft cooling tower. In **induced draft cooling towers,** an inlet fan is located on the top of the cooling tower. This creates a low-pressure situation. Axial fans are used for this type of draft.

Natural draft cooling towers do not utilize any form of fans. Heat is removed using a natural draft. The draft is created based upon the height and dimensions of the stack. These stacks are designed based upon the need of the facility. They offer distinct advantages due to their lack of fans when it comes to energy consumption as well as maintenance.

Cooling towers can be constructed in different ways depending upon the owner requirements and availability of laydown space and resources. The coordination between trades and the manufacturer is critical to ensure the units are installed correctly and the system is ready for the tower when delivery occurs. Factory assembled cooling towers are constructed and shipped in modular sections. The number of these sections is based upon the mode of transportation available to take the tower to the site. Small towers may ship as one unit, while larger towers may ship as several modules. These packaged units are then joined on site. Field-erected towers are constructed on site. The prefabricated pieces are shipped to the site for final assembly. The manufacturer typically provides the labor and supervision for the final assembly.

**FIGURE 14.8** Cooling tower.

## *EVAPORATIVE AIR COOLERS*

An evaporative air cooler cools the air by means of the evaporation of water (Fig. 14.9). When water evaporates into the air, it requires heat. The latent heat is taken from the air molecules, dropping the actual temperature of the air. These systems work well in areas were the maximum humidity level is 60% or less. The lower the humidity the more evaporation occurs. Evaporative air coolers can also be used to humidify dry air during the cool weather months.

**FIGURE 14.9** Evaporative coolers in commercial application.

## *DECENTRALIZED HVAC EQUIPMENT*

Decentralized HVAC equipment may be used in certain buildings where performance requirements are less demanding. These systems also offer a low initial cost and simplified installation that make them appealing. A decentralized system provides cooling to individual rooms or spaces rather than the entire building. The decentralized systems are modular units that are factory designed and assembled into a package. The units include fans, filters, a heating source, cooling coil, refrigerant coils, refrigerant controls, and a condenser unit. The cooling and heat removal operations all occur within the unit. These units are designed and built to meet specific performance specifications.

These air-conditioning systems are also referred to as direct expansion (DX) systems. This is because the cooling is delivered by directly exchanging heat with a refrigerant-type cooling coil. The system does not use chilled water as an intermediate cooling medium. They are designed to serve small spaces. The unit is located either with or directly adjacent to the space being conditioned. Installation of these units is simplified as there is no ductwork or pipes involved. These units are often individually metered at the unit. The challenges with these systems include the fact that they consume more energy than centralized systems and have a shorter equipment life. These systems come in different types that are installed and provide cooling in different ways.

Window air-conditioning units provide cooling only when and where needed. Their components are completely contained in a single box that is fitted in the window or cut-out slot in the wall of the room. These units are available in capacities between 0.5 and 3 TR. (A Ton of Refrigerant (TR) is a unit of power that indicates the heat-extraction capacity of refrigeration and air-conditioning equipment.)

Window unit split air-conditioning systems are split into two parts. The outdoor unit houses the compressor, condenser, and expansion valve. The indoor unit houses the evaporator or cooling coil and the cooling fan. The units are connected by a refrigerant pipe that transfers the refrigerant between the units. The separation distance between the two units is typically limited to about 100 ft. However, this ability to locate the units in different locations provides great flexibility in tight spaces.

The Variable Refrigerant Flow (VRF) Split System is essentially a split system; however, it possesses the ability to provide cooling to multiple rooms from a common compressor. The refrigerant flow is automatically adjusted in response to the heating or cooling load demand of the spaces. These systems are complex and contain microprocessor-based electronics. This means installation may require more than one specialty trade.

## PACKAGED AIR CONDITIONERS

Packaged Air Conditioners are preassembled units that provide space heating, cooling, and ventilation to small and medium spaces. These units are used to cool more than two rooms or a larger space in a residence or office. Packaged air-conditioning systems are available in various capacities ranging from about 5 TR to up to about 100 TR. As the size of the tonnage increases, so will the airflow and ductwork required to cover the spaces.

Package-type terminal air conditioners are relatively small systems typically below 7.5 TR. These units require no external ductwork. These units are mounted at floor level through a sleeve in the wall of the building.

Single-package rooftop systems are a single rooftop-mounted unit that contains all mechanical elements of the HVAC system. These units include a supply fan and filter system that connects to the building ductwork to distribute air. A connection to the air distribution ductwork is possible. The typical capacity for a rooftop-packaged unit is 5–130 TR.

Heat pumps contain a four-way valve in the refrigeration piping that allows the refrigeration cycle to reverse (Fig. 14.10). This enables heat to be extracted from outside air and released into the building, enabling these units to provide both heating and cooling. In warmer temperatures the heat pump removes heat from the building and transfers it outside via the condenser coil. In colder temperatures it removes heat from the outdoor air and transfers it to the building via an evaporator coil which acts as a condenser coil in heating mode. As temperature drops, the unit is required to work harder to pull heat from the air which affects the efficiency of the unit. When this happens, the heat pump will use supplemental electric resistance heaters to heat the air to the desired temperature.

Decentralized HVAC systems can also include convection heating and cooling units. These units condition the air by convection, fan power, induction, and/or radiation. These types of units are powered by electricity, steam, or water. They include baseboard radiation, cabinet heaters, convectors, draft barriers, fan coil units, finned tube radiation, induction units, radiant panels (excluding electric), radiators, unit heaters, unit ventilators, and wall heaters. These units can be used in all building types and categories.

**FIGURE 14.10** Home air-conditioning system.

A type of convection unit that is becoming popular in buildings is the chilled beam system. These units can operate as active or passive systems. They are a fin-and-tube type of heat exchanger that is contained in a casing that is suspended from the ceiling. Chilled water is passed through the tubes of the heat exchanger. In passive systems, warm air from the space will rise to the ceiling. The chilled beam will cool the air surrounding it, which will then sink back to the floor. This creates a convective air circulation to cool the space without the use of a fan. An active system requires a primary air supply that passes through nozzles. This induces air from the space up through the cooling coil. The induction process allows the active beam to provide greater cooling capacity. The air supplied to the beam also assists the building with meeting the code requirements for outside air ventilation. Active chilled beams are more commonly used.

Fan coil units contain a fan which draws air from a building into the unit. The air is then blown over a cooling or heating coil. The temperature of the air leaving the unit is either raised or lowered. These units are typically used to supplement central HVAC systems in applications where there are multiple small spaces requiring individual control. These units typically have a chilled water coil for cooling and either a hot water coil for heating or an electric heating element. Chilled water is provided from a chiller located in the central plant, and hot water from a boiler. Each unit must be supplied with a small amount of outside air to ensure adequate ventilation. These units do not work well in open spaces as adjacent units may conflict.

An induction system is comprised of a large central air handler that delivers primary air to terminal units. These terminal units use nozzles rather than fans to push air into the room. The air pushed through the nozzles causes a vacuum that draws secondary air from the room. This air is then mixed with the primary air before being returned to the room. Coils, most often water coils, regulate the temperature of the air. Induction units operate most efficiently when used for cooling. Challenges with this system include general deterioration, moisture problems, and the noise from the terminal units.

Radiators are connected to a central heating system through pipes (Fig. 14.11). Hot water from the system flows through the pipes and to the radiators. The air surrounding the radiator is heated via convection and is moved around the room. The hot water flows through the radiator and on to the next radiator in the chain. As the water flows through the system it begins to cool. The water returns to the boiler where it is heated and begins the process over again.

Expanding on the concept of the radiator, radiant heating systems supply heat directly to the floor or to panels in the wall or ceiling of a house. The heat is generated using either electric heating coils or hydronic-heating tubing.

**FIGURE 14.11** Radiator.

For flooring applications, electric systems are only cost-effective if they are used in combination with a material that has significant thermal mass, like a thick concrete floor that can store the heat and then gradually release it over time. This makes the hydronic systems more cost-effective. They pump heated water from a boiler through tubing in the floor.

Wall- and ceiling-mounted radiant panels are another application (Fig. 14.12). These are typically aluminum panels. These can be heated using electricity or with hydronic tubing. Due to concerns around leakage in the wall and ceiling, electrically heated panels are used more often. Radiant panels heat quickly and because they are individually controlled, they can be adjusted when no one is using a room.

**FIGURE 14.12** Ceiling-mounted radiant heater unit.

## *HUMIDITY CONTROL EQUIPMENT*

Maintaining the appropriate level of humidity in commercial and residential settings is an important component of indoor air quality. The use of humidifiers and/or dehumidifiers may be required depending on the time of year and climate.

Residential whole house humidifiers treat the air that is distributed through the ducts. The humidifier works with the heating system to ensure the appropriate level of humidity is maintained. Commercial humidifiers are installed within a central HVAC system typically between the hot air ducts and the cool air return ducts. A bypass humidifier draws warm air from the heat ducts of a furnace. These units are installed on the return plenum or the supply of a forced air-handling system. Fan-powered humidifiers work in the same manner, but they have a fan which blows air across an internal water panel. Lastly, steam humidifiers add humidity to the air even if the furnace is not operating. The system blower captures the steam and pushes it through the ventilation system.

High humidity in a building can cause issues with the indoor air quality, especially with mold. ANSI/ASHRAE Standard 62-2001, Ventilation for Acceptable Indoor Air Quality, and the United States Environmental Protection Agency (EPA) both recommend that indoor relative humidity be maintained below 60% to minimize the risks of microbial growth. To ensure the humidity remains within these limits, dehumidification systems may be required. The two common types of dehumidification systems are mechanical refrigeration based and desiccant based.

Mechanical refrigeration dehumidifiers are the most common type. These systems operate by passing moist air over a refrigerated coil using a fan. The cold evaporator coil of the refrigeration device condenses the water. The water is removed from the air and reheated by the condenser coil. The dehumidified air is then released into the room. This system operates well under conditions of high ambient temperatures with a high dew point temperature.

If the temperature and moisture levels are lower, a desiccant-based system is more economical than mechanical refrigeration systems. Typically, a desiccant dehumidification system is utilized for applications below 45% relative humidity. Desiccant dehumidifiers utilize chemical attraction instead of condensation to remove water from the air. A desiccant wheel with a grid of small air passages is impregnated with silica gel. The desiccant wheel rotates at different speeds to remove moisture from air that is moved through it using a blower. The size of this wheel and the speed at which it rotates determine the amount of water removed.

## HVAC AIR DISTRIBUTION

The distribution system used to deliver the heated or cooled air created by the HVAC system plays a key role in indoor air quality, energy efficiency, and thermal comfort. Components of this system include ductwork of various materials, sizes and shapes, fans, and terminal units (Fig. 14.13). Each of these components requires the coordination of different trades and specialty contractors to install. It is the responsibility of the construction manager to coordinate and schedule this work. Each of these components is discussed in the following section, highlighting key pieces of equipment, components, and installation details.

**FIGURE 14.13** Various types and sizes of ductwork.

### Insulation Requirements

Insulating supply ducts that deliver conditioned air to buildings is an energy-saving practice. It is also required by energy efficiency codes. The goal is to keep mechanically warmed or cooled air as close to the desired temperature as possible. The codes also address leakage from the duct system as it moves the conditioned air (Fig. 14.14).

Ductwork insulation requirements are identified in ASHRAE 90.1-2019 Section 6.4.4.1.2, Duct and Plenum Insulation. The section states that all supply and return ducts and plenums installed as part of an HVAC air distribution system shall be thermally insulated in accordance with the values shown in Tables 14.1 and 14.2.

**FIGURE 14.14** Insulated rectangular duct.

**TABLE 14.1** Minimum Duct Insulation R-value,[a] Cooling and Heating Only Supply Ducts and Return Ducts

| Duct location | | | | | | | |
|---|---|---|---|---|---|---|---|
| **Climate zone** | **Exterior** | **Ventilated attic** | **Unvented attic above insulated ceiling** | **Unvented attic with roof insulation[a]** | **Unconditioned space[b]** | **Indirectly conditioned space[c]** | **Buried** |
| **Heating-only ducts** | | | | | | | |
| 1,2 | None | None | None | None | None | None | None |
| 3 | R-3.5 | None | None | None | None | None | None |
| 4 | R-35 | None | None | None | None | None | None |
| 5 | R-6 | R-3.5 | None | None | None | None | None |
| 6 | R-6 | R-6 | R-3.5 | None | None | None | None |
| 7 | R-8 | R-6 | R-6 | None | R-3.5 | R-3.5 | R-3.5 |
| 8 | R-8 | R-8 | R-6 | None | R-6 | None | R-6 |
| **Cooling-only ducts** | | | | | | | |
| 1 | R-6 | R-6 | R-8 | R-3.5 | R-3.5 | None | R-3.5 |
| 2 | R-6 | R-6 | R-6 | R-3.5 | R-3.5 | None | R-3.5 |
| 3 | R-6 | R-6 | R-6 | R-3.5 | R-1.9 | None | None |
| 4 | R-3.5 | R-3.5 | R-6 | R-1.9 | R-1.9 | None | None |
| 5,6 | R-3.5 | R-1.9 | R-3.5 | R-1.9 | R-1.9 | None | None |
| 7,8 | R-1.9 | R-1.9 | R-1.9 | R-1.9 | R-1.9 | None | None |
| **Return ducts** | | | | | | | |
| 1 to 8 | R-3.5 | R-3.5 | R-3.5 | None | None | None | None |

[a]Insulation R-values, measured in (h-ft$^2$,°F)/Btu, are for the insulation as installed and do not include film resistance. The required minimum thicknesses do not consider water vapor transmission and possible surface condensation. Where exterior walls are used as plenum walls, wall insulation shall be as required by the most restrictive condition of Section 6.4.4.2 or Section 5. Insulation resistance measured on a horizontal plane in accordance with ASTM C518 at a mean temperature of 75°F at the installed thickness.
[b]Includes crawlspaces, both ventilated and nonventilated.
[c]Includes return air plenums with or without exposed roofs above.

**TABLE 14.2** Minimum Duct Insulation R-value,[a] Combined Heating and Cooling Supply Ducts and Return Ducts

| | Duct location | | | | | | |
|---|---|---|---|---|---|---|---|
| Climate zone | Exterior | Ventilated attic | Unvented attic above insulated ceiling | Unvented attic with roof insulation[a] | Unconditioned space[b] | Indirectly conditioned space[c] | Buried |
| | | | | Supply ducts | | | |
| 1 | R-6 | R-6 | R-8 | R-3.5 | R-3.5 | None | R-3.5 |
| 2 | R-6 | R-6 | R-6 | R-3.5 | R-3.5 | None | R-3.5 |
| 3 | R-6 | R-6 | R-6 | R-3.5 | R-3.5 | None | R-3.5 |
| 4 | R-6 | R-6 | R-6 | R-3.5 | R-3.5 | None | R-3.5 |
| 5 | R-6 | R-6 | R-6 | R-1.9 | R-3.5 | None | R-3.5 |
| 6 | R-8 | R-6 | R-6 | R-1.9 | R-3.5 | None | R-3.5 |
| 7 | R-8 | R-6 | R-6 | R-1.9 | R-3.5 | None | R-3.5 |
| 8 | R-8 | R-8 | R-8 | R-1.9 | R-6 | None | R-6 |
| | | | | Return ducts | | | |
| 1 to 8 | R-3.5 | R-3.5 | R-3.5 | None | None | None | None |

[a]Insulation R-values, measured in (h-ft$^2$,°F)/Btu, are for the insulation as installed and do not include film resistance. The required minimum thicknesses do not consider water vapor transmission and possible surface condensation. Where exterior walls are used as plenum walls, wall insulation shall be as required by the most restrictive condition of Section 6.4.4.2 or Section 5. Insulation resistance measured on a horizontal plane in accordance with ASTM C518 at a mean temperature of 75°F at the installed thickness.
[b]Includes crawlspaces, both ventilated and nonventilated.
[c]Includes return air plenums with or without exposed roofs above.

This includes:

- Supply and return ducts/plenums in unconditioned spaces
- Supply and return ducts or plenums located outside the building
- Supply and return ducts or plenums located within exterior components (floors, walls, or roofs) where the duct or plenum touches the exterior component

Table 14.3 describes the insulating requirements for each of the components shown in Fig. 14.15. Sealing of ductwork prevents the leakage of conditioned air from the duct as it is being distributed to specific locations in the building. Sealing requirements are outlined in ASHRAE 90.1-2019, Section 6.4.4.2.1, Duct Sealing.

Locations on the duct system that should be sealed are:

- All transverse joints, longitudinal seams, and duct wall penetrations
- Air handlers
- Filter boxes

Prior to construction it is important to review the plans to ensure the insulation requirements are met. This review should be completed with the owner, the construction manager, the mechanical subcontractor, and the designer. The review should verify the duct insulation locations and R-values required by code are reflected in the plans. Sealing requirements of the ductwork should also be reviewed. Duct pressure class and sealing requirements should be reflected in the design documents. Identify all joints and penetrations that the code requires to be sealed and the method by which they must be sealed are understood.

**TABLE 14.3** Duct Components and Insulation Requirements

| Duct component | Insulation requirement and code |
|---|---|
| 1. Insulation of Unit Casing | No insulation requirement.<br>Factory-installed ductwork within unit casings is exempt from insulation requirements. |
| 2. Exhaust | No insulation requirement.<br>Insulation requirements are specified only for supply and return ductwork, not for exhaust ducts. |
| 3. Supply and Return in Vented Attic | Insulation R-value requirements for unconditioned spaces in the IECC or for a vented attic in ASHRAE 90.1 must be met. |
| 4. Supply and Return Outside of Building | Insulation R-value requirements for exterior or outdoor locations must be met. |
| 5. Supply and Return in Unconditioned Shaft | Insulation R-value requirements for unconditioned spaces in the IECC or for unconditioned spaces in ASHRAE 90.1 must be met. |
| 6. Supply in Unvented Attic | Insulation R-value requirement for unconditioned spaces in the IECC or for unvented attics in ASHRAE 90.1 must be met. |
| 7. Return in Indirectly Conditioned Celling Space | No insulation requirement.<br>ASHRAE 90.1 has no insulating requirements for indirectly conditioned spaces. In the IECC, such spaces would be considered conditioned spaces, also with no duct insulation requirement. |
| 8. Exterior Wall of Return Plenum | Insulation R-value must meet either the requirement for above-grade walls or for exterior ducts. If the lesser R-value (most likely the duct R-value) is chosen, the envelope of the building must be shown to comply with a performance-based option. |
| 9. Supply Outlet in Plenum | Insulation R-value to match the supply duct. |
| 10. Supply Runout in Return Plenum | No insulation requirement beyond what is integral to flexible duct, unless the length of flex duct exceeds 10 ft. |
| 11. Supply in Return Plenum | No insulation requirement in the IECC.<br>If following ASHRAE 90.1, ducts must meet the requirements for indirectly conditioned space. |
| 12. Supply in Conditioned Space | No insulation requirement in IECC.<br>These ducts are located in conditioned space. Insulation may be installed for sound-reduction purposes or to control condensate, but it is not required. |
| 13. Supply in Vented Crawl Space | Insulation R-value requirements for unconditioned spaces in the IECC or for an unvented attic in ASHRAE 90.1 must be met. |
| 14. Buried Supply | Insulation R-value requirements for locations outside the building must be met if adhering to IECC. If ASHRAE is being followed, these ducts must be insulated to the level required for buried ducts. |

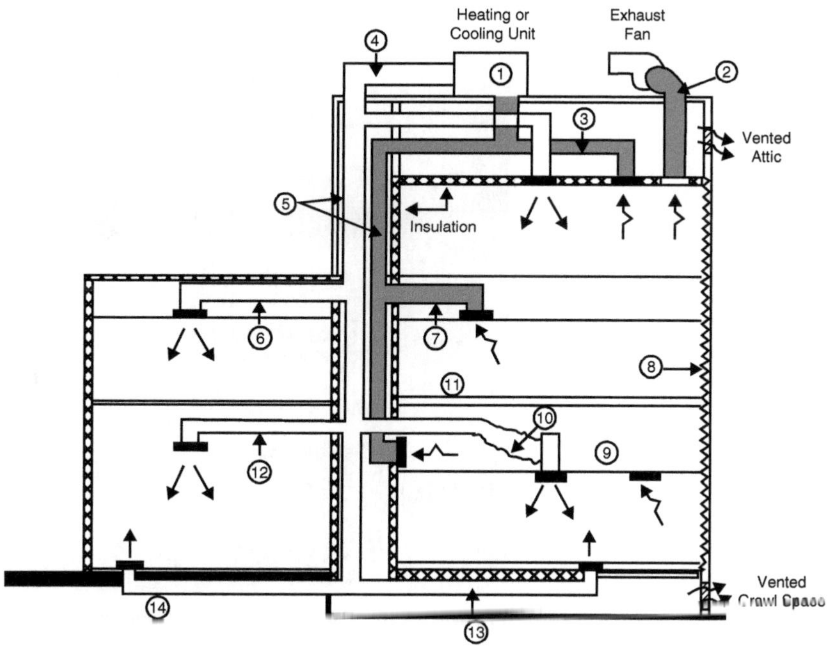


1. Insulation of Unit Casing
2. Exhaust
3. Supply & Return in Vented Attic
4. Supply & Return Outside of Building
5. Supply & Return in Unconditional Shaft
6. Supply in Unvented Attic
7. Return in Indirectly Conditioned Ceiling Space
8. Exterior Wall of Return Plenum
9. Supply Outlet in Plenum
10. Supply Runout in Return Plenum
11. Supply in Return Plenum
12. Supply in Conditioned Space
13. Supply in Vented Crawl Space
14. Buried Supply

**FIGURE 14.15** Duct components and insulation requirements.

## Pipe Insulation

The intent of the pipe insulation requirements is to reduce temperature changes while fluids are being transported through piping associated with heating and cooling systems, thereby saving energy and reducing operating costs.

Uninsulated piping systems that transport heated or cooled water, including piping systems with external heating should be thermally insulated to reduce heat loss or gain, allowing the fluid to be delivered at the intended temperature (Fig. 14.16). Both ANSI/ASHRAE/IES Standard 90.1-2019 (90.1-10) and the 2018 International Energy Conservation Code (IECC) require any piping associated with a heating or cooling system that carries fluids with an operating temperature of 105°F or more or 60°F or less to be insulated.

Any insulated piping in areas exposed to weather is required to be further protected from exposure to sunlight, moisture, and wind—all of which can damage the insulation materials over time, decreasing their effectiveness. This protection must shield the insulation from solar radiation. In addition, if the piping will be conveying chilled water, Standard 90.1-10 requires that a layer of vapor retardant material be installed on the outside of the insulating material. The 2018 IECC prohibits the use of adhesive tapes as shielding. An important thing to consider when developing installation and coordination plans for piping and ductwork is the thickness of the insulation and how that will alter the size of the pipe and duct.

**FIGURE 14.16** Piping insulation and labeling.

## HVAC PIPING AND PUMPS

Pumps and piping systems move water, refrigerants, and other fluids through various components in the HVAC system. These pumps and pipes come in a variety of types and sizes and are used for different applications.

### Hydronic Pumps and Piping

The term "hydronic" is used to describe a heating or cooling system that transfers heat by circulating a fluid through a closed system of pipes. Heat is transferred using liquids and pipes, rather than air and a ducting system. Hydronic piping is used to connect components for circulation, heat exchange, flow control, primary equipment and other piping components. These pipes are commonly carbon steel pipe, copper, or PEX tubing. Hydronic systems are able to convey more heat energy than ducting system, and they take up less space in the building. The main disadvantage is the lack of ventilation. Hydronic pumps come in four basic models:

- Base-mounted end suction pump
- Close-coupled end suction pump
- Vertical inline close-coupled pump
- Vertical inline split-coupled pump

When installing these pumps it is important to consider the floor space available, the space required for piping and supports, and any clearances needed. Coordination with plumbing connections and control requirements will require shop drawing review with the subcontractor installing and the associated trades.

### Condensate Pumps

Condensate pumps are used in HVAC systems to collect and remove condensate that cannot be removed by gravity. This water must be pumped to eliminate it from the system. Smaller pumps are used in residential applications while commercial pumps and motors are rated at much higher horsepower. In industrial applications, these pumps may double as feedwater pumps that return condensate to the boiler.

These pumps run intermittently and have a tank in which the condensate is stored. A float switch will trigger the pump to operate when the level of the condensate reaches a certain point. The pump will run until the level of condensate in the tank is lowered. Some pumps will have a two-stage switch that works as a failsafe mechanism for the system. The liquid in the tank will first rise and trigger the first stage and activate the pump. However, if the liquid level in the tank continues to rise, a second stage will be triggered. This stage may completely shut down the HVAC equipment or trigger an alarm.

The condensate tanks need to be considered in the installation. These tanks may be supported by their flanges or placed on the floor for small tanks. Large pump units are typically mounted on pads and draw their liquid from a tank below the floor. The smallest units use a drip pan and not tank at all. Regardless, the tank installation must also be coordinated as part of the condensate pump.

Steam condensate pumps are a critical part of the HVAC system. These pumps collect condensate from the boiler return line. This return system forces the condensate down the return line to a reservoir, where it is fed back into the boiler. The steam condensate pump supplies the necessary pressure to the condensate so it can re-enter the boiler.

## REFRIGERANT PIPING

Several HVAC systems utilize a refrigerant as part of their operation. This piping must be designed and installed on site, requiring coordination between trades, subcontractors, vendors, and manufacturers. For example, any condensing units, direct expansion coils in air handlers, and chiller systems.

The refrigerant used in HVAC systems is a compound typically found in either a fluid or gaseous state. The refrigerant absorbs heat from the environment and can provide air conditioning when combined with compressors and evaporators.

The most common refrigerants used for air conditioning over the years include:

- **Chlorofluorocarbons (CFCs). These include R12.** Production of these types of refrigerants ended in 1994.
- **Hydrochlorofluorocarbons (HCFCs). These include R22.** These are less damaging to the ozone than R12, but the EPA has mandated a phase out of R22 as a result of the Clean Air Act of 2010. R22 production ended on January 1, 2020. Only recovered, recycled, or reclaimed supplies of R-22 are available.
- **Hydrofluorocarbons (HFCs). These include R410A and R134.** These are being used in place of R22 and they are better for the environment because they do not contain chlorine.

## DUCTWORK

Ductwork is designed to distribute airflow from the HVAC equipment throughout building. This includes the return air and fresh air intakes. The intake and output of air will define how efficient the duct operates. If insulated and leak tested properly, the system will be efficient. Air is the only thing moved via ductwork. The shape, size, and material of the duct are determined by the function it is meant to perform. SMACNA HVAC Duct Construction Standards provides tables and details for constructing ductwork using different materials. This standard is commonly referenced in many construction specifications. Ductwork will require detailed coordination between piping and structural systems. The scheduling and coordination of each trade is the responsibility of the construction manager.

There are several codes and standards used in the design, construction, and testing of ductwork. Commonly referenced standards within construction specifications are listed. It is important for the construction manager to ensure that the mechanical contractor, vendor, and installers understand the installation and testing requirements for the system.

- ASHRAE Handbook Series Fundamentals: Ch. 2. Duct Design
- ASHRAE Handbook Series Equipment: Ch. 6. Duct Construction
- International Conference of Building Officials (ICBO): ICBO UMC Ch. 11
- National Fire Protection Association (NFPA):

  NFPA 90A Installation of Air Conditioning and Ventilating Systems
  NFPA 90B Installation of Warm Air Heating and Air Conditioning Systems
- Sheet Metal and Air Conditioning Subcontractors National Association (SMACNA):

  SMACNA HVAC Duct Construction Standards
  SMACNA Round Industrial Duct Construction Standards
  SMACNA Rectangular Duct Construction Standards
  IAQ Guidelines for Occupied Buildings Under Construction
- Underwriters Laboratories Inc. (UL): UL 181 Factory-Made Air Ducts and Air Connectors

## Ductwork Materials

**Galvanized steel** is a standard material used to fabricate ductwork (Fig. 14.17). Most of the ductwork that uses galvanized steel is hot-dip galvanized where the steel is coated with a thin layer of zinc. This protects the steel from corroding and is used where rust resistance is needed. Galvanized steel can be used in high-temperature applications up to about 392°F. These ducts are lined with a faced fiberglass duct liner or externally wrapped using an insulating duct wrap material. These ducts can be double walled. Double-walled systems help to control noise, improve thermal performance, and assist in reducing damage. These metals must meet the ASTM A653/A653M Standard Specification for Steel Sheet, Zinc-Coated (Galvanized), or Zinc-Iron Alloy-Coated (Galvannealed) by the Hot-Dip Process.

**FIGURE 14.17** Metal ductwork.

Stainless steel ductwork is primarily used in laboratories and research facilities due to the caustic or hazardous nature of the air being ventilated. Stainless steel ducting is used when corrosive substances are vented at high temperature.

Aluminum ductwork is light but exceptionally resilient against corrosion. Aluminum is an exceptional product for use in environments where moisture is common or in buildings where the interior is open to outdoor seasonal changes. Aluminum ducting is also available insulated or noninsulated from the main duct to the outlet system. Because of the ability of aluminum to be soft and bendable it is easy to manufacture special shapes and install. These ducts are not as common as galvanized steel due to their cost. Ductwork using aluminum must meet the standards outlined in ASTM B-209, Standard Specification for Aluminum and Aluminum-Alloy Sheet and Plate.

**Polyurethane and phenolic panels** use rigid insulation foam as the core material for these panels. They are manufactured with factory-applied aluminum foil finishes on both sides. The thickness of the aluminum foil is dependent upon whether the material will be used indoors or externally. Pentane is used as a foaming agent by most manufacturers instead of CFC, HFC, HCFC, and HC gasses. Ductwork using this material possesses low-vapor permeability and improved thermal resistance. Construction specifications reference the SMACNA HVAC Phenolic Duct Construction Standards.

**Fiberglass duct board** is a semirigid fiberglass insulation material often used for residential or light commercial heating and cooling applications. These panels are designed to offer in-built thermal insulation. Its interior surface can absorb sound, providing quiet operation of the heating or cooling unit.

**Flexible ducting** is manufactured with flexible plastic over a metallic wire coil in order to shape a tube. The flex ducts came in a wide range of configurations and are convenient for attaching the outlets that supply air to the rigid ductwork.

## Duct Accessories

As part of the overall distribution of air, ductwork has numerous accessories that work to direct air, reduce noise or assist with fire protection. Common accessories are detailed below.

Dampers used in HVAC ductwork are movable plates. These plates regulate airflow and redirect air to specific areas within a building. Dampers are typically used in zone control systems to manage airflow and regulate the temperature. Dampers are situated close to the mouth of the air distribution pipe. It will push or hold air from that point based upon demand. The pressure in the system will redistribute the airflow to other areas of the building that require air. These dampers operate in different ways based upon their application. Constant pressure-regulating dampers are easily installed in residential applications (Fig. 14.18).

Volume Control Dampers aid in the adjustment of the airflow volume to different sections of ductwork. These dampers can be automatic or manual. The automatic control in a simple system is provided by zone dampers while the variable air volume (VAV) controls the more complex systems.

**FIGURE 14.18** Constant pressure regulating damper.

Smoke and Fire Dampers are located in the duct system that passes through a fire curtain or firewall. These dampers are designed to seal off a duct immediately if smoke or fire is detected. The damper will remain sealed until a technician manually reopens them.

Turning Vanes are installed with ductwork to reduce the resistance and turbulence due to the flow of air. The air is directed through the vanes which enable it to change direction easily.

### Plenums

Plenums serve as the central collection and distribution units in HVAC systems. The return plenum in a building carries air from the return vents back to a centralized air handler. The supply plenum directs the air from the central system to the various locations that require heating or cooling.

Silencers within air duct systems are specifically designed to control airborne noise in ducts, openings in buildings, enclosures, or from equipment. The noise reduction capabilities make silencers ideal for commercial use at hospitals, conference rooms, hotels, and office buildings in which a silent environment is preferred. Space limitations, fan requirements, and pneumatic pressure drops are all factors that influence the incorporation of noise control products into the existing system construction.

## *HVAC FANS*

Fans are one of the key components required as part of the HVAC system. The function of the fan is to produce air movement to the space that is being cooled or heated. There are basically four types of fans that are commonly used in HVAC equipment. They are the propeller fan, centrifugal fan, vane-axial fan, and tube-axial fan.

The propeller fan is a disk-type wheel mounted on a plate. Connected to the plate is a direct drive or belt-driven motor. The fan is noisy and is only used where noise is not a factor. It is typically used as an exhaust and condenser fan.

Centrifugal fans increase the speed of an air stream with a rotating impeller. The speed increases as the air reaches the ends of the blades and is then converted to pressure. These fans are able to produce high pressure which makes them ideal for use with high-temperature, moist, or dirty air streams. The fans are categorized by their radial, forward curved, or backward curved blade shapes. Forward curved fans can create more air pressure, are noisy, and require more power to run. Backward curved fans have blades that can be curved or straight. These fans are good for high volumes of airflow and variable resistance.

Axial fans move an air stream along the axis of the fan. The blades of these types of fans generate an aerodynamic lift that then pressurizes the air. Propeller axial fan uses only a motor and a propeller to drive the airflow. The tube axial fan includes a venturi that surrounds the fan propeller and is designed to reduce air leakage. A vane axial fan has vanes trailing the rear of the propeller, which straighten out the swirling airflow as the air is accelerated by the motor. Axial fans are inexpensive, compact, and light but they are noisy.

Air terminal units are used to control the temperature of one space. These are either constant air or variable air volume models. The constant air volume units supply airflow at a constant rate, while the supply air temperature is varied to meet the thermal loads of a space. These systems are small and serve a single zone. A variable air volume terminal unit is a zone-level flow control device. These systems vary the airflow at a constant temperature.

Vent covers provide control over the air intake/exhaust rate and the direction of airflow. A vent cover also serves as an aesthetic seal over the holes and outlets for ventilation. The three main cover types are mostly grilles, registers, and diffusers.

Grilles are the least complex of the covers. The main function of a grill is to allow air to pass in and out of a room through it. Grilles are used for both air intake and air exhaust outlets (Fig. 14.19). They have no moving parts, so they are easy to maintain. Registers are the same as a grill, but with an adjustable damper (Fig. 14.20). Registers are only used to supply air to indoor spaces. Diffusers are similar to registers; however, they differ with respect to how they direct the airflow (Fig. 14.21). Diffusers are facing all around, unlike a register that only allows to flow in a single direction. Diffusers are typically located on ceilings.

**FIGURE 14.19** Grille.

**FIGURE 14.20** Register.

**FIGURE 14.21** Diffuser.

Exhaust hoods are a specialty component in HVAC systems. These are installed in kitchen applications as well as laboratories.

Commercial kitchen exhaust hoods are classified as either Type 1 or Type 2. Type 1 hoods are equipped to handle grease removal from the air in commercial kitchens. Type 2 hoods are designed for use in commercial kitchens without grease-producing appliance like grilles or fryers.

Fume hoods are used to limit the exposure of toxic fumes and vapors in the surrounding environment. These hoods are often used in laboratories to keep people safe around dangerous chemicals. A common hood is the constant air volume ducted hood, which vents unhealthy air out of the lab through a chimney. There are many other types of fume hoods. The most popular types are:

- Non-bypass Hoods
- Bypass Hood
- Low Flow/High Performance Bypass Hood
- Reduced Air Volume Hood
- Canopy Fume Hood

## HVAC CONTROLS

HVAC controls vary from fully integrated building automation systems to the wall thermostat (Fig. 14.22). The controls are the electronic devices within an HVAC system that govern the operation of the various equipment. At the heart of this system is data communication between devices, hardware, and computer systems.

Commonly referred to as Direct Digital Control or DDC, these automated controls govern the operation of several devices based upon an output signal that is generated from computer calculations. These calculations are based upon input signals and the desired environmental criteria set by the building owners and occupants. These systems collect and share information to ensure that the indoor environment is managed and the thermal comfort maintained. These systems are used in new and retrofitted commercial buildings.

While the exact make and model of the components in a DDC system will vary, there are foundational components that are standard. Controllers are devices that process input information and calculate output signals to control the environment. Sensors are the components that supply the input needed to the controllers. Common sensors include temperature, airflow, and humidity. The controllers will respond to the input from the sensors by raising the temperature if the sensor indicated the

**FIGURE 14.22** Thermostat.

air temperature is too low, or if airflow is too low, a damper control may be opened to allow for greater airflow. The controllers must consider the response time associated with implementing the information received from the sensors. This is accomplished through the programming within the controller using control loops. The control loops define the response that the controller sends to the equipment to alter the functioning. Each controller must be able to communicate with other controllers and computers that are collecting the data used to make control decisions in the building.

## SOLAR ENERGY HEATING EQUIPMENT

Commercial and residential developers and architects are seeking ways to improve the energy use within the buildings they design. One of the most expensive challenges associated with the efficient operation of any building is the management of the air quality. The goal of an HVAC system is to maintain air temperature and humidity at a certain level that is comfortable for humans to work or live in, and in many cases for machinery to operate in. Because the HVAC system is one of the largest users of energy in commercial and residential buildings the design of this system considers alternative energy sources to improve the overall energy efficiency. This reduces cost and assists the building in achieving energy-efficient ratings, such as LEED® and ENERGY STAR®, and moving a step closer to a net-zero energy status.

Alternative sources of energy are growing choices by many business and homeowners as they seek to reduce their consumption and cost. Solar energy is a popular source as it is clean and produces less strain on the environment. Solar HVAC systems are used in both commercial and residential applications. These systems must first collect and capture solar radiation. There are several ways to accomplish that task.

A solar collector is a device that collects solar radiation from the sun. They are primarily used for active solar heating systems. These collectors are generally roof mounted. There are many different types of solar collectors. All of these are constructed in the same basic manner. There is some material used to collect and focus the sun's energy and use it to heat water.

Flat plate collectors are metal boxes that have some sort of transparent cover on top of a dark-colored absorber plate. The sides and bottom of the collector are insulated to retain as much solar energy as possible. Solar radiation heats up the plate and the heat is transferred to water or air that is held between the plate and the transparent cover. The absorber plates are typically constructed from copper or aluminum.

Evacuated Tube Collectors use a series of evacuated tubes to heat water for use within a building. These tubes utilize a vacuum to capture the sun's energy. The unit has a metal tube which acts as the absorber plate. This tube is connected to a heat pipe that carries the water heated by the sun under pressure. The pressure allows for the thermal energy transfer to the water.

A concentrating solar collector uses reflective surfaces to concentrate sunlight onto a small area. The concentrated light is absorbed and converted to heat. The light can also be converted to electricity if run through a solar photovoltaic device. Concentrators can increase the power flux of sunlight hundreds of times.

The solar air heaters work by drawing in fresh outside air and circulating through a maze of black aluminum that has a special selective coating that absorbs much more heat than it emits. This heated air is then blown it into the dwelling via a solar-powered fan.

These collectors are used to concentrate the energy for use in solar-powered systems. Two common systems used in the building industry include:

1. Solar Hybrid
2. Solar AC with super-heated refrigerant

The solar hybrid HVAC system utilizes the solar energy generated by photovoltaic panels to supplement the electrical demand of the compressor. This reduces the need to draw electricity from the supply electrical grid of the building. Because the electrical demand is very large from HVAC systems, many solar panels are required to reduce the electricity required to power them. While the initial investment in these systems is high, the cost savings that can be realized over the life of the building are substantial.

Emphasis in solar-powered HVAC systems has increased due to the rising demand for energy-efficient air-conditioning systems for both residential and commercial applications. Solar-powered air-conditioning and minisplit systems are entering the market after solving technical challenges to the operation. The first is that the sun only shines for about half of the day and without a storage system, the solar cooling system will not operate at night. The second is the amount of energy that is required by typical heat pumps to generate cool air.

Technology advances have improved performance and energy storage as well as the overall operation of minisplit systems. Minisplits are heating and cooling systems in which the temperature in individual rooms or spaces can be controlled. These systems consist of an outdoor compressor and an indoor air-handling unit. They are easy to install and are good for home or small office renovations where connection to the building ductwork is not feasible. In most installations, the system only requires a small hole in the building envelop to run a conduit to connect the inside and outside units. The conduit houses the power and communication cables, the condensation drain line, and copper tubing. Many of the minisplit heat pumps are equipped with variable speed compressors so they can adjust their electricity use based on demand.

The heat pump used with photovoltaic-generated electricity must be a high-performance unit. The performance of a heat pump will vary depending on the weather and the amount of heating and cooling required. The measurement of heat pump efficiency is more realistic if determined on a seasonal basis. The Seasonal Energy Efficiency Ratio (SEER) is the rating system applied to the heat pump. The SEER rating is calculated by dividing the amount of heat removed from the air (in Btu) by the total amount of energy required by the air conditioner in watt-hours. The SEER rates the seasonal cooling performance of the heat pump. The higher the ratio is, the more energy efficient the unit is. SEER values in the teens are considered to be good when it comes to heat pump efficiency. These minisplit systems have achieved SEER values of 35, making them extremely efficient to operate.

The efficiency of a solar-powered air conditioner is challenged by the fact that photovoltaic panels only convert energy during daylight hours. For the air-conditioning system to run at night some type of battery storage is required. Battery storage is expensive and inadequate for all-night cooling operation.

The primary benefit of the hybrid solar-powered air conditioner is the elimination of the batteries. These systems are integrated into the power grid of the building. This also eliminates the need for special contracts with the local utility as the solar unit does not feed any excess power to the grid. When the photovoltaic panels stop converting energy, the system uses the electricity from the building's power grid. This also reduces the number of panels that must be installed in the array. The sizing of the photovoltaic array to supply the correct amount of electricity to power one of these solar air conditioners makes it an attractive alternative to stand-alone solar air-conditioning systems. In addition, there are systems available that are designed to integrate with the HVAC and other systems in the building. This allow for the solar energy to supplement the operation of other parts of the electrical demand when the HVAC system is not using the power.

Solar direct-expansion (DX) air conditioning takes the concept of solar hybrid air conditioning to the next level. The technology used in this method involves the superheating of refrigerant using solar heat captured in an evacuated-tube collector. These units are available for both residential and commercial split systems and package units. The SEER value for these units is as high as 37.5.

The technology involves altering the way the compressor heats and compresses the refrigerant. The standard air-conditioning system uses a vapor expansion compression cycle where the compressor does the majority of the work and consumes the most electricity. The system uses solar radiant energy to superheat refrigerant in the air-conditioning system. The refrigerant is diverted to an evacuated-tube solar thermal collector superheating it and allowing for more efficient condensing. This decreases the workload of the compressor, increasing its efficiency and decreasing the cost to operate the HVAC system.

According to the manufacturer it is the increased temperature difference between the condenser coil and the ambient air around it that allows the entire coil face to increase the heat exchange efficiency of the entire system—allowing cooler air to be generated in your home or office for less money.

This type of system was subjected to a test in Florida. A 5-ton commercial split system was installed in an office building. The system initially received a 15 SEER rating. A solar thermal collector superheating system was added to the installation. The result yielded a SEER improvement to 31.5 by an independent engineer. In addition to improving the efficiency, the capacity of the system is also increased due to the additional refrigerant required.

## GEOTHERMAL SYSTEMS

Geothermal heat pumps, also known as ground source heat pumps, provide highly efficient renewable energy for heating and cooling. The technology is applicable for both residential and commercial buildings.

A geothermal system takes advantage of the fact that approximately 5–6 ft beneath the earth's surface the ground maintains a constant temperature of 55–70°F depending on the location. It is warmer than the air at the surface in the winter and cooler than the air above it in the summer. The geothermal heat pump transfers heat stored in the earth or groundwater into a building during the winter and then transfers the heat out of the building into the ground in the summer. The ground acts as both a heat source and a heat sink. A geothermal HVAC system can save 25–50% on HVAC costs compared to conventional systems using air-source condensing units for cooling and fossil fuels for heating.

The entire geothermal system includes three principal subsystems:

1. **Earth Connection Subsystem**
   This subsystem is a series of pipes that are connected in a loop that is buried underground near the building that is being conditioned. Depending on the space available onsite and the size of the building being conditioned, the loop can be buried in either a horizontal or vertical orientation. These pipes circulate either water, or a mixture of water and antifreeze that absorbs or releases heat to the soil depending upon whether it is operating to heat or cool the building.
2. **Heat Pump Subsystem**
   The ground source heat pump during the heating process removes heat from the fluid circulating in the loop, concentrates it, and transmits it to the building. During the cooling process the heat is removed from the building and transferred to the circulating fluid then back to the ground.
3. **Distribution Subsystem**
   The heat pump is generally connected to a conventional system of ductwork and fans that will distribute the heated or cooled air throughout the building.

The efficiency of a geothermal system is greater than an air-sourced heat pump due to the fact that the system does not have to heat or cool the air. It is simply moving the heat from one place to another. Advanced models of geothermal systems are available with two speed compressors and variable fans that provide addition energy savings as they are only operated when the demand exists. When comparing a geothermal system to a traditional air-sourced heat pump, they last longer, require little maintenance, and run quieter. However, they are more costly to install than an ASHP system. Dual-source heat pumps offer an energy-efficient combination of an ASHP with a ground-source heat pump. These systems are more efficient than regular ASHP systems and more affordable than a full geothermal installation.

### Types of Geothermal Systems

Geothermal systems are either open- or closed-loop systems of some variety. The type of installation used depends upon a few factors:

- Climate
- Soil conditions
- Amount of land available
- Installation costs

**Closed-Loop Systems:** The closed-loop system is comprised of a continuous loop of buried pipe. The pipe material is polyethylene and they are joined by thermal fusion. The process involves the heating of the pipe connections and then fusing them together. This forms a joint stronger than the original pipe section. PVC pipe should never be used in a closed-loop system.

Once the pipe loop is installed in the ground, it is connected to an indoor heat pump. This forms a sealed underground loop through which a water-and-antifreeze solution is circulated. The system continuously recirculates the fluid in a pressurized pipe system. Most closed loops are trenched horizontally in areas adjacent to the building (Fig. 14.23). However, where adequate land is not available, loops are vertically bored. Types of closed-loop systems include horizontal, vertical, and pond/lake.

**FIGURE 14.23** Residential closed-loop system installed in horizontal trench.

The horizontal installation is a cost-effective means most applicable to residential applications (Fig. 14.24). It works well for new construction where sufficient land is available and the soil conditions are appropriate. Loop lengths and depth will vary with soil conditions and climate. The horizontal installation requires trenches between 4 and 6 ft in depth. The length of the pipe depends on the application but a standard rule of thumb is that 500–600 ft of pipe is required per ton of system capacity.

The vertical installation is most often used when the scale of the geothermal system is larger, for example, in commercial building applications. In these circumstances land is generally limited. The vertical installation is also a solution to poor or shallow soil conditions where burying horizontal loops in trenches is not feasible. It is more costly to install the vertical system; however, they require less land and minimize overall surface disturbance.

The pond/lake installation is the least expensive option of the closed-loop systems. They require access to an adequately sized body of water. This system uses a supply line pipe that runs underground from the building to the water source. The piping system is coiled into circles that are located at least 8 ft under the surface to prevent freeze. The criteria for the water source depend upon the size of the building being fed. Factors to consider when evaluating the water source include the volume, depth, and water quality.

**Open-Loop System:** Open-loop systems require access to a source of water. This type of system takes advantage of the fact that groundwater is at a relatively constant temperature throughout the year. Groundwater from an ordinary well or a surface body of water is used as the heat exchange fluid.

**FIGURE 14.24** Buried geothermal pipe indicator installed on residential application.

The water is pumped into the heat pump unit and the heat is extracted. Once the water has circulated through the system, the water returns to the ground. It can return through a source well, recharge well, or surface discharge. This option is only viable if there is an adequate water supply and if all local regulations regarding groundwater or surface water discharge are met.

The use of a geothermal system in commercial applications has become a popular option for heating and cooling. The major differences between residential and commercial systems are the number of heat pumps and the vertical borehole system. In a commercial facility, a series of heat pumps work to concentrate the low-grade heat from the vertical borehole system. The heat is transferred to the building into the HVAC distribution system using a heat exchanger. This HVAC distribution for commercial and institutional applications is typically an air system. A water-to-air system is often used for a larger multizone commercial building. Radiant-in-floor heating and hydronic fan coils for cooling are less commonly used options.

Commercial geothermal heat pump systems have been installed in various types of facilities to include schools, police stations, and office buildings. They can work in any large facility. These systems help many facilities attain LEED certification or help them on a journey toward zero energy.

Most commercial systems are vertical borehole loops that are installed under a parking lot or a recreation field. The heat is removed from these loops and transferred to the building using a series of heat pumps depending on the size of the facility. The heat is then distributed through the facility via ductwork or by more energy-efficient means. These can include large-scale radiant-in-floor heating and hydronic fan coils for cooling. The entire system is flexible and can be modified to fit any size and type of commercial building.

When selecting the equipment for the system, there are some critical items that must be appropriately sized. The heating and cooling loads for the building must be calculated. The loads will determine the heat pumps required. The climate will determine if the fluid in the pipes needs to be a water/antifreeze mixture. It will also determine if the equipment size should be based on the heating or cooling load. When the heating dominates, optimal heat pump sizes may be 65–75% of the design heating load. It also helps with energy efficiency if variable speed systems are selected. This will accommodate the heating requirements and provide the necessary cooling.

The sizing of the ground loop, whether horizontal or vertical, is the most important part of the geothermal system. There are common parameters associated with vertical ground loops associated with borehole depth and length of pipe. The boreholes can range anywhere from 150 to 450 ft in depth. Typical systems require between 150 and 200 ft of pipe per ton of load.

Geothermal systems possess long-term reliability. The average life of a heat pump is about 20 years and the underground piping infrastructure can last up to 50 years. These systems are also efficient. They can move between three to five times the energy they consume.

## HVAC COMMISSIONING

HVAC system commissioning involves the testing of the equipment and components of the system to ensure they function as designed and specified and that they meet the required performance criteria set by codes and regulations.

The commissioning process can begin as early as the predesign phase, but must be completed as part of the final acceptance phase. The opportunity for construction managers and contractors to be involved in the commissioning process occurs at all phases. The earlier they are involved in the process the smoother it typically proceeds.

During the predesign phase, the construction manager and general contractors are engaged in developing a preliminary commissioning plan that meets the needs of the owner. The involvement at this stage is ideal and may not occur on every project, given the nature of the contract and project delivery method. During predesign, the owner will identify the phases of the building that will be commissioned and the systems that will be commissioned.

During the design phase, the focus of commissioning is centered around the creation of the construction documents. The goal is to ensure that the owner's project requirements related to commissioning, the maintainability, functionality, and best practices are incorporated into the construction documents. The inclusion of specialty trade contractors during this process is highly recommended.

This phase should define the selected HVAC systems that meet the heating, cooling, and ventilation loads for the building. Included in the documents should also be the requirements for the commissioning process and their associated specifications. This will include the specification describing the commissioning activities required of the mechanical, electrical, controls, and testing and balancing contractors as well as any vendors or equipment supplier. The owners should also conduct a prebid meeting to answer questions that contractors may have about the commissioning process.

The construction phase of the commissioning process verifies that the equipment and systems are installed in accordance with the design documents. They also have to ensure the equipment is operable, maintainable, and ready for functional testing.

The commissioning team consists of various stakeholders on the project. It includes the owner, commissioning authority, design team, as well as the general contractor and the HVAC subcontractors. Typically, the testing that is required of the commissioning process is conducted by a third-party vendor or a testing and balancing specialty contractor.

The construction commissioning phase begins with a complete plan for the commissioning process. Checklists and forms should be developed to account for each required commissioning activity. A schedule for the commission activities needs to be included in the construction schedule and adequate time should be allotted for the activities.

The construction commissioning process begins with a review of all required submittals and shop drawing reviews. These reviews assist in the development of commissioning check sheets. These reviews ensure the systems are in conformance with the owner's requirements, the contract documents, and the overall functionality requirements. Any issues identified during this review should be recorded in a log that is then forwarded to the appropriate stakeholder for resolution. It is critical to document these issues as part of the project record.

Commissioning authorities visit the construction site throughout the construction process to inspect the progress with respect to the HVAC systems, equipment, and components being commissioned. The goal is to verify the construction is progressing according to the construction document. It is also to identify and document any quality issues that may lead to functional issues.

There are several types of prefunctional tests on HVAC equipment during construction. Installation verification tests, static tests, and equipment startup tests. Contractors must document that all equipment is installed in the correct location and is operable in accordance with the contract

documents. This should be coordinated with the commissioning authority. Installation Verification check sheets completed by the contractor shall be verified through site observations. Completion of all installation verification tests signifies that those systems are ready for startup activities.
Typical systems to review in the verification process include:

- Piping systems and component installations
- Duct systems and component installations
- HVAC equipment installations
- Building Automation and Control equipment and component installation

The installation verification shall include:

- Material or equipment name or description
- Model number
- Type and size
- Serial number
- Installation and location meet the design requirements
- Maintenance access verification
- Installation complete and ready for start up

The installation verification is a vital step in the commissioning process. Issues can be identified and corrected during equipment installation, resulting in faster and more efficient resolution. The items listed about may seem obvious, but checking them may uncover deficiencies. Some things to pay attention to during this process are the make and model as well as the energy-efficiency ratings on electric motors.

When verifying the make and model of the equipment, be sure to use the approved submittal for construction, as it contains the correct information regarding the equipment. Verification of the equipment make and model may seem trivial; however, errors in shipping or ordering may have taken place. An example of where this step is critical is when there are multiple pieces of a similar type of equipment. This could be a VAV box, compressor, or fan coil. If the facility is large, these pieces of equipment may be installed in the wrong location. The installation verification process is important as it can have a large effect on the performance of a building.

Electrical motors are used everywhere in commercial HVAC building systems. It is important to ensure that these motors meet not only the approved-for-construction submittals, but also the local building codes. For example, if the incorrect fan motor is installed in an HVAC rooftop unit, the building may not meet local energy codes, not to mention the additional expense to operate.

The contractor must perform hydrostatic and pressure tests as described in the contract documents. The hydrostatic and pressure tests must be verified by the contractor or code authority according to the contract documents or according to local codes.

- Hydrostatic testing of all piping systems to the appropriate test pressure and duration
  - Testing requirements for piping systems are covered in ASME B31.3 code as well as local codes
- Pressure testing of all ducts for leakage determination
  - Testing requirements for pressure testing ductwork are outlined in ANSI/SMACNA 006-2006—HVAC Duct Construction Standards and local codes

HVAC equipment startup is to be conducted by the subcontractor or vendor indicated in the contract documents. The systems and associated equipment has a startup checklist provided by the manufacturer. The contractor or vendor with the contractual obligation to provide startup services shall conduct them in accordance with the Contract documents and the manufacturer's requirements. It is critical to start up the equipment according to the manufacturer's instructions as failure to properly start up equipment may void the manufacturer's warranty. It is important to record the date of equipment startup as the 1-year warranty may begin on that date. Startup reports should be submitted as part of the HVAC subcontractor's project documentation.

The startup process will vary depending on the system and equipment. The verification may be as simple as checking to see if a fan is spinning in the proper direction to checking the refrigerant charge in chillers and verifying correct amperage drawn from pumps. Typically, startup tests at a minimum will include:

- Manufacturer's name
- Model number
- Type and size
- Serial number
- Actual settings or adjustments
- Actual electrical characteristics
- Actual load
- Operating data as required by the manufacturer

The testing and balancing process is a critical step in HVAC installation and the commissioning process. The system should be tested and balanced by a certified test and balance technician.

The HVAC system must be balanced to ensure that the correct air and water flows are present during the normal operation of a system. To balance the airflow systems, the balancing technician will alter fan rate and adjust equipment to ensure the desired airflow occurs. Water flow in an HVAC system is also balanced whether from a chiller, cooling tower, or boiler. The technician will adjust the pumps until the water is flowing at the appropriate rate.

The only way to ensure that a system is balanced properly so that the occupants will be comfortable is to have the system balanced by a technician following the guidelines of the two national balancing agencies: the National Environmental Balancing Board (NEBB) or the Associated Air Balance Council (AABC).

An important part of the building commissioning process is the Functional Performance Testing (FPT). FTP is the process of putting the HVAC system and associated equipment through every possible condition it will ever experience. The goal is to make sure a piece of equipment works as intended and can provide the expected performance. To accomplish this, each system and associated equipment has specific testing protocols which must be completed. The commissioning team monitors the performance and ensures the equipment performs over the desired range of loads and functional levels.

Commissioning documents that must be collected and delivered to the owner include:

- Final approved submittals
- Final approved shop drawings
- Record documents
- Service procurement information
- Operating manuals for each piece of equipment, systems, and components installed
- Maintenance requirements and procedures
- Recommended spare parts lists
- Warranty information

These documents should be reviewed, approved, and delivered to the owner.

# CHAPTER 15

# ENERGY-EFFICIENT BUILDING SYSTEMS

*The building sector accounts for over 12% of water use, 40% of $CO_2$ emissions, 65% of all waste outputs, and more than 70% of electricity consumption. There is a large potential to reduce energy consumption by constructing buildings as energy efficiently as possible. To accomplish this, buildings are designed and constructed using a whole-building approach. This approach ensures that all parts of the building are designed to work together to include the building envelope, the heating and air-conditioning systems, the lighting, the water distribution systems, as well as any equipment and building control systems. The design of efficient systems, coupled with improved construction techniques, assists builders to meet new energy requirements and certification standards and codes.*

## ENERGY-EFFICIENT BUILDINGS

Energy-efficient buildings can be defined as buildings that are designed and constructed to provide a significant reduction in the energy used to operate and maintain the structure. From the very beginning of each project, a team addresses all aspects of the building's construction. The team includes architects, engineers, developers, and owners, and even the people who will occupy and work in the building. Together, they set goals for energy efficiency, performance, and creative use of space. By adopting a collaborative approach, buildings can be up to 70% more efficient than conventional commercial buildings.

The United States does not have a national energy or building code. State and local governments adopt one of the national model energy codes or their own state- or local-specific code. Energy codes are applied to and related to the overall building. It does not include any appliances or equipment that are installed in the building; however, there is some overlap in areas such as lighting. Equipment and appliances have their own energy standards.

Ensuring that builders are complying with the energy codes is the responsibility of local permit or code officials who will conduct field inspections to ensure that code compliant plans are followed during construction. The US Department of Energy works to develop education and training materials that can be used to educate designers, builders, and inspectors on these codes. In addition, the DOE is developing methods to determine energy savings opportunities in both the residential and commercial sectors.

Many green or energy-efficient building programs that are adopted by owners not only address the building structure but also include site layout, orientation, and other features that reduce the heating and cooling loads on the structure. If an owner only addresses the building structure and not the entire site, ensuring the building is energy efficient is simpler. The design and construction can be tied directly to specific energy codes, like the International Energy Council Code (IECC) or the American Society of Heating, Refrigerating and Air-Conditioning Engineers (ASHRAE) 90.1 code that set the standard for the systems in the structure. If an owner considers the building from a much broader scope, codes and certifications such as ASHRAE 189.1 or Leadership in Energy and Environmental

Design (LEED) are more likely to be the standards to which a builder must comply. These energy codes are key to ensuring the building systems meet energy efficiency standards.

A key thing that builders need to understand when constructing energy-efficient buildings is how the efficiency will be measured upon completion. There are many ways a building's energy efficiency can be measured. A national code or standard could be applied as a benchmark with a percentage above that benchmark or a compliance with a unique method developed by the local municipality may be required.

States and cities have developed local regulations that are well beyond minimum national energy code standards. As an example, California, Washington, Vermont, and New York as well as cities such as Boulder, Colorado, the District of Columbia, and Scottsdale, Arizona are developing and adopting energy codes that are significantly more stringent than the IECC and ASHRAE 90.1 standards. Most notably, California's statewide energy code Title 24 has incentivized local governments in the state to develop and implement stretch codes that are even more stringent.

These codes utilize high-efficiency measures in lighting, heating, and cooling, and building envelope to maximize reductions of energy needed to power the building. National standards, benchmarks, and certification programs that are commonly used in measuring energy efficiency include:

- LEED Rating System
- IECC
- ENERGY STAR
- ASHRAE Standard 189.1 or 90.1
- HERS Index rating

## LEED Rating System

The Leadership in Energy and Environmental Design (LEED) certification is a green building rating system that was developed by the United States Green Building Council. There are various types of LEED certifications that ensure the quality of a building or neighborhood's green features. LEED works for all phases of building development from design and construction to the operation and maintenance of energy-efficient buildings.

In order to achieve a LEED certification, projects earn points across nine areas that focus upon key elements of green buildings. These areas are:

1. Integrative process
2. Location and transportation
3. Sustainable sites
4. Water efficiency
5. Energy and atmosphere
6. Materials and resources
7. Indoor environmental quality
8. Innovation
9. Regional priority

When addressing the concept of energy efficiency, each LEED program for commercial buildings includes points for a percent savings above the ASHRAE 90.1 code requirements, on-site renewable energy sources, heat island effects, enhanced commissioning, and other areas.

Other requirements related to energy specify the number of energy and atmosphere points that must be acquired in LEED to ensure it goes beyond minimum LEED certification. The certification is

**TABLE 15.1** LEED Rating Levels

| Points achieved | LEED rating level |
|---|---|
| 40–49 | Certified |
| 50–59 | Silver |
| 60–79 | Gold |
| 80–110 | Platinum |

achieved in levels, each corresponding to a certain number of points. Table 15.1 illustrates the LEED v4.1 rating system for commercial buildings.

The International Code Council (ICC) has developed model construction codes that all 50 states utilize, to include the International Building Code (IBC) and International Residential Code (IRC). The ICC also developed both the International Energy Conservation Code (IECC) and the International Green Construction code (IgCC).

The 2018 International Energy Conservation code established a baseline for energy efficiency in both residential and commercial buildings and has been adopted by most states. The code sets performance standards for the building envelope, mechanical systems, lighting systems, and service water heating systems. The US Department of Energy (DOE) recognizes the IECC as the minimum residential code but references ASHRAE 90.1-2007 as the minimum commercial energy code.

ENERGY STAR is not a code, but a voluntary program sponsored by the US Environmental Protection Agency (EPA). The program applies to homes and to commercial buildings and seeks to help owners improve the energy performance of these structures. The commercial buildings program seeks to benchmark buildings both new and existing against other buildings of similar use across the nation. Commercial buildings are eligible to receive the EPA's ENERGY STAR certification, while residential homes can receive an ENERGY STAR Label.

For commercial buildings to receive ENERGY STAR certification, an ENERGY STAR score of 75 or higher must be earned. This indicates that the building performs better than at least 75% of similar buildings nationwide. The objective of the ENERGY STAR score is to evaluate the energy performance for the whole building while reflecting the actual metered energy consumption. The score also considers the different sources of energy and then normalizes the energy use based on building activity. The program also provides a peer group against which energy performance can be compared.

The EPA awards a 1-100 ENERGY STAR score using the Portfolio Manager system where owners input information about the building size and operation, the actual energy use, regional weather information, and other data. Once the data is entered, actual and predicted source energy use intensity are calculated and used to determine the efficiency ratio. The efficiency ratio compares the actual use with the predicted use for the building. The ratio is then compared with the national distribution of similar buildings to determine the score.

ENERGY STAR requirements for a home require that energy-efficient systems are installed and their performance measured and verified by a third-party Home Energy Rater. The builder must construct the home to the strict energy efficiency requirement and must have the home certified by an Energy Rating Company. This company completes a review of the design and helps the builder choose the energy-efficient features that are appropriate to the local weather conditions that will meet the ENERGY STAR requirements. The company also conducts onsite inspections and testing during the construction process to ensure standards are met. Two typical tests that are performed are the blower door and duct blaster test. The blower door test is designed to measure the whole-house air leakage and the duct blaster test measures air duct leakage. The last thing the Energy Rating Company completes is a final inspection of the home to determine all ENERGY STAR requirements have been met in order to earn the ENERGY STAR label.

Home Energy Rating System (HERS) Index is a scoring system that was developed by the Residential Energy Services Network (RESNET). The scoring is accomplished by a certified home energy rater who assesses the energy efficiency of a home. The HERS reference home is a home built to the

International Energy Conservation Code standard and is awarded a score of 100 (Fig. 15.1). The reference home is a model home that is of the same size and shape as the actual home being assessed so the score is relative. The lower the number is, the more energy efficient the home is. Each 1-point deduction in the index corresponds to a 1% reduction in energy consumption when compared to the HERS reference home. For example, if a home is awarded a HERS index of 80, it is 20% more energy efficient than the reference home and conversely if the home earns a HERS index of 120 it is 20% less efficient than the reference home.

To calculate the HERS index rating, a certified inspector provides third-party verification of the construction and home performance according to specifications. The areas of the home that are included in the assessment include:

- Above- and below-grade exterior walls
- Floors over unconditioned spaces
- Ceilings and roofs
- Attics, foundations, and crawlspaces
- Windows and doors, vents, and ductwork
- HVAC systems, water heating system, and thermostat

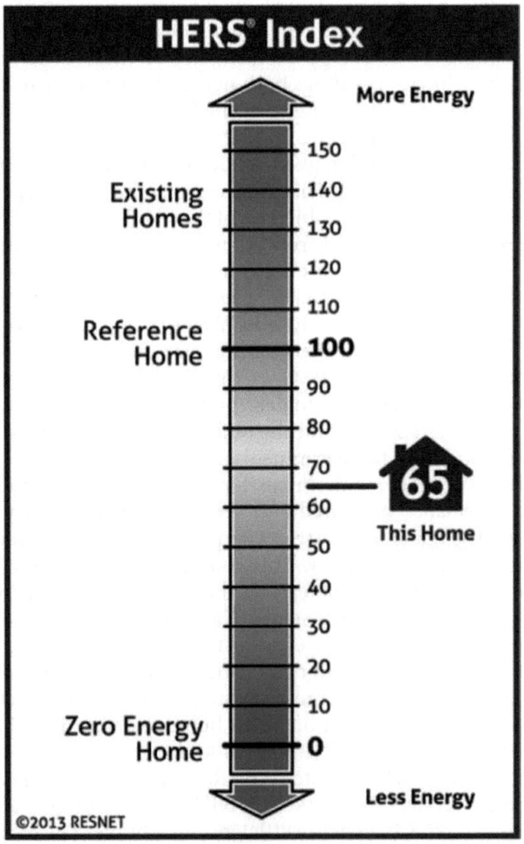


**FIGURE 15.1** Home Energy Rating System (HERS) index.

The ASHRAE Standard 90.1, Energy Standard for Buildings Except Low-Rise Residential Buildings, provides minimum energy performance requirements for all commercial and residential buildings of four stories or more. The standard includes the design and construction of the buildings as well as the operation and maintenance of the facility. The code specifically addresses performance and construction criteria for the building envelope, heating, ventilation and air-conditioning systems, service water heating, power sources, lighting, and other equipment installed in the building. It also contains criteria for determining compliance with these requirements. Every 3 years, ASHRAE reviews, updates, and publishes a new edition of the code. 2019 is the latest edition.

## ASHRAE Standard 189.1

The ASHRAE 189.1, Standard for the Design of High-Performance, Green Buildings Except Low-Rise Residential Buildings, is collaboratively developed and maintained by ASHRAE, the Illuminating Engineering Society of North America (IES) and USGBC and the ICC. The goal of the standard is to provide minimum requirements for the siting, design, construction, and plans for operation of high-performance green buildings. The standard seeks to better align green building codes across ICC and LEED, and serves as the technical content of the IgCC. The standard has been adopted by local jurisdictions across the nation. It provides the technical basis for mandatory building codes and regulations for high-performance green buildings.

This code addresses energy efficiency, a building's impact on the atmosphere, sustainable sites, water-use efficiency, materials and resources, and indoor environmental quality.

## Thermal Comfort

Thermal comfort can be defined as the condition of mind that expresses satisfaction with the thermal environment and is assessed by subjective evaluation. Thermal comfort in buildings is affected by the way the body regulates heat. This is through conduction, convection, thermal radiation, and evaporative heat loss.

Thermal comfort is the result of a combination of building systems that are designed to meet the needs of the building location, intended use, and energy efficiency goals. An efficient building envelope plays a large role in the thermal comfort of a structure as does the HVAC system.

A comfortable temperature for building occupants is maintained by providing a balance between the heat lost though conduction and ventilation with similar heating sources in the building. Thermal comfort is maintained by allowing the heat generated by the occupants to dissipate at a rate equal to their own personal thermal equilibrium. Heat gained or lost beyond the equilibrium of the body causes physical discomfort of the occupants. Uncomfortable working environments can lead to decreased productivity and unhappy occupants.

Commercial buildings and many residential buildings rely on HVAC systems to maintain a standard of thermal comfort for occupants. Various codes and standards are associated with HVAC systems as well as energy efficiency standards. HVAC systems control surface temperature, air temperature, relative humidity, air motion, and heat loss from buildings. Heat loss in a building depends upon factors as the materials and construction methods used in the building envelope and environmental factors such as the orientation of the building in relation to the sun and the amount of natural elements (wind, rain, snow) to which it is exposed.

To ensure thermal comfort is maintained and energy efficiency standards are met, a building must manage the following:

- Air temperature
- Surface temperature
- Humidity
- Absence of draughts

ASHRE standards that define thermal comfort and optimize energy efficiency in buildings are both ASHRAE Standard 55, Thermal Environmental Conditions for Human Occupancy, and ASHRAE 62.1, Ventilation for Acceptable Indoor Air Quality. Couple these with the requirements with building envelopes, green buildings, and other energy-efficient codes, and thermal comfort plays a large role in building construction.

ASHRAE 55 standard defines the occupied zone in a building as: "The region normally occupied by people within a space, in absence of known occupants, generally considered to be between the floor and 6 feet level above the floor and more than 3.3 feet from outside walls/windows or fixed HVAC equipment and 1 foot from internal walls."

ASHRAE 62.1 standard addresses the requirements for ventilation air in buildings. It defines ventilation air and the breathing zone as: "That portion of supply air that is outdoor air plus any recirculated air that has been treated for the purpose of maintaining acceptable indoor air quality. The breathing zone is the region within the occupied space between planes, 3 and 72 in. above the floor."

The challenge associated with thermal comfort is that it is related to the preferences of individual occupants as well as the energy codes. ASHRAE has defined six thermal comfort variables that impact the sensations of hot and cold that a building occupant feels. The variables interact to predict thermal comfort. They are a combination of environmental and personal factors. The environmental factors include the ambient air temperature as it is adjusted through natural ventilation and effects of solar gain; the radiant temperature achieved from the thermal mass in the floors and walls; the relative humidity from the water vapor in the air; and the rate at which air moves about and touches skin, the air velocity. These parameters are correlated to the thermophysical conditions of the building envelope and the HVAC system. The other two factors are personal to the occupants and include the metabolic rate based on the amount of energy expended and the clothing worn to retain or remove body heat.

The ASHRAE 55 standard analyzes the heat balance equation for the human body together with the six parameters to develop the predictive mean vote (PMV) index. The PMV index is expressed on a thermal sensation scale. This is a 7-point thermal scale that ranges from −3 on the cold side to +3 on the hot side. The PMV index defines thermal comfort conditions, not thermal sensations, so it can be used as an index to assess the thermal performance of the building HVAC system.

The predicted mean vote PMV scale for measuring thermal comfort is shown in Table 15.2 from the ASHRAE standard 55.

PMV can be used for the design of new buildings or for the assessment of existing buildings. Thermal comfort guidelines recommend that interior building spaces be controlled to maintain a PMV between −0.5 and +0.5. ASHRAE 55 specifies a vertical air temperature difference between the head and ankles of occupants to be no more than 5.4°F. The humidity level is not specified; however, the dew-point temperature is required to be less than 62.2°F. Air velocity should be less than 50 fpm during cooling mode and less than 30 fpm during heating mode. The PMV is viewed in relation to another variable, the Predicted Percent Dissatisfied (PPD). This indicates the percentage of people who are comfortable at the given temperature.

**TABLE 15.2** PMV Thermal Comfort Scale

| Value | Sensation |
|---|---|
| +3 | Hot |
| +2 | Warm |
| +1 | Slightly warm |
| 0 | Neutral |
| −1 | Slightly cool |
| −2 | Cool |
| −3 | Cold |

## *VENTILATION*

Critical to thermal comfort of occupants, indoor air quality, and the energy efficiency of a building, ventilation plays a major role in addressing these criteria. Ventilation moves outdoor air into a structure and then distributes the air throughout the structure.

When ventilating a building, ventilation rate, airflow direction, and air distribution are key components. Ventilation rate is the rate at which external fresh air flows into the building. Airflow can be controlled or uncontrolled and will depend on the location in the building. Air distribution refers to how air is moved to and from condition spaces within a building. Air distribution systems include all the components required to move the air, like fans, ductwork, filters, and dampers.

When improving the energy efficiency of a building, reducing air change rates helps to minimize the energy used to condition the indoor air. This concept poses a challenge to the quality of the indoor air if the amount of outdoor air is insufficient to replace air exhausted in a building by people, equipment, or exhaust fans.

Buildings can be ventilated using natural, mechanical, and mixed-mode or hybrid forms of ventilation.

In simple terms, natural forces such as wind or buoyancy pull fresh outside air through designed building envelope openings, like windows, solar chimneys, doors, or wind towers. The fresh air helps to force out the dirty, warm air inside the building. Minimum air change rate requirements and the need for exhausting of known sources of contaminants are the main mechanisms codes address.

There are two types of natural ventilation in buildings. Wind-based or cross-ventilation and buoyancy-driven ventilation.

Wind-based or cross-ventilation is a simple concept. Wind creates areas of high and low pressure as it moves across a structure. The windward side of a structure is under high pressure, while the leeward side and roof of the structure are low-pressure areas. Having various sizes and types of openings placed on both the high- and low-pressure sides causes the air to move through the building. Designing the location of these openings carefully can control the rate and direction of the air movement. This is a low-cost ventilation system, with limitations. This system depends on the direction in which the wind is blowing in addition to the fact that the wind may contain various pollutants and dust.

Buoyancy-driven ventilation or the stack effect works just like a fireplace. Warm air rises and colder air stays low, so the warm air is forced up. In a building structure, the natural tendency of warm air is to rise up seeking a place to escape. In building structures, this place is in the form of a natural ventilator. Cool air enters the structure through vents in the wall at floor level and replaces the warm air that has escaped. This method can work no matter which direction the wind is blowing.

### Mechanical Ventilation

Mechanical ventilation systems circulate fresh air through buildings using a combination of ducts, fans, vents, and other components. These systems can provide improved indoor air quality through filtration of pollutants, control of moisture, and removal of allergens. These systems provide more control over the amount of fresh air that enters a facility, to include appropriate airflow volumes required by code. These systems also provide a means to condition the outside air and provide dehumidification.

Commercial buildings use three common methods of air distribution:

- Partially mixed (most underfloor air distribution systems)
- Fully mixed (overhead distribution)
- Fully stratified (displacement ventilation)

**Partially mixed systems:** These systems conserve energy by conditioning the lower occupied level of a space and stratifying its upper-occupied level. Swirl diffusers or rectangular-shaped outlets delivering conditioned air from the plenum under the floor provide occupant thermal comfort. These systems commonly have a lower first cost than fully mixed systems and due to their low-supply air pressure, they help with energy efficiency.

**Fully mixed systems:** These systems deliver air to the space through overhead outlets and diffusers. The key with these systems is the pattern of air delivered to the space. For example to provide uniform temperatures a circular pattern is ideal for variable air volume (VAV) cooling. The cross-flow air pattern has longer throw, but its reduced induction means it may lose ceiling effect, which creates drafts in the occupied zone. Due to their flexibility, fully mixed systems can meet most air distribution challenges. These systems typically have the lowest first cost, making them more economical, but they may not help with energy efficiency

**Fully stratified systems:** These systems provide air to spaces through discharged cool supply air via an outlet placed at floor level. Low-velocity air (<80 fpm) is discharged horizontally across the floor; until it hits a heat source, this air moves with little mixing across the floor. This cooled air will mix with radiant heat from a source, and then stratify toward the ceiling.

Thermal displacement ventilation (TDV) systems offer energy savings and efficiency. They require less ventilation air to comply with ASHRAE 62.1, and they can use air-side economizers and warmer temperatures to match supply-air temperatures.

Various codes address ventilation and energy efficiency in commercial buildings. The IECC specifically focuses upon energy efficiency and ASHRAE Standard 62.1 focuses upon ventilation requirements. Table 15.3 highlights key requirements from both standards relating to HVAC and Ventilation requirements in buildings.

To meet the demands of ASHRAE and IECC codes, coupled with a need for performance above these standards to meet other green and energy-efficient building standards, HVAC systems need to be highly efficient. HVAC product manufacturers have developed products and systems that meet these energy-efficient needs.

**TABLE 15.3** Excerpts of Code Requirements for Ventilation and Energy

| Concept | International Energy Conservation Code, Section 503 for Commercial Buildings | ASHRAE Standard 62.1 |
|---|---|---|
| Venting and Exhausts | | Minimum exhaust rates are provided for multiple occupancy/building use categories. |
| Duct leakage or location<br>Limit of air migration from contaminated areas | • Duct and plenum insulation required (with exemptions) when located outside the conditioned space (R-6) or outside (R-8).<br>• All ducts, air handlers, and filter boxes must be sealed.<br>• Separate sealing requirements for low-, medium-, and high-pressure ducts.<br>• HVAC must be balanced prior to occupancy. | • Exhaust ducts must be sealed or negatively pressurized.<br>• Buildings with attached garages require vestibule, garage under negative relative pressure, and otherwise minimize entrainment of garage contaminants.<br>• Air classifications by degree of contamination are used with rules limiting the transfer of contaminated air into spaces with cleaner air.<br>• Environmental tobacco smoke (ETS) free areas must be positively pressurized relative to ETS areas. |

**TABLE 15.3** Excerpts of Code Requirements for Ventilation and Energy (Continued)

| Concept | International Energy Conservation Code, Section 503 for Commercial Buildings | ASHRAE Standard 62.1 |
|---|---|---|
| HVAC | • Ventilation required as per IMC.<br>• Mechanical systems must be capable of reducing outdoor air to minimum required.<br>• Demand control ventilation (DCV) required for spaces larger than 500 ft$^2$ and with occupant load of 40 people per 1,000 ft$^2$ of floor area.<br>• Airside economizers required (with exceptions) with capacity for 100% outdoor air and relief of excess outdoor air to prevent over-pressurization, and relief outlet located to avoid recirculation.<br>• No oversizing of HVAC allowed.<br>• Energy recovery required (with exceptions) for fan systems having supply air capacity of 5,000 cfm or more and minimum of 70% outside air.<br>• Required minimum recovered enthalpy of 50% of outdoor air–return air difference at design conditions.<br>• VAV required (with exceptions) for complex multizone systems with capacity to reduce outdoor air supply to defined minimums, including minimum required by IMC. | • OA intakes located at prescribed distances from known contaminant sources.<br>• Ventilation designed to provide minimum required outdoor airflow during all load conditions (specifically including VAV with fixed outdoor dampers at minimum flow).<br>• AHU-mounted outdoor requires rain and snow entrainment systems and bird screens.<br>• Systems with dehumidification required to have capacity for indoor<br>• RH of 65% or less under specified conditions.<br>• System must ensure that OA intake exceeds maximum exhaust whenever dehumidifying.<br>• Drain pan: Specified slope and otherwise designed for free flow of condensate.<br>• Specific requirements for access to HVAC for inspection and maintenance. |
| Ventilation | • Must meet provisions of the IMC for commercial buildings. | • OA ventilation may be supplied mechanically using a Ventilation Rate Procedure or an IAQ Procedure, or with natural ventilation with mechanical backup as needed.<br>• Outdoor air not meeting EPA standards to be treated to prescri bed levels for PM10, PM2.5, and ozone.<br>**Ventilation Rate Procedure**<br>• OA must be delivered to occupants' breathing zone at prescribed rates for all occupied spaces, during all hours of occupancy, and at full or part load conditions.<br>• Rates vary by type of occupancy, and include a rate per unit area and a rate per person for each space. |

(Continued)

**TABLE 15.3** Excerpts of Code Requirements for Ventilation and Energy (Continued)

| Concept | International Energy Conservation Code, Section 503 for Commercial Buildings | ASHRAE Standard 62.1 |
|---|---|---|
| | | • OA delivery to spaces is adjusted for breathing zone at prescribed rates.<br>• OA rates may be reset as conditions change: variable occupancy (demand control ventilation), variable ventilation effectiveness, and use of economizer.<br>**IAQ Procedure**<br>• For each contaminant of concern, a concentration limit and exposure period provided by a cognizant authority, and a satisfactory perceived air quality shall be used to establish the required OA ventilation rate.<br>• The OA ventilation rate shall be at least the highest needed to control any contaminant of concern.<br>**Natural Ventilation**<br>• Natural ventilation openings must be at prescribed maximum distances (based on ceiling height) from occupied space.<br>• Opening size must be a minimum of 4% of floor area.<br>• Openings must be accessible and open during occupancy. |

**Dedicated outdoor air systems**—These systems control humidity by conditioning outdoor makeup air separately from indoor return air. The removal of moisture from the outside air reduces the load on the cooling system and reduces energy use.

**Displacement ventilation**—This technique uses a low-velocity stream of fresh, cold air at floor level to displace stale air near the ceiling, improving indoor air quality.

**Electronically commutated motors**—A variable-speed component in the motor varies the speed of the HVAC blower to match the thermal demand in the building.

**Energy recovery ventilation** (ERV)—The ERV system captures energy from exhausted air and uses it to precondition incoming makeup air. ERVs save energy by reducing heating and cooling system needs while maintaining comfortable indoor air levels.

**Zero-degree heat pumps**—These heat pumps work effectively in cold climates (down to 0°F). Heat pumps move heat rather than creating it. This provides an energy efficiency of more than 100%.

**Desiccant dehumidification**—This technique absorbs moisture with a liquid desiccant solution and then evaporates the moisture using heat. This system removes humidity and precools makeup air, which reduces the air-conditioning load and saves energy.

**Demand controlled ventilation**—This dual system uses carbon dioxide sensors combined with economizers to maintain air levels based on the occupancy demand. This avoids overventilation of a space.

**FIGURE 15.2** Chilled beam system installation.

**Thermal energy storage**—This type of system shifts the timing of electrical use. The system provides cooling capacity by extracting heat from storage material, such as ice or chilled water that is used during the day. Electrical use is shifted to nighttime when electricity rates are lower.

**Chilled beam cooling**—These systems utilize natural convection and radiative heat transfer. The chilled beam delivers air directly to building spaces, reducing the need for air distribution through fans (Fig. 15.2). A limitation with the systems is that they may require a dedicated outdoor air unit and a tight building envelope to manage humidity.

**Adjustable speed drives**—This type of drive system reduces motor speed to accommodate varying HVAC loads and demands.

**Variable flow volume (VFV) systems**—These systems are ductless and use multiple compressors to provide part-load performance and zoned temperature control.

High-efficiency HVAC systems are only one part of the energy efficiency equation when it comes to thermal comfort. Once a building is complete and the occupants and operators are in control, there are many factors related to thermal comfort and energy efficiency that they control.

**Process loads**—These loads can create excessive internal heat gain and affect thermal comfort. While appliances are becoming more efficient with Energy Star and other codes, their energy use is not easily predicted.

**Plug loads**—These loads are a function of the work space and are hard to predict.

**Lighting control system**—Automating these systems is helpful, but individual lighting impacts the building internal heat gain.

**Thermostat setting**—In many situations, the temperature setting is locked, but some systems may provide occupants access to adjust setpoints.

## BUILDING ENVELOPE

No building can be considered energy efficient or green unless the building envelope is thermally efficient. A thermally efficient building envelope uses less energy to maintain thermal comfort as it minimizes heat from escaping the building in the winter and prevents too much heat from entering the building in the summer. It also prevents moisture migration and air leakage overall. This is accomplished through high R-value insulation, energy-efficient windows, appropriate sealants and waterproofing, and good design and construction practices.

The building envelope is a critical element in the energy efficiency of the built environment. A simple definition of the building envelope centers on the function it performs. It is the exterior of the building that protects occupants from natural forces. However, it plays a much more complicated role that involves the thermal comfort and health of the occupants and the energy efficiency of the building. The building envelope is a system that performs several functions that include structure, temperature control, and separates the conditioned from the unconditioned space. The envelope plays a role in almost all aspects of building performance as it interacts with every major building system. It is a balance between performance and function.

What is included in the envelope? The basement walls, exterior walls, floors, ceilings, roofs, windows, doors, and any other building element assemblies that enclose conditioned space or provide a boundary between conditioned space and exempt or unconditioned space (Fig. 15.3). The envelope contributes to the building performance by providing thermal, acoustic, and visual comfort, as well as providing aesthetics, accessibility, and safety.

It is the physical barrier that keeps out drafts, moisture, heat, light, and reduces noise. It is the roof, the walls, and the foundation all working together to complement each other. The envelope is a system of layers constructed of different materials each with different properties. Walls are composite structures of materials like wood, brick, insulation, and glass. There are penetrations in walls for windows and in roofs for ventilation. Each penetration must be carefully designed so the enclosure can perform the necessary functions of the envelope. These functions include managing the movement of moisture, thermal control, airflow, and providing structural support.

Every component of the envelope is exposed to the natural elements. The foundation is in contact with the earth and potential moisture migration. The roof experiences UV light, rain, snow, and wind, while the walls are exposed to wind and rain. Managing the movement of moisture is a key role of the envelope and directly impacts occupant health and comfort and the longevity of the structure. All the envelope surfaces are in contact with moisture so managing how the moisture that intrudes dries out is important.

Thermal control is directly related to thermal comfort. The ability of a building to prevent heat from escaping in the winter and cooling air from leaving in the summer is a measure of the energy

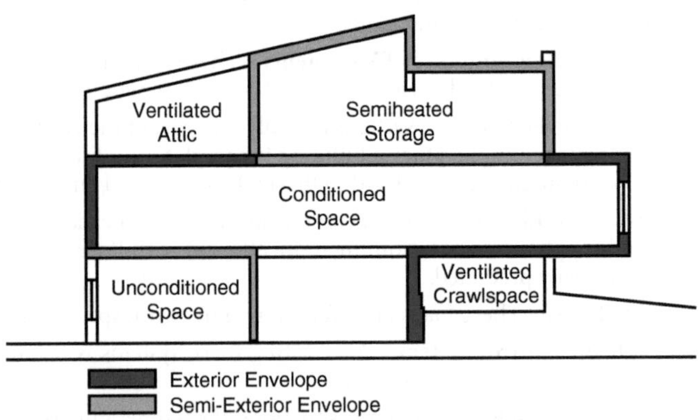


**FIGURE 15.3** Building envelope components.

efficiency of the structure. The lower the heating and cooling demand, the lower the cost of remaining comfortable. This is also related to how air flows through the structure and the envelope system. Air movement is important for air quality, thermal comfort, and the amount of energy a facility consumes. Airflow through the envelope or penetration in the envelope requires attention during construction and operation of a building.

The design and construction of a building envelope is a balancing act. Too much natural ventilation penetrating the envelope compromises the ability to maintain thermal comfort, reduce mold growth, and minimize energy consumption, but the airflow allows for a higher quality of indoor air and reduces the need for mechanical ventilation.

A tight building envelope will ensure that moisture penetration is minimized, temperature is controlled, and energy consumption is regulated. This requires careful construction practices, increases in insulation, and the use of energy-efficient windows that are sealed correctly. These types of structures will require additional mechanical ventilation, given the reduction in natural ventilation that is allowed.

**An energy-efficient building must be** virtually airtight and have a well-insulated building envelope to achieve any of the energy performance levels required by the existing energy codes. Special attention must be paid to the design and construction of the components of the envelope. An understanding of the parameters measured to determine efficiency is key in determining building methods and the requirements for building completion in the commercial and residential sectors.

## Wall Systems

Wall systems perform different functions as part of the building system (Fig. 15.4). These walls are part of the foundation or superstructure of the building. They also perform functions between the interior and exterior environments, such as the flow of air, moisture, heat, and sound as well as finishing the surface. Utilities such as power, water, sewer, and security are also distributed through these walls.

Controlling moisture is critical to the success of below-grade walls. Design and construction of these walls must focus on the details of moisture control and waterproofing. Particular attention must be paid to connections and penetrations through the walls. When a below-grade wall fails to control moisture it is generally caused by improper design, poor installation, or improper construction practices.

Successful waterproofing operations require collaboration and coordination among all parties involved. Prior to beginning the waterproofing operation, the waterproofing subcontractor, the manufacturer, designer, general contractor, and any other trade that is working in the area must

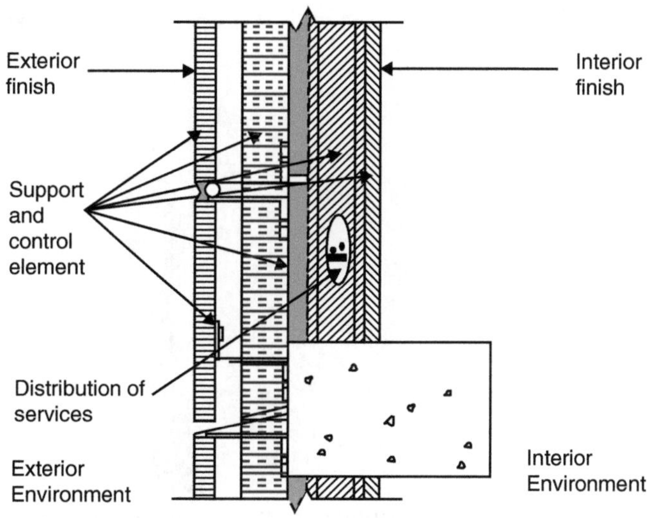


**FIGURE 15.4** Typical wall system functions.

coordinate. Specific features and characteristics required for installation and the entire system are to be reviewed. Schedule is important to this operation as the waterproofing is applied after an appropriate curing process has occurred on the wall. Manufacturers' specifications are to be strictly adhered to and comprehension of the required conditions for application or installation must be understood by all parties.

Standard best practices related to waterproofing involve inspection of the surface on which it is being applied. Ensure the surface is free of deleterious materials, such as dirt, coatings, or previous waterproofing attempts. Inspect the concrete and ensure it is free of honeycombs, large shrinkage cracks or tie remnants, and meets the manufacturer's specifications for application.

The application of a face-applied waterproofing system is completed on the exterior side of the foundation wall prior to backfilling soil. The waterproofing membrane should be continued unbroken from the face of the footing to the top of the footing to above-grade level. The membrane is typically lapped behind the water-resistant barrier of the above-grade wall control layer. If there are additional layers required in the system, these should be installed per manufacturer's recommendations.

Below-grade waterproofing systems are not just the waterproofing membrane applied to the wall. These systems are comprised of multiple elements which include vapor barriers, insulation, filter fabrics, drain board, and stone used for drains. Not every component is necessary for each situation. The goal of the waterproofing system not only is to prevent moisture from penetrating the wall, but also to alleviate the hydrostatic pressure that can be applied to the wall.

Many excellent waterproofing systems are available; however, to ensure success the system installed has to meet the design criteria and goals. Good planning and design, coupled with quality construction, inspection, and testing, will ensure a functioning system.

Keeping water away from below-grade walls is as important as keeping water out. Proper drainage is critical in the soils surrounding the foundation. Draining water away from the walls reduces the hydrostatic pressure the walls must endure, resulting in a lesser possibility of cracking or water penetration. Drainage systems that can be installed include drainage pipes, French drain systems, proper landscaping, and even properly located and installed downspouts and gutters (Fig. 15.5). Water needs to drain away from the foundation.

**FIGURE 15.5** Foundation drainage system.

The substrate surfaces should be dry and cured according to the waterproofing manufacturer's specifications prior to waterproofing application, or until the manufacturer is satisfied by various moisture test methods, such as a poly barrier condensation test. Finally, an adhesion pull test should be administered to the initial application of waterproofing to verify the bond to the substrate.

## Exterior Above-Grade Walls

Common exterior wall types associated with commercial buildings include a cavity wall and a barrier wall. Each of these wall types is commonly constructed with basic layers that each play a role in the energy and thermal efficiency function of the wall (Fig. 15.6). These layers include:

1. Exterior Cladding (Natural or Synthetic)
2. Drainage Plane(s)
3. Air Barrier System(s)
4. Vapor Retarder(s)
5. Insulating Element(s)
6. Structural Elements

According to the International Building Code, Chapter 14 Exterior Walls, exterior walls should provide a building with a weather-resistant exterior wall envelope. This includes flashing and

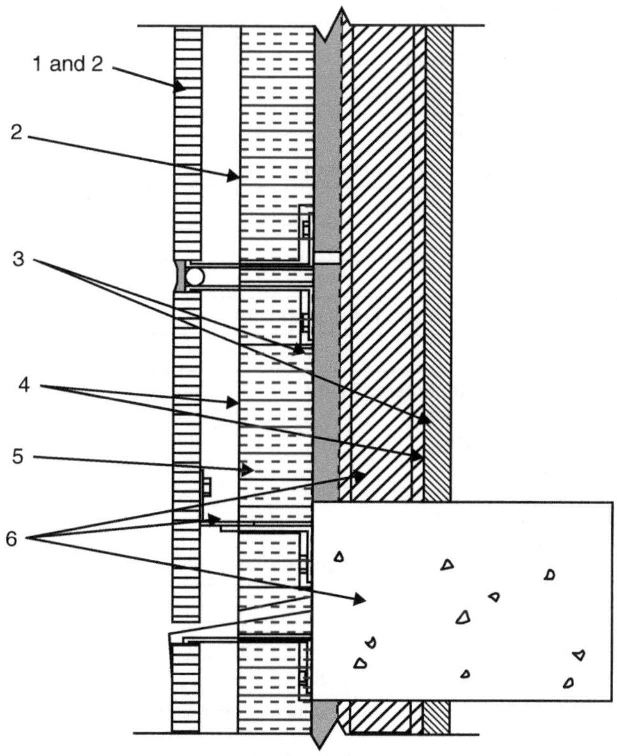


**FIGURE 15.6** Exterior wall components.

requires that the wall be designed and constructed to prevent the accumulation of water within the wall assembly. The wall should provide a water-resistive barrier and a means for draining water to the outside that enters the wall. The exterior wall assembly must protect against condensation within the wall.

IBC Code1403.2 applies to water-resistive barriers and specifies that not fewer than one layer of No. 15 asphalt felt, complying with ASTM D226 for Type 1 felt, shall be attached to the studs or sheathing with flashing in such a manner as to provide a continuous water-resistive barrier behind the exterior wall veneer.

Wall coverings serve as protection from the weather and must be applied at minimum thicknesses. Table 15.4 is an excerpt from the ICC code, Table 1404.2—Minimum Thickness of Weathering Coverings, for common materials used in building coverings.

Vapor barriers are an integral part of the exterior wall system. IBC Code 1403.3 addresses the requirements of the materials used as vapor barriers. The location of vapor barrier application interior or exterior sides of walls depends upon the building zone in which the structure is located. Vapor retarders are classified based upon the manufacturer's certified testing or a tested assembly. The classes of vapor retarders are:

**TABLE 15.4** Minimum Thickness of Weathering Coverings

| Covering type | Minimum thickness (in.) |
|---|---|
| Aluminum siding | 0.019 |
| Anchored masonry veneer<br>Stone (natural)<br>Architectural cast stone | <br>2.0<br>1.25 |
| Asbestos—Cement boards | 0.125 |
| Asbestos shingles | 0.156 |
| Exterior plywood (with sheathing) | 0.313 |
| Fiber cement panel siding | 0.25[c] |
| Hardboard siding | 0.25 |
| Porcelain tile | 0.25 |
| Steel (approved corrosion resistant) | 0.0149 |
| Structural glass | 0.344 |
| Stucco or exterior cement plaster<br>Three-coat work over:<br>Metal plaster base<br>Unit masonry<br>Cast-in-place or precast concrete<br>Two-coat work over:<br>Unit masonry<br>Cast-in-place or precast concrete | <br><br>0.875[b]<br>0.625[b]<br>0.625[b]<br><br>0.5[b]<br>0.375[b] |
| Vinyl siding | 0.035 |
| Wood shingles | 0.375 |
| Wood siding (without sheathing)a | 0.5 |

[a]Wood siding of thicknesses less than 0.5 in. shall be placed over sheathing conforming to code.
[b]Exclusive of texture.
[c]As measured at the bottom of decorative grooves.

- Class I: Sheet polyethylene, nonperforated aluminum foil with a perm rating of less than or equal to 0.1.
- Class II: Kraft-faced fiberglass batts or paint with a perm rating greater than 0.1 and less than or equal to 1.0.
- Class III: Latex or enamel paint with a perm rating of greater than 1.0 and less than or equal to 10.0.

Flashing material is key to preventing moisture from either entering the wall system or redirecting moisture to the exterior of the wall. IBC Code 1404.4 section applies to flashing material. Flashing is installed at any location where moisture could enter the wall. This includes exterior door and window assemblies, penetrations of exterior wall assemblies and exterior wall intersections. Exterior wall intersections include roofs, chimneys, porches, decks, and balconies, as well as built-in gutter systems. Flashing with projecting flanges shall be installed on both sides and the ends of copings, under sills and continuously above projecting trim.

Anchored veneer on exterior walls presents a different situation for water penetration and diversion. For example with masonry units, flashing and weep holes are located not more than 10 in. above finished ground level above the foundation wall or slab. At other points of support including structural floors, shelf angles, and lintels, flashing and weep holes shall be located in the first course of masonry above the support. There are specific details for each type of exterior building cladding detailed in the IBC specification. The key is to keep moisture from penetrating or moving it as rapidly as possible to the exterior.

## Cavity Wall

A cavity wall system is considered one of the best solutions for exterior walls and the preferred method of construction in many climate zones in the United States (Fig. 15.7). This type of wall system is a

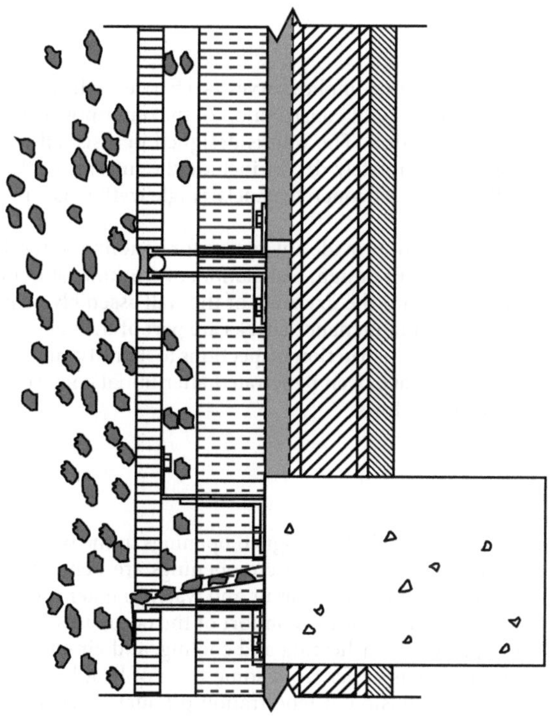

**FIGURE 15.7** Cavity wall.

double wall that consists of an inner layer that is tied to an exterior layer. The cavity between the walls can be filled with insulation and is normally between 2 and 4.5 in. wide. This type of wall system relies on a partially or fully concealed air space and drainage plane to resist wind-driven rain, absorb and reflect sound, and provide good thermal performance. It is commonly constructed of concrete masonry units and other architectural masonry units.

A typical drained cavity wall includes an exterior cladding element. This is the first line of defense against rainwater and its function is to shed or absorb the rainwater that penetrates before it enters the concealed spaces of the wall.

The function of the drainage cavity is to collect and control rainwater that penetrated through the cladding. This water is re-directed to the exterior of the building. Venting of the cavity can help with this process. The venting provides pressure equalization, preventing negative pressure that may draw water into the interior of the building along wall ties.

This wall system uses an internal drainage plane that is designed to function as the protection against uncontrolled rainwater penetration. This layer of the wall divides the wet and dry areas of the exterior wall. Prevention of condensation and potential mold growth are critical function of this layer.

The wall can also have an insulating layer depending on the climate where the wall is being constructed. This will determine the location of the insulation.

While the cavity wall is a commonly used wall in building construction if the wall is not properly designed and constructed, it can cause major disruptions to the operation of a building if repair is required after occupancy. The most common concerns are associated with the corrosion of the wall ties and structural connections as well as interior mold growth which can go completely undetected for years before manifesting themselves in a way in which it can be detected. Repair and remediation of these interior systems are costly and disruptive. To help alleviate these types of concerns, a quality assurance program should be developed to ensure the correct design and installation occurs at the time of original construction.

## Barrier Wall

Barrier wall as the name implies serves as a barrier to leaks and water entry (Fig. 15.8). A barrier wall relies on the weather tight construction of the outermost exterior wall to resist rainwater penetration and moisture migration. The wall is designed to resist all water penetration and keep all the layers behind the wall completely dry to include the interior of the building. Materials used to construct barrier walls are precast concrete panels, single wythe masonry walls, and exterior insulation and finish systems (EFIS).

These walls do not provide any means for handling water that penetrates the wall due to defects or damage. Any defect in design, installation, or workmanship in these types of walls will result in rainwater penetration into the dry zone of the wall assembly or potentially the conditioned space of the structure. Routine maintenance checks are part of the operation of the building with a barrier wall as well as conducting construction inspections to identify any defects. Defects causing infiltration can be found when walls have been penetrated after initial construction to install things like signs, lighting fixtures, or new windows.

## Fenestrations

Fenestration refers to all the designed openings in the building envelope. This includes the windows, the doors, and any skylights. These openings are penetrations through the building envelope and provide possible entry points for air and water to enter the building. These openings in the building envelope are necessary for the comfort of the occupants and can promote natural lighting, points of ventilation, passive solar heating and cooling, and air circulation. They also improve air quality. All fenestration products suffer from some form of heat loss due to radiation, conduction, convection, and air leakage. Inclusion of fenestration products decreases the efficiency of the building envelope

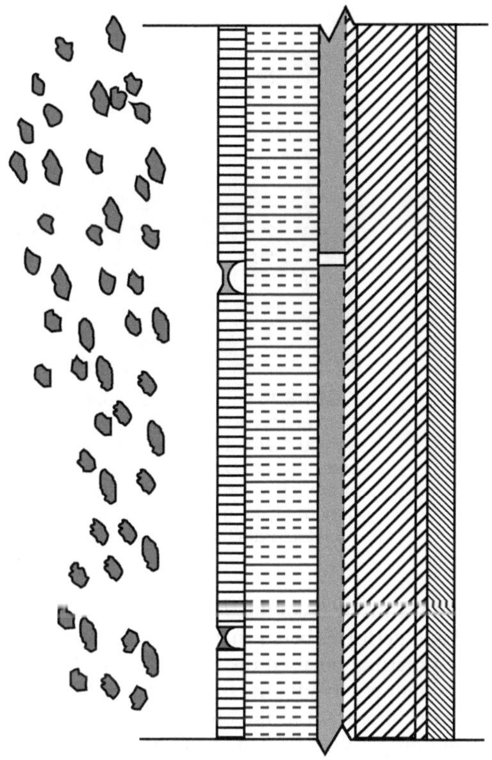

**FIGURE 15.8** Barrier wall.

because a component of the insulation is removed from the wall. This reduces the effective R-value of the wall and its ability to resist the flow of heat within the building.

The fenestration industry has made great progress in energy efficiency due to the integration of smart technology related to window, sensors, and other controls. There are fenestration energy codes for commercial and residential buildings that set the standards for the performance of all fenestration products.

Windows are the most common type of fenestration product. Successful window design includes integration with adjacent wall components to ensure an energy-efficient envelope. Focused detailing is required to integrate building envelope barrier systems with the window frame and glazing system. This includes water, air, and vapor barriers as well as any insulation, cladding, or fire protection. (See the Example Application.)

When coordinating with the building envelope, it is important to know if the glazing system being installed is a rain screen or a barrier system. A rain screen provides internal drainage within the glazing pocket. A barrier system assumes no water will ever penetrate the glazing perimeter and provides no drainage system in the design.

Energy efficiency with glazing assemblies is directly related to the details associated with the glazing, the window frame, and the perimeter. Thermal performance is affected primarily by how well the glazing can control radiative heat transfer through solar radiation. This can be accomplished with low emissivity (low-E) coatings on the glass, tinting, and the use of fritted glass (Fig. 15.9). The window frame thermal performance is determined by the material used, the geometry, and the design.

**FIGURE 15.9** Structural glazed lobby curtain wall of Asplundh Cancer Pavilion, Abington, PA. Fritted glass was selected to reduce thermal heat load. Twenty different types of glass were used on the exterior façade. Each had a different color, substrate (clear or tinted), and custom silk-screened pattern (light gray or white, gradated and solid lines). (Photo courtesy of the Architectural Glass Institute ©Joe Garvin.)

The lower the thermal conductivity of the frame material, the better the thermal performance is achieved. Wood performs better than aluminum and vinyl. The design of the frame would refer to the inclusion of a thermal break in the frame design. The thermal break involves inserting a material of low thermal conductivity into the frame. This reduces the frame strength and stiffness.

Insulation is important to the building envelope integrity around window penetrations. Placing the insulation properly around the window perimeter and façade face will reduce energy loss from around the windows.

Maximizing energy efficiency of windows is accomplished through choosing the correct window for the application and installing it correctly. Low conductivity materials in the frame, low-E coatings on the glazing, and triple-pained glass are standard characteristics of energy-efficient window.

Critical to all energy efficiency and the integrity of the building envelope is the correct installation of the window and ensuring the window perimeter detail designs are followed. Flashing plays an important role in this process. Improper flashing installation is a construction challenge. The sill, jamb, and head of the window must be flashed according to standard details.

Sill flashing should be of a durable metal and sloped to the exterior of the building with a drip edge over the face of the cladding wall. Install the flashing with a 1″ minimum upturn leg at the interior with end dams that are soldered to be water-tight. The horizontal portion of the flashing should not be penetrated with fasteners. The flashing should be attached to an angle that is inboard of the windowsill and then fastened through the upturned leg of the sill flashing and into the inboard leg of the sill frame.

Jamb flashings are typically a flexible membrane material and can also be metal. If the jamb flashing is detailed as part of the air barrier system, it must be continuously supported by a substrate that is designed to withstand the design air pressure. The jamb flashing and wall waterproofing must be lapped and fully adhered wherever they intersect. The jamb flashing must be fully sealed to the window frame, most likely through mechanical methods. Lastly the jamb flashing must be shingled over the sill flashing end dams and back leg to ensure that any water that runs down the jambs spills into the sill pan.

Head flashing should be of a durable metal and sloped to the exterior of the building with a drip edge, over the top of the window frame. The flashing material should extend several inches beyond the window frame. The end dams should be soldered watertight and the flashing sealed to the inner face of the windows and to the jamb flashings. Provide a minimum of 4 in. upturned leg and counter flash with the wall waterproofing membrane adhered to the vertical leg of metal flashing. If the window is installed in a concrete opening that will not allow the head flashing to extend beyond the opening, use dual sealant joint to capture water and direct it to the jamb flashing.

Windows may need to be laboratory tested, especially if the project requires custom windows. The test should be completed on a mock-up window construction, and the structural capacity of the window, the air infiltration, and the water leakage should be evaluated. This testing should be completed prior to the production of the window units for the project.

Even if the windows are stock orders, construct and test a field mock-up that is representative of the window installation on site. Tests should be conducted for window and wall air infiltration and water leakage (Fig. 15.10). The production windows should also be tested in the field for quality assurance of window fabrication and installation. This should be completed early in the construction process to catch installation issues, flashing errors, or manufacturer defects.

To ensure the windows are installed correctly, be certain to review the shop drawings carefully. These drawings should show all adjacent and related construction work and detail the sequencing of the work. Careful attention should be paid to flashings, window attachments, interior finishes, and coordination work between trades.

**FIGURE 15.10** Testing of the glazing assembly to verify performance criteria.

## *EXAMPLE APPLICATION*

Wall penetrations present a challenge to thermal control. Best practices in thermal control require insulation layers that are minimally penetrated, an effective air barrier system, good control of solar radiation, and interior heat generation management. The insulating value of a wall is major focus in energy-efficient design and any reduction in the overall U-value of a wall is the goal.

Both glass and metal are poor insulating materials, but if detailed correctly, glazing systems can be designed to provide good thermal performance. By minimizing the proportion of metal framing members exposed to the outdoors, providing thermal breaks, and using double or triple glazed window units, high thermal performance can be achieved. The use of appropriate insulation in areas of the wall also helps to improve the thermal performance. Glazing systems also have to incorporate fire prevention. Fire-prevention measures are mandated by code and will vary in different locations.

The following images (Figs 15.11–15.13) illustrate the progression of a glazing assemble from shop drawing detail, to revisions and installation.

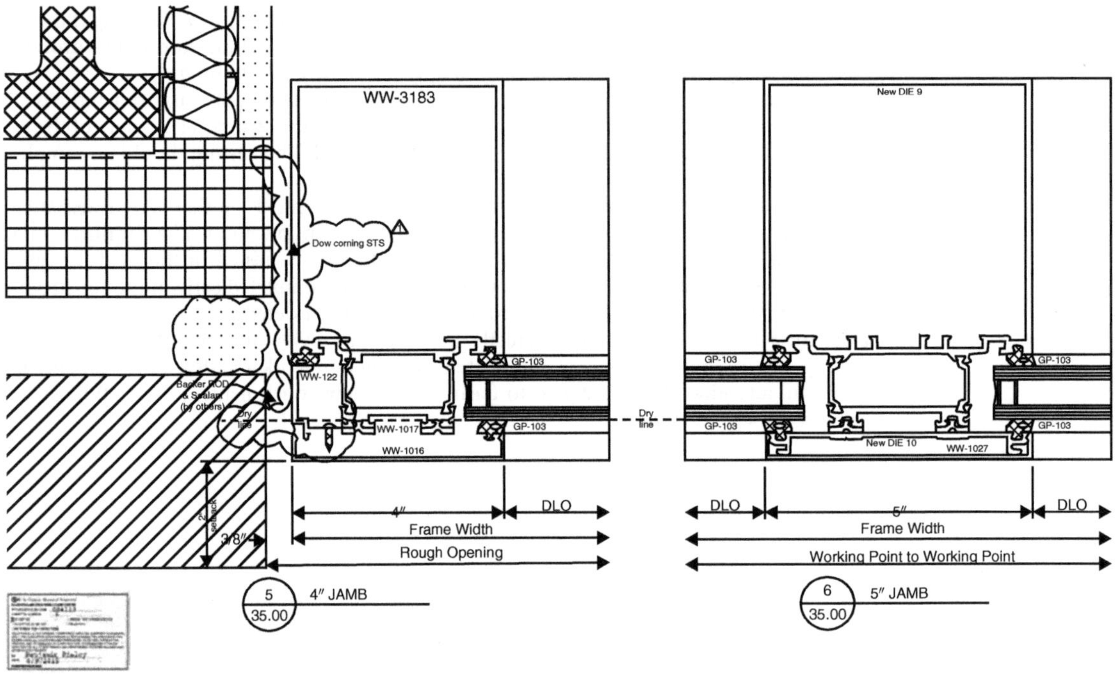


FIGURE 15.11 Shop drawing submission for typical glazing system jamb.

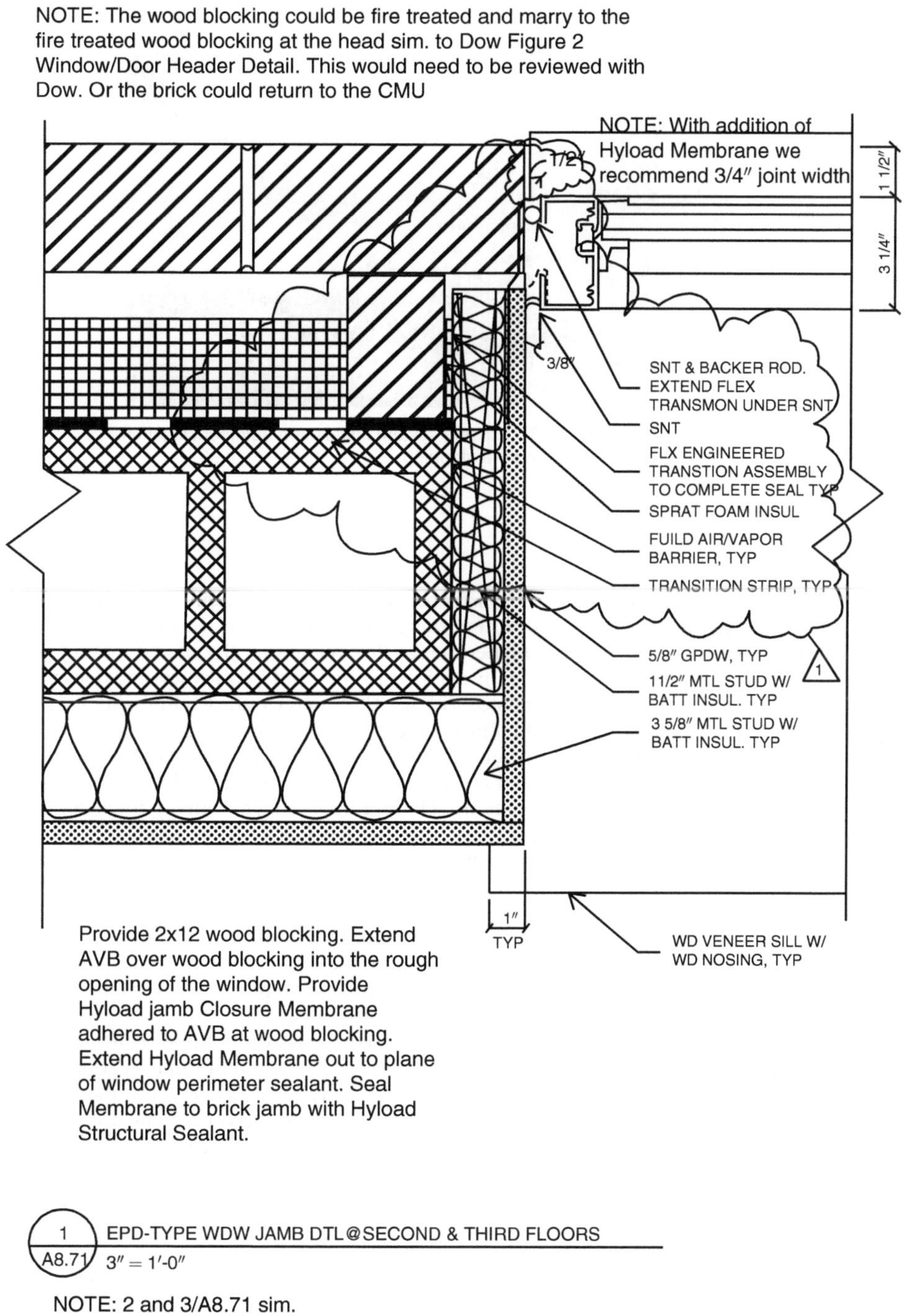


**FIGURE 15.12** Shop drawing comments to address thermal control as a fire requirement.

**FIGURE 15.13** Glazing system installation before installation of insulation and fire stopping.

### Insulation

Thermal insulation reduces the heat transfer between objects of differing temperatures when in thermal contact and prevents the loss of heat energy. Insulating materials are rated based upon their thermal resistance or R-value. The higher the R-value, the greater the effectiveness of the insulation. The R-value depends upon the type of insulation material, and its thickness and density. The R-value of insulation is determined in accordance with the US Federal Trade Commission R-value rule (CFR Title 16, Part 460) in units of °F $ft^2$ hr/Btu at a mean temperature of 75°F or 24°C.

Insulation is a key component in building envelope systems. Advances in thermal insulation have improved insulation production, materials, and function in building envelopes. Thin insulating materials and integrated insulation materials are solutions improving the energy efficiency of commercial and residential building envelopes.

Vacuum-insulated panels (VIPs) provide excellent energy conservation in building envelopes especially when space is limited. VIP is composed of a rigid thermal insulation panel evacuated and sealed by a vacuum with a gas-tight film. VIPs conduct very little heat due to the absence of air. VIPs are also coated with a low-emissivity facing that reduces heat lost to radiation.

These panels are extremely thin, making them a great choice when space is limited. VIP panels 1 in. thick possess an R insulation value of 50. Conventional fiberglass batt insulation would require a thickness of approximately 15″ to achieve the same R-value.

Structural insulated panels (SIPs) are prefabricated insulated structural elements for use in building walls, ceilings, floors, and roofs. These systems provide more uniform insulation than traditional stick built framing and offer an energy savings of 12–14%. An SIP panel is commonly constructed of a 4–8″ thick foam board insulation that is placed between two sheets of oriented strand board (OSB). By combining the structural elements of the OSB and the insulating properties of the foam board together, these panels produce energy-efficient envelop systems.

## LIGHTING

A critical area in building is lighting. Lighting consumes close to 35% of the electricity used in commercial buildings in the United States. The amount of waste heat that lighting generates impacts other building systems such as HVAC. With good design and appropriate selection, lighting energy use in most buildings can be cut at least in half while maintaining or improving lighting quality. Upgrading lighting systems with efficient light sources, fixtures, and controls can reduce lighting energy use, improve the visual environment, and affect the sizing of HVAC and electrical systems. Such designs and upgrades typically pay for themselves in energy savings alone within a few years, and they offer more benefits in terms of the potential for smaller and less costly cooling systems and more productive work environments.

Installing lighting systems that are efficient and controlled can reduce the size requirements of the HVAC system as well as the electrical system. When developing a lighting plan that is energy efficient, it must be tested to ensure that all systems work as intended. This requires a good commissioning plan.

Energy-efficient lighting design includes several important considerations that must be balanced. There must be an appropriate amount of light for the tasks being performed without glare. To reduce the need for electric lights, take advantage of daylight as much as possible and install automatic controls to turn off the lights or dim them. Be sure to use the most efficient light source application. Consider using high-performance fluorescent systems as the primary light source in most commercial buildings and replace the incandescent bulbs with compact fluorescent bulbs (Fig. 15.14). Other energy-saving lights include light-emitting diode and halogen incandescent.

**FIGURE 15.14** Energy-efficient lightbulbs.

Energy-efficient lighting systems use less energy in comparison to traditional lights. These highly efficient systems use 80% of the energy they obtain to produce light and only 20% is given off as heat. This saves on both lighting energy and energy required to cool a building. Energy-efficient lighting last longer than traditional lighting. These lights can last about 50,000 hr. This translates to 6 years if you leave your lights on every day for 24 hr.

Linear fluorescent lamps, T5HO lamps, are now used in many high bay applications in place of High Intensity Discharge (HID) lamps. T5HO lamps are good for providing general ambient lighting for a space. They are excellent for surface lighting and possess higher levels of optical performance. Indirect/direct linear fluorescent pendants and wall-mounted uplights are typical applications for these sources.

Compact fluorescent lamps (CFL) are commonly used as substitutes for incandescent lamps due to their significantly longer life and better energy efficiency. The self-ballasted, "screw-in" retrofit CFL lamps cannot be dimmed and their performance is not as good as the separate lamp and ballast combination. CFL lamps are used in recessed luminaires, wall- and ceiling-mounted fixtures, and even track lighting and task lighting.

Inductive fluorescent lamps are white light sources with very good color rendering and color temperature properties. These lamps have an extremely long life (over 100,000 hr) and provide an instant-on capability. They are powered by a small generator attached to the lamp by a short fixed-length cable. The generator induces a current in the lamp that causes it to glow. These lamps do not have any electrodes to wear out. They are often used in place of low- to medium-wattage high-intensity discharge sources because of the instant-on capability and reduced maintenance associated with the longer lamp life.

High-intensity discharge lamps (HID) are efficient lamps for lighting large areas or great distances (Fig. 15.15). Pulse-start metal halide lamps provide better color stability and longer life than previous technologies. PAR metal halide lamps with ceramic arc-tube enclosures are commonly used for accent lighting and highlighting in large spaces, and commonly used in retail applications. The small size of the metal halide arc-tube allows for excellent optical control.

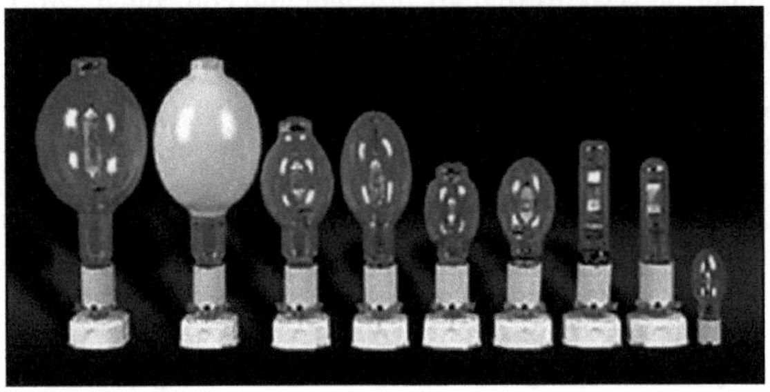

**FIGURE 15.15** HID Metal Halid lamps. Retrieved from https://www.wbdg.org/resources/energy-efficient-lighting

LED lamps are the newest addition to the list of energy-efficient light sources. LED lamps emit visible light in a very narrow spectral band; however, they can produce "white light." This is accomplished with either a red-blue-green array or a phosphor-coated blue LED lamp. LED lamps last 40,000–100,000 hr depending on color. The use of LED lamps has expanded in the past decade. According to the study, "Adoption of Light-Emitting Diodes in Common Lighting Applications" released by the US Department of Energy in 2020, from 2016 to 2018, installations of LED products have increased in all applications, increasing LED penetration to 30.0% of all general illumination lighting. The study also found that in the outdoor sector, parking garages are estimated to have the highest penetration of LED lighting in 2018. LED lighting has a higher penetration in outdoor applications compared to indoor applications. The study also found that the cumulative installed penetration of LED and connected lighting as of 2018 provided approximately 1,328 trillion British thermal units (tBtu) in source energy savings in 2018, which is equivalent to an annual cost savings of about $14.7 billion.

Energy-Efficient Ballasts regulate the current to the lamp and provide sufficient voltage to start the lamps. Fluorescent Ballasts are available in different types. The rapid start ballasts are the most common ballast and offer long lamp life and provide energy savings. Program rapid start ballasts are more expensive but are excellent for use for energy efficiency and long lamp life. Instant start ballasts are the least expensive ballast and have a higher efficiency, but the lamp life is shorter. The lamp life is even shorter if the frequency of starts is higher due to lighting controls. These are best used when lights are on continuously for long periods of time. Dimming electronic ballasts are used to provide low levels of light when needed and can be used to achieve energy savings by dimming lights when there is adequate daylighting.

Light fixtures are units that can provide energy efficiency based upon their components. Light fixtures are also called luminaires and they consist of one or more of these components:

- Lamp(s) and lamp socket(s)
- Ballast(s)
- Reflective material
- Lenses, refractors, louvers, blades, or other shielding

The energy efficiency of the luminaire is a function of the energy efficiency of each component. Table 15.5 is adapted from the Whole Building Systems Guide report on energy-efficient lighting and highlights examples of common types of luminaires, their benefits, cautions, and applications.

**TABLE 15.5** Common Luminaires

| Type of fluorescent luminaire | Description | Benefits | Cautions | Applications |
|---|---|---|---|---|
| Indirect/Direct Linear Luminaire | Primarily indirect, pendant or wall mounted, T8, T5, or T5HO lamping | Soft, even illumination, good visual comfort, easily dimmed | Choose spacing for good ceiling brightness uniformity | High and low bay areas and classrooms |
| Indirect/Direct Decorative Luminaire | Typically compact fluorescent or induction lamping | Significant energy savings, performance comparable to incandescent | Select diffuser for good brightness uniformity on glowing elements | Small offices, lobbies, waiting areas, atriums, and corridors |
| Linear Strip Luminaire | Surface mounted or pendant mounted with or without side reflectors, typically T8 lamping | Energy-efficient, small size, low-cost, easily dimmed | Best when concealed | In coves or wall slots, on top of cabinets, stacks or lockers, and mechanical rooms |
| Task Luminaire | Linear wall mounted "under shelf" or "arm type" | Task lighting allows for lower ambient lighting levels | Provide appropriate task/ambient contrast ratios | Any task surface (desks, counters, workbenches, etc.) |
| Indirect Recessed Luminaire | Recessed (light is directed up toward top of housing and reflected back down), typically 2′ x 2′ or 2′ x 4′, T8 or CFL biax lamping | Optimized for fewer lamps than typical recessed lensed troffer luminaires, good visual comfort | Does not brighten ceiling, consider minor supplemental lighting (such as wall sconces) | Corridors, open/private offices (can replace standard troffer in many applications) |
| Recessed Wall Washer | Linear or round can-type, Linear or CFL lamping | Significant energy savings, performance better than incandescent | Best when paired or in groups, choose spacings carefully | Select wall surfaces in many room types |
| Recessed Downlight | Round can-type, CFL lamping | Significant energy savings, performance comparable to incandescent | Does not brighten the ceiling, can create light "scallop" on walls | Localized infill lighting, often combined with other luminaire types |
| Wall Sconces | Wall mounted, decorative, CFL lamping | Significant energy savings, performance comparable to incandescent | Select diffuser for good brightness uniformity on glowing elements | Lobbies, corridors, conference rooms, etc. |

When developing the lighting system for a building, consider ongoing maintenance for the life of the building. This will include a detailed commissioning plan to ensure the system is performing as required and continued checks. The plan should also consider an option for re-lamping and proper disposal of old lamps and ballasts.

A building lighting system should be designed to provide the quality and quantity of light that is appropriate for the task performed in the space. Characteristics of the space and light that need to be considered are color, daylight availability, glare, and the distribution of light. Attention should also be paid to the exterior building lighting. Here the light should be directed where needed, reduce glare, and distribute illumination evenly. The quantity of light that is required for a space can be determined from Chapter 10 of the IESNA Lighting Handbook. In that section there are recommendations for illumination levels based on applications, space, and other characteristics.

The quality of light is critical to designing efficient lighting. Luminance ratios and glare are two important quality characteristics of light that ensure light level is balanced across a surface to avoid eye strain. High-frequency electronic ballasts are also important to visual performance because they reduce eyestrain and fatigue. Frequencies in the 20 kHz range and higher provide smooth, non-flickering lamp operation.

When determining the lights to use on the outside of the building ensure the appropriate design illumination level is used. In general, an average of 1 foot-candle (or less) is sufficient. Consider cut-off luminaires or similar type of fixture that do not spread light into the sky above the horizontal. Consider whiter light sources as they produce a better quality of light by which to see than the sodium lamps. Always provide controls on outdoor lights. Things like timers, motion sensors, or other controllers that can dim the lights when appropriate.

Lighting a building efficiently starts with using as much daylight as possible. The next step is to choose the lamp/ballast/fixture combination that will maximize efficiency while balancing all the other criteria. Some possible light sources include linear and compact fluorescent, high-intensity discharge, induction lamps, and light-emitting diodes. The color quality, service life efficacy, and applications of each of these types of lights' sources will vary.

### Daylighting Systems

Lighting systems that use daylight to supplement electric light can potentially cut energy use, decrease the peak demand, and improve the indoor environment. The sun provides 140 lumens (lm) of light for each watt (W) of heat energy, which compares well to most efficient lighting systems. To use daylight as a source of light, careful planning must occur to avoid glare and to be sure the electric lights are controlled and adjusted in response to the daylight. The electric lights should be reduced when enough daylight is present. Implementing a lighting control system ensures the correct lighting levels at all times and reduces the chance of users overriding the control system and eliminating energy savings.

## *APPLIANCES AND EQUIPMENT*

Energy-efficient appliances are designed to utilize minimum energy to complete the required task. The goal is to have these appliances maximize a small amount of energy to complete the task they are designed to accomplish. Energy-saving appliances may achieve the ENERGY STAR rating. This rating indicates that the appliance uses less energy than standard appliances of similar type.

Energy use in both commercial and residential buildings is attributed to appliances. Choosing energy-efficient appliances can reduce energy consumption, lower utility costs, and sustain an energy-efficient building.

**Refrigerators** use much less energy than they did in the past because of improvements in insulation and compressors. Refrigerators that are ENERGY STAR rated are 9% more efficient than models that meet the federal minimum energy efficiency standard.

**Dishwashers** waste energy and water; however, the technology has improved dramatically. New ENERGY STAR-certified models include innovations that reduce energy and water consumption and improve performance.

- **Soil sensors** test how dirty dishes are throughout the wash and adjust the cycle to achieve optimum cleaning with minimum water and energy use.
- **Improved water filtration** removes food soils from the wash water allowing efficient use of detergent and water throughout the cycle. The final clean-water rinse assures your dishes come out sparkling.
- **More efficient jets** use less energy to spray detergent and water over the dishes when cleaning.
- **Innovative dish rack designs** maximize cleaning by strategically situating the dishes.

**Commercial dishwashers** can clean and sanitize a large quantity of kitchenware in a short amount of time by utilizing hot water, soap, rinse chemicals, and significant amounts of energy. Size requirements for commercial dishwashing machines can be calculated by estimating the number of individuals served by the food service establishment. This information is a key determinant of the type of dishwasher that is most suited for a facility. ENERGY STAR-rated commercial dishwashers are about 40% more energy efficient and 50% more water efficient than standard models.

**Clothes washers** waste energy and water if they are old and inefficient. ENERGY STAR-certified clothes washers use about 25% less energy and 33% less water than models that simply meet the federal minimum standard for energy efficiency. ENERGY STAR also rates commercial clothes washers.

**Ceiling fans** will not replace air conditioning entirely, they only decrease the need for energy-intensive cooling. When a ceiling fan is in use, a thermostat can typically be set back by 4 degrees and still feel just as cool and comfortable. ENERGY STAR-certified units for a 20% boost in efficiency over standard models.

**Heat pumps** have the power to cool a building in the summer and heat it in the winter. They are able to remove excess humidity from the air, thereby using less energy than air conditioners during the summer. Heat pumps are also more efficient than other heating systems as they move heat and do not create it. Heat pumps have the capability to lower electricity usage in a facility.

ENERGY STAR-certified heat pump has a seasonal energy efficiency ratio (SEER) rating of up to 20.5 and a heating seasonal performance factor (HSPF) rating of up to 13.0.

The only constraint on the heat pump is that efficiency drops when the outside temperature falls below 45°F. Below freezing the heat pump will turn off and the system will use electric resistance coils for the heat source.

For commercial applications, a heat pump is an efficient, dependable technology that can be considered for use along with other heating and cooling systems. Used in combination with other systems, they can greatly enhance the efficiency of a heating and cooling system, at a lower installed cost.

**Smart Thermostats** enable you to program the HVAC system to heat or cool only when necessary. In residential settings the savings realized on heating and cooling costs is about 10% when dialing back the thermostat by about 10 degrees for 8 hr a day. These types of controls can be used in commercial settings also. Advanced models work seamlessly with the heat pump for maximum energy savings. Many thermostats have apps that can be downloaded on a smartphone to control temperature and even receive maintenance reminders like when to replace the filter.

**Water heaters** use a large amount of energy in both residential and commercial applications. ENERGY STAR certifies gas storage water heaters, gas tankless water heaters, heat pump water heaters, and solar water heaters. Choose an ENERGY STAR-certified water heater instead of a standard model and use up to 8% less energy.

ENERGY STAR-certified commercial water heaters include gas-fired storage and tankless units that use around 15% less energy than conventional commercial units by using energy-efficient heat exchangers. The ENERGY STAR label is now available for electric heat pump water heaters as well. These units transfer ambient heat to the water using a refrigerant-based system, using less than half the energy of an electric resistance unit. ENERGY STAR commercial electric heat pump water heater is expected to save at least 10 MWh.

## CONTROLS

The integration of control systems in buildings is a central component of energy-efficient buildings systems. The term "smart building" has been used to describe structures that use automated processes to control systems such as heating, ventilation, air conditioning, lighting, security, and others like elevators and shading. Sensors, microchips, and actuators are used to control the systems in a building and collect data to efficiently manage systems. Building controls can assist in maintaining energy efficiency and many are connected to and responsive to a smart power grid. The connectivity with the power grid allows building to respond to spikes in peak energy demand.

Energy management systems interact with every aspect of a building and the occupants. Regardless of how energy efficient the installed equipment is in a building, the ability to improve the energy efficiency of the structure cannot be accomplished without optimizing the energy. To optimize the energy used in a building, the consumption must be measured.

### Energy Metering

To manage the energy efficiency of a building, the energy consumption must be measured. Monitoring and metering energy usage are tools to measure the overall building energy efficiency. The information provided enables building stakeholders to make decisions regarding means to improve the energy performance of the building.

Metering systems are continuous, automatic control systems that monitor the building operation and energy consumption information. This is typically shown through an interactive dashboard that uses graphs, tables, and charts to illustrate how, when, and where a building is using energy in addition to how much (Fig. 15.16). Building owners can use the information collected to make sound energy choices.

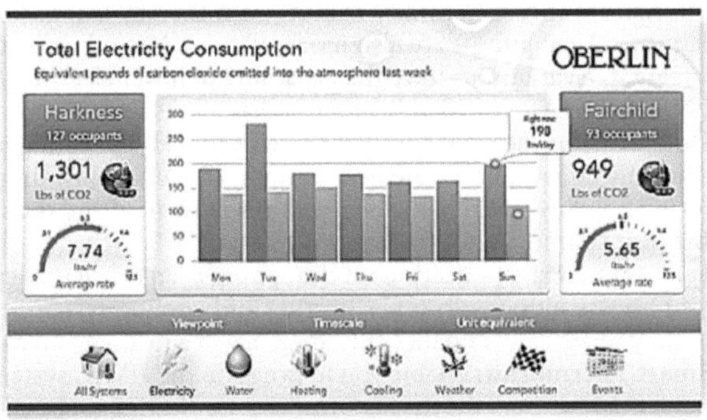


**FIGURE 15.16** Sample energy dashboard for buildings.

HVAC and lighting account for the majority of energy consumption in commercial buildings and a control system is a step in minimizing the energy use.

### HVAC Controls

These types of control systems measure ventilation parameters within a building and adjust the flow of outside fresh air flow in response to the changing indoor conditions. Variable air volume (VAV) systems are energy-efficient systems in which all major components are equipped with the ability to adjust in response to variations in load through controls.

The operation of these systems involves the heating or cooling of air from a central air handling plant which is then distributed to the VAV boxes. These boxes consist of a damper, which regulates air quantity, and a reheat element that adjusts the temperature of the air before discharging it to the space. A return air fan moves the air back to the air handler, where supply fan speed is a major control system within air handling units. Return air dampers are used to adjust the degree to which return air is recirculated and fresh air is introduced.

VAV boxes operate to increase the airflow in response to the temperature fluctuation in a specific building zone. Each zone is controlled by a separate thermostat which controls the VAV by opening and shutting the damper in response to temperature changes in the zone. Unlike a constant air volume system that provides a constant airflow at a variable temperature, a VAV system provides a varied airflow at a constant temperature.

While VAV systems provide airflow to a space when needed, many systems are designed to ventilate at a fixed minimum rate of 15–20 cubic feet per minute of outdoor air per person as that is the ASHRAE recommendation. Many building spaces are not fully occupied throughout the day and some may even be vacant. This will cause more outside air to be brought in than is necessary. This air must be either heated or cooled before entering the building, which requires the use of fans and motors.

A demand control ventilation system is an alternative. It regulates the quantity of outside air intake based upon actual building occupancy. The control system uses $CO_2$ sensors to monitor the level of human activity in the building and depending on the readings either increases or decreases the quantity of outside air intake. This reduces the heating and cooling system load and energy use.

The electric motors central to all HVAC systems consume a large amount of energy. Operating these systems more efficiently with a variable frequency drive reduces the energy used. A motor that is running less than full capacity is more energy efficient than if it was running at full speed. A Variable Frequency Drives (VFD) is an electrical device that is used to control the speed of a motor by adjusting the frequency of the electrical power supplied to that motor. A VFD can be used to control the speed of HVAC system components like fans, pumps, and compressors, thereby controlling the output of the system and increasing the energy efficiency. These can be used in peak-shaving operations of equipment energy use.

## Lighting Controls

Lighting systems can be controlled in a variety of ways. The system can be part of the building automation system or on a separate lighting control system. The lights can be turned off, on, or dimmed based upon the time day or occupancy. Types of controls include occupancy sensors, timers, and photosensors. Photosensors are excellent tools for both commercial and residential structures when it comes to outside lighting.

Many control systems enable lights to be dimmed or turned off in response to peak power demands, thereby saving energy. The most common types of lighting controls include dimmers, motion, occupancy, photosensors, and timers.

**Dimmer Controls:** These controls provide variable lighting by reducing the wattage and output of a lightbulb. This action saves energy and can increase the service life of the lightbulb. Care must be taken to ensure the bulb is compatible with the dimmer, as some compact fluorescent lamps and light-emitting diode (LED) bulbs may require special dimming ballasts and bulb holders.

**Motion Sensor Controls:** When motion sensor controls detect movement, they turn on lights and then shut them off a short time later. These sensors are useful in outdoor security and utility lighting and are often used in combination with a photosensor. This is because these lights are required when it is dark outside and there are people present.

**Occupancy Sensor Controls:** Occupancy sensor controls detect indoor levels of activity in specific locations. These sensors turn lights on when people enter a room and shut the lights off when all occupants have left, thereby saving money. These sensors must be placed in locations that will detect occupant activities in a room. These sensors use either sound (ultrasonic) or heat and motion (infrared) to identify occupant activity.

**Photosensor Controls:** Photosensor controls sense ambient light conditions and turn on outdoor lights once the ambient light is below a certain level. These controls work well outside and will prevent lights from operating during daylight hours.

**Timer Controls:** Timer controls are used to turn off and on both outdoor and indoor lights at specific pre-programmed times. Common types of timers are the manual timers and the in-wall programmable digital timer. Manual timers plug into an electrical outlet and typically control lamps, decorative lights, or night lights. The programmable digital timers are used to automate lighting both indoor and outdoor.

# CHAPTER 16
# SOFTWARE SUPPORT

*Historically, advancements in technology bring disruptive changes to the way we work and how business is conducted. It is no different in construction. There are numerous technology advancements driving efficiencies in the construction industry. These advancements include drones, GPS, 3D printing, and prefabrication of construction systems to support modular erection of a building or facility. The ideas described here focus on the information systems available to support advance technologies, and more specifically, on the software solutions now being used in the construction industry.*

## BUILDING INFORMATION MODELING (BIM)

One of the most disruptive technologies to be introduced to the construction since the industrial revolution has been the recent proliferation of software solutions in the area of Building Information Modeling or BIM. Also referred to as Virtual Design and Construction (VDC) or Digital Engineering, BIM is the digital representation of a physical building or facility. The term, BIM, commonly refers to the process of creating, utilizing, and managing project data with a collection of tools capable of representing project data in a 3D format.

Historically, designers communicate the project requirements to contractors by furnishing a set of two-dimensional drawings. These drawings typically illustrate the components or systems to be built from various perspectives, such as plan views, elevation views, or detail views. Individuals working with two-dimensional drawing need to be able to combine these perspectives into a mental image to fully understand what is to be built. This mental skill is necessary to effectively read plans, and it is not necessarily a skill easy for everyone to learn, especially for people who do not read plans on a regular basis.

Project teams are typically comprised of many different stakeholders, all with varying degrees of plan reading competence, so effective communication can become a challenge for large project teams working on complex projects. Through the use of BIM, designers use a 3D rendering (Fig. 16.1) to communicate project requirements. This reduces the likelihood of misunderstandings and enables project teams to spend more time focusing on project issues.

3D renderings can be interactive, helping the user better understand a project by allowing manipulation of the model; rotating around it, zooming into or panning around, and even slicing the model at a particular location to further interrogate it (Fig. 16.2).

Models are often referred to as "intelligent," meaning the various components making up the model contain vast amounts of **metadata**, providing users with further information about the model. The most common metadata included in a model is geometric data which the application uses to properly render the model. Geometric data includes dimensional measurements, length, width, and height of an object and its position relative to an X, Y, Z coordinate system.

**metadata**
*Properties or information providing details not relative to the data itself, "an underlying definition or description."*

**FIGURE 16.1** 3D rendering of a BIM model.
(Courtesy of InEight.)

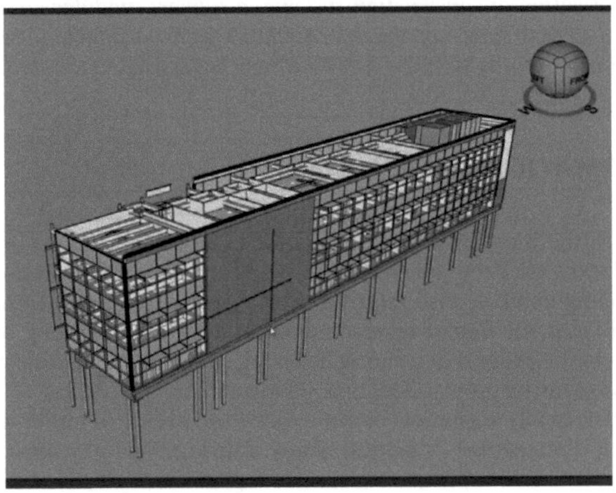

**FIGURE 16.2** 3D BIM model sliced to interrogate the model.
(Courtesy of InEight.)

Model elements can also contain many other types of metadata, most commonly specification-type data. For example, a concrete footing will be displayed in the model based upon dimensions and a coordinate position, but for the concrete footing there can also be other useful information provided, the required class of concrete, the bearing strength of the soil, or the size and type of rebar required. The amount of metadata included in a model depends on how much information the design team chooses to include when creating the model. As with conventional design documentation, providing a very detailed, information-rich model requires a lot of work; therefore, the amount and type of data included in a model is often driven by how the project team intends to use the model. What follows is an explanation of some of the more common uses of BIM models.

## Clash Detection

The final design of a facility is a collection of interrelated systems. A building typically includes a foundation system, a structural system, an enclosure system, mechanical and electrical systems, plumbing and fire protection systems, and others. These systems are often engineered by designers working independently in their particular specialized disciplines. The products of these team members are combined by a lead designer into a single collection or set of design documents. Combining all of these system designs requires a large amount of careful coordination between disciplines. As a result, it is not uncommon for coordination issues to arise or even go unnoticed in the process of producing the design documentation for a structure.

As an example, the mechanical design may size the HVAC ducts based on a required amount of airflow, while the structural design was based on the anticipated live and dead loads in the building. The owner makes a change to the design and desires a large open area in one part of the building, which in turn triggers a requirement for a larger air mover to maintain a certain level of airflow, which then increases the structural load requirements of the building. This results in the subsequent addition of a beam for structural support. These types of iterations of design requirements can easily lead to coordination issues and the result may be a **clash** between the previously designed air duct and the new structural beam. Worse yet, when the contractor attempts to build the project and comes across the issue in the field, the cost to resolve the clash rises exponentially and very often results in claims from the contractor to the owner due to "design errors."

**clash**
*The word means confrontation or encounter, a very appropriate word for the conflict.*

To mitigate this problem, one of the more common uses of BIM has been to leverage models for the purposes of clash detection. Clash detection is the process of taking various models and overlaying them on top of one another to identify clashes or locations within the model where physical elements are colliding or intersecting with one another. Identifying clashes early in the design process (Fig. 16.3) and not in the field by the contractor can save the project a tremendous amount of money for claims and lost schedule time. The alternative is attempting to redesign the project in the field to accommodate work already installed or, worse yet, requiring the reinstallation of work already in place.

> Owners are increasingly requiring designers and contractors to utilize clash detection tools on their projects.

## 4D BIM—Schedule Information

As described in Chapter 7, Project Schedules, one of the common ways for project teams to leverage a BIM model is by integrating it with the project schedule. The fourth dimension in 4D BIM is the element of time or it is the 3D model plus the project schedule. This is a way of visually representing

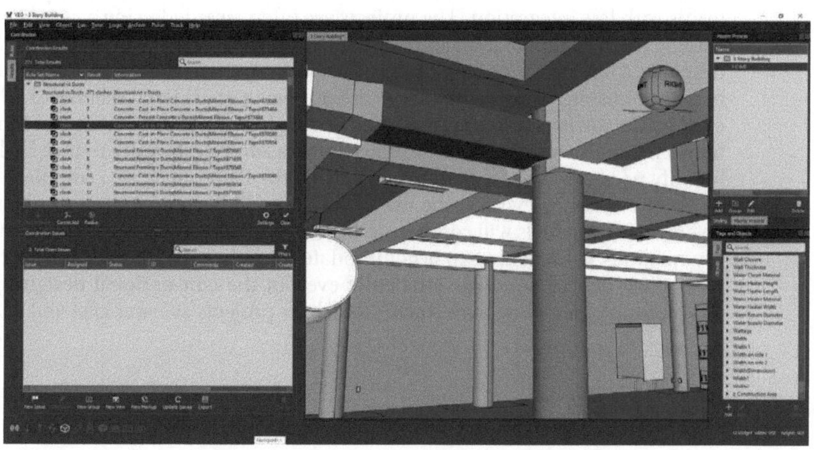

**FIGURE 16.3** Clash detection used to identify a structural member intersecting with an air duct. (Courtesy of InEight.)

the project over the scheduled time period it will take to complete the work. Such a model can be very impactful for communicating to interested parties how the work will be sequenced. Most 4D BIM software provides an effective animation of the work as it is sequenced by the project schedule. This helps the project team conceptualize how to build the work and identify potential issues with the sequencing of activities. For instance, the air handling equipment cannot be installed on the roof until the roofing work is completed. But an oversight in the schedule could have the roof equipment delivered and installed before the roof placement is completed. This type of scheduling error could easily be overlooked within a schedule for a complex project, but would be apparent using a model to visualize the sequence of construction activities.

### 5D BIM—Cost Information

Another valuable contribution to the construction process is the use of 5D BIM models, where the fifth dimension represents cost. Cost can be the estimated costs of work items or actual costs. Much of the benefit to using 5D BIM is that while managers work with complex cost estimates or control budgets, they will see those items with cost overruns "*light up*" in the model.

Actual costs, or as-built information, have also been referred to as 6D BIM. This sixth element captures information related to the specifics of the construction process, such as when any given model element was installed, by whom, what problems arose, and what materials were actually used.

### Challenges

The potential for leveraging BIM technology is great as it is underutilized relative to the modeling technology adopted in other industry sectors. The construction industry tends to be slow in its adoption of technology, even when the benefits of the technology are apparent. Use of a model requires more effort during the design of a project, but the investment results in a more efficient construction process and provides much more long-term value for the facility owner over the life of their asset.

Unfortunately, one of the challenges to implementing this technology is the extremely cost-competitiveness of the industry and the lack of perceived cost-benefit value the new technology can bring to a company.

The use of BIM technologies tends to be more prevalent among projects funded privately or where the project team has closely integrated designers and contractors, and not when low-bid or design-bid-build procurement methods are used. Publicly funded projects are characteristically project teams with less integration. The projects are often awarded using a competitive bidding process, precluding the selection of a designer or contractor most capable of delivering value through technology despite not having the cheapest price. This lack of a best-value selection process means the long-term benefits of using model technologies are not being considered relative to selecting contractors. However, private owners are coming to realize how BIM provides a powerful tool to analyze building systems, materials, and facility use. Smart owners can significantly reduce building operating costs and identify necessary system maintenance using the BIM model of a completed facility.

As the industry evolves, the use of models will become more prevalent, because the benefits of leveraging the models will continue to become more pronounced. The use of alternative project delivery methods is increasing to accommodate more best-value selection methods. Contractors thrive when they deliver buildings efficiently even in the conventional design-bid-build delivery system. BIM technologies allow a contractor to deliver projects at lower cost and with superior quality.

## ESTIMATING SOLUTIONS

Another category of software solutions is estimating tools. While the estimating process is discussed in detail in Chapter 6, Estimating Building Projects, this section will discuss the various ways technology solutions are used in supporting the creation of cost estimates.

Generally speaking, estimating involves analyzing a potential project and how it may be built most economically. The objective is to identify the most efficient work plan for use in preparing a cost-competitive bid proposal to maximize the likelihood of winning the work at a profit.

The difficulty lies in not only having the lowest bid price on bid day, but ensuring the cost estimate is based on a work plan and can be executed for a profit. Accordingly, estimating requires identifying and manipulating large amounts of design data, filling in missing or incomplete data from the design documents, and appropriately analyzing risk in order to develop bid price for the work.

Tight timeframes from when the contractor is provided with the bid documents to the time the proposal must be submitted is common. In addition, estimating tends to be an iterative process requiring the analysis of various approaches to a job to determine the most efficient work plan. As such, the value technology adds to the estimating process is improved efficiency in the creation, management, and processing of large amounts of data. This enables estimators to focus their time on understanding the requirements of the work and developing the optimal project work plan upon which to base the estimate.

## Quantity Takeoff

Takeoff is the process of using the plans and specifications to quantify all of the work elements required to deliver the project. Once a scope of work has been quantified, material prices can be requested, crew requirements and productivity rates can be determined, and the project costs estimated. Quantity takeoff can be accomplished in different ways. First, quantities are measured from the plans provided as part of the bid documents. In the past, this process was accomplished by an estimator using a full-size set of plans, a scale bar, and colored pencils to measure the scaled dimensions shown on the plans. Sometimes this is necessary to verify the plan dimensions specifically called out in the drawings. At other times a careful reading of the specification is required because specific plan dimensions were not indicated on the drawings. In either case, it is a time-consuming manual process to measure and count all the information needed to properly quantify the work.

Today, this process is commonly accomplished through the use of a **digitized** set of plans and a 2D takeoff software solution. The estimator opens the digitized plan set, often in the form of a **PDF file**, in the 2D takeoff tool and then digitally measures and counts the various quantities by pointing and clicking with the computer mouse. The takeoff software can contain a library of commonly measured items which it automatically summarizes as they are being taken off. This makes the manual process much more accurate and efficient and permits the estimator to produce more detailed information in less time.

**digitize**
*To convert data to digital form for use in a computer.*

**PDF file**
*Portable Document Format (PDF) file is an open standard file format. It supports easy sharing and presentation of files.*

A PDF file is essentially an image of a document. PDF files can be created in one of two ways, either as a raster image or a vector image. Raster-style image files contain a simple picture of the original source of the image. If a PDF has been created by scanning a copy of a physical document, this is most likely a raster-style image. By contrast, a vector-style image is the one commonly produced directly from a design/authoring tool. While it looks the same as a raster-style image, the image itself is based on the linework used in creating the design files. For purposes of performing a quantity takeoff, vector PDFs can greatly increase the efficiency and accuracy of the takeoff process by enabling the estimator to snap to existing linework or even derive information from the metadata contained with the linework such as line lengths or areas. Not all takeoff tools support the ability to utilize the additional data contained in a vectorized PDF file.

Another method of gathering quantity takeoff data is to use the original design file and an authoring tool capable of working with the file format. For example, CAD files can be used in AutoCAD to produce design quantities. To support construction contractors, it is common for the designer to produce PDF files from the design files rather than to make available the original design files. However, if contractors are able to gain access directly to the design files, they can then leverage all the data the designer created to generate the needed takeoff data.

If a 3D model is available, it may be possible to use the model elements and associated geometric data to determine dimensions along with other relevant design data. In a competitive bidding environment, typically the level of detail in the model is insufficient for deriving quantities. This is because a model at this point in project design will likely have been created for the purposes of communicating the intent of the design, not necessarily for the conveying design details. However, if a model is available, even a conceptual one, it can still serve as a good data source.

Lastly, design data may be included in the bidding documents in the form of tabular data. These are typically included as lists or tables to support details for various building systems, such as a cable schedule including lengths and terminations, a table of foundations including sizes and types, or a fixture schedule. These tables can be included in the PDF document itself and various tools can be used to export this data into a spreadsheet. There are also times when a spreadsheet file containing all the data is provided with the bidding documents. The most common spreadsheet application is Microsoft Excel and it is arguably the most commonly used tool in estimating.

## Cost Estimating

Quantifying the work can often be the single largest effort in developing a cost estimate. Once the project has been quantified, it then needs to be estimated. This can be done with a tool built specifically for cost estimating. These tools provide features intended to enhance the contractor's ability to derive a cost for the project. There are some software solutions that offer both quantity takeoff and cost-estimating features in a single product. For many general contractors, simply applying a unit rate to the various takeoff quantities is sufficient to produce an accurate cost estimate. For estimates requiring a more detailed analysis of the work, as is the case for many self-performing contractors, the costing solution must include more features and flexibility related to subdividing work into component activities, analyzing the various components, applying productivity rates to the quantities of those components, determining activity durations, and applying them with crew costs to establish an estimated project cost (Fig. 16.4).

Many estimating software solutions are geared to a specific sector of the market, meaning the solutions available for a sitework contractor are different from the solutions available for a plumbing or HVAC contractor. While the process of estimating is fundamentally similar across all sectors of construction, many estimating solutions tend to be "purpose-built" for specific types of contractors because certain software features tend to be more meaningful than others based on the type of work the contractor is estimating. For instance, a general contractor who subcontracts out the majority of the work would find more value in a solution with capabilities to manage quotes received from subcontractors, while a self-performing contractor is more interested in the ability to estimate based on current labor rates and job-specific productivity factors. Similarly, a construction manager participating in design and constructability reviews values a program that can quickly create conceptual estimates and easily revise the estimate as various design alternatives are being produced and evaluated. Specialty contractors, such as electrical or mechanical contractors, tend to use purpose-built solutions with catalogs of product information specific to their industry or with the ability to integrate to various industrial suppliers.

A software cost-estimating solution provides the ability to determine cost of the work using the estimating principals described in Chapter 6, Estimating Building Projects, together with many other features. There can be features to estimate all direct costs associated with the work, as well as the ability to estimate all of a company's indirect costs incurred in completing the work. Project indirect costs are typically driven by the project's total duration, so deriving a preliminary project schedule is an important consideration. For general contractors, indirect costs of an estimate are the required staffing of the project with appropriate management personnel, providing site facilities and job trailers, and many times, even providing services to support the trades such as a crane service or elevators. Another type of project indirect cost is the financial cost associated with bonding and providing insurance for the project, plus permit fees or regulatory requirements and taxes. Escalation costs for long duration projects and interest expenses for projects requiring the contractor to be in a cash-out-of-pocket position before receiving payment are other examples of indirect costs included in an estimate. Comprehensive estimating software includes features to assist with the proper estimation and inclusion of all these indirect costs.

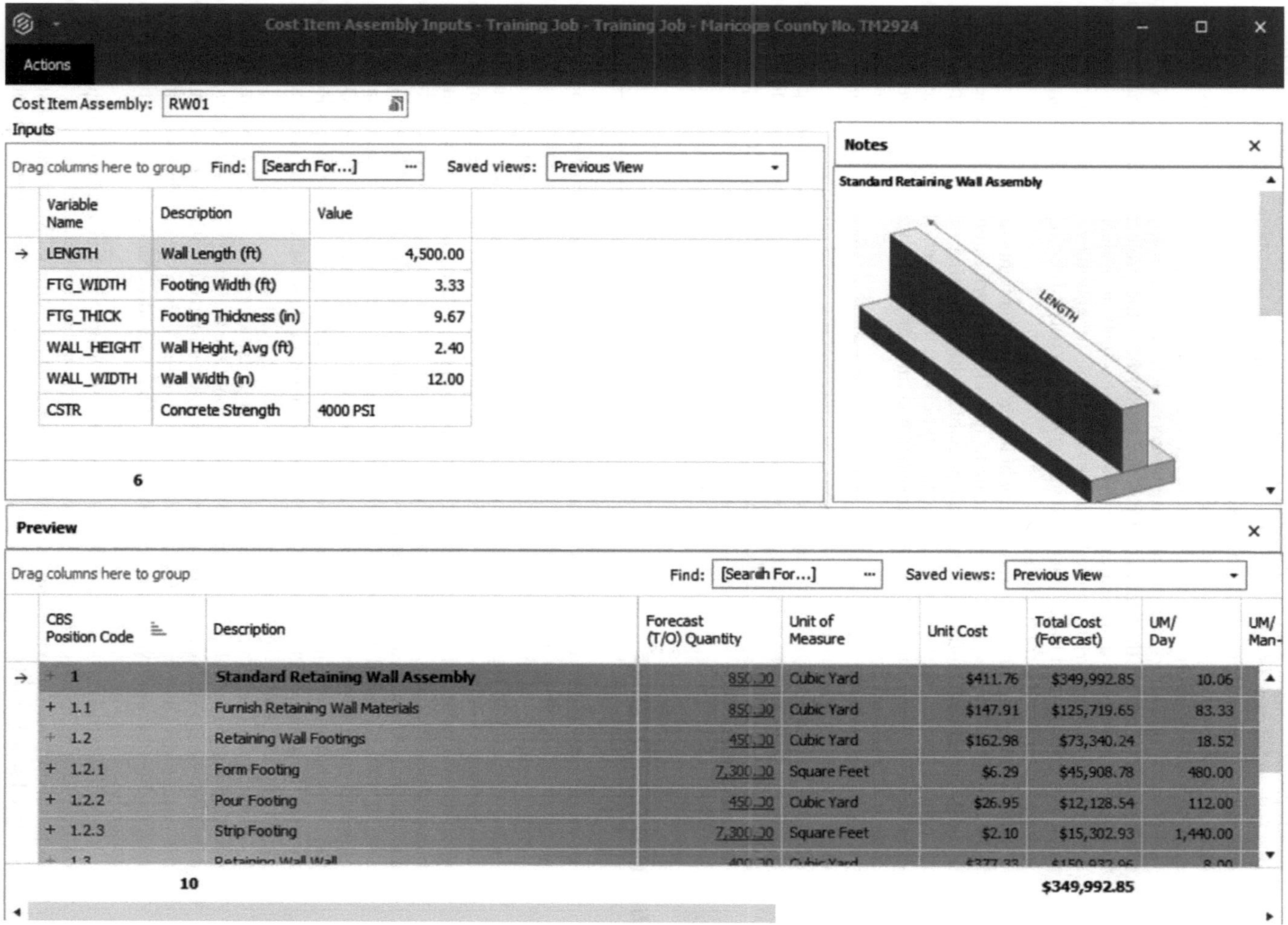


FIGURE 16.4 Cost estimating software.
(Courtesy of InEight.)

The estimating software needs to support the contractor's ability to determine an appropriate price for the project. This is not insignificant. An effectively priced proposal must consider all the reasonably anticipated costs for preforming all work tasks and it must include an appropriate amount for the contractor's overhead and profit. However, the bid price must not be so high as to reduce the chance of winning the work, specifically when the bids will be evaluated on the basis of lowest bid price. As such, choosing an appropriate markup amount or fee is a crucial question.

For most contractors, one of the main drivers of determining an appropriate fee is a fair and reasonable assessment of the amount of risk for performing the project. Some estimating solutions include a risk register. This feature permits the assessment of project-specific risks and assists in determining an appropriate amount of money to include for contingency. Contingency can either be carried as a line item in the estimate and drawn down as needed throughout the project, or it can be included as part of the contractor's fee.

Finally, some bid proposals require the bidder to furnish a lump-sum bid amount. Others require the contractor to provide **unit prices** on a schedule of bid items form. For such unit price bids, the estimating software should permit the estimator to ensure each bid item price covers the direct cost of doing the work, plus a proportional amount for indirect costs, overhead, and profit.

**unit prices**
*A price per unit of work performed. Usually used when item quantities are uncertain.*

Sophisticated estimating software enables a contractor to spread markup across bid items in various ways, considering quantity under- and overruns, and various alternative approaches to doing the work. Such pricing strategies enable bidders to put more money in bid items they will be paid for, and not put overhead and profit on bid items they expect to underrun or be deleted. This can provide the bidder with the best possible chance of winning the work.

## PLANNING AND SCHEDULING

There are many ways a contractor can lose money on a job, but one of the surest is poor planning, or no planning. As discussed in Chapter 7, Scheduling Projects, construction projects often involve a complex sequence of interrelated activities, where the delay of any particular activity can have a cascading effect on all the related activities, resulting in an overall delay to project completion. Most construction contracts require the work to be completed by a specified date and impose penalties upon the contractor for failure to meet the completion date. There can also be intermediate completion or milestone dates. The old adage, "time is money," is appropriate because a contractor estimates a project based on a reasonable presumption of the overall project duration. When the actual duration exceeds the anticipated duration, there is often insufficient margin to cover additional indirect costs of project supervision and temporary site facilities, such as job trailers and utilities. Other tangential time-dependent costs impact the contractor, like lost opportunities, because the project team cannot move to the next project. All of these factors underscore the importance of the planning and scheduling process in a contractor's success.

Planning can generally be defined as the development of a work plan. It is often a collaborative effort among individuals who are knowledgeable about the various project work tasks. The use of in-depth knowledge enables planners to define the work required and estimate reasonable durations for the various tasks and how the tasks relate to one another. This is the schedule logic for determining relationships between various tasks and their sequences and dependencies; when tasks cannot be commenced until other tasks have been completed.

Scheduling then takes the task durations and relationships to calculate planned dates for the work. Critical Path Method (CPM) scheduling is the most common software solution and there are many different options in the marketplace. For complex projects, some CPM scheduling solutions enable advanced scheduling based on different work calendars and resource or crew availability—a process referred to as resource-leveling. The concept behind resource-leveling is that if you have two activities planned to be performed at the same time, both of which require two cranes, you would need four cranes in total during the time period when both those activities will be performed concurrently. If, however, you have only two cranes on the project, then you would need to increase the duration of one or both activities to account for insufficient crane resources.

Today there is so much data being captured digitally, construction planners arguably have more data available than they know what to do with. The challenge, in using data efficiently, is sifting through the noise and ignoring the nonproductive information. More importantly, it is necessary to produce "actionable information"[1] from the data.

The project management team often focuses on managing cost and schedule. So some of the most meaningful information will be relative to keeping a project moving on time and under budget. Using historical and project data to forecast how much work will cost or how long it will take to complete various work tasks is a type of data analysis known as *predictive analytics*. This cutting edge approach to scheduling is really more of a knowledge-driven philosophy. For the first time, through the help of artificial intelligence (AI) and machine learning, the computer can actually start to guide the planner or scheduler through the schedule development process.

> The AI approach to planning results in an ability to identify and take corrective action early, before problems become critical issues threatening the successful completion of the project.

As an example, planning software can be used to more effectively develop the construction schedule. By leveraging data from past projects, the planning software can intelligently determine and make suggestions to the planner about what other related activities there may be and the typical duration of these activities (Fig. 16.5). Validating the planned assumptions against benchmark data results not only in a more accurate plan, helping instill confidence with the project team, but also helps the planner identify significant areas of risk within the plan.

For instance, if a scope of work for a task has historically seen large variations in its duration, it may represent schedule risk if the planned duration cannot be assured. A company's historical data shows foundation drilling for buildings of a particular size may take anywhere from 2 weeks to 3 months to complete. With this information, planners will understand that if they plan the foundation drilling to last 1 month, the duration may be longer because the assumed duration is on the low side of the durations experienced for other similar projects. However, if the planned duration for foundation drilling is scheduled to take 3 months, then the project will not be completed on time. Armed with this information, the estimators will understand the foundation drilling activity schedule risk and the project team can take measures to mitigate the risk. This could include a more thorough geotechnical investigation of the site, requiring the foundation contractor to be capable of accelerating the work by adding a second crew if necessary, or possibly even changing the building's foundation design to reduce the need for drilling.

Building construction necessitates a solid work plan because many of the work tasks for building systems are subcontracted to specialty contractors, and the general contractor must coordinate all the trades to a single project schedule. Where the available space on a project is very limited and transitions from one work task to another can be complex, it is very easy to encounter delays due to coordination issues.

Complicating things further, a delay in any one activity can have a cascading impact on all activities. If any of the activities are on the critical path, the entire project will be delayed. Therefore, a solid work plan: realistic and executable is crucial to the success of a construction project.

Often delays are caused by risk events or simply uncertainty surrounding the work itself (e.g., through quantity growth). Accounting for such risk and uncertainty is paramount to developing a realistic and achievable schedule forecast. The concept of developing what is known as a risk-adjusted plan, accounting for risk through the inclusion of contingency, is changing the science of CPM scheduling for the better.

The tools used in the development of a CPM plan should be flexible and support collaboration. They should allow for easy iteration of various approaches to the work in order to identify the optimal plan. Once a project team has developed a solid work plan, the project management team needs to track progress through the execution of the project and use the plan as a tool to determine how well

[1] Actionable information means having the necessary information available to deal with the situation at hand.

**FIGURE 16.5** Planning software used to develop a preliminary project schedule. (Courtesy of InEight.)

the project is progressing. This provides stakeholders with periodic updates on schedule and budget performance and identifies possible issues impeding the successful completion of the project.

## DOCUMENT MANAGEMENT

Effective management of project documents is critical to the success of any project. All projects have copious sets of documents defining the work. These are referred to as the contract documents. Contract documents are commonly the drawings and project specifications. They often incorporate other documents by reference, such as standard specifications or building codes.

In addition, by the terms of the contract, many documents must be produced: submittals, shop drawings, product data sheets, requests for information, pay requests, change orders, punch lists, and warranty and maintenance manuals.

Lastly, there are many other documents produced throughout the execution of the project, some of which may be required by company policy rather than dictated by the contract, but all are related to or necessary for the execution of a project. Some examples of these types of documents are time cards, daily logs, inspection forms, project photos, emails, and correspondence.

Invariably, even the best project team must contend with many issues during the construction process and differing interpretations of contractual requirements. When those interpretations result in a significant monetary impact, litigation to settle disputes among parties to a construction contract is all too common. When such disputes arise, the contract documents often determine the rules for resolving such disputes, and project documentation often becomes the key supporting element in either pursuing or defending contract claims. As such, contractors intuitively know the importance of effectively collecting and maintaining project documentation.

Prior to the proliferation of computers and digital storage of project documentation, it was common for contractors to maintain vast collections of boxes containing all the documentation generated throughout the course of a project. Documentation was often transmitted by mail for pen and ink/hard copy signature or in the case of time-sensitive information, by courier, which over the course of a project, added considerable delay to the necessary approvals and decision processes needed for the work to proceed. Today, the majority of the industry has migrated to digital storing of project files for accessibility and ease of distribution. While this has greatly increased the ability to share information more quickly and begin building the work faster, it has also introduced some complexity. Why? Because the tools used to manage project documentation are still somewhat disconnected since each stakeholder has typically invested in their own document management solution. As a result, email remains the most common way for project teams to share information and documentation with each other.

Portable Document Format, or PDF, has essentially become the industry standard file format for document storage. A common, industry standard file format is necessary because when files are shared digitally among different project stakeholders, the file format produced by software on one individual's computer needs to be readable by different software on the recipient's computer.

A common example is when a designer uses Autodesk AutoCAD to produce a design file and sends it to the contractor. The contractor may not own a license for AutoCAD or necessarily know how to use it, but still needs to access the information to understand the plan requirements. To complicate matters further, even designers sharing files with each other may not be able to open each other's files unless they are using the same software product. Worse, the design files might require the users to have the same version of the product. This interoperability challenge is relevant to all facets of digital information sharing, not just document management, but it becomes a predominant issue for document management software solutions due to the need to share project information.

Some of the most crucial pieces of documentation are the project plans and specifications of the contract documents which define the required work and level of quality. Almost gone are the days of the contractor spreading out the plans on the hood of a pickup truck. Today, contractors rely on the use of tablet or laptop computers to access and manipulate plan documents. Some want the added flexibility and functionality to access plans in the field on cell phones. Digital access to plan documentation from every device enables contractors to more easily find information and even mark up the plans and specifications, generating further documentation and enhancing documents with explanatory and clarifying information.

## Storage and Distribution

For many contractors, the most basic document management solution is storing digital files on a computer. Paper documents are still a common format, so scanners can be used to digitize the physical documents, thereby generating an image type or PDF file format. The digitized versions are then stored in a file folder on the computer and are accessible to email recipients. A slightly more sophisticated approach would be for a company to store the files on a **file server**, which makes it easier for multiple individuals in a company to share one common set of project documents. This approach to document management is the most basic and inexpensive solution, and it is very uncommon for a contractor to not employ at least this level of document management.

**file server**
*A computer responsible for the central storage and management of data files so other computers on the same network can access the files.*

Many contractors utilize non-construction-specific document management solutions, such as Dropbox, OneDrive, or Google Drive, which are commercial online portals allowing users to upload, manage, and share permission-based access to documents. These are often subscription-type services and charge a monthly fee for access to the portal. They can work well when everyone on a project has a subscription to the same service. For contractors, this makes the process of distributing the necessary project documentation to other project stakeholders very efficient, such as providing subcontractors and suppliers during the proposal process with a hyperlinked location URL where they can access copies of the plans and specs.

There are a number of construction-specific document management solutions available to help contractors better manage project documentation. These solutions tend to be online, subscription-based services, commonly referred to as a **Software as a Service** (SaaS) offerings. They drive efficiencies by providing features specific to the construction process. Most generic document management solutions do not effectively address these construction documentation issues (Fig. 16.6).

**Software as a Service**
*A method of software delivery and licensing in which software is accessed online via a subscription, rather than bought and installed on individual computers.*

## VERSION CONTROL

One construction-specific challenge is document versioning. It is very common for construction documents to go through multiple design iterations before they are finalized for construction and sometimes there are revisions after project execution has begun. Many times, issues arise or discrepancies are identified in the documents. These will necessitate drawing modification or change. The challenge for a contractor is to always work with the most up-to-date version of project plans. This can be especially problematic for subcontractors whose work depends on having the latest version of the drawings but who are further removed from the design iteration process.

In litigation, which version of the drawing and when it was distributed can become an important piece of evidence describing who knew what and when they knew it? Under current case law, just because a new version of a document existed, a contractor may not be obligated to incur the cost to redo work built out of compliance if the work was built in accordance with a previous version of the drawings and the contractor was not furnished with the new version.

Effective document management software provides a contractor with the ability to keep the latest version of the document available for the project team as well as to retain prior versions of project documentation. It can also notify applicable users when a new version of a document is available. The system can track acknowledgment of such notifications and record when the documents were accessed or downloaded and by whom.

**FIGURE 16.6** Document management software used to organize project documents. (Courtesy of InEight.)

### Workflow Management

Another issue addressed by document management solutions is the management of workflows. Contracts often define the process for how various parts of a project can be executed. For instance, before a contractor orders the structural steel for a project, it is usually a requirement to submit shop drawings and obtain approval from the structural designer. Failure to do this properly and within the time frames stipulated in the contract can result in the owner or the owner's representative deeming the work was done incorrectly or out of compliance with the contract, resulting in costly rework for a contractor.

Document management solutions can include features to track the submittal process, giving users the ability to manage a submittal log. This includes the due dates and lead times of various items and to see where any particular submittal may be in the process. The submittal log keeps together all the relevant documentation for a particular submittal and restricts which users can perform certain actions in the system, such as only permitting the designer to approve submittals. Contract requirements are different from project to project, so document management software should be flexible enough to provide companies with the ability to define workflows specific to the requirements of each project.

The management of project documents is a very important consideration for any project team. The solution chosen for any particular project can be as simple as defining a location to store all project-related documents. Or, it can be as sophisticated as implementing a solution to distribute project documents, manage specified workflows, and restrict user access based upon their project team role. The appropriate selection of a software will likely come down to the size and complexity of the project as well as contractual requirements for document production. It is necessary to also factor in the various people involved in the project, their need to work with documentation, and the willingness of the project team to implement and utilize a particular solution.

## *PROJECT COST MANAGEMENT*

Project management software varies widely in capabilities, the workflows supported, and process automation. Further, there are many overlaps in the features products offer and the depth to which they implement those features. With various solutions in the areas of job cost accounting, budget management, scheduling, forecasting, field data collection, and resource management, an effective project cost management solution often entails a complex combination of interrelated solutions. Each of these have their own strengths and purposes in the context of an overall solution intended to help the project team monitor progress and manage the successful completion of a project. The following discussion addresses the components of a comprehensive project cost management solution and how features can be leveraged to positively influence the execution of a project plan.

### Job Cost Accounting

Chapter 4, Construction Accounting, discusses how various aspects of accounting are unique to the construction industry. Effective project cost management requires more than just keeping track of how much money has been spent. It requires comparing actual costs to the anticipated costs budgeted as a part of the project's work plan. There are many commercially available accounting solutions, but very few accounting solutions are sufficiently comprehensive to provide a project team with all the tools needed to effectively manage a project. The ones optimized for the construction industry include features specific to job costing, construction accounting, and project reporting.

Accounting solutions, even construction-focused solutions, are typically very effective at tracking complex costs, such as payroll costs, managing payables and receivables, providing tools to manage cash flow, and reporting on financial results at a company or even at a project level. However, accounting solutions tend to fall short in helping project management account for things not commonly characterized as accounting transactions. As an example, effectively reporting on committed costs that have yet to be incurred can be challenging for accounting solutions.

Take for instance the situation where a contractor orders the fabrication of a long lead-time component such as a precast grade beam, so it will be ready in time to keep the project on schedule. The contractor will not contractually incur an expense until the element arrives on site for installation. Further, the contract with the supplier permits the contractor to cancel the order any time prior to delivery and allows the contractor to pay for only costs incurred in fabricating the precast element up until the time the order gets canceled. This committed cost can be difficult to determine but can represent a significant risk to the project if the possibility of a design change exists and the order may have to be changed or canceled. Being able to understand and track the potential cost, or the amount of money the project team may be liable for at any given point in time is often difficult for conventional accounting systems to handle because this type of transaction does not represent an actual cost. In many ways, it may be more important for the project team to track the potential costs or the risk associated with various aspects of a project than it is to simply report on actual expended project costs.

To report on the financial health of the project at any given time during construction there are nuances that must be specifically addressed. It's important for the company's accounting system to have the ability, through integration with add-on software, to express cost and risk in meaningful ways to project management.

## Control Budget

Software solutions to assist with managing the project budget are becoming more commonplace. Effective management of a budget starts with the establishment of a control budget. Much like establishing a schedule baseline, the control budget is a static version of the budget established at the start of the project and permits the project team to analyze the variance between budgeted values and costs actually incurred. Many of these systems can integrate with an accounting system to track progress against the budget by importing the as-built data back into the system as costs are incurred (Fig. 16.7).

Control budgets should be based on a detailed Work Breakdown Structure (WBS). The WBS must be detailed enough to permit a good analysis of actual project results while not being so detailed that cost cannot be reasonably tracked or offer meaningful information acquired.

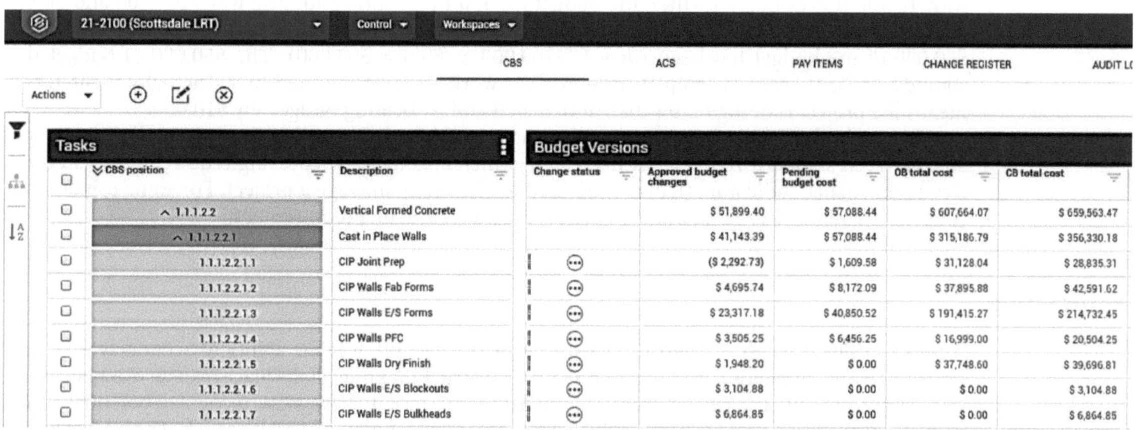


**FIGURE 16.7** Tracking project performance relative to a project's budget. (Courtesy of InEight.)

Consider the case of work done by crews in the field being entered on a time card, which is processed by payroll, resulting in actual labor costs compiled by WBS code and compared against the project budget. For a building project requiring the installation of 54 foundation footings of various sizes and dimensions and 450 lineal feet of spread footings, the project team needs to define a WBS to support an appropriate level of detail for tracking actual costs and managing the budget. Breaking the work down and creating a separate WBS item code for each of the 54 footings on the project would be too much detail, as it would require field personnel to individually track the amount of time crews worked on each.

If the project team were to encounter issues with the foundation footing work, this level of detail may permit them to very specifically determine which specific footing of the 54 footings is having issues, but the amount of work necessary to track at such a level of detail may not be worth the effort. Conversely, if all work for 54 footings were under one WBS Foundations item code, the lack of granularity may not help the team to determine which specific area of the project requires attention.

The level of detail necessary in a WBS is, therefore, subjective and is often a judgment call. A good rule of thumb for the person tracking the work in the field is to not require tracking of their time to more than a few items in the course of any single day's work. More than a few items and it is likely the field staff will not correctly record the work against the appropriate WBS item codes. Still, providing too few WBS item codes may make it easier for field staff to record their work, but it may not provide a sufficient level of work progress or cost detail to support good management decision making.

## EARNED VALUE MANAGEMENT

Of particular importance in tracking a project budget is being able to determine at any time how much scope of work has been completed. Many accounting systems tend to determine percentage of completion based on a cost-to-cost method. This means that if an item has a budget of $100,000 and you have incurred $40,000 of cost, the percentage of completion would be 40%. This may be technically accurate but requires constant re-estimation of the budgeted value of the work to be meaningful. Re-estimating is tedious and can make subsequent reporting less accurate if it is not done consistently and frequently.

A more meaningful way of determining percentage of completion is to use the physical quantities installed to date. If there are 450 linear feet of spread footing to install and 225 LF have been installed, then 50% is actually complete and any actual costs can be compared to the budget in a meaningful way. Taking this example further, the budget for 450 LF of spread footing was $100,000 and 50% is complete based on 225 LF being installed. As opposed to $40,000 of actual costs incurred to date, $50,000 of the budget has been *earned* ($100,000 × 50% = $50,000). The $50,000 of budget that should have been spent in completing 50% of the work is more than the $40,000 actually spent, which means the project is doing better than anticipated and is "beating budget" by $10,000.

This method of evaluating actual performance to the budget is called Earned Value Management and has been standardized per ANSI/EIA 748 which also defines reporting requirements and other Key Performance Indicators (KPIs) to be used to effectively manage a project. For some capital projects, the use of Earned Value Management is even a contractual requirement.

### Forecasting

Another critical element of a Project Cost Management solution is the ability to accurately forecast a project's cost at completion. Many activities on a construction project take place over an extended duration, which presents the project team an opportunity to monitor progress, recognize potential issues, and take corrective measures before they have a major impact on the overall project. The budget analysis described above enables the project team to determine which budget items may be in jeopardy of not meeting budget based on what has already been accomplished. However, a reasonable forecast of the

| CBS position | Description | % Complete | Forecast final cost | Forecast final unit cost | Forecast final MHrs | Forecast final labor cost/MHr | Forecast final man hours/Unit | Forecast final productivity factor |
|---|---|---|---|---|---|---|---|---|
| 1.1.1.2.2 | Vertical Formed Concrete | 3.25 % | $ 586,842.34 | $ 324.29 | 9,671.96 | $ 58.40 | 5.34 | 1.30 |
| 1.1.1.2.2.1 | Cast in Place Walls | 5.86 % | $ 322,102.52 | $ 670.97 | 5,205.60 | $ 61.88 | 10.84 | 1.31 |
| 1.1.1.2.2.1.1 | CIP Joint Prep | 0.08 % | $ 30,444.89 | $ 2.47 | 0.00 | $ 0.00 | 0.00 | 0.00 |
| 1.1.1.2.2.1.2 | CIP Walls Fab Forms | 16.38 % | $ 67,770.38 | $ 6.43 | 1,303.61 | $ 51.99 | 0.12 | 0.61 |
| 1.1.1.2.2.1.3 | CIP Walls E/S Forms | 6.13 % | $ 145,075.55 | $ 4.56 | 2,978.45 | $ 48.71 | 0.09 | 1.35 |
| 1.1.1.2.2.1.4 | CIP Walls PFC | 1.26 % | $ 26,960.50 | $ 34.07 | 0.00 | $ 0.00 | 0.00 | 0.00 |
| 1.1.1.2.2.1.5 | CIP Walls Dry Finish | 0.02 % | $ 45,019.34 | $ 0.99 | 911.44 | $ 49.39 | 0.02 | 0.88 |
| 1.1.1.2.2.1.6 | CIP Walls E/S Blockouts | 4.13 % | $ 631.86 | $ 2.61 | 12.10 | $ 52.22 | 0.05 | 5.00 |
| 1.1.1.2.2.1.7 | CIP Walls E/S Bulkheads | 4.59 % | $ 6,200.00 | $ 28.47 | 0.00 | $ 0.00 | 0.00 | 0.00 |

**FIGURE 16.8** Forecasting a project's cost at completion. (Courtesy of InEight.)

costs for the work remaining can help the project team incorporate the potential impact of necessary corrective measures. The team can more easily determine the parts of the project having the largest impact on the overall project. This can guide the project team on where to focus attention (Fig. 16.8).

## Field Execution

The overall productivity of construction is largely driven by labor productivity. Accordingly, much attention is paid to ways to improve the efficiency of the industry's labor force. Today, there are many different tools used to facilitate the collection of information from the field and drive efficiencies by automatically feeding the data to other systems, removing redundant data entry, and providing management with more timely information on the progress taking place in the field.

Prior to the ability to collect "hours worked" information electronically, it was common for a foreman to complete paper time cards each day for the crew's activities. The cards would then be delivered to the office where an individual would enter the information into a payroll system to generate pay checks and feed other systems, such as a job cost system, and to produce cost reports. While this process enables workers to be paid each week, it often results in long lead times between when the work was accomplished and when project management personnel would receive reports summarizing the actual costs incurred in doing the work.

If there were issues with the work being performed within budget, it would not be uncommon for such issues to persist for a week or more prior to being discovered, depending on the efficiency of the reporting system being used. On larger, more complex projects, this issue of delayed reporting could result in easily correctable issues going uncorrected for long periods of time because no one realized there was an issue.

The electronic collection of information has provided companies with the ability to monitor performance in a more real-time manner than was available in the past. Today, foremen use laptop computers, tablet devices, or even mobile phones to easily enter time expended by field personnel and equipment (Fig. 16.9), record progress, and even log issues as they arise in the field (Fig. 16.10). This real-time flow of data gives management immediate access to the efficiency of field execution activities data and the forecasted effects on the overall project schedule and budget.

While many of the construction industry technological improvements have resulted from collecting data directly from the field, of greater importance than simply tracking work is the ability to provide crews in the field with information to help them make better decisions quickly. This is critical because the individuals closest to the work are in the best position to improve efficiency.

An example of one such tool to improve field productivity is the concept of daily planning software. Daily planning software enables a foreman to choose the people and equipment to be

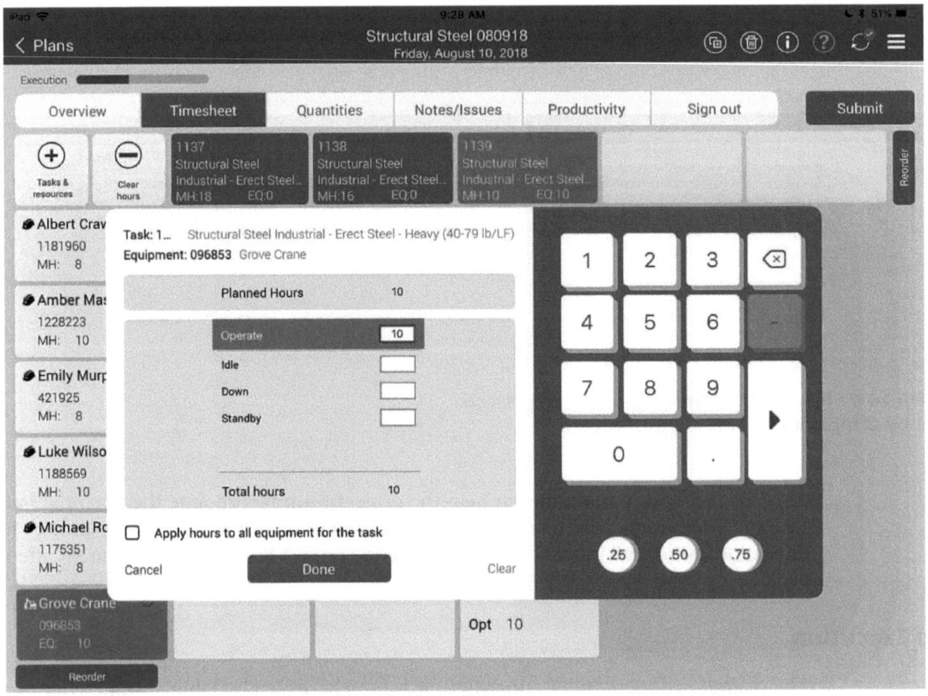


**FIGURE 16.9** Time card application used to record work hours from a foreman in the field. (Courtesy of InEight.)

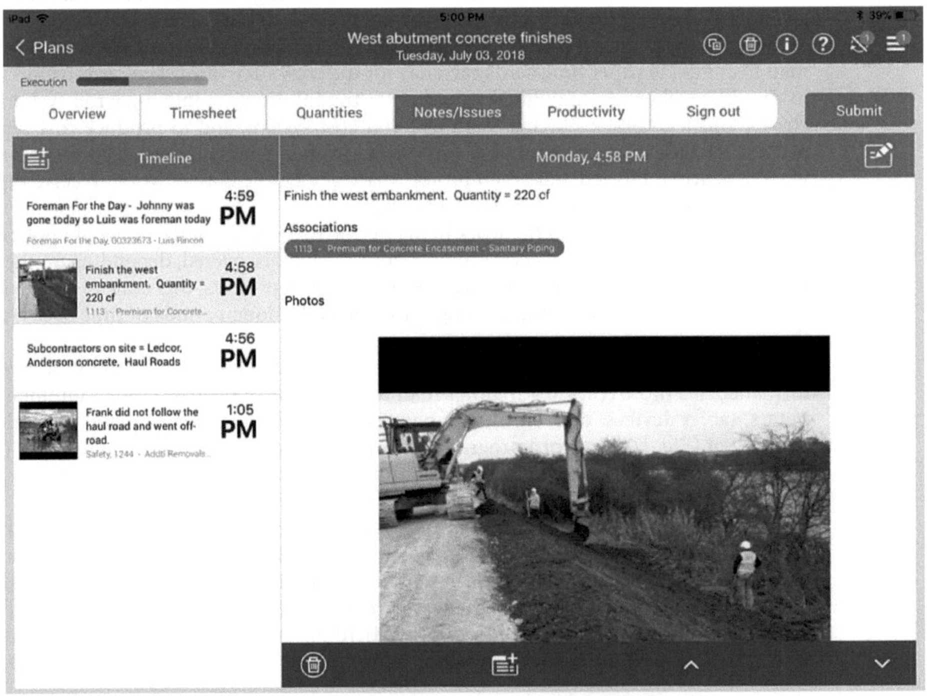


**FIGURE 16.10** Field issues being tracked real time with a mobile application. (Courtesy of InEight.)

employed. Based on the forecasted quantity to be installed, the foreman can compare the anticipated cost of doing the work with the budgeted value of the work. Giving the people who build the work immediate feedback on their planned approach helps them easily understand if the process used will make or lose money and lets them identify issues and take corrective actions before the work task have started.

## THE FUTURE OF TECHNOLOGY IN CONSTRUCTION

Technology will continue to evolve and improve how we build and how we manage the process of construction. Many companies have transitioned from viewing information technology as a cost center to viewing information technology as an investment. What is particularly challenging for firms is understanding what technologies are worth investing in and which ones are destined to fade away over time because technology changes at such a fast pace.

In response to this concern, companies are moving away from the traditional paradigm of owning and operating their technology infrastructure. Instead, they are adopting products delivered in a Software as a Service (SaaS) model. SaaS permits companies to avoid the capital costs of hardware and software and staffing and training the IT personnel needed to operate and maintain those systems. Companies pay instead for using SaaS systems on an as-needed basis.

Benefits of this approach, in addition to freeing up capital, include improved security and scalability of the technology because the host data centers are purpose-built for scalability, redundancy, and recovery at costs much less than a company maintaining its own infrastructure. The pay-as-you-go model does a much better job of supporting a growing company, letting them gradually scale their technology solutions as their business needs change. This greatly reduces the risk of being locked into technology decisions made years earlier. Features and enhancements to SaaS software are generally deployed automatically and eliminate the need for lengthy compatibility testing cycles. Also, SaaS solutions are commonly deployed as cloud-based solutions, enabling the contractor to greatly minimize any reliance on hardware and computers other than the most minimal configuration needed to run a web browser. Note: not all SaaS solutions are cloud-based solutions and not all technology solution providers offer a SaaS solution; however, the trend in the industry continues to move in that direction.

Another construction technology trend is in response to the lack of interoperability of various systems. For any particular category of software solutions in the construction industry, project management, estimating, and scheduling, there are numerous solution providers. And construction companies have many options when it comes to implementing the right solutions for their business. What does not currently exist is reliable interoperability among the various types of systems. Where a company may have a very good BIM solution and a very good estimating system, the ability for those two different solutions to integrate with each other and share data is often limited. In many instances, where cross-product integrations do exist, they can be incomplete and result in manual processes for reproducing the data shared between the systems. This is especially problematic because executive management requires a meaningful interpretation of project data from across all technology solutions at any point in time. This requires that the project systems work together because any issue occurring on one project is likely to touch all of the various information systems in use.

The trend in the industry is consolidation of the various software solutions onto a single platform to make interoperability seamless. The result is builders can leverage a single solution capable of managing every aspect of a project's lifecycle (Fig. 16.11).

While it is very difficult to predict how technology may influence the construction industry in the coming years, to be successful, a builder must embrace technology. Owners are increasingly facing budget and regulatory challenges and are consequently requiring their project teams to be more efficient. Delivering projects on time and within budget, while providing value to the project owner, will require contractors to find new ways to drive efficiencies in their operations and effectively mitigate project risk to build work profitably.

**FIGURE 16.11** Dashboard providing an executive view of project information. (Courtesy of InEight.)

Many of the most impactful technology innovations are disruptive, meaning they fundamentally change established business processes. Compared to other industries, construction lags in the adoption of new technology. Still, many firms are successfully implementing new software solution to drive efficiencies and add value to their business processes. As a result these companies are gaining a competitive advantage and will enjoy longevity.

# CHAPTER 17
# PRODUCTIVITY

*Productivity is defined as the ratio of input of a resource to the output. In construction, productivity is referred to by many different terms such as the performance factor, performance rate, and the unit person-hour rate. Productivity is more adequately defined in construction as the progress achieved per unit of time [Eq. (17.1)].*

$$Productivity = \frac{Output}{Labor + Equipment + Material} \tag{17.1}$$

*Factors that influence productivity in the construction industry are present at multiple levels. Factors at a high level are related to the influence of the owner and the contracting methods. Lower levels of impact relate to what occurs with the management of the jobsite and direct impacts on labor. Effectiveness and efficiency are vital to labor in the construction industry. Labor productivity is a concern of owners and construction managers who are responsible for cost and schedule control of constructed facilities.*

## INTRODUCTION

Data from various sources shows that US construction industry productivity has a long history of decline that continues today. In fact, construction is one of the few industries that is less productive now than it was 60 years ago. In a recent report, McKinsey Global Institute goes so far as to refer to lagging construction productivity as an "intractable productivity problem."

Good project management in construction must vigorously pursue the efficient utilization of labor, material, and equipment. Improvement of labor productivity should be a major and continual concern of those who are responsible for cost control of constructed facilities.

Material handling, which includes procurement, inventory, shop fabrication, and field servicing, requires special attention for cost reduction. The use of new equipment and innovative methods has made possible wholesale changes in construction technologies in recent decades. Organizations which do not recognize the impact of various innovations and have not adapted to changing environments have justifiably been forced out of the mainstream of construction activities. Table 17.1 highlights major factors that impact the productivity of the construction industry.

**TABLE 17.1** Broad Factors Impacting Construction Productivity

| Category | Factors |
|---|---|
| Project conditions | Weather |
| Market conditions | Material shortages<br>Lack of experienced construction managers |
| Design | Owner Changes |
| Construction management | Ineffective communication<br>Lack of planning and scheduling<br>Poor supervisor training |
| Labor | Lack of qualified labor<br>Restrictive union rules |
| Approvals | Slow approvals of documents<br>Permit issuance<br>Inspections |

## *INFLUENCE OF THE OWNER ON PRODUCTIVITY*

Designers and owners have a large role to play in productivity. As a design increases in complexity, the time it takes to build that design increases. This affects the productivity and the cost to construct. In addition, changes to the design during construction increase waste and the amount of rework required. Productivity can be improved through constructability reviews, value engineering, and prefabrication of elements.

### Design and Procurement Methods

Contracting methods also impact productivity. Harshly worded contracts continue to impede productivity and impact the bottom line of projects. When each party seeks to provide as much legal insulation as possible, it makes it more difficult for team members to engage directly with each other, increasing the likelihood of errors and discrepancies that lead to inefficiency.

The move to a more collaborative approach among all parties involved in the design, procurement, and construction process is producing greater efficiency in project outcomes. Having subcontractors involved early in the design as well as having a strong relationship with the project team is a step to increased productivity.

Once the construction of a project begins, it is much harder to improve productivity that is related to design and procurement. Nontraditional forms of contracts, such as integrated project delivery and design-build, provide the natural benefits of improved project coordination, site logistics, communications, teamwork, and ethics that improve productivity. With the increased trust and improved relationships there are fewer changes and miscommunications of those changes when they occur.

### Preconstruction Planning

During planning and execution of construction projects, project managers and estimators make assumptions with respect to construction means and methods, the availability of resources, and the schedule. Not all these assumptions are documented and discussed with the project team before the construction begins. Budget and schedule decisions based on poor assumptions can cause impacts such as rework, delays, and extra material cost. Ineffective project planning and design errors that are not corrected prior to beginning construction greatly impact worker productivity.

In addition, variability in the "**flow of work**" impacts how work is accomplished. The type and method of construction used to complete components of a project determine the schedule and efficiency of the work. Prefabrication and off-site production of building components is a construction methodology that is improving productivity.

**flow of work**
*The handing off of work activities from one crew to the next.*

## Changes of Project Scope

Construction projects may incur a major scope change during the building period. Almost any change in project features will have a negative impact on the productivity and result in inefficiencies. Changes in the work to be performed often lead to disputes between the owner and contractors, especially if there are errors and omissions in designs. This type of disruption in the original plan of work increases the project cost due to rework that is performed out of sequence, resulting in decreased labor efficiency for the general contractor and subcontractors. The decreased productivity and efficiency are caused by the manpower involved to implement the change, not necessarily the change itself.

**Management Issues:** The way management supervises employees, plans the work, and manages material has significant influence upon labor productivity.

**Supervision:** Supervision impacts how individual employees engage with work and are motivated to do their best. Good supervision has a direct influence on productivity and is the result of good leadership practices. A challenge occurs when supervision is diverted from scheduled work to analyze and plan contract changes, expedite delayed material, manage added crews, or other changes not in the original scope of work and schedule.

## Communication

Clear communication is an important productivity tool on every project. Workers need to understand the work that is being accomplished and the importance of productivity. Soliciting suggestions from workers to improve processes, work methods, and standard practices and incentivizing them to do so is an effective management instrument. Listening to the feedback provided and taking it into consideration in the way work is planned is important.

Communicating through various methods is standard on the connected jobsite of today. The use of a project documentation or other collaboration software to facilitate open communication and collaboration is an effective tool for sharing changes and other important information occurring on the jobsite. Ensure all parties who need to know about jobsite occurrences, especially changes in design are informed. The absence of proper communication and poor document control lead to rework. Such rework costs the US construction market billions of dollars each year.

Site supervision must also set realistic goals for productivity, given the capabilities of the workforce. It is important that workers have a goal to strive toward every day. Before beginning any project, a clear and detailed plan of how the work will be accomplished must be determined. This plan should include input and feedback from workers. Include detailed schedules for not only the activities to be performed but also workers, materials needed, and all other logistics.

## Materials Management

Material management is the system for planning and controlling all efforts necessary to ensure that the correct quality and quantity of materials specified for a project are received in a timely manner, obtained at a reasonable cost, and available on site when and where they are needed.

> The availability of materials and their ease of handling on site plays a large role in productivity.

The management of material is a worldwide problem and has been cited as a major cause of productivity loss. Repeated handling of materials, improper organization and markings, obstructed access to materials, insufficient supply of materials, and inefficient distribution methods are examples of material mismanagement. These actions cause delays or disruptions of the normal material workflow to a work area or lay-down yard. Material mismanagement may be the result of the substitution of materials due to contract changes, defects, or delays at the work site.

Project planning and control requires management of materials. Materials represent a large expenditure on construction projects and minimizing their cost as well as ensuring they are present on site when needed impact efficiency. The negative impact of shipping delays and poor procurement procedures are important to comprehend. Materials purchased early may tie up capital and interest charges may incur on the inventory. These materials may be damaged or stolen prior to installation causing delays and additional costs that impact project efficiency.

The availability of materials may greatly influence the schedule and just-in-time delivery where materials arrive on site when they are needed can improve productivity and reduce waste.

## FACTORS RELATED TO MANAGEMENT AND LEADERSHIP

Construction is a team process involving specialists collaborating from different organizations. To achieve high productivity, the well-being of workers must be considered by leadership. Leaders must control the flow of work activities and resources from the design phase to construction. Effective management and leadership skills coupled with effective resource planning have a large potential for affecting productivity. The leader of the team can affect the productivity of the design and construction. Who is responsible for leading the work is also dependent upon the contractual arrangement adopted for the project.

### Job Satisfaction and Motivational Factors

Worker motivation is defined as the driving force that stimulates individuals physiologically and psychologically to pursue goals to fulfill their needs or expectations. Workers are motivated both intrinsically by their own goals and extrinsically by the goals of the company or project of which they are a part. Key motivators to workers in the construction industry are a combination of one or more of the following: salary, job security, high achievement, recognition, the nature of the work itself, responsibility, and personal advancement and growth. Naturally, the factors that motivate each individual worker will be highly different. The key question is whether there is a relationship between financial incentives, motivation, and productivity. Research has shown that factors such as money, autonomy, supervision, safety, working conditions, policies, and interpersonal relations prevent dissatisfaction but do not motivate a worker. These factors prevent a decline in performance, but they do not increase productivity. Moral and attitude of workers can be lowered due to a variety of issues including poor site conditions, multiple contract changes, disputes, overtime, and disruption of work flow. Motivators such as the work itself, recognition, advancement, possibility of responsibility, and achievement can have a positive effect on job satisfaction which will lead to improved productivity.

### Organizational Factors

**Investment in technology and innovation**: The application of new technology and innovative practices has become critical for construction companies. Owners are increasing pressure to improve productivity and quality as well as reduce costs and speed up schedules. Innovations in technology, including material and information technology, have improved construction labor productivity in the past decade. Tools, machinery, software, and the automation and integration of information systems have increased management's capability to manage and improved the workplace. Investing in technology and innovation is essential to productivity, cost effectiveness, and the sustainable development of any construction firm. Information technology and the evolution of virtual design and construction practices can assist project teams in obtaining accurate information that leads to faster and more accurate decisions on site.

Construction productivity is influenced by the equipment and tools available. The type and quantity of the equipment as well as the operating condition and age of the equipment impact performance. Tools and equipment must be available in sufficient quantity and quality to meet the needs of the project. Labor productivity has been shown to improve as the type of equipment and technology associated with it have improved.

## Labor Issues

Labor productivity is a measure of the overall effectiveness of an operating system that is using the labor. It is a measure of how equipment and capital are used to convert the efforts of labor into the production of an output. It is not a measure of the capabilities of the labor force alone. Consider as an example, replacing a hand tool with a power tool. This will increase the output of work performed for the same hours, resulting in higher labor productivity. The productivity of labor is also influenced by direct impacts to the labor force, the work conditions, and the time spent on nonproductive activities. Each jobsite is "unique" in type and location. Labor customs are also unique and inefficiencies may arise due to the standard practices of a location. These may include added or extended breaks, unique customs, or other customary practices in the area.

As a schedule compression technique, **overtime** is often preferred because it can produce a higher rate of progress without the coordination problems of shift work or the additional workers needed for overstaffing. However, overtime introduces additional problems, including fatigue, low morale and a higher cost per unit, a higher accident rate, and the impact on pacing. Workers tend to pace themselves for longer work day or week and reduce overall productivity. Several studies have documented the capacity for overtime to reduce labor productivity.

**overtime**
*Federal overtime provisions are contained in the Fair Labor Standards Act (FLSA). Unless exempt, employees covered by the Act must receive overtime pay for hours worked over 40 in a workweek at a rate not less than time and one-half their regular rates of pay.*

Common reasons for overtime include the following:

- Aggressive schedule, scope change, makes up labor-related delays and missed milestones.
- Alternative to overtime are either overstaffing or shift work.
- Overtime is the most commonly used method of schedule acceleration and compression.

Common causes of schedule acceleration:

- Late start, such as late access to the site and late issue of notice to proceed.
- Delay of any sort, to include weather, late delivery of owner purchased items, delay by subcontractors.
- Late responses to RFIs, and late approval of submittals and shop drawings.
- Incomplete prerequisite work by other contractors resulting in delay.
- Change orders when the scope of work is added without extending project duration.

The "stacking of trades" occurs when multiple tradespeople are working within physically limited space on a project. This results in congestion, the inability to locate or even use tools, crew interference, added safety hazards, and the prevention of optimum crew sizes. This may be compounded when managers hire too many workers for the estimated scope and duration of the work. This strategy may be to compensate for absenteeism and turn over; however, it results in overstaffing of the project and a decrease in work productivity.

Occupancy and operation issues on projects can cause a decrease in productivity if they are not properly anticipated and planned. Joint occupancy of a space where one or more trades must share a work area that was not anticipated in the original bid requires coordination and longer periods of time to complete the work. Working in a space where the owner has taken beneficial occupancy also reduces productivity. This work is being completed over, around, or in close proximity of the owner's personnel and there are restrictions on noise and dust and even access that impact the work.

Working in areas that are still operating under normal conditions, like a hospital, or office building, will lower efficiency due to the added care and time it takes to follow safe working protocols. Inefficiencies may result due to work stoppages, the need for protective equipment, or special work permits.

Work in "confined spaces and hazardous conditions" can result in lost time due to restrictions on access, exposure limitations for workers, or even the time lost when getting to and exiting the work area. Additional safety precautions and training are also required in hazardous conditions.

The **planning** of the work plays a large role in the effectiveness of the labor. If work plans are altered due to issues caused by other trades, having a plan to keep a crew actively working is important to productivity. Planning must also consider the location of the work.

The **physical location** of the actual jobsite may cause inefficiencies if not properly assessed and planned, especially if the site is in a remote location. Issues associated with remote work include the time it takes to access tools, break areas, material laydown yards, and other resources. This nonproductive time has to be accounted for in the planning process.

Each project will require a period of "orientation" for workers to become familiar with the site conditions. They must learn the scope, layout, safety procedures, and work practices of the site. Crew turnover causes the same challenges. Time is lost to these activities and productivity is lowered until the crew is comfortable.

Inefficiencies happen at various times during projects when the flow of work is interrupted. This can be planned activities like mobilization/demobilization and holidays or surprise activities like a suspension of work. During these times productivity may drop as time is lost when crews move from one area or assignment to another. Suspensions of work for any reason will cause a decline in productivity due to the break in the schedule. Lowered productivity leading up to and following major holidays is due to workers discussing the time off and loss of previous momentum before they get back into their routine. Some projects may see a decline in productivity due to unique schedules or shift work. Rotating or staggered crews may be necessary in certain remote locations or under tight schedules. However, these practices impact the efficiency of the work.

There may be times when work is completed on site in shifts. This is when work is performed at any time other than the first shift or the morning shift of a work day. Work on second and third shifts are less efficient and may even be based on a shorter work period. The reduced daylight hours and problems with hand-offs and accurate communication between each shift results in less productivity.

The size of the crew is important. There is an optimal size for each work situation and each crew. When the optimal crew size is altered by adding or deleting crew members, the flow of work is disrupted. When workers are added or deleted from a crew, it breaks up the original team effort and rhythm of the crew and results in loss of productivity.

The consistency of a crew is also affected by absenteeism and worker turnover. There is a great deal of time and money lost associated with high turnover and absenteeism on projects. Construction projects in certain areas with low manpower and high demand for labor will usually be impacted more than others. Extreme weather conditions like cold or heat will also increase absenteeism and turnover. New crew members are not familiar with the work or the jobsite. These individuals require the time of other experience crew members to acclimate. This stoppage of productive work decreases productivity for each crew member.

There are many activities that occur on site that are nonproductive. These activities may or may not be paid by the owner; however, they do consume labor resources.

The time it takes to access a jobsite is nonproductive time. Standard delays like security checkpoints or surprise delays like interference with access to worksites all result in loss of productive time. Security checks are common on government sites and other types of projects like hospitals, education facilities, or other municipal sites. Interferences to the planned access to work areas can be a function of blocked stairways, roads, walkways, insufficient material-lifts, or congested work sites.

Construction activities are affected by the weather and the change of seasons. Working in extreme heat and cold or areas of intense changes in temperature affects the ability of a worker to function at the same level of productivity under normal conditions. Seasonal changes not only impact the temperature but also the available daylight hours in which work can be performed. Rain impacts the ability of many trades to complete work due to safety challenges. Others that live in wet regions of the country must work in rainy conditions at times or sacrifice the project schedule. While work can occur in the rain, inefficiencies associated with rain gear, visibility, safety, morale, discomfort, hazards, and other issues are present.

## MEASURED MILE ANALYSIS

The measured mile approach is a way for a contractor to measure productivity on a construction project. It is most often used as a method for measuring the lost productivity. The method involves comparing the cost of "impacted" work with the cost incurred to perform the same or similar "unimpacted" work. It creates a control period that is based upon the normal rate of production for the project. It then compares the rate against a production rate when the contractor is being delayed, disrupted, suspended, or otherwise negatively impacted. The baseline values of productivity come directly from the construction project when it is proceeding as planned. The method compares the typical production on an unimpacted day to the production on a delayed day to determine the level of impact on the contractor.

Using the measured-mile method to determine lost productivity is straightforward if the contractor has kept productivity records by location, type of work, and crews. The key is the documentation. Recommended steps to conduct a measured mile analysis include:

1. Identify and define the impacted work and define the unit of measurement for the work.
   - Inability to access an area of site due to permits not being secured by the owner. The amount of days delayed is the unit.
2. Identify the impacted and unimpacted time periods and the location for the analysis.
   - Selection of the unimpacted period and location for the project is important for the analysis. Construction projects take place in phases, at different times and even in different locations of the site. In the above example, the contractor may be able to achieve a higher production after accessing the prohibited location.
3. Evaluate the difference between the two periods and select a representative unimpacted period.
   - Ensure the unimpacted selection of work is representative of the project as the measured-mile method assumes all work on the project is performed at the same rate as the unimpacted segment.
4. Locate and assemble job-cost records, identifying labor-hours, equipment, and material used.
   - Record keeping is critical to calculate and support any lost productivity claim. Review records for all unimpacted work periods. Field personnel need to maintain the records in generally the same manner for the impacted and unimpacted sections.
5. Determine the units on which the analysis will be based—Time or Money.
   - A time approach is based on the total crew hours required to complete a work task.
   - A money approach is based on the total cost to complete a task, including labor costs, equipment rental, operating costs, and consumables that vary with time.
6. Develop an unimpacted benchmark productivity measurement.

## EARNED VALUE ANALYSIS

Earned value analysis is a method used to measure project progress. The method compares the actual work completed at any time during the project to the original budget and schedule. The analysis allows for the forecasting of the final budget and schedule and provides a measurement of productivity.

The performance of an earned value analysis involves the calculation of several variables that are based upon project information related to the schedule and budget. Access to information related to the status of the project from a financial and time perspective is critical. Accurate documentation of work status and cost are the foundation for the analysis.

### Determine Percent Complete

Determine the percent complete of each project activity. The level of accuracy to which these values are determined will improve the results. Use information from weekly progress meeting to determine the work put in place.

### Determine Planned Value (PV)

Planned Value is defined as the approved amount of the activity to be completed in a given amount of time. It is the value that should have been earned based on the schedule. Planned Value is also referred to as Budgeted Cost of Work Scheduled (BCWS).

### Determine Earned Value (EV)

Earned Value is defined as the amount of the task activity is actually complete. It is the work of the work accomplished to date. Earned Value is also known as Budgeted Cost of Work Performed (BCWP).

### Identify the Actual Cost (AC)

The Actual Cost is the actual cost of the work performed. The cost needs to include all labor, materials, equipment, fixed costs, and subcontractor costs. The Actual Cost is also known as Actual Cost of Work Performed (ACWP).

### Calculate Schedule Variance (SV)

The Schedule Variance represents the schedule status of the project. It indicates how far ahead or behind the project is compared to the baseline schedule. The schedule variance is the difference between the EV and the PV and is calculated by

$$SV = EV - PV \qquad (17.2)$$

A negative schedule variance indicates the activity is behind schedule. A positive schedule variance indicates the activity is ahead of schedule.

### Calculate Cost Variance (CV)

The Cost Variance represents the cost status of the project. It indicates how far ahead or behind the project is compared to the baseline budget. The cost variance is the difference between the EV and the AC and is calculated by

$$CV = EV - AC \tag{17.3}$$

A negative cost variance means the task is over budget. A positive cost variance indicates the activity is under budget.

### Determine Other Status Indicators

There are other variables that may be calculated to provide more information regarding the status of the project. These variables are calculated using the values of EV, PV, and AC.

- **Schedule Performance Index (SPI).** This is the ratio of the EV to the PV [Eq. (17.4)], and expresses the schedule variance in percentage terms, for example, SPI = 0.9 means the project is 10% behind schedule.

$$SPI = \frac{EV}{PV} \tag{17.4}$$

- **Cost Performance Index (CPI).** This is the ratio of the EV to the AC [Eq. (17.5)], and expresses the cost variance in percentage terms, for example, CPI = 0.8 means the project is 20% over budget.

$$CPI = \frac{EV}{AC} \tag{17.5}$$

### Estimate at Completion (EAC)

The expected budget at the end of the project is based upon the variances that have occurred up to the point of analysis. This calculation [Eq. (17.6)] assumes that the project will continue along the same path and the original budget value (BAC) is the same.

$$EAC = AC + BAC - EV \tag{17.6}$$

**Estimate to Completion (ETC):** The expected cost to finish the rest of the project if nothing changes:

$$ETC = EAC - AC \tag{17.7}$$

**To Complete Performance Index (TCPI):** Represents the efficiency level of the project that will make the project finish on time, which is the required CPI necessary to finish the project right on budget [Eq. (17.8)]. For example, TCPI = 1.10 means you need to find 10% efficiencies to finish on budget.

$$TCPI = \frac{(BAC - EV)}{(BAC - AC)} \tag{17.8}$$

The results of the analysis are a snapshot based upon the date of the data gathered. The calculations enable the efficiency of the activity to be determined. The calculation provides an indicator of the activity progress. Interpreting the results is simple. If the schedule variance value is negative, the activity is behind schedule. If the cost variance is negative, the activity is over budget.

## WORK SAMPLING

Work sampling is a series of observations of work in progress taken randomly over a period of time. The observations are used to infer the productivity of the workers engaged in the activity. Work sampling estimates the percentage of time a worker is productive relative to the time the worker is involved in the activity. This method provides information on the amount of time workers spend performing productive, supportive, and nonproductive work. The method also allows for the monitoring of labor productivity on a variety of different trades at the same time.

The details of how a work sampling study is conducted may vary but the fundamental activities are the same. To begin a work sampling study, the observer starts at a random time and follows a planned route through the jobsite. As each trade or crew is encountered, the observer records the activity in which the worker was engaged in the moment they came into contact. The observer will categorize the worker into either being engaged in productive, supportive, or nonproductive work.

**Productive work** is defined as direct, hands-on activity that advances the project, such as constructing formwork.

**Supportive work** is defined as any activity used to support the productive, such as cutting lumber for the formwork.

**Nonproductive work** is defined as wasted time spent on the jobsite that can be recovered as productive time such as waiting for lumber, starting late, or quitting early on a work activity.

Observations continue to be recorded until the number of work samples required has been collected. Work samples should be taken throughout the entire day as productivity will vary based upon the time of day. In addition, most workers require time to set up at the beginning of the day and put tools and materials away at the end of the day. These activities must be accounted for or the calculated productivity will be inflated. As the number of work samples increases so does the reliability of the data collected. Once all the samples are collected, the data is compiled and analyzed and recommendations for productivity improvement are made.

## TIME AND MOTION STUDIES

Time and motion studies are a measurement technique for recording the time it takes to perform a certain task. Time studies are a direct and continuous observation of a task. During the observation, time is tracked using a particular method, such as a stopwatch, cell phone timer, or video camera, to record the time to accomplish a task. Time and motion studies are performed to eliminate waste.

The time spent on performing activities has a direct impact on resources. The more time it takes to accomplish an activity, the more it costs to complete. In this way time and motion influence determine the overall productivity of the work. These studies are completed to create a baseline against which processes, crew sizes, and equipment used can be evaluated.

The completion of a time and motion study is simple and the goal is to determine the best method to complete an activity in a specific time period.

The work process that is being studied is divided into standard steps that follow the typical work sequence. The observer must understand the standardized process being studied and have a uniform method for collecting data. Generally the only equipment required for the study is a stop watch and a form on which the data can be recorded. The time to complete each step of the process is recorded. Several observations are made to determine the average time it takes workers to complete the task. The data collected is analyzed to determine areas of inefficiencies.

In general these methods involving direct observation of workers indicate that the time it takes to complete activities includes time not dedicated to actual work.

The inefficiencies in the processes are most often attributed to poor planning, inadequate material handling, poor work practices, and poor utilization of space.

Other productivity study methods are primarily used to prove lost time in the case of construction claims that end up in court. These are the Total Cost Method and the Modified Total Cost Method. The Total Cost or the Modified Total Cost methods are only used when other more precise methods of proof for loss of productivity are not available. It is also appropriate when only owner-caused impacts affected contractor performance and when there is no evidence that the contractor contributed to the cost increases. These methods are based completely on the argument that the entire cost overrun is attributed to the alleged delay.

The Modified Total Cost Method is a variation of the Total Cost Method, but allows for the contractor to account for their own contributions to delays and inefficiencies. The method subtracts the planned costs of work from the actual cost to complete the work. This method considers the contractors total cost as a starting point, rather than an end, and reduces damages from there.

Industry organizations have completed productivity studies over the years. The two that are most referenced include the National Electrical Contractors Association and the Mechanical Contractors Association of America. Key takeaways from the NECA study include:

- Placing an electrical crew on scheduled or unscheduled overtime reduces labor productivity and increases labor cost.
- A greater number of hours worked beyond the regular 40 hours per week are related to higher productivity losses.
- There is a direct correlation between labor inefficiency and the duration of overtime use.
- Despite productivity losses related to overtime, placing an electrical crew on overtime schedule is more efficient than overmanning.

The MCAA Labor factors provide guidance about the loss of productivity for a set of 16 different factors that impact the productivity of the labor force. A percent of loss in productivity factor that corresponds to each of these factors was developed. The percentage of loss is expressed in a range of values from minor to average to severe. Logistics, trade stacking, crew inefficiencies, and weather were noted to have the greatest impact on productivity.

## *LEAN CONSTRUCTION APPROACHES*

The idea behind lean production is to eliminate waste and increase efficiency. Lean production is commonly associated with manufacturing processes developed by Toyota. Lean construction builds on the lean principles developed for manufacturing settings. Lean construction focuses on adding value to raw materials, such as lumber and structural steel, as it proceeds through various processing steps to become a finished product. Due to the limited supply or high cost of construction materials, equipment, and labor, lean construction focuses on avoiding, eliminating, or reducing waste associated with those resources. The overarching principles of lean are to:

- Eliminate waste
- Pull materials through the production system to meet specific customer demands
- Synchronize and physically align all steps in the production process reducing wait time for people or equipment and virtually no staging of partially completed products
- Clearly document, update, and constantly report the status of all process flows to all involved
- Pursue perfection by continuous improvement

The systematic elimination of waste is a constant process in lean construction.

## Practical Lean Tools

**Offsite Prefabrication:** Construction sites are inherently chaotic and subject to the environment, weather, site constraints, and other challenging factors can negatively impact productivity. However, shifting assembly of building components from the project site to a covered and conditioned space offsite can increase productivity and quality while eliminating waste. It is becoming increasingly common for wall sections, HVAC ductwork, complete bathroom units, and many other building components to be prefabricated offsite and then shipped to the project site for final installation.

**Just-in-time Delivery:** Just-in-time (JIT) delivery is a system for producing or delivering the right amount of building components at the time they are needed for production. For example, JIT delivery of structural steel involves picking columns and beams directly from the delivery truck and installing them. This eliminates the need for a site laydown yard and rehandling the steel for final installation.

**Big Room Meetings:** Proper planning requires input and feedback from personnel of all trades associated with a project. Big room meetings provide an opportunity for all players working on a project to provide input to project planning and status reporting. The room where the meeting is held is large enough to hold the construction team, as well as the designers, and the owner. Conducting regularly structured meetings with all stakeholders present encourages collaborative planning and real-time resolution of project issue. These meetings allow multiple trades to coordinate their work in an efficient manner. They can also resolve requests for information requiring contractor and designer input because the whole team is present.

**Last Planner™ System/Pull Planning:** In traditional planning a project schedule is typically developed by the general contractor in isolation. There is little feedback on the schedule from subcontractors and negligible collaboration between the various subcontractors involved. This practice results in individual subcontractors being assigned a time frame in which to perform their work. The subcontractor has not agreed to the duration or the timeframe; however, they focus on optimizing their own operations to fit that window of time. The resulting schedule is often less than optimal and results in inefficiencies and may create conflict between subcontractors.

The Last Planner™ system (LPS), developed by members of the Lean Construction Institute, is an effort to improve on traditional planning. LPS uses coordinated planning among all trades. The collaborative scheduling requires the input of those supervising the work that is being completed. This is often the superintendents and foreman.

Unlike traditional planning, which typically involves a general contractor imposing a schedule with little feedback from subcontractors, LPS builds a schedule leveraging the feedback from the source best suited for providing accurate schedule input. Foremen and superintendents, the people best suited to organize and control field operations, are the last planners in LPS. Figure 17.1 illustrates one view of the LPS.

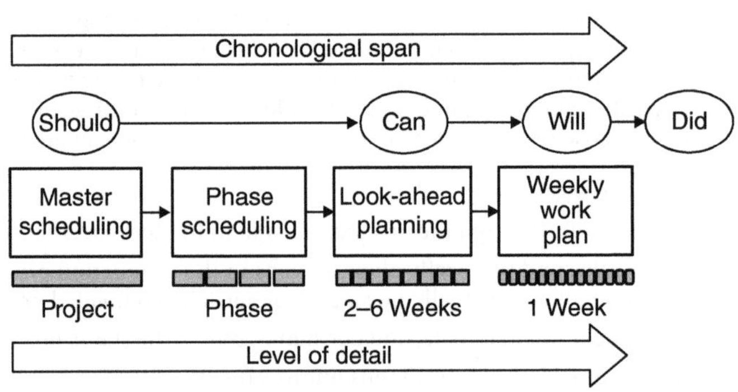


**FIGURE 17.1** Last Planner™ flow.

# CHAPTER 18
# QUALITY MANAGEMENT

*Quality is often defined as delivering a project without defects. But there is no strict definition of quality. All owners have their views about quality and for every project there can be wide variations in expectations. To deliver quality a contractor must understand the project owner's requirements. Developing this understanding is critically important for project success. Quality Control involves processes used by a contractor to track and control work quality as tasks are being performed. Quality Assurance programs are used to systematically monitor and evaluate contractor performance.*

## *IMPORTANCE OF QUALITY*

When the Pennsylvania Railroad (the largest corporation of the 19th century) began to specify mild steel for its rails, because they were longer-lasting, the steel companies told the railroad "if they did not take the rails offered, they would not get any." Companies with such attitudes do not survive in the marketplace today. A reputation for delivering quality projects on time is a key factor in developing project owner satisfaction and growing a business. A culture of always providing quality will improve profitability by eliminating the cost of removal and replacement of faulty work.

## *QUALITY CONTROL*

The first element in the delivery of quality is Quality Control (QC). A contractor's QC efforts seek to ensure compliance with the project specifications for material and workmanship quality. QC includes all testing and procedures used to monitor and document performance. Just as importantly, it provides the data to adjust processes as construction proceeds.

> At its core, QC is about real-time process control.

A contractor's QC plan identifies the work item characteristics to be tested or verified and states the frequency of testing. QC activities provide the data for applying statistical procedures to identify variance. QC activities are, therefore, a continual process. With the information generated by the QC activities, management can continuously improve work processes and deliver quality more efficiently. A poor system of documentation will diminish the value of QC efforts.

The American Society for Quality™ (ASQ) describes a Quality Control Plan as a document specifying quality standards, practices, resources, specifications, and the sequence of activities relevant to a particular product, project, or contract. The ASQ outline for a QC plan advises inclusion of the following information:

- Objectives to be attained (characteristics selected to represent each type of work or the specification)
- Steps in the processes (necessary sampling for assessment and monitoring)
- Allocation of responsibilities
- Specific documented standards, practices, procedures, and instructions to be applied (an owner may specify sample lots, with samples providing statistical representation of the materials and work in a lot)
- Suitable testing, inspection, examination, and audit programs at appropriate stages
- A documented procedure for changes and modifications to a quality plan
- A method for measuring the achievement of the quality objectives
- Other actions necessary to meet the quality objectives

Many owners/agencies have their own QC plan requirements: For example, specification direction may be very succinct, as:

> The CONTRACTOR shall submit a Quality Control Plan outlining how the CONTRACTOR shall assure that the materials and Work are in compliance with the CONTRACT Documents. The initial plan shall be submitted to the agency for review and approval at least 30 days prior to the beginning of any construction activity.

With such an open-ended directive the submitted plan will probably have to be revised unless further direction is requested. While such arguments can be very open ended, the following sections of a plan are appropriate.

## QUALITY CONTROL PLAN

Project Quality Control Plans describe the measurements, inspections, and process checks used to monitor quality. They should address:

- Organization
- Personnel and essential certification—Identification of the Quality Control Manager (QCM) and authority
- Communications
- Assurance
- Subcontractors and Suppliers
- Specifications
- Inspections and Tests—Sampling descriptions: required tests and frequency by task
- Usefulness—Description of data analysis to establish conformance
- Documentation—Tools used to confirm quality trends
- Nonconformances
- Project Completion Inspections

### Organization

Provide an organizational chart at the beginning of the construction quality control plan. This serves to delineate who is responsible for what activities. During project execution the chart functions as personnel responsibility reference. Because many owners require specific qualifications for certain duties, the organizational chart should reference resumes and qualifications of each listed member. In addition, it should identify all certifications held by each member.

## Personnel

There should be a designated quality manager in charge of project quality. The owner must know this person's (1) **responsibilities** and (2) **authority**. Project specifications can be very specific about QC leadership and might state:

> The contractor is required to place a competent representative onsite to oversee the QC system. The representative must have full authority to act for the contractor on QC matters. Responsibilities include workmanship, methods, and techniques to ensure the performance of all work by properly qualified and conscientious craft workers.

The U.S. Army Corps of Engineers (USACE), the Naval Facilities Engineering Command (NAVFAC), and other government agencies require the contractor's Quality Control Manager to have a Certificate of Completion for the *Construction Quality Management* (CQM) *for Contractors* course. Without certification the individual will not be acceptable for performing the quality control responsibilities outlined in the contract.

**Sample QC Plan Personnel Section**

**Quality Control Coordinator**, shall:

**A.** Assist the Superintendents and Project Managers as time and qualifications permit, including complete delegation of their tasks.

**B.** Establish and enforce the preparation and filing of all documentation required by the QC Plan.

**C.** Assist the Superintendent and Project Manager with the initiation of the Quality Control Plan and process.

**D.** Assist with the administration of the overall Quality Control Plan.

**E.** Prepare Monthly QC Status Report to be reviewed in conjunction with the Superintendent and Project Manager reviews.

**F.** Administer Subcontractor weekly QC meetings.

**G.** Coordinate/Conduct Owner, Third Party Inspectors and A/E inspections.

The plan demonstrates the allocation of responsibilities and authority during the different phases of the project. The quality manager has overall responsibility for the quality of the project whereas project superintendents are primarily responsible for the day-to-day field operations. Such a division of responsibility should be clearly stated in the plan. Other QC personnel should be listed and their certifications to perform necessary tests detailed.

## Communications

Quality information/activities must be discussed and communicated. This part of the plan should include information describing: (1) the creation and collection of data, (2) the types of reports and distribution of quality control information, (3) the record keeping of test results and/or inspection forms, and (4) the conduct of meeting to discuss quality issues.

## Assurance

With Design-Build (DB) contracting it is common for quality assurance staff requirements to be included in the contractor's contract. Retaining an independent quality assurance agent through a direct contract with the owner removes the potential conflicts of interest. However, this approach requires additional contract management and administrative effort (incurring additional costs) on the part of the owner.

## Subcontractors and Suppliers

The prime contractor must identify each major subcontractor and supplier. The performance requirements for each subcontractor's work should be clearly identified including independent quality inspection requirements. For fabricated products, the QC plan is developed and executed by the fabricator, instead of the contractor. The responsibility for inspection and testing of off-site materials must follow all applicable owner specifications. A certification system is commonly used for manufactured products, but the basis for the certification acceptance is the manufacturer's QC plan. Structural steel element and precast concrete product QC is typically performed at the production facility by representatives of the owner. The subcontractor and supplier representatives responsible for quality management will interface with the prime contractor and should be identified in the QC plan.

## Specifications

Often the owner's project specifications fail to identify the building codes or industry standards applicable to specific work items. All pertinent building codes or industry standards should be listed with the applicable version identified.

## Inspections and Tests

All mandatory inspections and/or tests together with the applicable standards or procedures, responsible personnel, sampling points, and test frequencies must be itemized and the required documentation identified.

While QC plans are typically required by the project owner, their purpose is to establish a structure for the contractor to monitor and control of delivery of work items. Therefore, it is necessary for QC results to be available real time so process changes can be made immediately and the delivered item is always with the specification limits.

## Usefulness

QC is a contractor's tool for controlling delivered quality. As such, the sampling plans should be designed based on the number of samples needed to provide a valid representation of the population which comprises the work.

## Documentation

Quality control charts give QC personnel and superintendents a way to quickly assess the suitability of the produced work item. Statistical control charts are effective visual quality monitoring tools. They provide rapid and real-time process control feedback. These provide graphical illustrations of changes in quality characteristics over time. Statistical quality control monitoring using range and average values are straightforward tools for tracking delivered quality. The average values illustrate the central tendency of the population and the range shows the variability. Something is wrong with a process and needs to be corrected when there is movement toward an established quality limit. Quality limits can be established at specification limits or can be more restrictive in order to provide advance notice regarding the need for corrections.

When an Alternate Project Delivery (APD) contracting mechanism is used, quality management documentation is critically important. Some will claim performing the work and complying with performance objectives is more important than documentation. However, documentation is needed for two purposes, first to control the work processes by use of the data in the documentation, and second, when disputes arise, records are needed to provide the information necessary to identify what went wrong or to verify how the required quality was actually provided.

With the latest construction management software packages and the use of the cloud it is possible to provide digital documents to all project personnel. Everyone should have immediate access to the up-to-date plans, drawings, and specifications.

### Control of Nonconformances

The Quality Control Plan must state the procedures to use for the documentation of changes and modifications to performance criteria. Nonconformance Report (NCR) systems should be detailed. The NCR system is a protocol for recording specification deviations and identifying installed work items not in conformance with project quality requirements.

> There must be a prescribed system for tracking deficiencies.

As a minimum, NCRs should state:

- The reason for the NCR: the deviation is in respect to what part of the specification, code, or other requirement.
- How did this deviation happen?
- What corrective action was taken or will be taken to correct the deviation this time?
- In what manner should the process be changed to prevent the issue from reoccurring?

### Project Completion Inspections

To ensure the construction work meets project specifications, final inspections are conducted by the Owner's Team and regulating authorities. These must be completed satisfactorily before the project is turned over to the Owner and an Occupancy Certificate issued by local authorities. Such inspections may include systems testing. The QC plan should sketch the proposed timing of the inspections and present a corrective action plan for any items found not in conformance with the project specifications.

## *QUALITY ASSURANCE*

Quality Assurance (QA) is a management step beyond Quality Control. It is concerned with fulfillment of quality requirements. However, the terms QA and QC are often used interchangeability. QA consists of the processes used to document and verify the quality of the constructed work. The project specifications, in most cases, explicitly describe required quality standards for items of work. By reference, these specifications become part of the contract between the project owner and the contractor.

To ensure desired or prescribed standards of quality are met *Quality Assurance* (QA) programs are established for systematically monitoring and evaluating contractor performance. QA programs are used by owners to verify the quality of work carried out by contractors. The American Society for Quality™ (ASQ) defines Quality Assurance as the "part of *quality management* focused on providing confidence that *quality requirements* will be fulfilled."

> Quality assurance focuses on the processes and procedures used to deliver quality and verifies compliance with contract documents.

Contractor QC activities provide sampling and testing data for control of work processes. In the case of a Design-Bid-Build (DBB) contract, the contractor's role in the QA is normally limited to developing and carrying out its own QC plan. To verify the contractor's QC data and how accurately it represents the quality of the work, QA will sample and test the same work at lesser sampling frequencies.

## International Organization for Standardization

To help companies and owners efficiently maintain an effective quality system, the International Organization for Standardization (ISO) sets standards on quality management and quality assurance. ISO 9001:2015 "Quality Management Systems" specifically applies to quality management and quality assurance. Because the standard is not specific to any one industry, it can be applied to quality management process for construction projects. It requires the establishment of a documented quality management system. The documentation provides evidence of work accomplishment according to plan. The standards offer recognized procedures, record keeping, technologies, and resources for managing quality work. They promote the adoption of a process approach. The first step is to identify all of the organization's processes and their interrelationships. A process is a set of activities used to transform inputs into outputs. This is exactly what is done on a construction project. The planning element of ISO 9001 encourages more specific strategic planning which is always necessary for complex building projects.

## Core QA Elements

If an owner uses the Design–Bid–Build delivery method there are six core elements to a QA program:

1. **Contractor quality control:** processes/tests used by a contractor to ensure the delivered project complies with the quality requirements of the contract. This would include QC of construction tasks, materials, and the work by subcontractors.
2. **Owner acceptance:** evaluates how well a contractor met specified requirements.
3. **Owner independent assurance:** unbiased and independent evaluation of all the sampling and testing procedures used in the acceptance program.
4. **Dispute resolution:** a procedure to resolve discrepancies between the owner's acceptance testing and contractor's QC testing results when differences could impact payment.
5. **Qualified laboratories:** capable as defined by appropriate accreditation groups or by specific owner established programs.
6. **Qualified personnel:** capable as defined by appropriate established programs or certifications recognized by the owner.

While the elements do not change when the project is delivered using DB method, the sharing of quality delivery risk between the owner and the contractor changes dramatically.

Recognizing the impact of different contracting methods, the Transportation Research Board (TRB) of The National Academies of Sciences, Engineering, and Medicine has defined Quality Assurance as: "All those planned and systematic actions necessary to provide confidence that a product or facility will perform satisfactorily in service … . QA includes the elements of quality control (QC), independent assurance, acceptance, dispute resolution, laboratory accreditation, and personnel certification. The use of the term QA/QC or QC/QA is discouraged and the term QA should be used."

## Quality Assurance and APD Methods

With alternative delivery contracting methods being utilized more frequently and continuing to evolve, establishing an effective structure to support a strong quality management program will be the key to the success of APD projects. When an owner uses an APD method, a portion of the quality assurance elements is often transferred to the contractor. This transfer of responsibility can cause concerns about how quality activities will be executed. With an APD method the project owner must establish the structure of the quality assurance organization (QAO) early in project development because the typical QA organizational structures used for Design–Bid–Build projects are not suitable. With many APD methods, the contractor performs in a manner not covered by the owner's usual standard specifications. The shifting of risk and responsibilities causes owner organizations to operate outside of their customary "comfort zone." The QA organization for an APD project must be designed considering the distinctive characteristics of the project. Often the most important characteristic is delivery speed. Nevertheless, government agencies must operate within laws and rules of their funding source. Many APD methods transfer project design to the contractor so the ability to manage design QA must be clearly defined in the QAO plan.

Based on the use of APD methods for transportation projects, five possible quality assurance organizational structures have been identified (*Guidebook on Alternative Quality Management Systems for Highway Construction*, NCHRP Report 808, Transportation Research Board).

1. **Deterministic** (the traditional approach used with DBB projects). The owner retains responsibility for all project quality roles and activities. The construction contractor responds to the owner's specifications and has no responsibility for design quality.
2. **Assurance.** The owner is responsible for all aspects of quality except for design and construction QC. Whether used with separated contracts for design and construction (a DBB situation) or with a single contract (the case of a DB situation) the owner must first accept the design and then release it for construction.
3. **Variable.** The owner uses different approaches for acceptance of project design and construction. With a DB contract, the owner might yield design acceptance to the contractor.
4. **Oversight.** Design and construction QC and acceptance are assigned by an owner to outside parties. The owner defines the required quality but does not dictate how to produce the work.
5. **Acceptance.** The owner retains responsibility for verification testing and final acceptance. The acceptance approach is used primarily with public-private partnership (PPP) projects.

With the construction manager (at risk) form of APD the contractor provides ideas and preferred construction approaches to the designer, but is not responsible for the project design. In such a case, design QA activities remain the responsibility of the owner. If the DB method is used, the owner transfers delivery of the project including design to the DB team. The contractor team might be responsible for design QA or it could participate with the owner in a shared QA plan. A third option would be for the owner to use an independent third party as its representative for QA of design and/or construction. As the contractor's responsibility for the project delivery increases, greater responsibility for QA follows. With PPPs the concessionaire assumes a significant portion of the performance risk of the facility during the operation term. Consequently, there is a strong incentive for concessionaire to be very involved in the quality assurance program.

### Building Code Requirements

Construction of a building is usually controlled by local and international building codes intended to safeguard public health and safety. These codes govern both building design and construction. Consequently, quality assurance plans for building elements must adhere to the applicable building code for the elements.

Building work will typically require a permit and is subject to inspection by building officials. Many items of work need to remain accessible for inspection until approved, potentially impacting a contractor's schedule. A building inspector views the work as it progresses and makes a final inspection upon completion. Inspections by building inspectors provide another element of oversight and assurance of construction quality.

## *STRUCTURING FOR SUCCESS*

Project quality is the result of a variety of initiatives:

1. Qualified employees: The workforce must be qualified, so hiring practices are important. The selection of employees should be based on appropriate minimum qualifications.

> The leading cause of quality failures is defective workmanship.

2. Safety: There is a proven correlation between construction safety and construction quality. Rework necessitates removal of work (demolition), creates schedule pressure, impacts the allocation of resources, and leads to unstable work processes.
3. Training: Employee training is a significant element contributing to delivered quality. To deliver quality a contractor must invest in workforce training. Management should review workforce abilities and invest in skill improvement.
4. Procurement: Identify and use suppliers who deliver quality products.
5. Personnel policies: Reduce turnover and retain skilled employees. Base incentive pay eligibility for benefits on delivered quality.

Every construction project is unique because of location, delivery schedule, weather uncertainty, logistical complexities, facility features, and a new set of team members. To deliver quality and sustain success, companies must continually improve task delivery. The ability to deliver quality requires more than establishing policies and procedures.

> The ability to deliver quality requires establishment of a "quality culture."

Rework costs and reduced contract pay from the owner for nonconformance with quality requirements adds on average between 2.3% and 9.4% to total project cost.

## *TOTAL QUALITY MANAGEMENT*

W. Edwards Deming, the influential quality management thinker, preached the use of statistical methods to continually improve product quality. Using Deming's methods, which are termed Total Quality Management or TQM, it is possible to both deliver quality and reduce costs because nonconformance with quality standards causes waste, rework, and possibly litigation. A fundamental premise of Deming's ideas is the need for continuous incremental improvement. Building on Deming's work, eight quality management fundamentals have been advanced.

1. Customer focus (the ultimate judge)
2. Employee empowerment (involvement)
3. Process centered (from input to output)
4. Integrated system (think horizontal, eliminate silos)
5. Strategic and systematic approach (organizational vision)
6. Fact-based decision making (collect and analyze data)
7. Communications (organizational)
8. Continuous improvement (process improvement)

## Customer Focus

In the context of TQM the term "customer" includes both the external customer (the project owner) and the internal customers (the subcontractors and suppliers) who come together to deliver the project. In the case of many projects, customers can also include the local community, political groups, and government regulators.

## Employee Empowerment

Employee motivation and decision-making involvement is vital to delivering project quality. Each task must be owned by a responsible employee.

> The key to achieving quality is empowerment of the workforce.

Management must facilitate open discussion of ideas and give consideration to employee recommendations. Employees are members of the production team. They are knowledgeable problem solvers. Their input should be welcomed, encouraged, and not met with criticism or adverse treatment by management.

## Process Centered

Management must seek to remove organizational hindrances from production process. The emphasis is on "production processes" rather than individual tasks. The approach looks at internal inputs and the contribution of external suppliers. Consideration must be given to such simple things as tool use, tool storage, and availability. Think in terms of people and how the complete and integrated production process is designed.

## A Unified System

A long-term effort is required to change the culture and consistently deliver quality. Participation by all members of the organization is required. To succeed in delivering quality all managers and employees must be involved. Companies have multiple internal groups supporting the actual work of building a project. These groups must all participate in improving the processes used to deliver the project. The goal is the development of seamless horizontal processes.

## Strategic and Systematic Approach—Vision

To consistently deliver quality a company must have a strategic plan—with quality as the first priority. Instead of casting blame on workers, quality management seeks the cause of quality problems in the construction processes. It is necessary to view processes not from a perspective of what they should be doing, but to ask, "What are they actually doing and can the process be changed?"

## Fact-Based Decision Making

Decisions must be made based on facts. Consequently, it is necessary to expend effort gathering data and evidence before decisions are attempted. The decision-making steps are:

- Collect the data.
- Check the data.
- Analyze the data.
- Use effective statistical techniques.
- Implement the decision.

## Communications

Communication must be both vertically, up and down the leadership chain, and horizontally between all internal departments or groups. Communication from customers is needed and from employees with ideas. Horizontal communications link the different functional elements of the company.

## Continuous Improvement

Continuous improvement cannot be just a "buzz" word. It must be the strategic plan. Deming preached the: *Need for consistency of effort to improve quality and productivity, and thus constantly decrease costs*. The focus is on seeking improvement, continuous improvement. Every process can be improved. To deliver quality projects, it is necessary to purposely seek and benefit from opportunities for improvement.

Application of these eight TQM fundamentals does not eliminate the need for quality control activities and inspections. The data from inspections is used to identify and remove the cause of defects. Process control now becomes problem solving for continuous improvement, using prior project experiences to make improvements.

Building construction has multiple repetitive activities (processes): column formwork and placement of concrete are examples. These processes are affected by a number of factors, including the season of the year, daily weather/climate conditions, access to the site, lifting hook availability for moving the forms, and other physical conditions. To deliver quality it is necessary to apply the eight fundamentals to each part of the process. Management must take time to observe and ask questions:

- From where is the form material purchased?
- Is it delivered on time and in good condition?
- Would it be better to fabricate the forms off site?
- Are the necessary tools and equipment available to the crew?
- Does the workforce have the needed skills?

**FIGURE 18.1** Concrete formwork.

- Has appropriate and necessary lifting equipment been provided (Fig. 18.1)?
  - ➢ Are the forms always carefully removed and cleaned for reuse?
  - ➢ Are all parts of the forming system properly stored for future use?

One construction contractor who undertakes large commercial projects applied the fundamentals to its processes and, by asking the questions, found it was better to utilize prefabrication to reduce costs, shorten project schedules, and achieve a reduction in waste and scrap.

There is a strong relationship between delivered quality and safety. The Experience Modification Rate (EMR) for the building contractor who applied these fundamentals to its construction processes was well below 1.

The Construction Industry Institute (CII) has studied the costs of quality deviations and believes a company seeking to deliver quality needs to focus on how it operates. Control and improvement of process is more important than inspection of the produced end product. Inspection is necessary to provide the data, but improvement of processes is the goal. The cost for delivering a concrete mixture of sufficient quality to meet the project owner's specification is driven by the cost of the most expensive component of the mixture, the cement. However, good quality aggregates and appropriate mixture proportioning will reduce the required quantity of the expensive mixture component. Purchasing department decisions also affect quality. It is not always best to purchase aggregate from the supplier with the lowest cost. It is necessary to look at the total cost of the final product. This goes to the communication fundamental: to deliver quality there must be horizontal communication within the company. The question is about aggregate quality in terms of reducing the cost of producing quality concrete.

## DELIVERING QUALITY

The eight TQM fundamentals and seven principles of the International Organization for Standardization (ISO) provide a structure for consistently meeting customer expectations. The seven ISO principles are:

1. Customer focus (TQM Fundamental 1)
2. Leadership: vision and direction
3. Engagement of people (TQM Fundamental 2)
4. Process approach (TQM Fundamental 3)
5. Improvement (TQM Fundamental 8)
6. Evidence-based decision making (TQM Fundamental 6)
7. Relationship management

Lean production is another approach used to improve delivered quality. Lean is not just about the elimination of scrap material. Like TQM it is about managing and improving the construction process. Believing individual processes cannot be considered in isolation; "Lean" is applied using a comprehensive view of the production system. With Lean any activity which fails to add value for the customer is considered waste.

Construction companies can apply the basic tenants of these three approaches and achieve excellent results. However, senior management must provide the leadership for development of a quality culture. Without management leadership, project quality will not be delivered even under the most detailed and structured programs.

> Consistent quality performance happens when senior leadership nurtures a culture cognizant of and responsive to customer expectations.

## EMPLOYEE EMPOWERMENT AND TRAINING

Management must lead and create the culture but managers have to relinquish power and entrust employees to instantly respond to changing situations.

> A quality culture empowers those doing the work with the ability to make decisions.

Employee empowerment is based on trust and ownership. It is necessary to invest in those performing the work. Each person must be provided training so they possess the vital task-specific knowledge necessary. A company noted for consistently delivering quality has this posted on its website: "*We exist to provide a place where inspired People can do their best work—building great projects, producing quality materials, and developing enduring customer relationships.*" Yes, this company spells "People" with a capital P.

According to a survey of large construction companies by the U.S. Chamber of Commerce over 70% of the companies were experiencing trouble finding qualified labor. Construction workers change employment for many reasons but the industry as a whole has one of the highest employee turnover rates. The rate varies because of many factors but the average is 20–25% per year. Employee training programs are effective in reducing turnover and in producing quality work. To increase project profits and improve project quality a company needs to retain its skilled employees. Only by investing in people can the necessary workforce be recruited and retained.

The greatest number of process improvement ideas originates with those doing the work. This is because the workers are the ones experiencing the problems. Their ideas can first be tested in a limited manner and additional data collected. If evaluation of the data confirms improvement in the process, the idea is then fully implemented.

## *IMPROVEMENT*

Continuous improvement seeks efficiencies by application of incremental steps. However, technological advances or even rethinking the purpose of a process should not be neglected. Sometimes it is necessary to challenge the basic assumptions of a process. One facility's owner reviewed its standard requirements for confirming its quality requirements were being met by the contractors. It discovered many documents served only to increase project delivery time. By reducing mandated record-keeping requirements by over 40%, project delivery was accelerated and essential rigorous requirements were maintained. An important area for contractors to examine is delivery and storage of required materials (Fig. 18.2). Are the company's practices creating a need to handle material repeatedly and as a result causing damage to materials?

## *QUALITY MANAGEMENT PLANS*

Development of a Quality Management Plan involves everyone in the organization. Nevertheless, for a plan to deliver results senior management must be the driving and motivating force. A lack of management commitment always results in a loss of productivity, delivery of subquality work, and accidents. Consequently, Quality Management Plans lay out a strategic approach to bring people and processes together with unified goals for quality, safety, and productivity. Attention to safety will increase productivity. An accident, any accident, diminishes productive work. Haste in completing a task leads to delivery of substandard work, lower quality, and accidents.

Well-structured quality plans transfer quality delivery ownership to the employees. The plan should match the company's performance to customer expectations and effectively communicate objectives and responsibilities.

**FIGURE 18.2** No order in the storage of materials around the building's perimeter.

**Sample Quality Management Plan: Overview Section**

This document is established to provide specific Quality Control on the "Grumpy Owner" project. XYZ Company will provide supervision to ensure all Systems are in compliance with the project specifications, drawings, and codes in respect to the furnished equipment, materials, workmanship, construction, finish, functional performance, and identification.

An important purpose of all Quality Management Plans is to eliminate the need for taking corrective actions to meet quality standards. One of the foremost tools for delivering quality and avoiding rework is Building Information Modeling (BIM).

**Sample Quality Management Plan: Installation Controls**

Building Information Modeling

This project shall be fully modeled in 3-Dimensions utilizing the latest software to perform clash detection between trades, ensure efficient use of space, and provide installation drawings to more efficiently manage production in the field. This process shall be performed in cooperation with all trades and will ultimately result in a mutually accepted installation model for use in the field months prior to physical installation.

Interior Elevation Modeling

In addition to the aforementioned "3D" model, a 2-Dimensional elevated rendering will be completed for every wall in the facility. This 2D model will be utilized to further the clash detection effort performed in the 3D model and will ultimately serve as a "final view" of each wall to ensure all devices are located in accordance with the intent of the design and needs of the end-user.

Quality, production, and safety are linked goals. They cannot be successfully embedded in a company's culture unless the link is recognized. The Experience Modification Rate (EMR) of a company is a Key Performance Indicator (KPI) of a company's ability to deliver quality. Sophisticated owners recognize the correlation between quality and safety and if a construction company's EMR is greater than 1, these owners will not allow the contractor to submit a project proposal.

As part of its Quality Management Plan one company states:

*It is our belief that Safety and Quality go hand-in-hand and one does not have to be sacrificed for the sake of the other. As such, the Quality Control Manager will act as on-site Safety Representative during all phases of construction.*

When a company makes the cultural shift to quality, there will be new emphasis on safety.

# CHAPTER 19
# EQUIPMENT FOR BUILDING CONSTRUCTION

Construction sites of low- and mid-rise buildings (up to about ten stories) will typically use several or all of the following equipment types: tower cranes and mobile cranes (including loader cranes and boom trucks), concrete mixers and concrete pumps, material handlers, hoists and lifts, and landscaping equipment. The smaller models of some of these machines are also used during the construction of single-story buildings. As with equipment used for construction of high-rise buildings, the manufacturers of equipment to support construction of low- and mid-rise buildings constantly seek to improve their products for enhanced safety, performance, work range, comfort, environmental friendliness, and compliance with the latest regulatory requirements (e.g., emission standards). The equipment can be procured on a purchase or rental basis. Given the shorter machine service time required on low- to mid-rise building sites, as compared to high-rise projects, on-demand service of some equipment types may be more economic than having equipment permanently on site for the full duration of a project.

## *TOWER CRANES*

Both tower crane configurations—top slewing and bottom slewing—can technically be used for the construction of low- and mid-rise buildings. However, if the height of the constructed building enables the use of the height-limited *bottom-slewing crane* (Fig. 19.1), the use of that equipment is expected to be more economic. Unlike top-slewing cranes, whose erection and dismantling are a long (can be up to a week) and costly process, bottom-slewing cranes can be erected and put to work in a matter of hours. In addition, their transportation to the site is much simpler than for a top-slewing crane.

Bottom-slewing cranes can be towed by a truck to the site (Fig. 19.2), with the counterweights (all or in part) being hauled on a separate truck, whereas transporting a top-slewing crane to the site requires a convoy of trailer trucks. Moreover, because of the weight and size of the components of a bottom-slewing tower crane, there must be careful loading, transporting, and unloading planning. Yet, top-slewing cranes, which are the only type used for high-rise building construction, can be seen on mid-rise building construction sites, particularly when lifting capacities or the horizontal reach required are greater than those provided by bottom-slewing cranes. Figure 19.1 shows a typical bottom-slewing tower crane serving the construction of a mid-rise building. Note there is no counter-jib to carry counterweights for balancing the crane when it lifts a load. The crane's ballast is placed only at its base (Fig. 19.3), and therefore its carrying capacity and horizontal reach are limited.

Bottom-slewing tower cranes have the swing circle (turntable) located under a slewing platform, and both the tower and jib assembly (as well as the ballast and motors) rotate relative to the base chassis. The tower is essentially a two-part telescoping mast, and hence the term "telescopic tower crane" is often applied to this type of crane. The mast usually has an open lattice design (always for the larger

**FIGURE 19.1** Bottom-slewing tower crane in the construction of a mid-rise building.

**FIGURE 19.2** Bottom-slewing tower crane transported by a truck.

cranes, Fig. 19.1), but a closed tubular (hollow) section mast, similar to the telescopic boom of a mobile crane, is used for smaller and, increasingly, midsize cranes of this type. The tubular mast on the smaller models is often foldable instead of telescopic.

A major advantage of bottom-slewing **tower** cranes is their ability to erect themselves using their own motors (hence they are often referred to as "self-erecting"), whereas the erection of a top-slewing tower crane requires the assistance of other equipment, commonly a high-capacity mobile crane (Fig. 19.5). On the other hand, the revolving mast of the bottom-slewing crane prevents it from being braced to the constructed building for increased stability, resulting in the crane's limited service height. The base chassis of the bottom-slewing crane can be adapted to mounting on rails, providing the crane with traveling capacity to increase work area coverage. However, in most cases the crane has a fixed base in the form of a jacked (outriggered) chassis (Figs. 19.3a and 19.3b). A detachable wheeled undercarriage (trailer) serves for relocation on the jobsite and for transportation between sites. For on-site relocation, the crane's jib is usually fully or partially folded. Such relocations are infrequent, if at all. Under no condition is it allowed to tow the crane while loaded.

Since bottom-slewing cranes are always free standing, they do not necessarily have to be located near the constructed building (this is opposite to the case of a crane anchored to the building). If, however, a location near the façade is desired (commonly for better horizontal coverage of the building), room must be left for the slewing platform. This is because the ballast blocks of the crane are located at the rear of the slewing platform, always opposite the jib (Figs. 19.3a and 19.3b). For better

a b

**FIGURE 19.3** Base platform and ballast blocks of bottom-slewing tower cranes.

counterweight effectiveness, the distance of the ballast from the mast should be as great as possible (Fig. 19.3a). On the other hand, to allow crane placement near the building and for increased safety the projection of the platform's rear from the crane's footprint (as determined by the undercarriage) should be as small as possible. New models, therefore, have a compact platform, compensated for by a high stack of ballasting blocks (Fig. 19.3b).

As an option to operating the crane from a top cab or from a control stand at the tower's base, bottom-slewing cranes can be operated by wireless remote control from the ground or from the floor of the building being constructed. In such a case the crane is not equipped with a top cab (the crane shown in Fig. 19.1). Many new models offer a climbing cab that can move along the mast for optimal view.

Typical jib length of bottom-slewing tower cranes is 70–180 ft, maximum height is 70–130 ft, and maximum lifting capacity is 2,000–20,000 lb (1,000–7,000 lb at the end of the jib).

What makes a *top-slewing crane* uniquely useful in high-rise construction is the ability to anchor its fixed mast to the constructed building, allowing the crane to achieve any required height. This is possible because only the upper works, including the main jib, counter-jib, and operator's cab, rotate. In high-rise construction the crane is initially assembled to a certain height within its allowable free-standing height, and then jacked up, as required, with the installation of anchors at model-specific height intervals. Yet, the height of mid-rise buildings is less than the common free-standing height of a top-slewing crane (200–250 ft). Thus, when used in low- to mid-rise construction (Fig. 19.4), the crane is assembled to its final required height and no anchoring is required. Same as with a bottom-slewing crane, the free-standing top-slewing crane does not necessarily have to be located close to the façade of the constructed building. This is of utmost importance when considering the dismantling of the crane at the completion of construction. Dismantling of top-slewing cranes involves a similar procedure to their assembly, including the assistance of a high-capacity mobile crane. Therefore, providing accessibility and room for the dismantling mobile crane must be planned when determining the crane's location, considering the different geometry and spatial conditions since a completed building will be in place. Not having to locate the crane close to the constructed building may help with this planning.

Top-slewing tower cranes are commonly used as stationary machines mounted on a fixed base. For a free-standing crane in mid-rise construction, the base is usually a ballasted chassis, anchored to the ground (Figs. 19.5a and 19.5b). The dimensions and footprint area of this base must be considered

**FIGURE 19.4** Top-slewing tower cranes in the construction of mid-rise buildings.

a b

**FIGURE 19.5** Assembly and foundation of a top-slewing tower crane.

when planning the location of the crane on the site. The other common method for mounting the crane is with fixing angles sunk in an engineered concrete mass foundation. This method, in which the above-ground footprint area and dimensions are much smaller (namely, those of the mast's cross-section), is commonly used for cranes anchored to the building in high-rise construction.

Figure 19.4 shows the two types of top-slewing crane jibs: on the right in the figure is a pendant-free jib cantilevered from the crane's mast (earning the crane the name "flat-top" or "topless"); on the left a saddle jib supported by pendants connected to an "A-frame" (also termed "cat-head"). Note the greater height dimension of the jib on the right, to allow for the greater bending moment imposed on the jib because of its structural action as a cantilever. On the other hand, and notable in Fig. 19.4 as well, is the jib's lightweight profile design, owing to the unidirectional action of all forces in the cantilevered jib. Saddle-jib cranes have been the prevalent type for years the world over, but since the early 2000s flat-top cranes have been gaining growing popularity, mainly given the solution they offer for both crane overlapping and height restriction situations. In addition, they are easier and quicker to set up, and require reduced reach for the erecting mobile crane. A third jib type for top-slewing cranes is the luffing jib. Unlike saddle-jib and flat-top cranes, whose working radius is attained by a trolley traveling along the horizontal jib, with luffing-jib cranes (commonly referred to as "luffers") the working radius is attained by upward and downward motion of the inclined jib. This operation mode is useful in the construction of multistory buildings where a greater amount of crane service (lifting power) may be required in order to speed delivery of the project. Such a situation dictates the use of several overlapping cranes in a relatively small footprint area. The use of more than two trolley-jib cranes would usually be impractical because of the resulting shared work zones (namely, a slower activity rate to maintain safety and thus decreased productivity); using luffing-jib cranes solves this problem, as their jibs can be luffed up (raised) to avoid overlapping. Therefore, luffers, whose procurement is costlier than that of trolley jib cranes, are uniquely associated with high-rise construction and are not seen on low- to mid-rise building projects.

Regardless of their jib type, the masts of top-slewing tower cranes are assembled to the desired height from modular lattice-type sections. The crane is stabilized partly at its base, as explained above (Fig. 19.5b), and partly by ballasts on the counter-jib, termed counterweights (Fig. 19.4).

Top-slewing cranes are commonly operated from a top cab, which grants the operator a bird's eye view of the entire work arena. While this is almost critical in high-rise construction, it is usually not so important in low- to mid-rise construction, where the operator can often have a good view even if not located at the top of the crane. Thus, the use of cabin-less top-slewing tower cranes, although not as common as using cranes equipped with a cab, is becoming more widespread. One of the benefits of operating the crane by remote control is saving the need for the operator to climb the crane to reach the cab. In some countries there are regulations that mandate the provision of a mechanical climbing means for the operator. Where such regulations already exist, a common minimum threshold for mechanical climbing is 80–100 ft. Below the stated height no such means is required. As this is roughly the height limit of mid-rise construction, operating the crane from the cab would mean manual climbing. Thus, the use of a cabin-less crane spares the climbing.

In addition to their common use as stationary machines mounted on a fixed base, top-slewing cranes can be mounted on rails to provide mobility—a likely solution for projects of linear extent. This way their work envelope is increased, with a better coverage of the constructed building and the site around it; however, this is useful only as long as crane service demands in any given zone of the building are not extensive such that they render the crane stationary anyway, even if mounted on rails. Note that granting a top-slewing crane traveling capacity is possible only for free-standing cranes; hence, it is unique to low- and mid-rise construction.

Typical jib length of free-standing top-slewing tower cranes is 100–270 ft. Base dimensions are 13 × 13 ft to 20 × 20 ft, and maximum lifting capacity is 10,000–80,000 lb (2,000–13,000 lb at the end of the jib).

## MOBILE CRANES

Mobile cranes are traditionally associated with heavy civil and infrastructure construction. Yet, various machine types, particularly small and midsize models, are used in the construction of low- to mid-rise buildings. This section refers to mobile crane types serving on the site during extended periods of time, and not to machines brought to the site for one-off, specific lifting tasks, such as for the erection and dismantling of tower cranes.

It is common to see any of the following types of mobile cranes on low- to mid-rise building sites. They can be used for the full duration of a project or for short durations for specific lifting needs. Providing lifting services for the construction of the building's frame, the following types are common: lattice-boom crawler, telescopic-boom crawler, rough terrain, all-terrain, truck-mounted tower cranes, and boom trucks.

*Lattice-boom crawler cranes* (Fig. 19.6) have been for years the machines of choice in the United States, and they are still favored to a great extent by American contractors for mid-rise building construction though tower cranes have gradually been replacing them since the early 2000s. The full revolving superstructure (namely, undercarriage and boom) of this type of unit is mounted on a pair of crawler tracks. Units in the low to middle range of lifting capacity (maximum of about 20–100 tons at minimum radius) have good lifting characteristics and are capable of duty-cycle work (namely, work in which the crane is engaged in repetitive lifts of relatively short cycle time), such as handling a concrete bucket. The main lattice-type boom can reach 300 ft or more; further extension by a jib is possible with many models.

The crawler tracks provide the crane good travel capability around a jobsite. The large ground contact area allows the crane to operate even on low-bearing capacity soils.

> Before hoisting a load, the crane must be leveled and ground settlement considered.

When ground conditions are adequately firm, a crawler crane can move with the hoisted load. This ability to carry a hoisted load, "pick and carry," along with the crane's capacity to travel and work even in poor underfoot conditions, is the main advantage a crawler crane has over a wheeled (truck) crane.

Small to midsize crawler cranes are commonly relocated between projects by several truck trailers. As the size of the crane increases, the time and cost to dismantle and load the crane, to investigate haul routes, and reassemble the crane increases. These machines usually have lower initial cost per rated lift capability than other mobile crane types, but movement between jobs is more expensive.

**FIGURE 19.6** Lattice-boom crawler crane. (a) Main boom and extension jib. (b) Crawler tracks and revolving superstructure.

**FIGURE 19.7** Small-size telescopic-boom crawler crane in building construction.

Therefore, crawler-type machines should be considered for projects requiring long-duration usage at a single site.

*Telescopic-boom crawler cranes* (Fig. 19.7), commonly referred to as *telecrawlers*, combine the advantages offered in certain job-site conditions by the tracked undercarriage of the lattice-boom crawler crane and the telescoping boom of the wheeled mobile crane: (1) high on-site maneuverability; (2) quick rigging requiring less space; (3) ability to travel with the load on the hook; and (4) boom length adjustability while lifting. Furthermore, the compact dimensions of telescopic crawlers reduce the difficulty of crawler crane transportability between jobsites. Figure 19.7 demonstrates these benefits, as well as the high reach of the boom and the low ground pressure provided by the tracks. These provide versatile lifting services in building construction. Another typical application of telescopic-boom crawlers in building construction is installing prefabricated concrete elements.

*Rough-terrain (RT) cranes* (Fig. 19.8) are mounted on two-axle carriers. The operator's cab is commonly mounted in the upper works, enabling the operator to swing with the load; on some models, the cab is located on the carrier—a simpler design resulting in a lower cost. These units are equipped with unusually large wheels and closely spaced axles to improve maneuverability at the jobsite. They further earn the right to their name by their high ground clearance, as well as the ability of some models to move on slopes of up to 35 degrees. Most units can travel on the highway but have maximum speeds of only about 30 mph; in addition, the ride is rough. In the case of long moves between projects, rough-terrain cranes are transported on low-bed trailers (Fig. 19.8b).

Most units use joystick controls, enabling the operator to manipulate more than one function simultaneously. Small to midsize models are in the 10- to 60-ton capacity range and are typically employed as utility cranes. They are primarily lift machines but are capable of light, intermittent duty-cycle work. To utilize their maximum lifting capacity, the cranes must be stabilized on outriggers

a b

**FIGURE 19.8** Rough-terrain (RT) crane. (a) In operation. (b) In transportation.

**FIGURE 19.9** All-terrain (AT) crane demonstrating its roadability.

while in operation (Fig. 19.8a). Since these machines are commonly used for work requiring frequent on-site relocation, the small to midsize models can also be operated on their wheels, albeit with reduced loading. They can even move slowly with loads on firm, leveled ground, but only without a fly jib and with the boom in a straight-ahead (i.e., not revolved) position.

All-terrain (AT) cranes combine the advantages of truck cranes and rough-terrain cranes (Fig. 19.9). The undercarriage makes travel on public roads much more comfortable. The carrier has all-axle drive and all-wheel steering, crab steering, large tires, and high ground clearance. All-terrain cranes have dual cabs—a lower cab for fast highway travel and a superstructure cab with both drive and crane controls. Because of this the crane has both job-site mobility and transit capability. This feature is important when multiple lifts are required at scattered project sites or at multiple work locations on a single project. Since these cranes combine two features, they have a higher cost than an equivalent capacity telescoping truck crane or rough-terrain crane. But an all-terrain machine can be positioned on the project without the need for other construction equipment to prepare a smooth travel way, as truck cranes would require. In addition, the all-terrain crane does not need a lowboy to haul it between jobsites, as would a rough-terrain crane. Small to midsize models are in the 40- to 100-ton capacity range.

*Truck-mounted tower cranes* are a class of "hybrid" machines. They combine the main advantages of the tower crane and the wheeled mobile crane (Fig. 19.10). These machines have excellent lifting height and outreach, as would a normal tower crane. In addition, they have independent travel capacity to the jobsite and fast-erection and dismantling capability similar to a wheeled mobile crane.

a b

**FIGURE 19.10** Mobile truck-mounted bottom-slewing tower crane in prefabricated concrete building construction. (a) Full view. (b) Carrier and revolving superstructure.

They must have their outriggers extended and down before the mast is raised and, therefore, cannot travel with a load. Operation is either by radio control or from the height-adjustable elevating cab. Midsize models feature a mast height of up to 80 ft, but in most units the jib can be luffed up as much as 45 degrees to attain a total hook height of up to 160 ft (and twice as much in the largest models). Common uses are not only building frame construction, including prefabricated concrete construction, but also finishing works such as façade cladding and screen walls.

Not to be confused with the truck crane, whose carrier vehicle is a specialized truck, the carrier vehicle of the *boom truck* (also known as a truck-mounted crane) is a commercial truck. In terms of dimensions and capacities, boom trucks typically are smaller in size, shorter in boom length, and have lower lifting capacities than truck cranes. In the case of a boom truck, a truck manufacturer supplies the carrier vehicle and a crane manufacturer the lifting unit mounted on the truck. In that respect, boom trucks may resemble loader cranes; however, unlike loader cranes, boom trucks have no cargo-hauling capacity. These two crane types differ also in the common mounting of the boom: commonly at the rear of the truck in the boom truck and directly behind the truck cab in the loader crane. Boom trucks also typically feature higher lifting capacities than loader cranes (up to 40 tons at minimum reach for the midsize models).

## LOADER CRANES

Loader cranes are self-contained hauling and loading/unloading units, a truck fitted with a crane. When a loader crane arrives on the jobsite, it can discharge its load, either at a storage location or place the load directly at the required position on the building being constructed, there is no need for another crane. This makes loader cranes very useful when constructing low- to mid-rise buildings. These cranes often make it possible to complete the project without the need for an onsite tower crane. In addition, loader cranes can be used for short- or long-term services unrelated to their hauled load, yet another advantage when no other crane is on site.

There are two boom configurations: a straight boom (Fig. 19.11) and an articulated boom (Fig. 19.12). The straight boom extends telescopically. The articulated boom (a knuckle boom is the term often used) can also telescope some or all of its sections. Typical reach is up to 80 ft horizontal

**FIGURE 19.11** Straight-boom loader crane used in construction of a low-rise building.

**FIGURE 19.12** Articulated-boom loader crane used in construction of a mid-rise building.

and 100 ft vertical above ground. Most articulated-boom models feature below-ground reach. Typical lifting capacities are up to 20 tons at minimum reach and 2 tons at maximum horizontal reach. The largest machines extend these dimensions and capacities considerably.

The boom is commonly mounted directly behind the truck cab, facing rearward. This way, most of the truck bed is free for cargo. A less common configuration is a rear-mounted boom, leaving less space for cargo but allowing for increased lifting capacity, as the boom is supported directly by the rear axles. Operation is commonly from a stand-up station (most models have two such stations, one on each side of the truck).

On the construction site, loader cranes offer advantages by their two service modes. As independent hauling and loading equipment, they can save crane time by self-unloading their cargo (or picking up cargo from the site), as well as in lifting it directly to the building. They also have an advantage in being able to drop off or pick up cargo outside the work envelope of the site's crane. In the capacity of lifting-service providers augmenting site resident equipment, their hourly operation cost is lower than a truck crane or rough-terrain crane. This is also why on low- to mid-rise building projects there is a preference for using loader cranes frequently, on an on-demand basis, rather than having an on-site crane.

## CONCRETE MIXERS

Most projects have concrete delivered by special *truck mixers* (Fig. 19.13). The trucks are loaded at a central concrete plant and by the time they arrive at the site, the concrete mix is ready to be placed, hence the term ready-mixed concrete. Obviously, to be useful, a ready-mixed concrete plant must be within a reasonable distance from the jobsite. On projects requiring large quantities of concrete, particularly at remote locations, provisional concrete plants are often set up on the site.

Truck mixers of all types use *freefall* mixing. With the freefall mixing method, the concrete is mixed by lifting the ingredients with fixed mounted blades inside a rotating drum and then dropping the material by overcoming the friction between the mixture and the blades. The drums of the truck mixers are of the reversible type: in mixing position, the drum rotates in one direction, while for discharging, rotation is reversed. The drum has one opening for both loading and discharge.

Truck mixers are available in sizes up to approximately 20 cy. The common sizes are in the range of 8–12 cy, and the most popular size is 11 cy. Truck mixers are commonly of the *rear-discharge* type (Figs. 19.13a and 19.14), but in some areas, *front-discharge* truck mixers are used (Fig. 19.13b). Drums of the rear-discharge truck mixers can be mounted on any commercial vehicle, which is this type's main advantage. Yet, owing to their design and configuration, front-discharge truck mixers feature lower wheel pressure and hence provide a better solution for meeting road and bridge load regulations. Front-discharge mixers have additional advantages: (1) convenient forward-driving approach to discharge location and in-cab driver control of discharge operation, and (2) smoother drum rotation (no vibrations) due to elongated shape of the drum and even weight distribution, enabling higher road-travel speed and increased rotation velocity and discharge rate. As an alternate solution to

a b

**FIGURE 19.13** Concrete truck mixers. (a) Rear-discharge truck. (b) Front-discharge truck.

**FIGURE 19.14** Truck-mounted pump with a truck mixer casting concrete in a low-rise building.

meeting road load restrictions, truck mixers may be equipped with an extra, hydraulically operated, foldable trailer axle for additional weight distribution. This rear axle is folded up when the mixer is not carrying a load or when there are no wheel-load regulations.

When concrete is hauled to the site in large and heavy truck mixers, attention must be given to on-site travel limitations in terms of both maneuvering space and type of terrain. In planning a concrete placement, concrete discharge time must be considered. Discharge time is usually governed by the rate at which concrete is accepted into the pump hopper (Fig. 19.14) or directly into the formwork, or by the crane bucket cycle time (Fig. 19.3a). If a truck pump is used, time allocation for possible pump relocation should be made, including coordination with the plant regarding the schedule of concrete delivering truck mixers. For further considerations regarding production rates of the combined operation of truck mixers and concrete pumps, see the concrete pumps section.

For concrete delivered directly into the formwork, a chute is used. To use the chute, the opening of the truck mixer drum must be sufficiently higher than the accepting formwork. Truck mixers are normally equipped with built-in swing (and often extendable) chutes. These chutes are hydraulically operated for direct placing of the mix when the concreting location is within chute reach and the chute's slope is steep enough for the concrete to flow continuously without segregation.

For crane and bucket placing, buckets of various types and sizes may be used. The most common bucket sizes are in the 1.0–2.5 cy range. The weight of a fully loaded bucket, including self-weight, is about 4,800 lb for a 1.0-cy bucket and 9,200 lb for a 2.0-cy bucket. The bucket size selected should be such that when fully loaded it can be safely lifted to the required concrete placing locations by the job-site crane. Crane lifting capacity must always exceed the weight of the loaded bucket. Quite often, bucket size is determined on the basis of required production rates. With a larger bucket, there are fewer work cycles required to execute a given concrete placement. If a certain crane-bucket combination satisfies most of the project's production requirements but not all, then for extended-reach locations either a smaller bucket will be used or the bucket in use will be partially loaded. With any given crane-bucket combination, the production rate can be increased by using two buckets instead of one. With bucket loading time thus virtually eliminated, this is particularly beneficial in the case of low- and mid-rise construction, where bucket loading (and then unloading at the placement location) makes up the better part of the overall cycle time.

## *CONCRETE PUMPS*

Concrete pumping equipment comes in three primary configurations: (1) a pump with a separate pipeline; (2) a pump-and-boom combination; and (3) a pump and pipeline with a separate boom.

The most common configuration used on low- and mid-rise construction projects is the pump-and-boom combination (Fig. 19.14), also known as truck-mounted pump, truck pump, or boom pump.

The pump is mounted on a truck and equipped with a slewing boom along which a fixed-length delivery line is connected. The line is made of steel pipe, commonly of 5 in. diameter. The free end of the line has a flexible pipe, 9–12 ft long (termed end hose), connected to it. The end hose eases the task of the worker who must maneuver it and can also be used to slightly extend the boom's reach. The booms—hydraulically operated and articulated—come in lengths of 60–230 ft, but the more commonly used booms are in the range of 80–180 ft. Carbon fiber, a lightweight and strong material, is used today to manufacture considerably longer truck-pump booms. Most truck-mounted pumps have four to five section booms. For a given boom length, a greater number of sections provides the boom with increased maneuverability, which may be an advantage in confined areas.

The term "boom length" indicates the boom's maximum *vertical* reach, measured from the ground and, therefore, it is only a nominal length. In order to obtain the boom's *effective horizontal* reach, allowance must be made for loss of distance due to the following factors (Fig. 19.15).

1. Height above ground of the joint connecting the boom to the truck, which serves as the boom's slewing axis (usually 13 ft for most makes and models of truck-mounted booms)
2. Height of those boom sections extended vertically or diagonally as required to reach the height of the concreted element
3. Distance from the slewing axis to the front of the truck or to the truck side
4. Dimension of the extended outriggers by which the truck is stabilized in the direction of the boom
5. Distance from the building (or scaffolds, peripheral fence, etc.) including minimum required safety boundary

The boom's *effective vertical* reach cannot be greater than the nominal length minus the length of the end section (in which configuration the horizontal reach is minimal). It is important to consult specific boom charts and truck dimensions when selecting a truck-mounted pump for a given concreting operation.

The truck-mounted pump would be optimally positioned to enable maximum coverage of the concreting area for each operation. Because the length of the boom is limited, coverage from one location may be impossible, requiring relocation of the truck in the course of the operation. Relocation

**FIGURE 19.15** Allowances for loss of distance when calculating effective horizontal reach of truck-mounted pump.

interrupts work continuity and may cause delays, depending on the time it takes to fold the boom, retract the outriggers, move the truck, extend the outriggers, and unfold the boom to continue concrete placing. If relocation was not preplanned, a line of waiting ready-mixed concrete trucks may result.

High placing rates are one of the main advantages of pumping, but the theoretical outputs are seldom realized. Placing rates are determined in practice by the dimensions of the cast element, the strength of the formwork, the size of the concrete placing crew, and the rate at which the concrete truck mixers can deliver concrete to the pump.

> In a properly organized placing operation, one truck will always be waiting to position itself at the pump as soon as the previous truck finishes discharging concrete into the pump.

This is important in maintaining a constant pressure head in the pipeline and eliminating cavitation in the pump housing. If possible, there should be room for two trucks to be positioned at the pump hopper—one discharging and the other ready to discharge. This requires adequate access routes for trucks to reach the site. It also requires space for two trucks to be in the discharge position and then depart. These requirements are often a challenge in tight urban settings. Thus, even under the best of conditions, the placing rate is first and foremost dictated by gross truck discharge cycle times.

Crew size is a limiting factor only when the element being cast or the access to the work area for the workers limits the size of the placing crew, such as during wall concreting. Otherwise, the size of the crew will be determined so it does not control the placing operation. Another factor impacting the rate of placing is the formwork, especially for vertical elements. The lateral pressure of the concrete on wall or column forms is a crucial factor. The rate of placing and the imposed hydrostatic pressures from the concrete are commonly much higher with concrete pumping than with crane and bucket placing. This factor must be considered when designing the formwork and planning the concrete placement.

The prevalent maximum *theoretical* output for the currently available range of boom pumps is 80–200 cy/hr, but in the case of the largest pumps the outputs can reach 300 cy/hr. *Effective* outputs, distinguished from theoretical outputs, are project specific and can vary widely, so it is difficult to provide average output values, even for a given pump. Assuming the pump does not control output and the work is conducted under normal conditions, the following outputs can be used as an initial guide for planning: common building elements of regular dimensions 40 cy/hr; thick slabs (in excess of 20 in.) and similar elements 60 cy/hr; and mass concreting of large elements (i.e., dams, raft foundations) 80 cy/hr.

The main advantages of concrete pumping are high placing rates, work convenience, and the ability to deliver the concrete while bypassing project site obstructions. A concrete pump, however, is good for concrete transportation only; hence, most building construction projects will require equipment, such as cranes, for the transportation of many other materials and building elements. Such on-site cranes can then be used for concrete placing, as well as other lifting tasks. Thus, concrete pumping should be considered for on-site concrete transportation and placing mainly in the following cases and types of projects: (1) projects such as single-story building, for which concrete pumps may satisfy all or most material transportation needs, with no or only minor requirements for any additional transportation equipment; (2) for high-rate concreting of structural elements of large dimensions, such as thick slabs and large beams; (3) sites with accessibility limitations (e.g., to avoid obstacles); and (4) projects on which material transportation is a bottleneck to work progress.

## MATERIAL HANDLERS

Telescopic material handlers, commonly referred to as *telehandlers*, are an equipment class particularly suitable for the construction of low- to mid-rise buildings, as long as site conditions permit their mobility. With a typical maximum vertical reach of 60 ft—about six stories—and a horizontal reach

**FIGURE 19.16** Telescopic material handler.

of up to about 40 ft, their versatility and wide range of available attachments often make telehandlers a viable alternative to using a tower crane, a concrete pump, and landscaping equipment such as wheel loaders. Figure 19.16 shows a material handler used as a forklift, which is the most common use of this equipment type. Note the high ground clearance of this forklift, which renders it a rough-terrain machine suitable for working on uneven surfaces typical of construction jobsites. As its name implies, the telehandler uses a telescopic boom to reach horizontally and vertically much farther than the traditional forklift.

Other common attachments are buckets of various types, extendable lifting arms and lifting hooks, spreader bars, pipe grapples, and sweepers and cleaners. For increased maneuverability and off-road capability, telehandlers have four-wheel drive, and most models feature multiple steering modes—four-wheel, two-wheel, and crab steering.

Typical capacity and dimensions of telehandlers are maximum lifting capacity 5,000–15,000 lb, maximum lifting height 20–60 ft (but up to 80 ft for the largest machines), maximum forward reach 10–40 ft, measured from the front of the front tires (60 ft for the largest machines), capacity at maximum height 4,000–10,000 lb, capacity at maximum forward reach 1,000–3,500 lb, and maximum travel speed 15–20 mph. Some models are equipped with a pair of front stabilizers for increased reach and lifting capacity. The largest machine on the market today (2020) has a maximum lifting height of 167 ft, a maximum forward reach of 113 ft, and a maximum lifting capacity of 13,200 lb.

## MOBILE LIFTS

This is a unique class of machines, compared to the other machines reviewed in this chapter: it is the only equipment type designated to carry people rather than construction materials and elements. Mobile lifts of various kinds and configurations provide variable height access to specific areas above ground level for work crews and other site personnel (e.g., for inspection). For service on low- to mid-rise building sites, this class of machines includes the following equipment types: boom lifts, scissor lifts, and mast climbing work platforms. Boom lifts and scissor lifts are also widely known by the terms *mobile elevated work platforms* (MEWPs) and *aerial work platforms* (AWPs). Another type, single- or double-cabin mast-based hoists are used to move crews between the ground and a building's floors in high-rise buildings. Although technically feasible in lower buildings as well, such hoists are usually not cost-effective for low- and mid-rise buildings, in which workers commonly use the building's stairs for vertical movement.

*Boom lifts* come in two configurations (Fig. 19.17), telescopic boom and articulated boom, both offer vertical and horizontal reach (unlike scissor lifts, which offer only vertical reach). Mobility of the

FIGURE 19.17 Boom lifts in low-rise construction (right: articulated boom; left: telescopic boom).

FIGURE 19.18 Boom lift transported on a lowboy.

self-propelled machine around the jobsite is provided by the wheeled chassis and high ground clearance. For transportation between jobsites, the lift is hauled on a truck or lowboy (Fig. 19.18). The operator can control the vertical and horizontal position from the ground or the upper work platform (which, combined with its guardrails, is commonly referred to as a "basket"). Most models are limited to a vertical height of 60–90 ft (but the largest telescopic models extend as high as 210 ft). Their maximum horizontal reach is in the range of 50–70 ft (but the largest models have a 100-ft maximum outreach). Common platform capacity is in the range of 600–1,000 lb, rated for two to four occupants plus tools. Counterweights and outriggers are available on larger models to provide stability. Diesel or gasoline powered, all electric, and hybrid models are offered by most manufacturers.

Some manufacturers offer crawler boom lifts, commonly with nonmarking rubber tracks. These machines can manage soft grounds and delicate floor finishes requiring contact area larger than that provided by a wheeled chassis. In addition, they have the ability to climb slopes with ease. In working mode, the machine is stabilized by outriggers. Typical working height is in the range of 40–100 ft, and typical outreach is 30–50 ft.

A document developed by several trade associations, based on ANSI standards, details best practices for workplace risk assessment and AWP equipment selection. As stated, it is aimed at aiding in the identification of workplace hazards, offering guidance for conducting a thorough workplace risk assessment and implementing control measures, and assisting in the selection of the most appropriate AWP equipment for the work involved. Key selection factors are: (1) Required elevation, (2) required horizontal reach, (3) required capacity, (4) adequate ground support, (5) suitable driving conditions, (6) working indoors, (7) work area access, and (8) other factors, such as proximity to electrical conductors, weather conditions, and presence of pedestrian or vehicular traffic in the work area.

*Scissor lifts* have traditionally been used mostly in interior works, such as finishing (Fig. 19.19) and system installations; these are the industrial type, low ground clearance machines designated to travel on smooth surfaces (e.g., warehouse floors). Construction-type, high ground clearance machines (often referred to as rough terrain equipment) for exterior works (e.g., building frame, exterior finishing) are also available. If a smooth surface is available, the industrial-type machine can be used also for exterior works (Fig. 19.20). The movement of the machine on the floor as well as the height extension of the "scissors" for vertical positioning can be controlled from the upper deck. While scissor lifts offer only vertical reach, they have two advantages compared with boom lifts: they are suitable for indoors work and they can manage particularly narrow spaces (the slimmest machines can fit through a doorway). It is common to see several machines working close to each other in a confined space (Fig. 19.19). Typical maximum platform height is in the range of 20–40 ft for the industrial-type scissors and 30–50 ft for the construction-type, RT scissors. Common platform capacity is in the range of 500–800 lb for industrial scissors and 800–1,500 for construction scissors.

A similar lift is the vertical mast lift, in which the height reach is attained by a telescopic mast rather than scissors. This small-footprint configuration is helpful in working over and around assembly lines, shelving, ductworks, or partitions. Typical working height is in the range of 20–30 ft, and common platform capacity is 500 lb.

*Mast climbing work platforms* (Fig. 19.21), also known as mast climbers, have a power-driven work platform that climbs a vertical tower or a pair of towers. The tower is composed of modular stacked units resting on either a mobile chassis or stationary beams. A drive unit moves the platform up and down. It can be either a gas-powered hydraulic system or an electric motor system. Multiple mast towers can be arranged with bridging to create a long, continuous scaffold platform across the face of a building. Mast climbers provide access for façade works such as installation of exterior insulation and finish systems, glazing, and metal framing, as well as inspection, restoration, and maintenance. The entire assembly can operate as a free-standing unit up to a manufacturer-specified height. For higher operational needs, they must be anchored to the structure being accessed.

**FIGURE 19.19** Scissor lifts in interior finishing works.

**FIGURE 19.20** Scissor lift positing worker for assembly of slab formwork.

**FIGURE 19.21** Mast climbing work platforms.

## *LANDSCAPING EQUIPMENT*

In the context of this chapter, landscaping equipment refers to small-size earthwork machines commonly used for shaping the grounds of low- and mid-rise building projects (Fig. 19.22). These machines can handle the volume of work required on such projects, while their dimensions make them useful for operating in tight and confined spaces without compromising maneuverability.

a b

**FIGURE 19.22** Compact landscaping equipment. (a) Skid-steer loader. (b) Track excavator.

**FIGURE 19.23** Compact excavators hauled by a small truck and tow-behind trailer.

Because of their size, these compact machines can be hauled between sites by small trucks (Fig. 19.23), with easy access and unloading at the site. The range of such machines offered is extensive, whether they are made by the manufacturers of regular-size earthwork machines or as specialized equipment. Figure 19.22 shows two compact earthwork machines in typical landscaping work on a single-story house (a) and a mid-rise building jobsite (b). The size of these compact machines grants them extra flexibility and agility lacking with larger machines. These machines are commonly diesel-powered but hybrid (dual mode) and fully electric models with overnight rechargeable batteries are recently offered, particularly suitable for working in noise- and emissions-sensitive environments (e.g., urban settings near residential areas or hospitals). A wide selection of attachments often earns these machines the name "tool carriers." (Note: this section addresses only ride-on machines. Some sources use the term "mini" to refer to both ride-on and walk-behind machines, but other sources use that term to refer only to walk-behind machines. Therefore, the term "compact"—rather than "mini"—is used here.)

*Compact loaders* come in two undercarriage configurations: wheel and track. To increase their maneuverability, compact wheel-based loaders commonly feature either an articulated chassis or *skid steering*. In skid-steer loaders (Fig. 19.22a), the right and left wheel pairs are propelled separately. To turn the machine, the wheel pairs are turned at different speeds. The machine thus turns by skidding (namely, the wheels are dragged across the ground), hence the term "skid steering." The wheel pairs can also be turned in opposite directions, namely, forward on one side and backward on the other, like crawlers; this way, the turning radius is kept to a minimum. Some models feature both all-wheel steering and skid steering. All-wheel steering has an advantage when tearing up finished surfaces or leaving marks on driveways, as would be with skid steering, is to be avoided.

A wide variety of attachments are available for quick mounting on most models, wheel or track, to increase versatility regarding landscaping requirements. Such attachments, other than a loader bucket, are pallet forks, plate compactors, grader blades, and brooms. *Wheel* loaders feature faster mobility but need relatively level and firm ground. Typical machine dimensions for skid-steer loaders are 8–12 ft length (with bucket on ground) and 4–6 ft width, maximum speed is 7–12 mph, maximum rated operating capacity is 1,300–4,000 lb, and full lifting height is 8–12 ft. *Track* loaders, with either metal or rubber tracks, are slower but can handle challenging underfoot conditions. Typical machine dimensions are 10–13 ft length (with bucket on ground) and 5–6.5 ft width, maximum speed is 8–10 mph, maximum rated operating capacity is 1,000–3,600 lb, and full lifting height is 9–12 ft.

*Compact excavators* are commonly track machines (Figs. 19.22a and 19.23), but wheel-based models are offered. Steel, rubber, or a mix of steel and rubber are the materials used for the tracks. In landscaping work, compact excavators are used mainly to dig holes and trenches. The machine is equipped with a long digging arm, which in some models is extendable for additional digging depth and reach. The type of a digging bucket used depends on the type of soil at the specific jobsite. The machine is also equipped with a dozer blade, for backfilling, leveling, and grading. The blade also adds stability while digging. As with compact loaders, various attachments are available for quick mounting, such as hydraulic concrete breakers, earth augers, trenchers, and frost rippers. Some models feature retractable tracks for reduced machine width, an advantage in particularly tight workspaces such as doorways, as well as for ease of trailer loading. Typical machine dimensions are 10–20 ft length (with digging arm on ground = transport length) and 4–7 ft width, maximum digging depth is 7–16 ft, maximum reach at ground level is 12–23 ft, maximum dumping height clearance is 8–16 ft, and maximum travel speed is 2.5–3.5 mph.

## SAFETY

Construction sites are inherently hazardous workplaces. Working in the presence of heavy machines contributes significantly to the hazardous nature of the site. Among the various types of construction

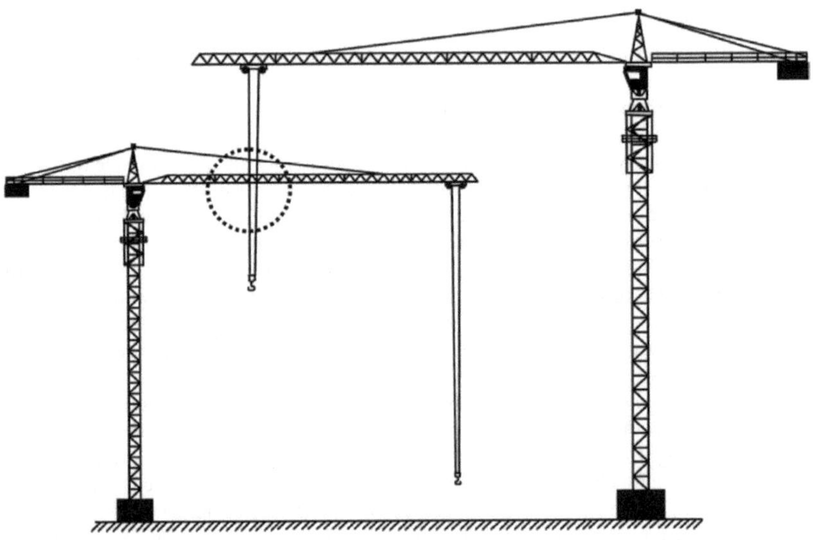

**FIGURE 19.24** Typical hazardous scenario of two overlapping tower cranes.

**FIGURE 19.25** Typical hazardous scenario of work zone shared by two overlapping tower cranes and a boom of truck-mounted concrete pump.

**FIGURE 19.26** Tilted truck-mounted concrete pump with collapsed outriggers.

equipment reviewed in this chapter, cranes are perhaps the focus of accident prevention efforts, given their work regime—hoisting and transporting a variety of loads near and above people, which further increases the safety risk on the site. When two or more tower cranes overlap and work in shared zones (Fig. 19.4), an additional risk develops as the result of a potential contact between one crane's cable or hook block and the other crane's jib (Fig. 19.24). Even more hazardous is the case of shared work zones of overlapping cranes and the boom of a truck-mounted concrete pump (Fig. 19.25).

Yet, the presence of other machines should not be overlooked when taking measures to secure a safe working environment. Figure 19.26 illustrates an accident involving a truck-mounted concrete pump following the collapse of one of its outriggers. The truck tilted to the side of the heavyweight extended placing boom. The collapse resulted in a sudden, uncontrollable downward motion of the boom; a worker standing next to the boom was struck by it in his head and killed instantly.

The risk of unstable ground yielding under the pressure of outriggers due to an ill-prepared work plan is not unique to concrete pumps; it is common to all wheeled-type equipment requiring the use of mat-supported outriggers for stabilization. If ground-bearing capacity is not carefully considered in relation to outrigger pressure, and the area size and thickness of the mats are insufficient, an accident is likely to result. Ground-machine interaction must also be considered when designing foundations for tower cranes, particularly given the duration of this equipment on site for periods of time spanning different ground conditions (e.g., hard and dry versus flooded by rainwater).

A considerable number of workers struck by moving equipment are recorded each year. Vehicles such as hauling trucks, mobile cranes, earthmoving machines, and material handlers are commonly equipped by back-up alarms, yet these do not always prevent an accident. It is imperative to acknowledge that equipment employed on low- and mid-rise building sites is not potentially less hazardous than high-rise construction equipment. The likely perception of workers operating in a more protected environment due to the lack of the height dimension must be countered by an overall site safety plan. Such a plan should include strict preventive measures covering the fleet of machines operated on site either resident or per-call equipment.

# CHAPTER 20
# SAFETY

*The construction industry is one of the most dangerous professions. According to the Occupational Safety and Health Administration (OSHA), one in five worker deaths annually is in construction. Construction industry leaders must strive to safeguard their employees by following safe workplace guidelines and regulations. Safety standards for construction are detailed in the OSHA regulation 29 CFR 1926, Safety & Health Regulations for Construction.*

*This chapter will highlight these requirements, outline suggestions for developing robust corporate and site-specific safety plans, and discuss jobsite processes and procedures to assist in achieving the goal of "zero incidents." Also discussed are the most commonly cited OSHA standards in construction with tips on mitigating these risks as well as hazard mitigation strategies for the jobsite.*

## *OSHA REQUIREMENTS*

Construction sites are required to follow OSHA regulations, procedures, and requirements. These include keeping records related to the safety of operations on the jobsite. Employers with more than ten employees are required to keep a record of serious work-related injuries or illnesses. Minor injuries that only require first aid do not have to be reported.

### Recordkeeping Requirements

The Log of Work-Related Injuries and Illnesses (Form 300) is used to classify work-related injuries and illnesses and to record the extent and severity of each incident. This form is used to record the details about what and how the incident happened. The Summary of Work-Related Injuries and Illnesses (Form 300A) shows the total number of injuries and illnesses that occurred during the year for each category of illness and injury. The log and summary are expected to be kept for each site and made accessible to all employees. Form 300 (Fig. 20.12) and Form 300A (Fig 20.13) are include at the end of the chapter.

The information provided by these records helps all stakeholders evaluate the safety of the workplace, understand the hazards present, and implement worker protections to reduce and eliminate hazards, preventing future workplace injuries and illnesses.

The records must be maintained for at least 5 years. Also required is the posting of a summary of the injuries and illnesses that were recorded the previous year. This occurs from February through April. The regulations also require, if requested, copies of the records must be provided to current and former employees, or their representatives.

### Reporting Requirements

Employers must report any worker fatality within 8 hr and any amputation, loss of an eye, or hospitalization of a worker within 24 hr (29 CFR 1926.22).

### Contractor Requirements

OSHA regulations state under the general safety and health provisions [29CFR 1926.20(a)(1)] that a contractor may not require any employee to work in conditions that are unsanitary, hazardous, or dangerous to their health and safety. The contractor is also responsible for implementing protocols to prevent accidents and they must initiate and maintain safety and health programs. The safety program must include provisions for regular inspection of the jobsite, materials, and equipment. All tools, materials, and equipment used on the site must be in good operating order and inspected prior to use.

The regulations state that the contractor is required to provide each employee with personal protective equipment [29CFR 1926.20(f)(1)]. In addition, the contractor must provide training related to the hazards employees will encounter on the worksite [29CFR 1926.20(f)(2)]. Employees must be able to recognize and avoid unsafe conditions as well as understand the methods to control or eliminate the hazard. Detailed requirements for the training are provided in the specific topic-related sections of the regulation.

These standards are simply a baseline and jobsite safety should achieve levels beyond the OSHA baseline. Safety must extend to the culture and climate of a company. Successful construction firms establish a safety culture that starts at the top of the company and is practiced by all. The foundation of the culture is the espoused beliefs, attitudes, and values that the company has surrounding safety. These are the norms of how safety is practiced within the organization.

The true test of how a safety culture is embraced beyond the board room is the actual safety climate. How do employees perceive the consistency between stated and practiced safety policies and procedures? Consistency between the culture and the climate is what distinguishes highly functioning companies and the jobsite must also illustrate these principles. Here, the integration of safety climates is from different entities, like the owner, subcontractor, or local labor practices. Here is where the safety challenge lies: engaging not only your employees, but all the stakeholders.

Safety is often overmanaged and under lead. To lead, management must be committed to safety through all phases of the construction life cycle. A focus on safety begins with the project initiation, planning, and design phases and continues to the actual construction, performance, and monitoring phase. Safety is the responsibility of all stakeholders and requires a commitment to education, training, and implementation of practices and protocols.

## *MANAGING SAFETY*

Managing safety within a construction operation requires the development of a safety and health program for the organization. The goal of this program is ensuring the well-being of the workforce and protecting them while they are performing work. These programs are tools that seek to foster improved employee morale, productivity, and retention. A proactive approach to identifying and mitigating jobsite hazards before they can cause injury or illness is the focal point of these programs.

Instead of relying on post incident or accident data, these programs seek to proactively engage workers to collaboratively identify and solve safety issues before they become a problem. While the goal is worker safety, this type of collaboration and fostering of communication builds a level of trust among the workforce that leads to other business improvements.

A corner stone to successful safety and health programs is the involvement of the field workers in the development of the program. Construction worksites typically involve more than one employer. In this case a project-specific safety and health plan should be developed that includes input from all stakeholders involved. This provides improved insight into the work being accomplished by each entity and ensures their concerns are heard and needs are met. On these sites, it is critical to underscore the importance of collaboration, communication, and coordination with all parties involved to ensure a foundational understanding of the safety culture and climate promoted on site.

These plans are not meant to be stagnate. Each program should be seeking ways to continuously improve and develop metrics they can use for the assessment of the effectiveness of the program.

When estimating and planning a project, safety must be considered as part of the core elements and should be budgeted for during the bidding process. While the safety plan is an investment designed to keep the workplace injury- and incident-free, there are many other benefits to a robust safety culture and climate.

Companies that have a strong commitment to the health and safety of their workers and have worked to foster a positive safety culture have reaped other benefits that improve their bottom line. Examples include improved productivity and quality of work, increased employee morale and retention, and a positive reputation in the industry.

## Leading Safety

Leaders in construction companies set the tone for the types of behaviors tolerated. The safety and health of all workers must be a core value to the organization. On the jobsite, these firms are committed to identifying and eliminating hazards, by investing the appropriate resources to demonstrate their commitment. These resources are beyond the funds required for personal protective equipment (PPE) and other safety precautions. It involves providing employees with training and the time necessary to complete these trainings. It also requires that senior leadership demonstrates their commitment to safety by leading by example.

In leading by example, top executives must demonstrate their commitment by establishing a written health-and-safety policy that they sign and implement. They must ensure that safety is considered in all decisions to include the means and methods by which they do their work. Safety should be included in labor rates and selection criteria for all subcontractors and vendors.

Another key to managing safety is to ensure that policies surrounding implementation are communicated often and in multiple ways. The policy clearly states and emphasizes application to everyone on the jobsite and all stakeholders are held accountable. A safety lead should be established for the company that is responsible for implementation, training, and performance. Roles and responsibilities for the adherence to the program must be established to ensure continued success.

Including employee input in the development of the health-and-safety program is paramount to success. Not only because they know the most about the hazards involved with the work they do, but they also have the most to lose if a program fails. Engagement of the field workers demonstrates the level of commitment of the company and establishes a foundation of trust that extends far behind the safety-and-health program. This engagement can be in fun and interesting ways. Having workers express their commitment to safety is also important (Fig. 20.1). Workers will be more likely to provide input and report any safety concerns and hold their fellow workers accountable if they have input to the design of the program.

Encouraging the reporting of safety incidents, unsafe conditions, and near misses will enable these issues to be addressed before an employee is seriously injured. This line of communication with workers is the best way to improve your program and identify gaps in the policies. Learning from near misses is critical, but these teachable moments might not be possible if workers are not comfortable with reporting or if they do not know the process.

Encourage workers to review all documentation associated with the health-and-safety program. These include items such as injury and illness data, incident investigation reports, and job hazard analysis forms and processes. This ensures they understand the hazards and how to mitigate them and creates a sense of ownership. It will also create a level of trust so they will report issues they see without fear of reprisal or retaliation.

A health-and-safety plan must be meaningful and that requires that not only are workers involved but that it is reviewed periodically for relevance. While jobsite safety plans will vary based upon the unique requirement of each project, the corporate health-and-safety plan should be a consistent document that portrays the values of the company and demonstrates their commitment to the health and safety of the employees.

A health-and-safety plan must contain various elements that assist with deployment and implementation. Employees must be involved; they must be educated and trained in hazard recognition and in how to conduct incident investigations and reports. They must also know that they are

**FIGURE 20.1** Artist's capture of what safety means to participants.

empowered to stop any work that they feel is not safe. They have the ability and the authority to do so. The components and recommendations for the development of a corporate health-and-safety plan are discussed in depth.

## Corporate Health-and-Safety Plan

A corporate health-and-safety plan must contain certain key elements to demonstrate the critical role safety plays in the company. The plan must include a clear policy statement from leadership that emphasizes the importance of safety. Roles and responsibilities of all stakeholders must be clearly defined, to include expectations of behaviors and work practices. The process for planning safety and a set of procedures to follow for all work being pursued. This should include prequalification of subcontractors and vendors as well as a detailed review of specifications related to safety. Detailed project site communication protocols to include a safety training and orientation for all site personnel must be included. Specific focus should be given to field implementation of the plan and require the sharing of all safety-related information among personnel. Emergency management plans and processes as well as incident investigation processes should also be detailed. In addition, all workers need to understand their rights on the site. They should be informed that they have the ability and authority to shut down any work they deem unsafe without fear of repercussions.

The plan should also outline a process for evaluation of all procedures, employee training, and work practices.

## EDUCATION AND TRAINING

Before a worker can set foot on a jobsite, they must be fully aware of all hazards present. A worker who does not understand the hazards is a danger to all on the site as their ignorance may unknowingly put everyone at risk. It is the construction manager's job to make sure that every worker is aware of the dangers present and the steps they must take to remain safe.

### Hazard Identification and Assessment

The identification and assessment of hazards requires a process designed to collect and review information related to the risks. This involves the inspection of the jobsite regularly and the investigation of injuries, incidents, and close calls to determine the root cause of the issue. Was the issue caused by gaps in the health-and-safety program? Are there trends in the data regarding the hazards that are identified?

When evaluating jobsites, safety hazards may evolve or be introduced over time as the work on site progresses. A hazard assessment plan includes time for regular jobsite inspection where hazards are identified and addressed prior to any future incidences occurring. This is the responsibility of one person who will assess the jobsite hazards on a regular basis. This means at a time interval that is most relevant for the type of work. For example, during an excavation operation, trenches will be inspected at a minimum before each shift, but may need to be evaluated more often, especially after a rain event. Consideration must be given to the likelihood of the hazard occurring during an operation. OSHA has developed excellent safety checklists that can be used as a guide to assist in jobsite inspections. When conducting inspections all items that impact the health and safety of the work are to be included. Check lists are a means to ensure there is consistency in the inspections and how they are reported.

Begin the site inspection with a focus on the commonly cited sources of injuries, such as falls, caught in/between, struck-bys, and electrocutions, then seek to capture other hazards that are specific to that site. Be sure to include the activities of onsite contractors, subcontractors, and temporary workers. The mitigation or elimination of the hazard is desirable if possible. The data collected by the inspections can assist in identifying means to control or reduce the hazards.

While daily inspections will identify hazards for routine operations, there are also hazards associated with nonroutine operations. For example, demobilization of the site, concrete placement, critical crane lifts, or equipment installations all present unique hazards. Procedures must be defined to identify and respond to hazards in these nonroutine situations.

The identification of hazards and the types of incidents that can occur on site is important in the determination of the controls to be implemented to keep workers safe.

### Hazard Prevention and Control

When determining how to control and mitigate hazards, involve workers in identifying measures of protection. Utilize the hierarchy of controls (Fig. 20.2) to identify possible controls to protect workers from the hazard as well as how to protect workers if an emergency occurs.

What controls are currently being used? Evaluate the effectiveness that these controls supply to determine if different controls would be more appropriate. Consider new technologies that may provide more effective and reliable protection while being less costly.

When determining the appropriate controls, review what has worked in the past and what control measures are currently being used in similar situations on other jobsites. Determine if these controls

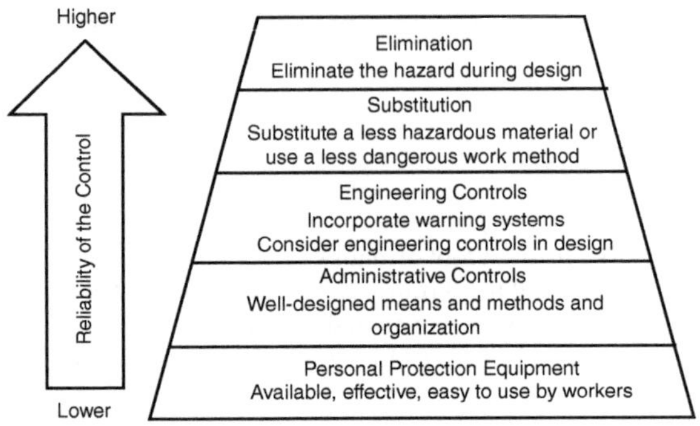


**FIGURE 20.2** Hierarchy of controls.

would be effective to protect the workers on the jobsite. As mentioned, be sure to seek input from workers. They may be able to suggest and evaluate potential controls based on their knowledge of the jobsite, equipment, and work processes.

While the ultimate goal of a hazard mitigation plan is elimination of the hazard, there are steps that can be taken to select controls that are feasible, effective, and permanent.

Consider the danger present on jobsites related to falls and how the hierarchy of controls shown in Fig. 20.2 can be used to work to eliminate fall hazards. Starting at the bottom with a lower reliability of control method, personal protective equipment, and work to highest level of control, which is elimination.

Personal protective equipment in the form of a fall protection harness is a control method that can be distributed to employees who will be working at heights over 6 ft. Employees must receive training in the proper wearing, use, and inspection of the harness prior to deploying it on the jobsite. The harness will protect individuals in the event of a fall, but the danger of falling during their work is still present.

Administrative controls seek to mitigate risk through designing thoughtful work practices. For falls on jobsites this could include strict housekeeping rules to minimize tripping hazards, written policies pertaining to ladder use, or detailed requirements for leading edge guardrail installations. These are work practices and methods that coupled with the appropriate PPE can reduce the level of hazard exposure.

Engineering controls incorporate methods designed into the building, equipment, or work process to minimize worker exposure to hazards. This may include preinstalled anchorage points on roofs so workers have clear locations where they can attach their fall protection equipment. The redesign of a construction process to minimize the time workers are required to work at height is another example (Fig. 20.3).

Substitution requires that a less dangerous work method be employed on site. This could be substituting work that was planned to be accomplished using scaffolding systems over a long duration with prefabricated systems that require a shorter duration of working at heights.

Elimination of a hazard means completely removing it so there is nothing to be concerned about. This may not be possible. For example, the full elimination of a fall hazard would mean not putting any workers at heights. Scaffolding erection presents major fall hazards to the employees completing the process. During the erection process there is no means to safely tie off in many situations. Preassembling the scaffolding sections on the ground and then having them lifted into place is a possible way to eliminate working at heights without protection.

Prevention through design is another method of control that can be utilized in construction. It falls in the categories of substitution and elimination. Prevention through design considers construction safety in the design of a project. The designer makes decisions based upon the safety risks presented to the construction workers. They seek to eliminate elements of the design that place the workers in high-risk situations. In Fig. 20.4 the connections for the fall protection line were designed into the building.

**FIGURE 20.3** Fall protection anchor point.

**FIGURE 20.4** Fall prevention line.

## DEVELOP AND UPDATE A HAZARD CONTROL PLAN

The hazard control plan describes the implementation of the controls selected. Effective plans address serious hazards first, with the overall goal to ensure effective long-term control of hazards. Control plans at a construction site may need to be updated and modified often as the project develops and the site conditions and hazards change. Awareness of these changing conditions is paramount.

The daily routine of a construction site is filled with safety challenges and a good safety-and-health plan will identify mitigation strategies for these. The construction site also sees many nonroutine tasks during certain phases of the work. During excavation of a foundation, the site may be susceptible to collapses, while during steel erection, the site is subjected to overhead lifts. Tasks that workers do not normally do raise concern from the perspective that they may not be familiar with the hazards that are present. This work needs to be approached with caution. Prior to beginning the work, a walk-through of the actual site and a review of the steps that need to be taken to complete the work should be completed. A thorough job hazard analysis (JHA) should be developed for these types of activities. This JHA should be reviewed by all the workers involved with the task prior to work beginning with a focus on receiving input on the plan from all involved. Others on the jobsite should be notified of the work taking place, when the work is scheduled and any precautions that must be taken while on site during the activity. It may be as simple as having a large crane onsite for a few hours to complete a heavy lift or it may be a longer activity that involves people and equipment working in an open trench that is greater than 6 ft deep. Regardless, all people on site need to be aware of the presence of the hazard and the precautions they must take.

### Conducting a Job Hazard Analysis

Prior to conducting a job safety analysis, be sure to consult the OSHA standards. Complying with these regulations and incorporating the requirements into the job hazard analysis will ensure the health-and-safety program meets federal standards. In addition, be sure to check the contract and with the owner of the project as there may be additional safety guidelines and requirements with which to comply.

A hazard is a condition that if left uncontrolled may result in illness or injury. Identifying hazards and then determining means to control or eliminate them helps to prevent harm to employees.

A tool that can be used to assess the potential for a hazard to arise during a work task is a job (or task) hazard analysis. This technique focuses on the interaction between the worker, the task, the tools, the equipment, and the work environment to identify hazards. Once hazards are identified steps can be taken to eliminate them or reduce them to an acceptable level of risk. Sample forms used for job hazard analysis are shown in Figs. 20.5 and 20.6. Figure 20.5 highlights a component of general safety that applies to most construction sites, while Fig. 20.6 is a task-specific hazard analysis for elements of pile driving.

Reviewing standard work practices and training all employees in safe practices will reduce the occurrence of incidents and illnesses. It will also result in more efficient work practices leading to an increase in worker productivity. The job hazard analysis once standardized is an excellent tool to train new employees in the steps required to safely perform their work.

The job hazard analysis process like the safety plan must involve front-line employees, given their knowledge of the work performed. This also ensures future buy in to the solutions developed and minimizes the chance of missing key elements in the work process. To begin, have a team of employees review the incident reporting for the company over the past several years to identify any reoccurring issues and hazards that are commonly encountered. For example, if there are trip and fall injuries on several jobsites over multiple years, a stricter set of jobsite housekeeping policies and procedures may need to be developed and training implemented on each site. Compare the reports to the processes and procedures outlined in the company safety policy. Are these hazards addressed in the plan? Are the incidents caused by poor work practices or inadequate training in the identification of the hazards? Seek to identify areas of concern and work to develop plans to control or eliminate hazards.

**JOB HAZARD ANALYSIS** General Safety

| Job Description | Hazard Identification | Hazard Controls |
|---|---|---|
| Personal Protective Equipment (PPE) | • Falling Objects<br>• Improper Visibility of Workers to Pedestrians and Equipment Operators<br>• Slips and Punctures<br>• Cuts and Burns | 1. All employees are required to wear hard hats while on site.<br>2. All employees are required to wear Class 3 High Visibility Safety Vests.<br>3. All employees are required to wear steel toe boots and gloves while handling any material or operating any equipment. |
| Fall Protection | • Wet or Slippery Surfaces<br>• Ledges | 1. All employees are required to wear properly fitted safety harnesses for all work over 5 feet.<br>2. All edges of work areas will be marked with caution tape.<br>3. All employees will remain 3 feet from all work edges. |
| Ladders/ Scaffolding | • Fall hazard<br>• Unstable Base<br>• Dropped Tools and Materials<br>• Overloading<br>• Hoisting Equipment and Materials<br>• Inclement Weather | 1. Employees will wear safety harnesses and be tied off to a safety rail when working above 5 feet.<br>2. Scaffolds will be inspected daily and will be able to hold 4 times the desired load.<br>3. Employees will keep 3 points of contact with the ladder at all time. |

| | Employee Instructions: |
|---|---|
| | 1. Hard hats, safety glasses, work clothes, gloves, and work boots shall be worn at all time. |
| | 2. All employees and materials will remain 3 feet from all work edges. |
| | 3. All employees are required to wear Class 3 High Visibility Safety Vests. |

***All crew members must print/sign their name below to acknowledge their understanding of the JHA.***

| | | |
|---|---|---|
| | | |

THA conducted by: ______________________________ Date conducted: ____________________

Manager / Supervisor Signature: ______________________ Date: ____________________________

**NOTE TO MANAGER, SUPERVISOR OR FOREMAN: WORK MUST NOT START UNTIL HAZARDS ARE DISCUSSED, CONTROLS ARE IN PLACE, AND ALL SIGNATURES ARE OBTAINED.**

**FIGURE 20.5** General safety job hazard analysis sample form.

## JOB HAZARD ANALYSIS

Pile Driving

| Job Description | Hazard Identification | Hazard Controls |
|---|---|---|
| Rope | • Easily cut or abraded<br>• Tail ends should not be cut off<br>• Workers should be cautious of a load that is quickly released<br>• Spliced or broken wire rope | 1. Frequent inspection of ropes needed for wear and tear<br>2. Rope tail ends should be tucked in beyond the splice at all times<br>3. Workers should never stand alongside any rope under heavy strain<br>4. Wire rope should be discarded when significant wear from outside wires is noticed |
| Slings | • Loads being picked up with a wrapped hoist line<br>• A sling with 6 or more closely spaced broken wires<br>• Rusted/ dirty slings | 1. Loads should always be picked up with slings<br>2. Slings are to be inspected before each use<br>3. Keep slings cleaned and lubricated<br>4. When handling mats, use a two-leg sling |
| Hooks | • If they are being used for drills, jets, siphons, or other semi permanent attachments<br>• Hooks showing signs of cracks or straightening | 1. Should only be used for temporary connections<br>2. Hooks need to be inspected often and regularly<br>3. Shackles are often preferred<br>4. Hooks should be turned so the pull of the line or chain is always into the hook |

| Employee Instructions: |
|---|
| 1. Hard hats, safety glasses, work clothes, gloves, work boots, and **life vests** shall be worn at all time. |
| 2. Employees will be required to lift no more than 50 pounds at one time. |
| 3. Rebar ends to be capped after placement. |

***All crew members must print/sign their name below to acknowledge their understanding of the THA.***

| | |
|---|---|
| | |
| | |

THA conducted by: ______________________ Date conducted: ______________

Manager / Supervisor Signature: ______________________ Date: ______________

**NOTE TO MANAGER, SUPERVISOR OR FOREMAN: WORK MUST NOT START UNTIL HAZARDS ARE DISCUSSED, CONTROLS ARE IN PLACE, AND ALL SIGNATURES ARE OBTAINED.**

**FIGURE 20.6** Pile driving job hazard analysis sample form.

Each project site will have its unique set of hazards, so while general procedures for standard tasks may exist, each project should be reviewed for hazardous tasks. Identify the hazards, rank their risk, and prioritize each for analysis.

Review the work that will be undertaken and break it into simple tasks. An excellent way to begin a job hazard analysis is to watch an individual perform the task and record the effort. The recording can be viewed by many workers and used to develop an accurate list of the steps required to perform the work. Review the list of tasks and answer the following questions to help guide the analysis.

- What hazards exist?
- What can go wrong?
- What are the consequences?
- What factors contribute to the hazard?
- How likely is it that the hazard will occur?

Consider the environment, the exposure hazard, and what events will cause or trigger the hazard. Evaluate the consequences if the hazard occurs and the contributing factors.

After reviewing your list of hazards with employees, consider what control methods will eliminate or reduce them. The most effective controls are engineering controls that physically change work environment to prevent employee exposure to the hazard. The more reliable or less likely a hazard control can be circumvented, the better. If this is not feasible, administrative controls may be appropriate or a lower level of PPE. The results of a job hazard analysis may be the way that employees complete their work. To determine if these new work processes are feasible, discuss the idea with employees who perform the work. Make sure the employees understand the reasons behind why the procedures are being changed and what they must do to comply with the new processes.

Once selected, hazard prevention measures need to be implemented on site. Initiate the simplest measures first, such as housekeeping, the wearing of appropriate PPE, before you initiate the more time-consuming and expensive measures. Make the simple measures part of a daily routine and soon they will be part of the workplace culture.

## EDUCATION AND TRAINING

OSHA requires that employers train all employees in the recognition and identification of hazards [29CFR 1926.20(f)(2)]. In addition to receiving training in hazard identification, all employees need to have a greater understanding of the health-and-safety program. This way all employees can contribute to the development, implementation, evaluation, and improvement.

The primary training that employees should receive should be focused upon their ability to complete their work safely. Understanding jobsite hazards and how to avoid creating additional dangerous situations is required prior to an employee taking on any task. In addition, they must comprehend the procedures for controlling the hazard as well as the process for reporting the hazard. Consider providing the OSHA 10-hr course for construction or a similar course to be supplemented by orientation training and toolbox talks to cover hazards on each specific site.

If the work involves unique hazards, like explosive gases, specialized training may be required. Other training may be needed that does not necessarily involve the identification of a hazard. If an employee serves a role in the health-and-safety program, such as incident investigation or emergency response, they may need specialized training to ensure they can fulfill their role.

Training can take many forms. It can be a formal classroom approach or an on-the-job intensive demonstration. Safety education can be in the form of on-the-job training, worksite demonstrations, daily toolbox talks, pretask hazard analysis review, or even peer-to-peer mentoring (Fig. 20.7). As long as it is occurring, the form does not matter.

Be sure to review the safety requirements of the local municipality in which the work is taking place. Many have minimum requirements for workers to even step foot on jobsites, like the

**FIGURE 20.7** Toolbox talk prior to work beginning.

completion of the OSHA 10-hr training. Many jobsites will not allow workers site access unless they can present a valid OSHA 10-hr card.

In addition to specific worksite hazards, all workers, managers, and supervisors need to understand the intent and the requirements of the health-and-safety plan. They must also understand the steps to take when implementing the plan and specifically what role they play.

Guidelines must clearly indicate who they should contact for information about the program. Most importantly, all employees must be clear on how to report hazards, injuries, illnesses, and close calls or near misses. They also need to be aware of the employer's responsibilities and their own rights as workers under the Occupational Safety and Health Act.

An emergency response plan for the site needs to be developed and communicated to all employees. The specific actions that are to be taken and notifications that must be made are key components of this plan. Clearly indicate the individuals to be contacted for emergencies to include phone numbers and a protocol for contacting them.

The training of all workers on their specific roles in the safety-and-health program will be important in the implementation, evaluation, and improvement of the plan. Beyond the identification of hazards and appropriate jobsite behavior and actions, these employees must know how to report injuries, incidents, and concerns. Workers who will be participating in investigations and program evaluation must understand their roles in the process.

## PROGRAM EVALUATION AND IMPROVEMENT

The developed program must be reviewed periodically to verify that the program is being implemented as intended and that it is working. Develop metrics and goals that indicate the program is meeting the safety targets. Evaluate the program to identify any gaps and areas for improvement. Define indicators to track the performance and progress on critical items as well as the methods to collect and analyze the data.

These indicators should include both leading and lagging data. Leading indicators track how well the program is implemented and the prevention of injuries and illnesses before they occur. Lagging indicators track data about worker injury after they occur.

Evaluation of the program is critical and the level that it is evaluated will depend on the hazards controlled. The focus of the evaluation is the effectiveness of the program and ensuring that it is operating as intended and health-and-safety goals are being achieved. These results of the program evaluation will help to identify opportunities for improvement and the change to correct any gaps in the program. To demonstrate and reinforce the commitment to safety, employers must take immediate action to correct a problem and work to prevent its recurrence.

## INCIDENT INVESTIGATIONS

Despite best efforts to eliminate hazards there may be times where incidents occur on a project site. These incidents, no matter how major or minor, must be investigated because they provide a clear indication of where hazards exist. These investigations help to identify potential weaknesses in the health-and-safety plan once the root cause of the incident is identified. Once the problem is known, preventative measures can be developed to keep it from happening again.

A clear plan needs to be developed that follows prescribed procedures for conducting incident investigations. The investigation must begin immediately when an incident occurs so that no information or time is lost. The plan should detail items and procedures such as:

- Who will be involved.
- The lines of communication during the investigation and access to documents.
- Materials, equipment, and supplies needed to conduct the investigation.
- Required reporting forms and templates.

As part of health-and-safety training, individuals identified as part of investigative teams must receive specific training on incident investigation techniques. This training should emphasize objectivity and impartiality throughout the investigation process. The team should include representatives of both management and site workers.

The plan defines what requires an investigation. Any injury, incident, or anything requiring more than normal first aid on site will be required to be reported to OSHA using the standard Form 300. These events should be investigated as part of the standard health-and-safety plan. In order to learn from events that occur on site, investigations should be conducted on any close calls or near misses that occur, even if they do not require an OSHA report. There is valuable information to be learned from each of these incidents. These types of investigations should seek to identify and analyze the root causes that lead to the incident. The investigation should serve as a tool to address the gaps in the health-and-safety program that allowed the incident to happen. The results of the investigation are to be communicated to all stakeholders to help to minimize the likelihood of the incident recurring.

## SUBCONTRACTOR MANAGEMENT

**Choosing Subcontractors:** Prior to soliciting bids, prequalify subcontractors using safety as a criteria. As part of the prequalification questionnaire, be sure to request safety data like their Experience Modification Rate, copies of their OSHA recordkeeping forms (300, 300A, and 301), as well as their company health-and-safety plan. This data provides an indication of their commitment to a safety culture and climate and can assist in the choosing of a subcontractor with similar values.

**Training Subcontractors:** As part of the contract and bid documents, include requirements for safety training of all subcontractors and vendors. This may include requirements that before they step foot on the site they must undergo site safety orientation and while on the site that they attend all safety briefings, tool box talks, and participate in the job hazard analysis of all work tasks in which they have a role.

One of the greatest challenges on jobsites is the presence of many different subcontractors at various times. To ensure that the safety culture and message is communicated clearly, subcontractors must be aware that they will be held to the same standards as all of the general contractor's employees on site.

Ensuring clear communication and coordination occurs requires preliminary planning and agreement on how communication will flow. All workers on site must understand the types of hazards that may arise from the work being done on site by all subcontractors, contractors, and vendors. Clearly communicate the procedures needed to avoid or limit the exposure of workers to these hazards. What is the chain of command and who is the contact person if a safety concern is experienced? Before entering the site, all contractors, subcontractors, and vendors must receive training and review existing jobsite hazards to ensure they are aware of what currently exists and how to handle them. In addition, all subcontractors and vendors must be informed of how to contact the general contractor to receive more information, report an injury, illness, or incident or if they have a safety concern.

## CONTRACT REQUIREMENTS—SITE-SPECIFIC SAFETY PROGRAMS

A key component of construction contracts is the requirement to write a site-specific safety plan for the work you plan to accomplish. In addition, many owners are now requiring as part of a bid package a written statement of the corporate safety culture or philosophy. Realize that the safety plan is a living document that should reflect the specific needs of the project, the contractual requirements, and company safety policies, procedures, and philosophy.

The required elements in a safety plan will vary depending on the size, complexity, and hazards encountered on the site. To begin, the construction means and methods should be reviewed to identify known hazards that are present on the project site. General guidance should be prepared for those hazards that may be unexpected or unknown at the time you develop the plan. Include the steps to be taken to mitigate the hazard, the responsible party for various actions during an emergency, and how and when that individual will act in the case of the emergency.

Site-specific safety plans should amplify and reflect the policies and philosophies embraced in the corporate health-and-safety plan. Suggested areas to address are outlined below.

### Safety regulations

A generalized statement highlighting compliance with all local, state, and federal safety regulations and a commitment to following OSHA and other applicable agency regulations for the site. The plan should state the company intends to take any steps necessary to protect its employees, and others, from accidents and hazardous conditions in the workplace. Specifically mention that OSHA regulation "29 CFR 1926, Safety & Health Regulations for Construction" will be followed. Each section of your safety plan should detail how you are going to provide this protection.

### Who is Responsible for Site Safety?

Identify the individual who is responsible for the overall jobsite safety. This site supervisor has safety and first-aid training and any other specialized hazard training required by the type of work. Clearly state that the supervisor has the authority to expend resources to eliminate imminent dangers to people and property. Emphasize that every employee has the right and authority to stop work, should a hazardous condition arise. The plan should also identify the corporate safety representative who is responsible for all safety training and the certification of workers with safety regulations.

## Safety Orientation Procedures and Policies

Outline a detailed process for how all employees are briefed on safety issues, safe worksite practices, and jobsite safety policies prior to setting foot on site. Identify what PPE is required to work on site and how employees will be trained in the proper use of all PPEs (Fig. 20.8). Describe how hazard analysis is conducted on site and how and when regular safety briefings and toolbox talks occur. Identify any specific safety training requirements of the project, like an OSHA 10-hr card, and how these will be checked for each employee.

Highlight the appropriate policy and procedures for employees to raise safety concerns to management and notify them they are encouraged to report these issues. Also clearly state the processes by which these complaints are reviewed.

**FIGURE 20.8** Proper wearing of PPE to include fall protection harness. (Photo Courtesy of Philadelphia D&M, Inc.)

## Subcontractors

Indicate that all subcontractors and vendors working on the site must follow all safety guidelines adopted by your company. They will receive the same training and be required to participate in the same briefings, job hazard analysis processes, and tool box talks as company employees. Also ensure that they have the opportunity to provide input into the jobsite safety program, especially procedures related to the work they will be performing.

## General Housekeeping

A strong indicator of the attitude a company has for safety is reflected in the orderliness of the jobsite. Insisting on a neat and organized worksite reinforces the importance of safety and set the tone for what will be tolerated on site. Identify who is responsible for the cleaning and maintaining a clean worksite and the frequency at which cleanup activities will occur.

## Who and When Safety Inspections are Conducted?

Clearly state that the supervisor of each worksite shall conduct continuous safety inspections of the work area and that any problems will be immediately noted and corrected.

## Required Personal Protective Equipment

Identify the PPE issued to each employee and what the employee is responsible for providing themselves. For example, the company may provide safety glasses and a high reflective vest, and the employee is responsible for purchasing their own steel-toed boots. The plan must clearly state the requirements for all PPE use and the process to replace worn, damaged, or lost PPE.

## Incident Reports

Every incident that occurs on site, no matter how minor, should be reported to the site safety supervisor. The incident should be documented in a designated file that contains all the information pertinent to the incident and the steps that were taken to eliminate the hazard that lead to the incident. Identify the individual responsible for conducting investigation into the incident (usually the site safety representative) and where the records will be kept. In addition, define how the information collected will be communicated to all employees and how the data collected will be used to improve the site safety.

## Work Zone Warning Devices

Safe work practices require that the jobsite detail the types of work zone warnings that are utilized to protect workers. These may include spotters for any vehicle backing up, flaggers for cranes movement, or the use of various audible alarms for overhead work. Standard safety equipment such as barriers and fall protection should also be highlighted to workers.

## Fire Protection

Clearly indicate the location and type of fire protection equipment provided on site. Outline procedures for safe fire response and identify the emergency personnel that should be summoned in case of a fire. Describe fire prevention practices like a fire-watch during welding operations or precautions when storing flammable liquids. Understand the use and location of fire safety equipment (Fig. 20.9).

**FIGURE 20.9** Fire extinguisher in active work zone.

## Equipment inspections

Jobsites have many different types and sizes of equipment. The equipment must be inspected on a regular basis that complies with manufacturer recommendations as well as federal, state, and local requirements or requirements of the owner.

Identify who is responsible for inspections of equipment and where the records of these inspections are kept. These inspections should be conducted on a regular schedule and the operator of the equipment should be trained in safe operating practices and procedures and should inspect the equipment prior to use. A process should also be defined for how damaged or faulty equipment is taken out of use.

## Control of Lockout/Tag-Out

Electrocutions are a common cause of construction site injuries and fatalities. A detailed procedure for the lockout/tag-out requirements for the site must be discussed with all employees. This should include clear guidelines for when it should be used and who has the authority to remove a lockout/tag-out tag. Establish clear lines of communication for the process and reference the formal lockout/tag-out policy adopted by the company. This is typically a separate document required as part of the contract.

## Hazard Communications

Every jobsite is required to keep onsite a Material Safety Data Sheet (MSDS) as part of the OSHA Hazard Communication standard. Indicate where the MSDS are filed and who to contact with MSDS questions. Provide reference to the hazard communications plan for detailed procedures related to hazardous materials and the training required.

### Medical Treatment

Spell out the specifics of who provides medical services, when and where these services will be provided, and how and when emergency personnel shall be summoned.

### Employee Periodic Plan Review

Identify who reviews this plan, when it is reviewed (at least annually), and who can approve and/or make changes to this plan. A comprehensive review prior to each project is most desirable.

### Hazard Control Measures

Information obtained from a job hazard analysis is useless unless hazard control measures recommended in the analysis are incorporated into the tasks. Managers should recognize that not all hazard controls are equal. Some are more effective than others at reducing the risk.

## OSHA FATAL FOUR

OSHA has defined the top four causes of injuries and deaths in the construction industry. They refer to these as the fatal four. These are falls, struck-by, caught-in/between, and electrocutions. It is these four areas that all construction site employees must be trained to recognize and be educated upon the steps to mitigate, minimize, and ultimately eliminate these hazards.

### Falls

Falls can be from height or at the same level or a trip and fall. There are safety precautions that can be taken to prevent falls, the most obvious being the use of fall protection and arrest equipment. In addition, jobsite safety mechanisms can be put into place as part of standard work plan to ensure the safety of those working, visiting, or passing by the jobsite. These include identifying perimeter protection of roofs or at heights, labeling floor openings and covering, and securing roof and floor penetrations (Fig. 20.10). Employees must receive training in the safe use of equipment that can be the root cause of a fall from height, like scaffolding and ladders. The jobsite should also be kept neat and all walkways should be clear of debris or other tripping hazards. Housekeeping on a project will promote a safe jobsite.

Construction sites use a wide range of tools and equipment. Some is large and cumbersome, like a bulldozer or an excavator, while some are smaller and held in your hand, like a nail gun, or any type of powered saw. It does not matter the size, both can cause serious injuries or deaths on site. At all times employees must be aware of their surroundings. They should never stand or work between areas that have access to moving objects where they could be pinned between it and fixed object. Workers should be visible to operators and be actively communicating with the operator if using heavy equipment to move material on the site. The wearing of approved PPE may include bright reflective clothing that can easily be seen by operators of equipment as well as regular vehicles on the site.

Employees working around equipment that has the capability to pull them into the operating gears need to be careful and not wear loose clothing or jewelry that could get caught in the equipment, machinery, or power tool.

### Trenches and Open Excavations

Trenches and open excavations are highly dangerous and very common on most construction sites at various stages of construction. The OSHA ruling for shoring or protecting a trench or excavation is a

**FIGURE 20.10** Fall protection railing.

**FIGURE 20.11** Trench box.

function of depth and sloping. Any trench or excavation that is deeper than 5 ft must have an adequate protective system in place if people will be entering the area. Protection may include sloping, benching, or other types of support of excavation options like trench boxes (Fig. 20.11).

## Electrocution

The possibility of being electrocuted on a jobsite comes in many forms, some not as obvious as others. As part of your sitework plan, be certain that all underground and overhead utilities are identified, located, and labeled. Be certain to communicate the safe distance requirements for all power lines on site and include these in project documentation for all stakeholders on the project.

Check all cords on all portable electric power tools and be sure they are operated according to the manufacturers' specifications. Implement a jobsite tool inspection procedure and ensure that it is followed. While checking for frayed cords on tools and with extension cords may seem like tedious work, if it is part of the culture and climate, employees will adopt the practice as part of their daily procedure. Identify all electrical hazards when working with ladders and scaffolding.

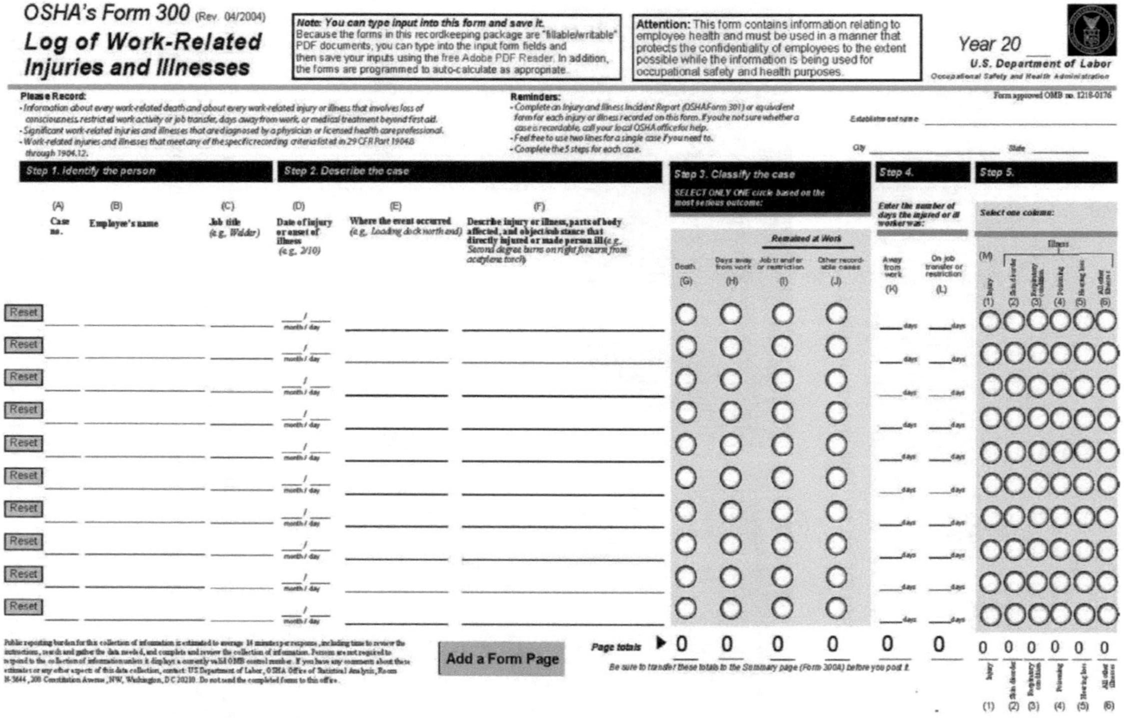

OSHA's Form 300 (Rev. 04/2004)

**Log of Work-Related Injuries and Illnesses**

**Note: You can type input into this form and save it.** Because the forms in this recordkeeping package are "fillable/writable" PDF documents, you can type into the input form fields and then save your inputs using the free Adobe PDF Reader. In addition, the forms are programmed to auto-calculate as appropriate.

**Attention:** This form contains information relating to employee health and must be used in a manner that protects the confidentiality of employees to the extent possible while the information is being used for occupational safety and health purposes.

Year 20 ___

U.S. Department of Labor
Occupational Safety and Health Administration

Form approved OMB no. 1218-0176

**Please Record:**
- Information about every work-related death and about every work-related injury or illness that involves loss of consciousness, restricted work activity or job transfer, days away from work, or medical treatment beyond first aid.
- Significant work-related injuries and illnesses that are diagnosed by a physician or licensed health care professional.
- Work-related injuries and illnesses that meet any of the specific recording criteria listed in 29 CFR Part 1904.8 through 1904.12.

**Reminders:**
- Complete an Injury and Illness Incident Report (OSHA Form 301) or equivalent form for each injury or illness recorded on this form. If you're not sure whether a case is recordable, call your local OSHA office for help.
- Feel free to use two lines for a single case if you need to.
- Complete the 5 steps for each case.

Establishment name ______

City ______ State ______

Step 1. Identify the person | Step 2. Describe the case | Step 3. Classify the case | Step 4. | Step 5.

(A) Case no. | (B) Employee's name | (C) Job title (e.g., Welder) | (D) Date of injury or onset of illness (e.g., 2/10) | (E) Where the event occurred (e.g., Loading dock north end) | (F) Describe injury or illness, parts of body affected, and object/substance that directly injured or made person ill (e.g., Second degree burns on right forearm from acetylene torch)

Step 3: SELECT ONLY ONE circle based on the most serious outcome: Death (G) | Days away from work (H) | Remained at Work: Job transfer or restriction (I) | Other recordable cases (J)

Step 4: Enter the number of days the injured or ill worker was: Away from work (K) ___ days | On job transfer or restriction (L) ___ days

Step 5: Select one column: (M) (1) Injury | Illness: (2) Skin disorder | (3) Respiratory condition | (4) Poisoning | (5) Hearing loss | (6) All other illnesses

Reset (×10 rows) ___ month / day ___

Page totals ▶ 0 0 0 0 0 0 0 0 0 0 0 0

Be sure to transfer these totals to the Summary page (Form 300A) before you post it.

Add a Form Page

FIGURE 20.12 OSHA form 300.

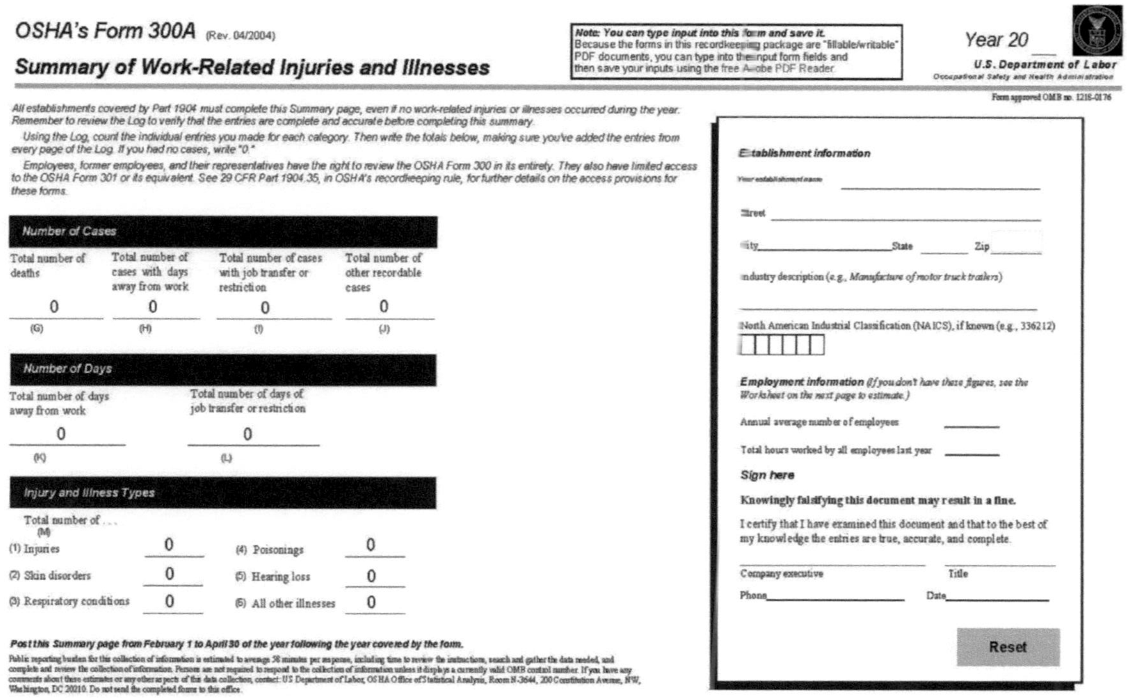
OSHA's Form 300A (Rev. 04/2004)

Summary of Work-Related Injuries and Illnesses

Note: You can type input into this form and save it. Because the forms in this recordkeeping package are "fillable/writable" PDF documents, you can type into the input form fields and then save your inputs using the free Adobe PDF Reader.

Year 20 __

U.S. Department of Labor
Occupational Safety and Health Administration

Form approved OMB no. 1218-0176

All establishments covered by Part 1904 must complete this Summary page, even if no work-related injuries or illnesses occurred during the year. Remember to review the Log to verify that the entries are complete and accurate before completing this summary.

Using the Log, count the individual entries you made for each category. Then write the totals below, making sure you've added the entries from every page of the Log. If you had no cases, write "0."

Employees, former employees, and their representatives have the right to review the OSHA Form 300 in its entirety. They also have limited access to the OSHA Form 301 or its equivalent. See 29 CFR Part 1904.35, in OSHA's recordkeeping rule, for further details on the access provisions for these forms.

**Number of Cases**

| Total number of deaths | Total number of cases with days away from work | Total number of cases with job transfer or restriction | Total number of other recordable cases |
|---|---|---|---|
| 0 | 0 | 0 | 0 |
| (G) | (H) | (I) | (J) |

**Number of Days**

| Total number of days away from work | Total number of days of job transfer or restriction |
|---|---|
| 0 | 0 |
| (K) | (L) |

**Injury and Illness Types**

Total number of . . .
(M)

| | | | |
|---|---|---|---|
| (1) Injuries | 0 | (4) Poisonings | 0 |
| (2) Skin disorders | 0 | (5) Hearing loss | 0 |
| (3) Respiratory conditions | 0 | (6) All other illnesses | 0 |

***Post this Summary page from February 1 to April 30 of the year following the year covered by the form.***

Public reporting burden for this collection of information is estimated to average 58 minutes per response, including time to review the instructions, search and gather the data needed, and complete and review the collection of information. Persons are not required to respond to the collection of information unless it displays a currently valid OMB control number. If you have any comments about these estimates or any other aspects of this data collection, contact: US Department of Labor, OSHA Office of Statistical Analysis, Room N-3644, 200 Constitution Avenue, NW, Washington, DC 20210. Do not send the completed forms to this office.

**Establishment information**

Your establishment name ______

Street ______

City ______ State ______ Zip ______

Industry description (e.g., *Manufacture of motor truck trailers*) ______

North American Industrial Classification (NAICS), if known (e.g., 336212)

**Employment information** *(If you don't have these figures, see the Worksheet on the next page to estimate.)*

Annual average number of employees ______

Total hours worked by all employees last year ______

**Sign here**

**Knowingly falsifying this document may result in a fine.**

I certify that I have examined this document and that to the best of my knowledge the entries are true, accurate, and complete.

Company executive ______ Title ______

Phone ______ Date ______

Reset

**FIGURE 20.13** OSHA form 300A.

# INDEX

Printed by Libri Plureos GmbH in Hamburg,
Germany

9 781260 456882